AN INTRODUCTION TO PRACTICAL BIOTECHNOLOGY

An Introduction to
PRACTICAL BIOTECHNOLOGY

[A Handbook on Practical Biotechnology]

by
HARISHA S.
M.Sc. PGDBI
Lecturer
Department of Biotechnology
Nagarjuna College of Engineering and Technology
Devanahalli, Bangalore

LAXMI PUBLICATIONS (P) LTD
An ISO 9001:2015 Company
BENGALURU • CHENNAI • GUWAHATI • HYDERABAD
KOCHI • KOLKATA • LUCKNOW • RANCHI
NEW DELHI

AN INTRODUCTION TO PRACTICAL BIOTECHNOLOGY

Printed and bound in India

Typeset at: Kalyani Computer Services, New Delhi

First Edition: 2006; *Reprint:* 2027, 2010, 2015, 2018, 2021; *New Edition* : 2026

ISBN : 978-81-7008-879-0

Branches

✆	Bengaluru	080-26 75 69 30
✆	Chennai	044-24 34 47 26
✆	Guwahati	0361-254 36 69
✆	Hyderabad	040-27 55 53 83
✆	Kochi	0484-405 13 03
✆	Kolkata	033-40 04 77 79
✆	Lucknow	0522-430 36 13
✆	Ranchi	0651-224 24 64

PUBLISHED IN INDIA BY

LAXMI PUBLICATIONS (P) LTD

An ISO 9001:2015 Company
113, GOLDEN HOUSE, GURUDWARA ROAD, DARYAGANJ,
NEW DELHI - 110002, INDIA

Corporate Office : B-33, Sector-5, Noida,
Gautambuddha Nagar (U.P.) - 201 301, INDIA
Telephone : +91-120-44 46 900
www.laxmipublications.com
info@laxmipublications.com

C– R/026/05
Printed at : Ajit Printers, Delhi

PREFACE

Biotechnology is a new discipline. It is constantly evolving as more information is added to the knowledge and it has become an important and integral part of life science courses now-a-days.

Biotechnology is an area of multi-disciplinary science subject involving a variety of distinct subject, where various living organisms and parts of organisms are used for the welfare of human beings.

It includes practical aspects of Biochemistry, Microbiology, Cell Biology, Genetics, Molecular Biology, Plant & animal tissue culture techniques, recombinant DNA Technology etc. As a teacher, I felt the need for a complete & comprehensive book on orientation towards application aspects of Practical Biotechnology.

Although there are number of Biotechnological procedures or protocols followed in the biological research, but there has been no attempt made for assembling the information at one point. Therefore, this book is the combination of all biotechnological protocols, needed for the students who are studying biotechnology either in B.Sc., M.Sc., or B.E.

The students obviously need a book that can provide a complete and comprehensive aspects of the subjects and various Biotechnological procedures applied in biology. This book aims to fulfil such a need of the students and teachers as well.

An Introduction to Practical Biotechnology : A Laboratory Manual provides exercises suitable for use in advanced general life science courses as well as applied biotechnology. The manual consists of 200 exercises arranged in 9 modules or parts. The protocols are self-explanatory and the user can easily follow the steps.

I will be glad to accept constructive criticism and suggestions from the faculties, students and readers to make this book a better one in subsequent editions.

I hope this book would be of use to all those engaged in Biotechnology research as a guide to make a start alone and carry the research work smoothly without any difficulty.

—Harisha S.

ACKNOWLEDGEMENT

It's my great pleasure to take this opportunity to thank everyone who helped me while writing this book.

First and foremost, I would like to thank Dr. Gopal M. Advirao, Chairman, Department of P.G. Studies and Research in Biochemistry, Kuvmepu University, for writing the foreword and adoring the book.

I express my sincere thanks and gratitude to Dr. Somashekhar.R for correcting the protocols and for his valuable suggestions.

A special thanks to Mr. Prabhakar. G .S. and Mr. Shivakumar K. for their timely help in checking the printout of protocols.

My special thanks to publisher for the neat execution of the work.

Author

FOREWORD

Biotechnology impinges on everyone's lives. It is one of the major technologies of the twenty-first century. It's huge, wide-ranging, multi-disciplinary activities include recombinant DNA techniques, cloning, genetics and the application of microbiology to the production of goods as prosaic as bread, beer, cheese and antibiotics. It continues to revolutionize treatments of many diseases, and is used to provide clean technologies and to deal with environmental problems.

This Handbook of practical Biotechnology, edited by Sri. Harisha.S. is true to its title in being both a handbook and very practical. The focus in the following pages is how to do things with one's own hands with a very brief preamble for each technique, which discusses the principles and theoretical basis of the technique. This unique laboratory manual is particularly appropriate for courses in molecular biology techniques, recombinant DNA techniques, tissue culture and genetic engineering. Also included is an especially helpful section to aid new instructors in avoiding potential pitfalls of specific experiment.

KEY FEATURES:

- Contain student tested, easy – to – follow protocols
- Presents background information that reinforces principles behind the methods presented.
- Provide both detailed descriptions of experimental procedures and a theoretical support section.
- Sequentially links experiments to provide a "Project" approach to study biotechnology.

This text will serve as an excellent aid in the establishment or instruction of introductory courses in the biological sciences.

DR. GOPAL.M. ADVIRAO.
Chairman,
Department of P.G. Studies and
Research in Biochemistry,.
Kuvempu University,
Davanagere-577 002.

Dedicated with profound gratitude

to my parents

and to my teachers

CONTENTS

1 GENERAL INSTRUCTION AND GENERAL LABORATORY METHODS

GENERAL INSTRUCTION

1. An observation notebook should be kept for laboratory experiments. Mistakes are not to be erased but should be marked out with a single line. The notebook should always be up-to-date and can be collected at any time.
2. *Index:* An index containing the *title* of each experiment and the page number should be included at the beginning of the notebook.
3. Write everything what you do in the laboratory should be in your observation notebook. The notebook should be organized by *experiment* only and *should not* be organized as a daily log. Start each new experiment on a new page. The top of the page should contain the *title of the experiment, the date,* and the *page number.* The page number is important for indexing, referring to previous experiments. Each experiment should include the following:
4. *Title/Purpose:* Every experiment should have a title and it should be descriptive.
5. *Background Information:* This section should include any information that is pertinent to the execution of the experiment or to the interpretation of the results. A simple drawing of the structure can be helpful.
6. *Materials:* This section should include any materials, i.e., solutions or equipment, that will be needed. Composition of all buffers should be included unless they are standard or are included in a kit. Include all calculations made in preparing solutions. Biological reagents should be identified as to original source, for example, genotype, for strains; concentration, source, purity, and/or restriction map for nucleic acids; and base sequence, for oligonucleotides.
7. *Procedure:* Write down exact procedure and flow chart, before what you are going to do before you do it and make sure you understand each step before you do it.
   - You should include everything you do including all volumes and amounts; many protocols are written for general use and must be adapted for a specific application.
   - Writing a procedure out helps you to remember and to understand what it is about. It will also help you to identify steps that may be unclear or that need special attention.

- Some procedures can be several pages long and include more information than is necessary in a notebook. However, it is good laboratory practice to have a separate notebook containing methods that you use on a regular basis.
- If an experiment is a repeat of an earlier experiment, you do not have to write down each step but refer to the earlier experiment by page or experiment number. If you make any changes, note the changes and why.
- Flow charts are sometimes helpful for experiments that have many parts.
- Tables are also useful if an experiment includes a set of reactions with multiple variables.

8. *Results:* This section should include *all* raw data, including gel photographs, printouts, colony counts, graphs, autoradiographs, etc. This section should also include your analyzed data, for example, transformation efficiencies, calculations of specific activities or enzyme activities.
9. *Conclusions/Summary:* This is one of the most important sections. You should summarize all of your results, even if they were stated elsewhere and state any conclusions you can make.

GENERAL LABORATORY METHODS

1. SAFETY PROCEDURES

(*a*) *Chemicals.* A number of chemicals used in this laboratory are hazardous. All manufacturers of hazardous materials are required by law to supply the user with pertinent information on any hazards associated with their chemicals. This information is supplied in the form of Material Safety Data Sheets or MSDS. This information contains the chemical name, CAS#, health hazard data, including first aid treatment, physical data, fire and explosion hazard data, reactivity data, spill or leak procedures, and any special precautions needed when handling this chemical. A file containing MSDS information on the hazardous substances used in the Molecular Biology course is kept in the lab. In addition, MSDS information can be accessed on World Wide Web on the Biological Sciences Home Page. You are strongly urged to make use of this information prior to using a new chemical and certainly in the case of any accidental exposure or spill.

The following chemicals are particularly noteworthy:

Phenol—can cause severe burns
Acrylamide—potential neurotoxin
Ethidium bromide—carcinogen.

These chemicals are not harmful if used properly: always wear gloves when using potentially hazardous chemicals and never mouth-pipette them. If you accidentally splash any of these chemicals on your skin, *immediately* rinse the area thoroughly with water and inform the instructor. Discard the waste in appropriate containers.

(*b*) *Ultraviolet Light.* Exposure to ultraviolet light can cause acute eye irritation. Since the retina cannot detect UV light, you can have serious eye damage and not realize it until 30 min to 24 hours after exposure. Therefore, *always wear appropriate eye protection when using UV lamps.*

(c) *Electricity.* The voltages used for electrophoresis are sufficient to cause electrocution. Cover the buffer reservoirs during electrophoresis. Always turn off the power supply and unplug the leads before removing a gel.

(d) *General Housekeeping.* All common areas should be kept free of clutter and all dirty dishes, electrophoresis equipment, etc. should be dealt with appropriately. Since you have only a limited amount of space to call your own, it is to your advantage to keep your own area clean. Since you will use common facilities, all solutions and everything stored in an incubator, refrigerator, etc. *must be labelled.* In order to limit confusion, each person should use his initials or other unique designation for labelling plates, etc. Unlabelled material found in the refrigerators, incubators, or freezers may be discarded. Always mark the backs of the plates with your initials, the date, and relevant experimental data, e.g. strain numbers.

2. PREPARATION OF SOLUTIONS

(a) *Calculation of Molar, % and "X" Solutions*

(i) A molar solution is one in which 1 litre of solution contains the number of grams cqual to its molecular weight.

Example. To make up 100 ml of a 5M NaCl solution = 58.456 (mw of NaCl) g × 5 moles × 0.1 litre = 29.29 g in 100 ml sol mole litre.

(ii) *Percent solutions.*

Percentage (*w*/*v*) = weight (g) in 100 ml of solution

Percentage (*v*/*v*) = volume (ml) in 100 ml of solution.

Example. To make a 0.7% solution of agarose in TBE buffer, weight 0.7 of agarose and bring up volume to 100 ml with TBE buffer.

(iii) *"X" solutions.* Many enzyme buffers are prepared as concentrated solutions, e.g. 5X or 10X (five or ten times the concentration of the working solution) and are then diluted such that the final concentration of the buffer in the reaction is 1X.

Example. To set up a restriction digestion in 25 µl, one would add 2.5 µl of a 10X buffer, the other reaction components, and water to a final volume of 25 µl.

(b) *Preparation of Working Solutions from Concentrated Stock Solutions.* Many buffers in molecular biology require the same components but often in varying concentrations. To avoid having to make every buffer from scratch, it is useful to prcpare several concentrated stock solutions and dilute as needed.

Example. To make 100 ml of TE buffer (10 mM Tris, 1 mM EDTA), combine 1 ml of a 1 M Tris solution and 0.2 ml of 0.5 M EDTA and 98.8 ml sterile water. The following is useful for calculating amounts of stock solution needed:

$C_i \times V_i = C_f \times V_f$, where C_i = initial concentration, or concentration of stock solution

V_i = initial volume, or amount of stock solution needed

C_f = final concentration, or concentration of desired solution

V_f = final volume, or volume of desired solution.

(c) *Steps in Solution Preparation*

(i) Refer to the laboratory manual for any specific instructions on preparation of the particular solution and the bottle label for any specific precautions in handling the chemical.

(ii) Weigh out the desired amount of chemical(s). Use an analytical balance if the amount is less than 0.1 g.

(iii) Take the chemical(s) in an appropriate size beaker with a stir bar.

(iv) Add less than the required amount of water. Prepare all solutions with double distilled water (in carboy).

(v) When the chemical is dissolved, transfer to a graduated cylinder and add the required amount of distilled water to achieve the final volume. An exception is in preparing solutions containing agar or agarose. Weigh the agar or agarose directly into the final vessel.

(vi) If the solution needs to be at a specific pH, check the pH meter with fresh buffer solutions and follow instructions for using a pH meter.

(vii) Autoclave, if possible, at 121°C for 20 min. Some solutions cannot be autoclaved, for example, SDS. These should be filter sterilized through a 0.22 μm filter. Media for bacterial cultures must be autoclaved the same day it is prepared, preferably within an hour or two. Store at room temperature and check for contamination prior to use by holding the bottle at eye level and gently swirling it.

(viii) Solid media for bacterial plates can be prepared in advance, autoclaved, and stored in a bottle. When needed, the agar can be melted in a microwave, any additional components, e.g. antibiotics, can be added and the plates can then be poured.

(ix) Concentrated solutions, e.g. 1M Tris-HCl pH = 8.0, 5M NaCl, can be used to make working stocks by adding autoclaved double-distilled water in a sterile vessel to the appropriate amount of the concentrated solution.

(d) *Glassware.* Glass and plastic ware used for molecular biology must be scrupulously clean.

Glassware should be rinsed with distilled water and autoclaved or baked at 150°C for 1 hour. For experiments with RNA, glassware and solutions are treated with diethylpyrocarbonate to inhibit RNases which can be resistant for autoclaving.

Plastic ware such as pipettes and culture tubes are often supplied sterile. Tubes made of polypropylene are turbid and are resistant to many chemicals, like phenol and chloroform; polycarbonate or polystyrene tubes are clear and not resistant to many chemicals. Micropipette tips and microfuge tubes should be autoclaved before use.

3. DISPOSAL OF BUFFERS AND CHEMICALS

(i) Any uncontaminated, solidified agar or agarose should be discarded in the trash, not in the sink, and the bottles rinsed well.

(ii) Any media that becomes contaminated should be promptly autoclaved before discarding it. Petri dishes and other biological waste should be discarded in Biohazard containers which will be autoclaved prior to disposal.

(iii) Organic reagents, e.g. phenol, should be used in a fume hood and all organic waste should be disposed of in a labelled container, not in the trash or the sink.

(iv) Ethidium bromide is a mutagenic substance that should be treated before disposal and should be handled only with gloves. Ethidium bromide should be disposed of in a labeled container.

4. EQUIPMENT

(a) *General Comments.* Keep the equipment in good working condition. Don't use anything (instrument) unless you have been instructed in the proper use. Report any malfunction immediately. Rinse out all centrifuge rotors after use and in particular if anything spills.

Please do not waste supplies—use only what you need. If the supply is running low, please notify to the instructor before the supply is completely exhausted. Occasionally, it is necessary to borrow a reagent or a equipment from another lab, notify to the instructor.

(b) *Micropipettors.* Most of the experiments you will conduct in this laboratory will depend on your ability to accurately measure volumes of solutions using micropipettors. The accuracy of your pipetting can only be as accurate as your pipettor and several steps should be taken to insure that your pipettes are accurate and are maintained in good working order. Then they should checked for accuracy following the instructions given by the instructor. If they need to be recalibrated, do so.

There are two different types of pipettors, Rainin pipetmen and Oxford benchmates. Since the pipettors will use different pipette tips, make sure that the pipette tip you are using is designed for your pipettor.

(c) *Using a pH Meter.* Biological functions are very sensitive to changes in pH and hence, buffers are used to stabilize the pH. A pH meter is an instrument that measures the potential difference between a reference electrode and a glass electrode, often combined into one combination electrode. The reference electrode is often $AgCl_2$. An accurate pH reading depends on standardization, the degree of static charge, and the temperature of the solution.

(d) *Autoclave Operating Procedures.* Place all material to be autoclaved in a autoclavable tray. All items should have indicator tape. Separate liquids from solids and autoclave separately. Make sure lids on all bottle are loose.

Make sure chamber pressure is at zero before opening the door.

5. WORKING WITH DNA

(a) Storage

- The following properties of reagents and conditions are important considerations in processing and storing DNA and RNA. Heavy metals promote phosphodiester breakage. EDTA is an excellent heavy metal chelator.
- Free radicals are formed from chemical breakdown and radiation and they cause phosphodiester breakage. UV light at 260 nm causes a variety of lesions, including thymine dimers and crosslinks. Biological activity is rapidly lost. 320 nm irradiation can also cause crosslinks.

- Ethidium bromide causes photooxidation of DNA with visible light and molecular oxygen. Oxidation products can cause phosphodiester breakage. If no heavy metals are present, ethanol does not damage DNA.
- 5°C is one of the best and simplest conditions for storing DNA. –20°C this temperature causes extensive single and double strand breaks.
- –70°C is probable excellent for long-term storage. For long-term storage of DNA, it is best to store in high salt (>1M) in the presence of high EDTA (>10 mM) at pH 8.5.
- Storage of DNA in buoyant CsCl with ethidium bromide in the dark at 5°C is excellent.

(b) *Purification.* To remove protein from nucleic acid solutions:

(i) Treat with proteolytic enzyme, e.g., pronase, proteinase K.

(ii) Phenol Extract. The simplest method for purifying DNA is to extract with phenol or phenol:chloroform and then chloroform. The phenol denatures proteins and the final extraction with chloroform removes traces of phenol.

(iii) CsCl/ethidium bromide density gradient.

(c) *Quantitation*

(i) **Spectrophotometric.** For a pure solution of DNA, the simplest method of quantitation is reading the absorbance at 260 nm where an OD of 1 in a 1 cm path length = 50 µg/ml for double-stranded DNA, 40 µg/ml for single-stranded DNA and RNA and 20–33 µg/ml for oligonucleotides. An absorbance ratio of 260 nm and 280 nm gives an estimate of the purity of the solution. Pure DNA and RNA solutions have OD_{260}/OD_{280} values of 1.8 and 2.0, respectively. This method is not useful for small quantities of DNA or RNA (<1 µg/ml).

(ii) **Ethidium bromide fluorescence.** The amount of DNA in a solution is proportional to the fluorescence emitted by ethidium bromide in that solution. Dilutions of an unknown DNA in the presence of 2 µg/ml ethidium bromide are compared to dilutions of a known amount of a standard DNA solution spotted on an agarose gel or Saran Wrap or electrophoresed in an agarose gel.

(d) *Concentration: Precipitation with ethanol* DNA and RNA solutions are concentrated with ethanol as follows: The volume of DNA is measured and the monovalent cation concentration is adjusted. The final concentration should be 2–2.5M for ammonium acetate, 0.3M for sodium acetate, 0.2M for sodium chloride and 0.8M for lithium chloride. The ion used often depends on the volume of DNA and on the subsequent manipulations; for example, sodium acetate inhibits Klenow, ammonium ions inhibit T4 polynucleotide kinase, and chloride ions inhibit RNA-dependent DNA polymerases. The addition of $MgCl_2$ to a final concentration of 10 mM assists in the precipitation of small DNA fragments and oligonucleotides. Following addition of the monovalent cations, 2–2.5 volumes of ethanol are added, mixed well, and stored on ice or at –20°C for 20 min to 1 hour. The DNA is recovered by centrifugation in a microfuge for 10 min (room temperature is okay). The supernatant is carefully decanted making certain that the DNA pellet, if visible, is not discarded (often the pellet is not visible until it is dry). To remove salts, the pellet is washed with 0.5–1.0 ml of 70% ethanol, spun again, the supernatant decanted, and the pellet dried. Ammonium acetate is very soluble in ethanol and is effectively removed by a 70% wash. Sodium acetate and sodium chloride are less effectively removed. For fast drying, the pellet can spun briefly in a Speedvac, although the method is not recommended for many DNA preparations as DNA that has been overdried is difficult to resuspend and also tends to denature small fragments of DNA. Isopropanol is also used to precipitate DNA but it tends to co-precipitate salts and is harder to evaporate since it is less volatile. However, less isopropanol is required than ethanol to precipitate DNA and it is sometimes used when volumes must be kept to a minimum, e.g., in large scale plasmid preps.

(*e*) *Restriction Enzymes.* Restriction and DNA modifying enzymes are stored at –20°C in a non-frost free freezer, typically in 50% glycerol. The enzymes are stored in an insulated cooler which will keep the enzymes at –20°C for some period of time.

2 TOOLS AND TECHNIQUES IN BIOLOGICAL STUDIES

1. SPECTROPHOTOMETRY

A spectrophotometer measures the relative amounts of light energy passed through a substance that is absorbed or transmitted. We will use this instrument to determine how much light of a certain wavelength(s) is absorbed by (or transmitted) through a solution. Transmittance (T) is the ratio of transmitted light to incident light. Absorbance (A) = – log T. Absorbance is usually the most useful measure because there is a linear relationship between absorbance and concentration of a substance. This relationship is shown by the Beer-Lambert law:

A = ebc

e = extinction coefficient (a proportionality constant that depends on the absorbing species)

b = pathlength of the cuvette. Most standard cuvettes have a 1 cm path and thus this can be ignored

c = concentration.

A spectrophotometer or calorimeter makes use of the transmission of light through a solution to determine the concentration of a solute within the solution. A spectrophotometer differs from a calorimeter in the manner in which light is separated into its component wavelengths. A spectrophotometer uses a prism to separate light and a calorimeter uses filters.

Both are based on a simple design of passing light of a known wavelength through a sample and measuring the amount of light energy that is transmitted. This is accomplished by placing a photocell on the other side of the sample. All molecules absorb radiant energy at one wavelength of another. Those that absorb energy from within the visible spectrum are known as pigments. Proteins and nucleic acids absorb light in the ultraviolet range. The following figure demonstrates the radiant energy spectrum with an indication of molecules which absorb in various regions of that spectrum.

The design of the single beam spectrophotometer involves a light source, a prism, a sample holder and a photocell. Connected to each are the appropriate electrical or mechanical systems to control the illuminating intensity, the wavelength, and for conversion of energy received at the photocell into a voltage fluctuation. The voltage fluctuation is then displayed on a meter scale, is displayed digitally, or is recorded via connection to a computer for later investigation.

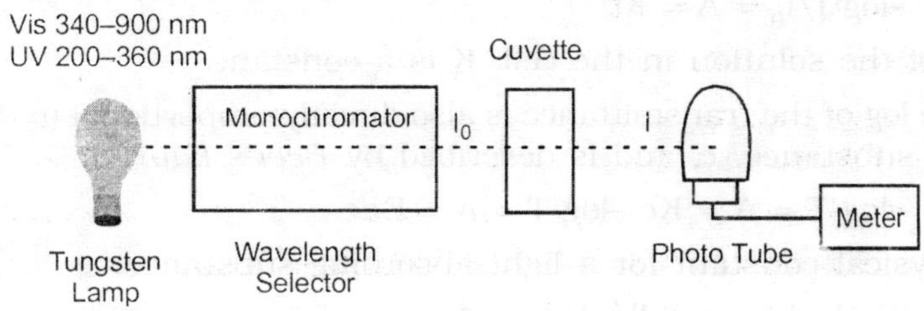

Fig. 1. Spectrophotometer construction.

Spectrophotometers are useful because of the relation of intensity of colour in a sample and its relation to the amount of solute within the sample. For example, if you use a solution of red food colouring in water, and measure the amount of blue light absorbed when it passes through the solution, a measurable voltage fluctuation can be induced in a photocell on the opposite side. If now the solution of red dye is diluted in half by the addition of water, the colour will be approximately 1/2 as intense and the voltage generated on the photocell will be approximately half as great. Thus, there is a relationship between the voltage and the amount of dye in the sample.

Given the geometry of a spectrophotometer, what is actually measured at the photocell is the amount of light energy which arrives at the cell. The voltage meter is reading the amount of light TRANSMITTED to the photocell.

We can monitor the transmission level and convert it to a percentage of the amount transmitted when no dye is present. Thus, if 1/2 the light is transmitted, we can say that the solution has a 50% transmittance.

Transmittance is the relative percent of light passed through the sample.

The conversion of that information from a percent transmittance to an inverse log function known as the Absorbance (or Optical Density).

The monochromator selects a particular wavelength. The sample and a blank are located in cuvettes. The light from the lamp passes through the cuvette and hits the phototube. The meter then records the signal from the phototube.

I_0 = incident light, has intensity I_0

I = light coming out of the cuvette (that contains light absorbing substance), has intensity I.

Quantitative Aspects of Light Absorption: The Lambert-Beer Law

Transmittance, T, is the amount of light that passes through a substance. It is sometimes called percent transmission:

$$T = I/I_0$$

$$\%T = I/I_0$$

I_0 is the intensity of the incident light and I is the transmitted light. The light absorbed by the substance at a particular wavelength depends on the length of the light path through the substance. The negative logarithm of the transmittance, the absorbance A, is directly proportional to the amount of light absorbed and to the length of the light path and is described by the Lambert Law:

$$-\log T = -\log I/I_0 = A = Kd$$

d is the length of the solution in the cell. K is a constant.

The negative log of the transmittance is also directly proportional to the concentration of the absorbing substance, *c*, and is described by *Beer's Law*:

$$-\log I/I_0 = -\log T = A = Kc \quad -\log T = A = Edc$$

where E is a physical constant for a light absorbing substance.

A = E*cd*, *d* is usually 1 cm

A = absorbance (sometimes called the optical density)

E = molar extinction coefficient

c = concentration of the light absorbing substance.

Operation

1. Turn on the spectrophotometer and allow 10 minutes for warm up of the instrument before use.
2. Adjust the wavelength to that specified for the procedure you are using.
3. Be sure the cover is closed on the cuvette holder and use the left knob on the front panel to adjust the dark current such that the meter is reading 0 transmittance. At this point, you are simply adjusting the internal electronics of the instrument to blank out any residual currents. This adjusts the lower limit of measurements. It establishes that no light is equivalent to 0 transmittance or infinite absorbance.
4. Insert a clean cuvette containing the blank into the holder. Be sure that the tube is clean, free of finger prints and that the painted line marker on the tube is aligned with the mark on the tube holder. Close the top of the tube holder. The blank for this exercise is the solution containing no dopachrome, but all other chemicals. The amount of solution placed in the cuvette is not important, but is usually about 5 ml. It should approximately reach the bottom of the logo printed on the side of the cuvette.
5. Adjust the meter to read 100% transmittance, using the right knob on the front of the instrument. This adjusts the instrument to read the upper limit of the measurements and establishes that your blank will give a reading of 100% transmittance (0 absorbance).
6. Remove the blank from the instrument and recheck that your 0 transmittance value has not changed. If it does, wait a few minutes for the instrument to stabilize and read steps 1–5. Periodically throughout the exercise, check that calibration of the instrument is stable by re-inserting the blank and checking that the 0 and 100% T values are maintained.
7. To read a sample, simply insert a cuvette holding your test solution and close the cover. Read the transmittance value directly on the scale.
8. Record the % transmittance of your solution, remove the test tube cuvette and continue to read and record any other solutions you may have.

It is possible to read the absorbance directly, but with an analog meter (as opposed to a digital read out), absorbance estimations are less accurate and more difficult than reading transmittance. Absorbance can be easily calculated from the transmittance value. Be sure that you note which value you measure!

Absorption Spectrum

Analysis of pigments often requires a slightly different use of a spectrophotometer. In the use of the instrument for determination of concentration (Beer-Lambert Law), the wavelength was pre-set and left at a single value throughout the use of the instrument. This value is often given by the procedure being employed, but can be determined by an analysis of the absorption of a solution as the wavelength is varied.

The easiest means of accomplishing this is to use either a dual beam spectrophotometer or a computer controlled instrument. In either event, the baseline must be continuously re-read as the wavelength is altered.

To use a single beam spectrophotometer, the machine is adjusted to zero first, with the blank solution and then the sample is inserted and read. The wavelength is then adjusted up or down by some determined interval, the zero is checked, the blank re-inserted and adjusted, and the sample re-inserted and read. This procedure continues until all wavelengths to be scanned have been read.

In this procedure, the sample remains the same, but the wavelength is adjusted. Compounds have differing absorption coefficients for each wavelength. Thus, each time the wavelength is altered, the instrument must be recalibrated.

A dual beam spectrophotometer divides the light into two paths. One beam is used to pass through a blank, while the remaining beam passes through the sample. Thus, the machine can monitor the difference between the two as the wavelength is altered. These instruments usually come with a motor driven mechanism for altering the wavelength, or scanning the sample.

The newer version of this procedure is the use of an instrument which scans a blank, and places the digitalized information in its computer memory. It then rescans a sample and compares the information from the sample scan to the information obtained from the blank scan. Since the information is digitalized (as opposed to an analog meter reading), manipulation of the data is possible. These instruments usually have direct ports for connection to personal computers, and often have built in temperature controls as well. This latter option would allow measurement of hanges in absorbtion due to temperature changes (known as hyperchromicity). These in turn can be used to monitor viscosity changes, which is related to the degree of molecular polymerization with the sample. For instruments with this capability, the voltage meter scale has given way to a CRT display, complete with graphics and built in functions for statistical analysis.

A temperature controlled UV spectrophotometer capable of reading several samples at pre-programmed time intervals is invaluable for enzyme kinetic analysis. An example of this type of instrument is the Beckman DU-70.

2. ELECTROPHORESIS

Electrophoresis is the migration of charged molecules in response to an electric field. Their rate of migration depends on the strength of the field; on the net charge, size and shape of the molecules, and also on the ionic strength, viscosity and temperature of the medium in which the molecules are moving. As an analytical tool, electrophoresis is simple, rapid and highly sensitive. It is used analytically to study the properties of a single charged species, and as a separation technique.

There are a variety of electrophoretic techniques which give different information and have different uses. Generally the samples are run in a support matrix, the most commonly used being agarose and polyacrylamide. These are porous gels and under appropriate conditions they provide a means of separating molecules by size. We will focus on those methods used for proteins. These can be denaturing or nondenaturing. Nondenaturing methods allow recovery of active proteins and can be used to analyze enzyme activity or any other analysis that requires a native protein structure. Two commonly used techniques in biochemistry are Sodium Dodecyl Sulfate Polyacrylamide Gel Electrophoresis (SDS/PAGE) and Isoelectric focusing (IEF). SDS/PAGE separates proteins according to molecular weight and IEF according to isoelectric point. This laboratory exercise will introduce you SDS/PAGE.

SDS/PAGE

The gel matrix used is a crosslinked acrylamide polymer. This electrophoretic method separates the proteins according to size and not charge due to the presence of SDS. The dodecyl sulfate ions bind to the peptide backbone both denaturing the proteins and giving them a uniform negative charge.

The gels we will be running use a discontinuous system, that is they have two parts. One is the separating gel which has a high concentration of acrylamide and acts as a molecular sieve to separate the proteins according to size. Before reaching this gel, the proteins migrate through a stacking gel which serves to compress the proteins into a narrow band so they all enter the separating gel at about the same time. The narrow starting band increases the resolution. This part of the gel has a lower concentration of acrylamide to avoid a sieving effect.

The stacking effect is due to the glycine in the buffer, the low pH in the stacking gel, and the higher pH in the running buffer. At the low pH, the glycine has little negative charge and thus moves slowly. The chloride ions move quickly and a localized voltage gradient develops between the two. As the gel runs, the low pH of the stacking gel buffer is replaced by the higher pH in the running buffer. This maintains a discontinuity in the pH and keeps the glycine moving forward (any glycine molecules behind would acquire a higher charge and speed up). Since there is no real sieving going on, the proteins (which have intermediate mobility) form a tight band, in order of size, between the slower glycine and the faster chloride ions. The separating gel buffer has a higher pH so the glycine molecules become more negatively charged and move past the proteins and the voltage gradient becomes uniform. The proteins slow down in the smaller pore size of the separating gel and separate according to size.

Exercise: You will be given protein molecular weight standards, several different solutions containing individual proteins, and a sample of the same serum you used in the protein quantitation lab. Your job is to determine the molecular weights of the individual proteins and the major components in the serum sample. You will run each sample on two gels, one you prepare yourself and a commercial precast gel and compare the results.

Before doing electrophoresis, you must know the amount of protein in each sample. Determine the protein concentrations of each of your samples using a protein assay before coming to lab to do any electrophoresis. For this exercise, the only sample of unknown protein concentration is the serum which you used for one of your unknowns last week. The amount of protein to be loaded depends on the thickness and length of the gel, and the staining system to be used. Using the Coomassie blue staining system, as little as 0.1 μg can be detected but more will be easier to see. As a guide use

0.5–5 μg for pure samples (one or very few proteins) and 20–60 μg for complex mixtures where the protein will be distributed amongst many protein bands. Overloading will decrease the resolution.

Protocol: The apparatus used in gel casting or running electrophoresis vary, make sure you look over the appropriate manuals before you operate.

Caution: UNPOLYMERIZED ACRYLAMIDE IS A NEUROTOXIN. BE CAREFUL! Do not pour unpolymerized acrylamide down the sink, wait for it to polymerize and dispose of in the trash.

TEMED is also not very good for you and very smelly, avoid breathing it. Open the bottle only as long as necessary or use it in the hood.

1. Make sure gel plates are clean and dry. Do not get your fingerprints on them or the acrylamide will not polymerize properly.
2. Prepare gel solutions (separating and stacking) except do NOT add polymerizing agents, APS and TEMED (this would start the polymerization).
3. Lay the comb on the unnotched plate and mark (on the outside, using a sharpie) about 1 cm below the bottom of the teeth. This will be the level of the separating gel. If available, use an alumina (opaque, white) plate for the notched plate as this conducts heat away from the gel more efficiently than glass. Set up the gel plates, spacers, and plastic pouch in the gel casting as described in the manufacture's directions. When everything is completely ready, add TEMED to the separating gel solution, mix well, and pour it between the plates up to the mark. Wear gloves if you pour directly from the beaker. You can also use a disposable pipette. Work quickly or the solution will polymerize too soon. Carefully layer isopropanol (or water saturated butanol) on top of acrylamide so it will polymerize with a flat top surface (i.e., no meniscus). Do this at the side and avoid large drops so as not to disturb the gel surface. When the leftover acrylamide in the beaker is polymerized, that between the plates will also be ready.
4. If you will run the gel on the same day, prepare samples while the acrylamide is polymerizing otherwise wait until you are ready to run the gel.

   (*i*) You will need a sample of each unknown plus the molecular weight standards. Prepare samples in screw cap microcentrifuge tubes. The protein content should be at 1–50 μg in 20–30 μL sample.

   Total sample volume that can be loaded depends on the thickness of the gel and the diameter of the comb teeth. For Bangalore genei apparatus ~ 30 μL/well. To prepare sample, mix 7–10 μL sample (depending on protein concentration) +20 μL 2X sample buffer containing 10% β-mercapto-ethanol (BME). Use the BME in the hood - IT STINKS! For dilute samples 40 μL sample +10 μL 5X sample buffer. Add 2 μL BME. Heat to 90°C for 3 minutes to completely denature proteins. It is important to heat samples immediately after addition of sample buffer. Partially denatured proteins are much more susceptible to proteolysis and proteases are not the first proteins to denture. (Heat samples to 37°C to redissolve SDS before running the gel if samples have been stored after preparation).

   (*ii*) If you wish proteins in the sample to retain disulfide bonds, do not add BME. If both reduced and non-reduced samples will be run on the same gel, leave at least 3–4 empty wells between samples as the BME will diffuse between wells and reduce proteins in adjacent samples.

(iii) *MW Stds:* 7 μL of Rainbow stds +10 μL sample buffer (do not make in advance). Heat to 37°C before use.

5. After the separating gel has polymerized, drain off the isopropanol. Add TEMED to the stacking gel solution, pour the solution between the plates and insert the comb to make wells for loading samples. The person putting in the comb should wear gloves. Keep an eye on this while it's polymerizing and add more gel solution if level falls (as it usually does) or the wells will be too small.

6. After polymerization, do not cut the bag; we reuse them. The gel may be stored at this point by taping the bag shut to prevent drying.

   When ready to run gel: Mark the position of each well as they are difficult to see when full.

7. Remove comb and rinse wells with running buffer. See manual directions for setting up the gels in the buffer chambers. The apparatus can run two gels simultaneously. There is a blank plate to use when running only one. Fill upper chamber with running buffer first and check for leaks. Adjust plates if necessary. Load samples using a micropipettor with gel loading tips (these are longer and thinner than the normal tips). This will be demonstrated. Do not load samples in the end wells. Make sure to write down which sample was loaded in each well.

8. Electrophoresis (takes 1–2 hrs)

   Connect the gel apparatus to the power supply and run at 15 mA/gel until the tracking dye (blue) moves past the end of the stacking gel. Increase the current to 20–25 mA/gel but make sure the voltage does not get above 210 V. Run until the blue tracking dye moves to the bottom of the separating gel.

   For BioRad apparatus do not exceed 30 mA regardless of the number of gels.

9. Disassemble the apparatus and carefully separate the gel plates using a flat spatula.

   Cut off the stacking gel and any gel below the blue tracking dye. Note the colour of each of the molecular weight standards as they will all be blue afterstaining. Wash 3X with distilled water.

   Place the gel in a plastic staining container and add Coomassie blue staining solution. Keep it in this 1 hr to overnight. Wash again with water.

   You can wrap the gel in plastic wrap and xerox or scan it to have a copy. The gel may also be dried.

Data Analysis

Measure the length of the gel (since you cut off the bottom, this is the distance traveled by the dye).

Measure the distance traveled by each of the molecular weight standards. Measure the distances of each unknown band.

For samples lanes with many bands (Serum in this exercise), measure all bands in those with just a few and the major bands in those that have many.

Prepare a standard curve by plotting log MW vs relative mobility (Rf, distance traveled by protein divided by distance traveled by dye). Use this and the mobility of bands from your fractions to determine the MW of the unknown proteins. (Review standard curves from the protein quantitation lab if necessary). MW of proteins which do not run very far into the gel or run near the dye front will not be accurate.

If you have reduced and unreduced samples, compare the number of bands and MW of each to determine the number of subunits.

Gel Solutions

1. Separating Gel: (15 mL, enough for two gels) 10% acrylamide.

   40% Acrylamide/bisacrylamide mix 3.55 mL.

   1.5 M Tris pH 8.8, 3.75 mL, H_2O 7.4 mL, 10% SDS 150 μL, 10% Ammonium Persulfate (APS) 150 μL (prepared fresh), TEMED 6 μL.

2. Stacking Gel: (5 mL) 5% acrylamide.

   Compresses protein sample into narrow band for better resolution.

   40% Acrylamide/bisacrylamide mix 0.625 mL.

   0.5 M Tris pH 6.8, 1.25 mL, H_2O 3.0 mL, 10% SDS 50 μL, 10% APS 50 μL, TEMED 5 μL.

3. 2X Sample Buffer (10 mL) store in freezer for extended time.

4. SDS must be at room temperature to dissolve.

5. H_2O 1.5 mL, 0.5M Tris pH 6.8, 2.5 mL, 10% SDS (optional) 4.0 mL, Glycerol 2.0 mL, BPB 0.01%, β-Mercaptoethanol (optional) 0.1 mL.

6. Running Buffer (5L)

   30 g Tris Base, 144 g Glycine, Dissolve in sufficient H_2O to make 1.5 L and put into final container.

   Add 1.5 g SDS (CAUTION: do NOT inhale dust).

When adding SDS, avoid making too much foam which makes measuring and pouring difficult.

Final pH should be around 8.3 but do not adjust or ionic strength will be too high and the gel will not run properly. If pH is way off, it was made incorrectly or is old and has some contamination.

Running buffer can also be made more concentrated (5X or 10X) and diluted as needed to save bottle space.

3. COLUMN CHROMATOGRAPHY

Column chromatography is one of many forms of chromatography. Others include paper, thin-layer, gas, HPLC, and so on. Most forms of chromatography use a two-phase system to separate substances on the basis of some physical-chemical property. One phase is usually a stationary phase. The second phase is usually a mobile phase (often a buffer in biochemistry) which carries the sample components along at different rates of mobility. The separation is based on how well the stationary phase retards the components vs. how quickly the mobile phase moves them along. Substances with different properties will thus elute (exit) from the column at different times. Some common types of column chromatography used in biochemistry are gel filtration, ion exchange and affinity. You will have the opportunity to use one or more of these during your projects. In this exercise you will use gel filtration chromatography.

(*a*) *Gel Filtration (permeation) Chromatography.* Gel filtration uses a gel matrix as the stationary phase. The matrix consists of very small porous beads. The large molecules of a sample solution do not get "caught" in the pores of the gel and will transit through the column more rapidly because they can go around the beads. They are said to be "excluded" from the matrix. Smaller molecules that can enter the gel pores must go through the beads thus taking more time to reach the bottom of the column. Medium size molecules can enter larger pores but not small ones. This form is also referred to as "molecular sieve" chromatography in that the components of a sample are separated according to their molecular size (and to a certain extent, molecular shape). The gel matrices are commonly made of crosslinked polysacchrides or polyacrylamide both of which can be made with varying pore sizes. The information supplied by the manufacturer will state the size of the beads, the approximate size of molecules that will be excluded, and the range of molecular weight range that can be separated. By using gels of different sizes and porosities, one can separate samples having a large variety of components.

A few useful definitions:

Bed volume (V_t) is the total volume inside the column.

Void volume (V_0) is the volume of solution not trapped in the beads.

Internal volume (V_i) is the volume of solution trapped in the beads.

Volume of the gel matrix (V_g).

$$V_t = V_0 + V_i + V_g$$

Elution volume (V_e) is the volume necessary to elute a substance from the column.

(*b*) *Ion Exchange Chromatography.* In this type of chromatography, the matrix is covalently linked to anions or cations. Solute ions of the opposite charge in the mobile liquid phase are attracted to the resin by electrostatic forces. There are two basic matrix types; anion exchangers bind anions in solution and cation exchangers bind cations. As the sample components go through the column those with the appropriate charge bind and the others are eluted. Proteins have many ionizable groups with different pK values thus the charge on the protein will depend on the pH of the buffer used. Thus, one must carefully choose the exchanger and pH of the buffer used for the mobile phase. Once all unbound substances have passed through the column the bound molecules can be eluted by changing the buffer. One way is to increase the ionic strength (either gradually using a gradient or all at once depending on whether you wish to fractionate the bound components elute them all at once respectively). The anions or cations in the salt will compete with the bound molecules and cause them to dissociate from the matrix. The higher the charge density on the bound molecules the higher salt concentration will be required to effectively remove them. Another option is to change the pH and thus the charge of the proteins. Problem: You use ion exchange chromatography with DEAE cellulose (an anion exchanger) to separate proteins with the following pI values: 3.5, 5.2, 7.1, 8.5. The proteins are loaded onto the column in a low ionic strength buffer, pH = 7.0. The column is then washed and eluted with a gradient of 0.05–0.50 M NaCl in the same buffer. What is the order of elution of the proteins?

(c) *Affinity Chromatography.* Affinity chromatography utilizes the specific interaction between one kind of solute molecule and a second molecule that is immobilized on a stationary phase. For example, the immobilized molecule may be an antibody to some specific protein. When solute containing a mixture of proteins are passed by this molecule, only the specific protein is reacted to this antibody, binding it to the stationary phase. This protein is later eluted by changing the ionic strength or pH. Alternatively, an excess of the molecule immobilized on the stationary phase may be used. For example, if the molecule you wish to purify binds glucose it can be separated from molecules that don't by using a glucose affinity column (The matrix contains immobilized glucose molecules). Only glucose binding molecules will bind to this matrix. The bound molecules can be eluted by adding glucose to the elution buffer. This will compete with the matrix-bound glucose for the binding sites on the protein and the proteins (now bound to free glucose) will dissociate from the matrix and elute from the column. This method is gentler but can only be used in some cases. This elution method is only feasible when the immobilized molecule is small, readily available, and cheap as is the case for glucose.

Exercise for Gel Filtration Chromatography

Determine the "bed volume" of the glass column by filing the column with water and measuring with a graduate cylinder.

Preparation of the Gel

1. You will use Sephadex G-100 for this experiment. The gel has a fractionation range for proteins of 4000–150,000 daltons. Sephadex is supplied as a dry powder and must be hydrated before use. The amount of water absorbed and the time required depends on the type of gel. Sephadex G-100 takes three days at room temperature or 3 hours in a boiling water bath. One gram of dry powder will make about 15–20 mL of gel. Weigh out the powder and add a large amount of water. Gentle stirring may be used but vigorous stirring will break the beads.

2. When the gel is ready, decant the water. Some of the very fine particles will also be decanted. This is not a problem. In fact, it is good to remove the "fines" as they will pass through bottom support screen of the column or clog the column and slow the flow. Replace the water with PBS and stir to equilibrate the gel with the buffer. Allow to settle and decant again.

3. Degas with gentle vacuum just before use.

Packing the column:

4. Close the outlet of the column. Stir the gel to create a slurry and carefully fill the column without creating areas of different densities. The most even packing will be achieved if you pour all the necessary slurry into the column at once. If necessary, stir the settling gel to prevent layers of gel from forming. Open the outlet and add buffer as the gel packs. Do not let the buffer drop below the top of the gel bed! If it is necessary to add more Sephadex, stir the top of the gel bed before adding more slurry.

5. If layers or air bubbles are still present in the column, invert the column and allow it resettle, doing this as many times as is necessary to obtain a well packed column.

6. Connect the column to the peristaltic pump and equilibrate the column by eluting 1 bed volume of PBS buffer at a flow rate of 1ml/min. Collect the eluent in a graduated cylinder.
7. Determine the void volume and check the packing. Blue Dextran is a large polysaccharide (average molar mass is about two million daltons). It is excluded from the beads and will be eluted in the void volume. Add blue Dextran solution to the top of the column and let it run into the gel. Immediately start collecting the eluent in a graduated cylinder. Gently put more buffer over the gel and run the peristaltic pump. Measure the amount of PBS eluted during the time it takes the Blue Dextran fraction to run the length of the column. This volume is the void volume. If your column was evenly packed, the Blue Dextran should run as a horizontal well-defined band through the column.
8. Prepare your protein mixture to 1 mg/ml concentration and add it carefully to the top of the column as you did for the Blue Dextran. For the best resolution, the sample volume should not exceed 1–2% of the column volume. You will run the following substances: Hemoglobin, myoglobin, cytochrome c, and vitamin B12. Vitamin B12 has a molar mass of 1355 D and should be completely included in the Sephadex beads. All of these substances are colored various shades of red or brown and so you should see them as they make their way down the column and in the collected fractions. Run the column at a rate of 0.5 mL/min. Rates that are two fast will decrease resolution and compress the gel. Start collecting one mL fractions and start the chart recorder as soon as you add the sample. The eluent passes through an absorbance detector (280 nm) and will detect the proteins as they elute.
9. Note the elution volume of each substance. A plot of log molar mass vs elution volume should be linear over the useful fractionation range (for roughly spherical proteins).

4. pH METER

Most biochemical experiments are done using buffered solutions since many reactions are very sensitive to the pH and some reactions use or produce hydrogen ions.

Buffer – a solution whose pH does not change very much when small amounts of acid (H^+) or base (OH^-) are added. This does not mean that no change occurs, only that it is small compared to the amount of acid or base added, the more acid or base added the more the pH will change. Buffer solutions consist of a conjugate acid-base pair (weak acid plus its salt or weak base plus its salt) in approximately equal amounts (within a factor of 10). Thus, buffers work best at pH within one pH unit of the pKa. The concentration of a buffer refers to the total concentration of the acid plus the base form. The higher the concentration of the buffer the greater its capacity to absorb acid or base. Most biological buffers are used in the range of 0.01–0.02 M concentration. The ratio of the two components and the pKa of the acid component determine the pH of the buffer.

$$\text{pH} = \text{pKa} + \log \frac{[\text{base form}]}{[\text{acid form}]}$$

If everything is behaving ideally, the pH should not depend on the buffer concentration or the presence of other ions in solution. In reality some buffers do better at this than others. It's best to check the pH of the final solution when preparing buffers from concentrated stocks.

Temperature will also affect pH since pKa values, like other equilibrium constants, change with temperature. Again, it's best to check the pH of the buffer at the temperature it will be used.

Some common buffers are given in the following table.

Agent	*pKa at 20°C*	*Comment*
Bicarbonate	6.3	Often used in tissue culture media, volatile, pH is affected by CO_2.
Citrate	6.4	Good buffer with sodium salt but sometimes binds metal ions.
Glycine	9.9	Often used in electrophoresis.
HEPES	7.5	Interferes with Lowry, forms radicals, not good for redox work.
Phosphate	pK2 = 7.2	May inhibit some enzymes and is a substrate for others, binds divalent cations, pH increases with dilution.
Tris	8.3	Interferes with Lowry and BCA assays, pH changes with temp. and dilution, may react with aldehydes.

pH meters should be calibrated regularly using commercially available reference buffers.

LIST a protocol for making 1 liter of phosphate buffered saline (PBS, 0.15 M NaCl, .02 M phosphate, pH = 7.2).

What is the pH of a 0.00043 N solution of HCl?

Weak Acid (dissociation is incomplete)

$$HAH^{+} + A^{-}$$

$$K_a = [H^+][A^-]/[HA]$$

$$[H^+] = K_a[HA]/[A^-]$$

$$\log [H^+] = \log K_a + \log [HA]/[A^-]$$

$$-\log [H^+] = -\log K_a - \log [HA]/[A^-]$$

$$pH = pK - \log [HA]/[A^-] \text{ or}$$

$$pH = pK + \log [A^-]/[HA]$$

Henderson-Hasselbach Equation

$$pH = pK - \log [HA]/[A^-] \text{ or}$$

$$pH = pK + \log [A^-/[HA]$$

Definition. A buffer is a mixture of a weak acid and its salt (or a mixture of a weak base and its salt).

pH Meter and pH Electrode

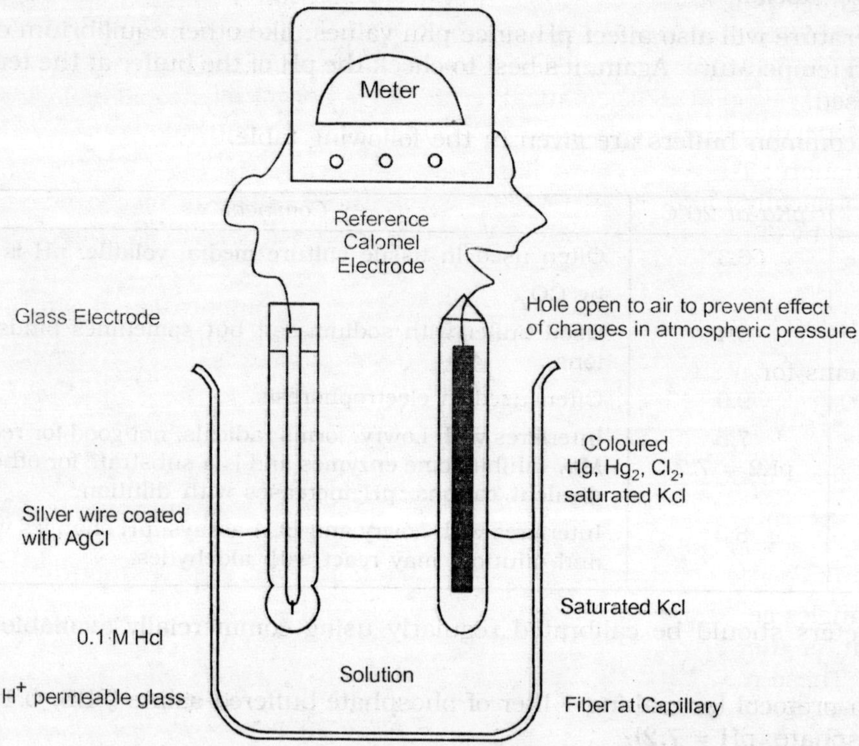

Fig. 2. Glass and reference electrodes of a pH meter.

Note: pH Meter : A pH meter measures the voltage between electrodes placed in a solution.

Most commonly used electrode is made from borosilicate glass which is permeable to H^+ but not to other cations or anions.

Inside is a 0.1 M HCl solution; outside lower H^+ concentration; thus passage of H^+ from inside to outside; leaves negative ion behind which generates an electric potential across the membrane.

$$E = 2.3 \times RT/F \times \log [H^+]_1/[H^+]_2$$

R = gas constant,

T = absolute temperature,

F = Faraday constant $[H^+]_1$ and $[H^+]_2$ are the molar H^+ concentrations inside and outside the glass electrode.

A reference electrode (pH-independent and impermeable to H^+ ions) is connected to the measuring electrode. Reference electrode contains Hg-Hg_2Cl_2 (calomel) paste in saturated KCl.

The concentration of 0.1 M HCl (inside the measuring electrode) may decrease by repeated use-therefore pH meter has to be standardized against solution of known pH.

5. CENTRIFUGATION

A centrifuge is a device for separating particles from a solution according to their size, shape, density, viscosity of the medium and rotor speed. In biology, the particles are usually cells, sub cellular organelles, viruses, large molecules such as proteins and nucleic acids. To simplify mathematical terminology we will refer to all biological material as spherical particles. There are many ways to classify centrifugation.

Introduction : The single most important advance in the use of centrifugal force to separate biologically important substances, was the coupling of mechanics, optics and mathematics by T. Svedberg and J.W. Williams in the 1920's. They initiated the mathematics and advanced the instrumentation.

Nowadays, any technique employing the quantitative application of centrifugal force is known as ultracentrifugation. The design of the basic instruments, the rotors and the optical systems for measurement.

Rotors

Rotors for a centrifuge are either fixed angles, swinging buckets, continuous flow, or zonal, depending upon whether the sample is held at a given angle to the rotation plane, is allowed to swing out on a pivot and into the plane of rotation, designed with inlet and outlet ports for separation of large volumes, or a combination of these.

Fixed angles generally work faster; substances precipitate faster in a given rotational environment, or they have an increased relative centrifugal force for a given rotor speed and radius. These rotors are the work-horse elements of a cell laboratory, and the most common is a rotor holding 8 centrifuge tubes at an angle of 34°C from the vertical.

Swinging bucket rotors (horizontal rotors) have the advantage that there is usually a clean meniscus of minimum area. In a fixed angle rotor, the materials are forced against the side of the centrifuge tube, and then slide down the wall of the tube. This action is the primary reason for their apparent faster separation, but also leads to abrasion of the particles along the wall of the centrifuge tube. For a swinging bucket, the materials must travel down the entire length of the centrifuge tube and always through the media within the tube. Since the media is usually a viscous substance, the swinging bucket appears to have a lower relative centrifugal force, that is it takes longer to precipitate anything contained within. If, however, the point of centrifugation is to separate molecules or organelles on the basis of their movements through a viscous field, then the swinging bucket is the rotor of choice. Most common clinical centrifuges have swinging buckets.

Cell biologists employ zonal rotors for the large scale separation of particles on density gradients. The rotors are brought up to about 3000 rpm while empty, and the density media and tissues are added through specialized ports.

Rotor Tubes

In using either a fixed angle or swinging bucket rotor, it is necessary to contain the sample in some type of holder. Continuous and zonal rotors are designed to be used without external tubes.

For biological work the tubes are divided into functional groups, made of regular glass, Corex glass, nitrocellulose, or polyallomer. Regular glass centrifuge tubes can be used at speeds below 3,000 rpm, that is in a standard clinical centrifuge. Above this speed, the xg forces will shatter the glass.

For work in the higher speed ranges, centrifuge tubes are made of plastic or nitrocellulose. Preparative centrifuge tubes are made of polypropylene and can withstand speeds up to 20,000 rpm.

Analytical/Preparative Centrifugation

The two most common types of centrifugation are analytical and preparative; the distinction is between the two is based on the purpose of centrifugation.

Analytical centrifugation involves measuring the physical properties of the sedimenting particles such as sedimentation coefficient or molecular weight. Optimal methods are used in analytical ultracentrifugation. Molecules are observed by optical system during centrifugation, to allow observation of macromolecules in solution as they move in gravitational field.

The samples are centrifuged in cells having windows that lie paralleled to the plan of rotation of the rotor head. As the rotor turns, the images of the cell (proteins) are projected by an optical system on to film or a computer. The concentration of the solution at various points in the cell is determined by absorption of a light of the appropriate wavelength. This can be accomplished either by measuring the degree of blackening of a photographic film or by the deflection of the recorder of the scanning system and fed into a computer.

The other type of centrifugation is called preparative and the objective is to isolate specific particles which can be reused. There are many type of preparative centrifugation such as rate zonal, differential, isopycnic centrifugation.

Ultracentrifugation/Low Speed Centrifugation

Another system of classification is the rate or speed at which the centrifuge is turning. Ultracentrifugation is carried out at speed faster than 20,000 rpm. Super speed ultracentrifugation is at speeds between 10,000 and 20,000 rpm. Low speed centrifugation is at speed below 10,000 rpm.

MOVING BOUNDARY/ZONE CENTRIFUGATION

A third method of defining centrifugation is by the way the samples are applied to the centrifuge tube. In moving boundary (differential centrifugation), the entire tube is filled with sample and centrifuged. Through centrifugation, one obtains a separation of two particles but any particle in the mixture may end up in the supernatant or in the pellet or it may be distributed in both fractions, depending upon it size, shape, density, and conditions of centrifugation. The pellet is a mixture of all of the sedimented components, and it is contaminated with whatever unsedimented particles were in the bottom of the tube initially. The only component which is purified is the slowest sedimenting one, but its yield is often very low. The two fractions are recovered by decanting the supernatant solution from the pellet. The supernatant can be recentrifuged at higher speed to obtain further purification, with the formation of a new pellet and supernatant.

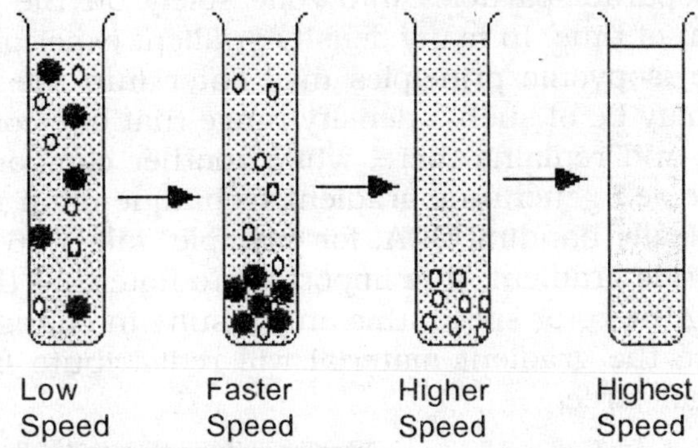

Fig. 3. Differential centrifugation.

In rate zonal centrifugation, the sample is applied in a thin zone at the top of the centrifuge tube on a density gradient. Under centrifugal force, the particles will begin sedimenting through the gradient in separate zones according to their size shape and density. The run must be terminated before any of the separated particles reach the bottom of the tube.

$$s = \frac{V}{C} = \frac{2/9\, r^2\, (d - d_0)}{\eta}$$

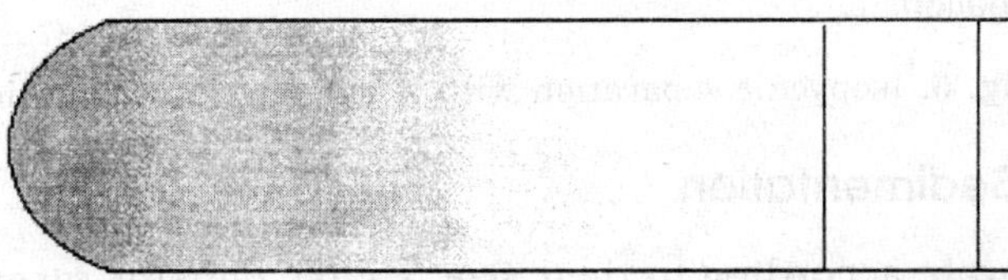

Fig. 4 (a).

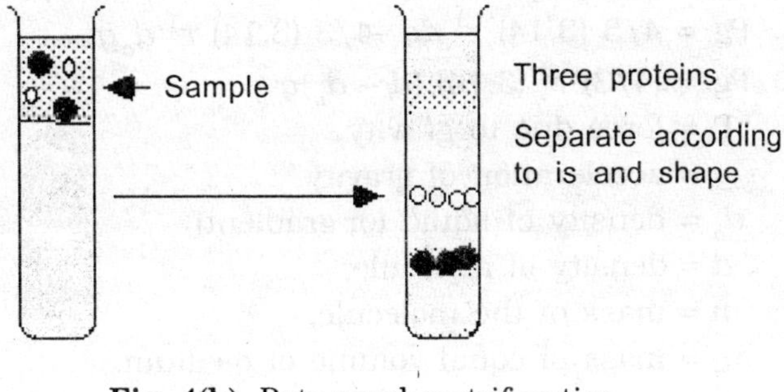

Fig. 4(b). Rate zonal centrifugation.

In isopycnic technique, the density gradient column encompasses the whole range of densities of the sample particles. The sample is uniformly mixed with the gradient material. Each particle will sediment only to the position in the centrifuge tube at which the gradient density is equal to its own density, and there it will remain. The isopycnic technique, therefore, separate particles into zone solely on the basis of their density differences, independent of time. In many density gradient experiments, particles of both the rate zonal and the isopycnic principles may enter into the final separations. For example, the gradient may be of such a density range that one component sediments to its density in the tube and remains there, while another component sediments to the bottom of the tube. The self generating gradient technique often requires long hours of centrifugation. Isopycnically banding DNA, for example, takes 36 to 48 hours in a self-generating cesium chloride gradient. It is important to note that the run time cannot be shortened by increasing the rotor speed; this only results in changing the position of the zones in the tube since the gradient material will redistribute farther down the tube under greater centrifugal force.

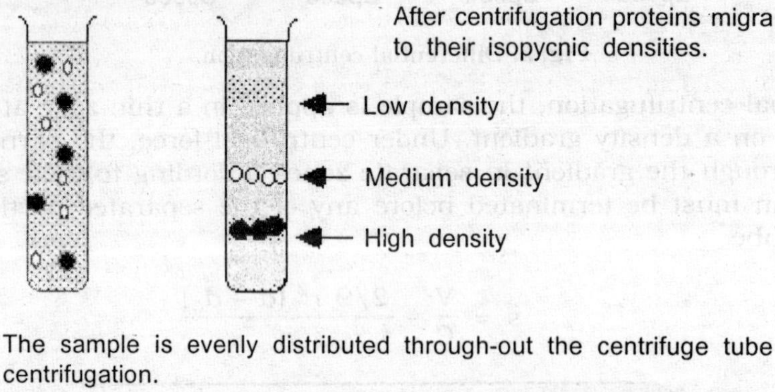

Fig. 5. Isopycnic separation with a self-generating gradient.

Basic Theory of Sedimentation

Molecules separate according to their size, shape, density, viscosity, and centrifugal force. The simplest case is a spherical molecule. If the liquid has the density of d_o and the molecule has a density of d and if $d > d_o$ then the protein will sediment. In gravitational field, the motor force (P) equals the acceleration of gravity (*g*) multiplied by the difference between the mass of the molecule and the mass of a corresponding volume of medium.

Equation 1

Equation 1. $Pg = (m - m_0)g$

Equation 2. $Pg = 4/3\ (3.14)\ r^3\ dg - 4/3\ (3.14)\ r^3\ d_o g$

Equation 3. $Pg = (4/3)\ r^3\ (3.14)\ (d - d_o)g$

P = force due to gravity,

g = acceleration of gravity,

d_o = density of liquid (or gradient)

d = density of molecule,

m = mass of the molecule,

m_o = mass of equal volume of medium.

In a centrifugal field, the gravitational acceleration (*g*) is replaced by the centrifugal force.

Fig. 6. Sedimentation of particles by gravity.

6. ULTRASONIC CELL DISRUPTION

The treatment of microbial cells in suspension with inaudible ultrasound (greater than about 18 kHz) results in their inactivation and disruption. Ultrasonication utilises the rapid sinusoidal movement of a probe within the liquid. It is characterised by high frequency (18 kHz^{-1} MHz), small displacements (less than about 50 m), moderate velocities (a few m s^{-1}), steep transverse velocity gradients (up to 4,000 s^{-1}) and very high acceleration (up to about 80,000 g). Ultrasonication produces cavitation phenomena when acoustic power inputs are sufficiently high to allow the multiple production of microbubbles at nucleation sites in the fluid. The bubbles grow during the rarefying phase of the sound wave, then are collapsed during the compression phase. On collapse, a violent shock wave passes through the medium. The whole process of gas bubble nucleation, growth and collapse due to the action of intense sound waves is called cavitation. The collapse of the bubbles converts sonic energy into mechanical energy in the form of shock waves equivalent to several thousand atmospheres (300 MPa) pressure. This energy imparts motions to parts of cells which disintegrate when their kinetic energy content exceeds the wall strength. An additional factor which increases cell breakage is the microstreaming (very high velocity gradients causing shear stress) which occur near radially vibrating bubbles of gas caused by the ultrasound.

Much of the energy absorbed by cell suspensions is converted to heat so effective cooling is essential.

Equipment for the large-scale continuous use of ultrasonics has been available for many years and is widely used by the chemical industry but has not yet found extensive use in enzyme production. Reasons for this may be the conformational lability of some (perhaps most) enzymes to sonication and the damage that they may realize though oxidation by the free radicals, singlet oxygen and hydrogen peroxide that may be concomitantly produced. Use of radical scavengers (e.g. N_2O) have been shown to reduce this inactivation. As with most cell breakage methods, very fine cell debris particles may be produced which can hinder further processing. Sonication remains, however, a popular, useful and simple small-scale method for cell disruption.

High Pressure Homogenizers

Various types of high pressure homogenizer are available for use in the food and chemicals industries but the design which has been very extensively used for cell disruption is the Manton-Gaulin APV type homogeniser. This consists of a positive displacement pump which draws cell suspension (about 12% w/v) through a check valve into the pump cylinder and forces it, at high pressures of up to 150 MPa (10 tons per square inch) and flow rates of up to 10,000 L hr^{-1}, through an adjustable discharge valve which has a restricted orifice. Cells are subjected to impact, shear and a severe pressure drop across the valve but the precise mechanism of cell disruption is not clear. The main disruptive factor is the pressure applied and consequent pressure drop across the valve. This causes the impact and shear stress which are proportional to the operating pressure.

The cell suspension is pumped at high pressure through the valve impinging on it and the impact ring. The shape of the exit nozzle from the valve seat varies between models and appears to be a critical determinant of the homogenization efficiency.

The location of an enzyme within the cells can influence the conditions of use of an homogenizer. Unbound intracellular enzymes may be released by a single pass whereas membrane bound enzymes require several passes for reasonable yields to be obtained. Multiple passes are undesirable because, of course, they decrease the throughout productivity rate and because the further passage of already broken cells results in fine debris which is excessively difficult to remove further downstream. Consequently, homogenizers will be used at the highest pressures compatible with the reliability and safety of the equipment and the temperature stability of the enzyme(s) released. High pressure homogenizers are acceptably good for the disruption of unicellular organisms provided the enzymes needed are not heat labile. The shear forces produced are not capable of damaging enzymes free in solution. The valve unit is prone to erosion and must be precision made and well maintained.

Use of Lytic Methods

The breakage of cells using non-mechanical methods is attractive in that it offers the prospects of releasing enzymes under conditions that are gentle, do not subject the enzyme to heat or shear, may be very cheap, and are quiet to the user. The methods that are available include osmotic shock, freezing followed by thawing, cold shock, desiccation, enzymic lysis and chemical lysis. Each method has its drawbacks but may be particularly useful under certain specific circumstances.

Certain types of cell can be caused to lyse by osmotic shock. This would be a cheap, gentle and convenient method of releasing enzymes but has not apparently been used on a large scale. Some types of cell may be caused to autolyse, in particular yeasts and Bacillus species. Yeast invertase preparations employed in the industrial manufacture of invert sugars are produced in this manner. Autolysis is a slow process compared with mechanical methods, and microbial contamination is a potential hazard, but it can be used on a very large scale if necessary. Where applicable, dessication may be very useful in the preparation of enzymes on a large scale. The rate of drying is very important in these cases, slow methods being preferred to rapid ones like lyophilisation.

Enzymic lysis using added enzymes has been used in the laboratory scale but is industrial purposes. Lysozyme, from hen egg-white, is the only lytic enzyme available on a commercial scale. It is used to lyse Gram positive bacteria in an hour at about 50,000 U Kg^{-1} (dry weight). Where costs are reduced by the use of inexpensive, lysozyme-rich, dried egg white, a major separation problem may be introduced. Yeast-lytic enzymes from Cytophaga species have been studied in some detail and other lytic enzymes are under development. If significant markets for lytic enzymes are identified, the scale of their production will increase and their cost is likely to decrease. Lysis by acid, alkali, surfactants and solvents can be effective in releasing enzymes, provided that the enzymes are sufficiently robust. Detergents, such as Triton X-100, used alone or in combination with certain chaotropic agents, such as guanidine HCl, are effective in releasing membrane-bound enzymes. However, such materials are costly and may be difficult to remove from the final product.

7. CONDUCTIVITY METER

Conductivity of any substance, solution is measured on the basis of Ohm's law (V = I·R).

$$\text{Conductance} = 1/\text{Resistance} = 1/R = G$$

A different type of electrode is used in the conductivity meter compared to the pH meter.

G is proportional to the area A of the faces or the electrode and inversely proportional to the distance between them. Resistance (R) is expressed in "ohms"; G is expressed in 1/ohms x cm; in modern instruments as "Siemens" (S)

What influences the conductivity in your experiment, measuring conductivity of a salt solution: The number of ions, their concentration, their charge, their size, ion mobility; therefore:

1. Conductivity is measured at very low concentration (mM).
2. Usually, a standard curve of that specific salt solution is generated, before you determine the (unknown) concentration.

8. RADIOACTIVE TRACERS

The use of radioactive tracers in cell research is an effective and safe means of monitoring molecular interactions. Radiation, its misuse can lead to increased environmental pollution, and at worst can lead to serious long term injury. It should be handled safely.

Radioactivity is caused by the spontaneous release of either particulate and/or electromagnetic energy from the nucleus of an atom. Atoms are composed of a positively charge nucleus, surrounded by the negatively charged electrons. In an uncharged atom, the number of orbital electrons equals the number of positively charged protons in the nucleus. In addition, the nucleus contains uncharged neutrons. A proton has a mass of 1.0076 amu (Atomic Mass Units), while a neutron has a mass of 1.0089 amu.

If the mass of a helium nucleus is examined, there is a difference between the expected mass based on its proton and neutron composition, and the actual measured mass. Helium contains two protons and two neutrons in its nucleus, and should have a corresponding mass of 4.0330 amu. It has an actual mass, however, of 5.0028 amu. The difference (0.0302 amu) is the equivalent energy of 28.2 MeV and is known as the binding energy. It would require 28.2 MeV to fuse two protons and two neutrons into a helium nucleus, and the fission of the helium nucleus would yield the same energy.

In addition, the electrons orbit the nucleus with precise energy levels. When the electrons are in their stable orbits, they are said to be in their ground state. If the electrons absorb energy (e.g. from photons), they jump to excited state. The energy difference between a ground state and an excited state can take the form of an electromagnetic radiation.

The number of protons in the nucleus of an atom is called the atomic number, while the number of protons plus neutrons is the mass number. The mass number is approximately equal to the atomic weight. In the representation of an atom used in the periodic table of elements, the atomic number is a subscript written to the left of the letter(s) designating the element, while the mass number is written as a superscript to the left.

The chemical identity of an element is determined by the number of protons in the nucleus of the atom. The number of neutrons may vary. Elements sharing the same number of protons, but having different numbers of neutrons are known as isotopes. For example, Hydrogen has one proton. All nuclei containing one proton are hydrogen nuclei. It may have one, two, or three neutrons. The isotopes of hydrogen would be written as ${}_1H_1$ ${}_2H_1$ ${}_3H_1$ (in all further references, the atomic number subscript 1 is left off for clarity). ${}_1H$ is the most stable form of hydrogen and is therefore the most abundant (99.985% of all forms). ${}_2H$ is also a stable form of hydrogen, but less stable than ${}_1H$, and constitutes about 0.015% of the total hydrogen found. It is known as deuterium.

${}_3H$ is unstable and constitues a very small fraction of the amount of hydrogen available. Termed tritium, this element readily reorganizes its nucleus, and is said to decay. The emission of its sub-atomic particles and energy is therefore known as radioactive decay, or simply radioactivity. Deuterium is a stable, but heavy isotope of hydrogen, tritium is a radioactive isotope of hydrogen.

Note that each of the three will chemically react as hydrogen. This is important for tracer work in cell biology. The substitution of either deuterium or tritium for hydrogen in a molecule will not effect any chemical or physiological changes in the activity of the molecule. Tritium will, however, tag the molecule by making it radioactive.

Radiation emissions have several forms. When an atom reorganizes its sub-atomic structure to a more stable form, it may emit neutrons, protons, electrons, and/or electromagnetic waves (energy). An alpha particle is 2 protons plus 2 neutrons. A beta particle is an electron. Gamma rays are electromagnetic energy waves similar to x-rays. The release of sub-atomic particles and energy, resulting in the change of one element to another is known as radioactivity.

Radioactive elements thus, by their very nature, self destruct. The loss of their sub-atomic particles is a spontaneous process, and once it has occurred, the element is no longer radioactive. With time a percentage of all radioactive elements in a solution will decay. Statistically, it is nearly impossible to predict which individual element will radioactively decay, but we can make a prediction about large numbers of the elements. That is, we can say that if we wait 14,000 years, half of the radioactivity in a sample of $_{14}C$ (a radioactive isotope of $_{12}C$) will be lost (1/2 remains). We then say that $_{14}C$ has a half-life of 14,000 years. After a second 14,000 years, half of the remaining half would have been lost, or 3/4 of the original amount. Based on this information, could you predict how long it would take for all radioactivity to have disappeared from the sample?

With a half-life of 14,000 years, radioactive carbon will be around for a very long time. This is why it is used for dating rocks and fossils. If one makes some assumptions about the activity of the carbon when the fossil was formed, and measures the current level, the age of the fossil may be determined.

The amount of radioactive material is measured by how many nuclei decay each second, and this value is known as the activity. It is measured in curies each radioisotope has three important properties; the type of particles emitted, the particle energy, and the half-life. The energy and kind of decay particle will determine the penetration of the radiation, and therefore determine the degree of shielding necessary to protect the user. The half-life determines both the remaining activity after storage or use, and the time that the isotope must be stored before disposal.

In cell biology, only a few of the many radioactive elements are used routinely. The primary elements used are $_{3}H$ (Tritium), $_{15}C$ (Carbon-14), $_{32}P$ 20(Phosphorus-32), $_{125}I$ 20(Iodine-125) and $_{131}I$ 20(Iodine-137).

Measurement of Dose

When alpha or beta particles, or gamma radiation pass through matter, they form ions. They accomplish this by knocking electrons from the orbits of the molecules they pass through. We can monitor the ionization effect by allowing the radiation to pass through dry air and measuring the numbers of ions formed. This is most often done by designing a chamber with an electrical charge capacitance, allowing the radiation to pass through the chamber and monitoring the amount of capacitance discharge caused by the formation of ions. The device is a Geiger-Mueller Counter and has many variations.

The ionizing ability is measured in roentgens, and a roentgen is the number of ionizations necessary to form one electrostatic unit (e.s.u.) in 1 cc of dry air. Since the roentgen is a large unit, dosage for cell research use are normally divided into milliroentgens (mR).

Curies measure the amount of radioactive decay, roentgens measure the amount of radiation transmitted through matter, over distance. Neither unit is useful in determining biological effect, since biological effect implies that the radiation is absorbed by the tissues that are irradiated.

The rad (radiation absorbed dose) is a unit of absorbed dose and equals 100 ergs absorbed in one gram of matter. The roentgen is the amount of radiation exposure in air, while the rad represents the amount of radiation exposure in tissue. The two are usually very close in magnitude, however, since for most biological tissues, 1 roentgen produces 0.96 rad.

Not all radioactive emissions have the same penetrating power, however. If radiation safety (monitoring of dose) is considered, then the rad is insufficient. A linear energy-transfer dependent factor must be defined for each type of emission. An alpha particle, for example, would not travel very far through tissue, but it is 10 times more likely to be absorbed than a gamma wave of the same energy dose. This factor is known as the Quality Factor (QF) or Relative Biological Effectiveness (RBE). The RBE is limited to work in radiobiology, the QF is used in other exposure monitor schemes. The use of the QF results in a new parameter, the rem. The rem is a unit of dose equivalent and is equal to the product of the QF × rad.

Detection of Radioactivity

Ionization chambers. The most common method of measuring radiation exposure is the use of an ionization chamber. Among the more common forms of ionization chambers are the **Geiger muller counter,** Scintillation counter and Pocket dosimeter.

The chambers are systems that comprise two electrical plates, with a potential established between them by a battery or other electrical source. In effect, they function as capacitors. The plates are separated by an inert gas, which will prevent any current flow between the plates. When an ionizing radiation enters the chamber, it induces the formation of an ion, which in turn is drawn to one of the electrical plates. The negative ions are drawn to the anode (+ plate) while the positive ions are drawn to the cathode (– plate). As the ions reach the plates, they induce an electric current to flow through the system attached to the plates. This is then expressed as a calibrated output, either through the use of a digital or analog meter, or as a series of clicks, by conversion of the current through a speaker.

The sensitivity of the system depends on the voltage applied between the electric plates. Since alpha particles are significantly easier to detect than beta particles, it requires lower voltage to detect the high energy alpha particles. In addition, alpha particles will penetrate through the metal casing of the counter tube, whereas beta particles can only pass through a quartz window on the tube. Consequently, ionization chambers are most useful for measuring alpha emmissions. High energy beta emissions are able to be measured if the tube is equipped with a thin quartz window and if the distance between the source of emission and the tube is minimal.

A modification of the basic ionization chamber **pocket dosimeter**. This device is a capacitor, which is charged by a base unit and which can then be carried as a portable unit. They are often the size and shape of a pen and can be thus carried in the pocket of a lab coat. When exposed to an ionizing radiation source, the capacitor discharges slightly. Over a period of time, the charge remaining on the dosimeter can be monitored and used as a measure of radiation exposure. The dosimeters are usually inserted into a reading device which is calibrated to convert the average exposure the dosimeter has had directly into roentgens or rems. Since the instrument works by discharging the built up charge, and the charge is upon a thin wire in the center of the dosimeter, it can be completely discharged by the flexing of that wire, as it touches the outer shell upon impact. When later read for exposure, the investigator will be informed that they have been exposed to dangerously high levels of radiation as there will be no charge left in the dosimeter. Besides causing great consternation with the Radiation Safety Officer, and a good deal of paper work, it also causes some unrest with the investigator. The dosimeters should be worn in a location where they can not impact any other objects. Since the dosimeters normally lack the fragile and vulnerable quartz windows of a Geiger tube, and carry lower voltage potentials, they are used for the measurement of X-ray and high energy gamma radiation, and will not detect beta emmissions.

Photographic Film

Low energy emissions are detected more conveniently through the use of a film badge. This is simply a piece of photographic film sandwiched between cardboard and made into a badge which can be pinned or clipped onto the outer clothing of the investigator. They can be worn routinely and collected on a regular basis for analysis.

When the film is exposed to radiation, it causes the conversion of the silver halide salts to reduced silver (exactly as exposure of the film to light). When the film is developed, the amount of reduced silver (black) can be measured and calibrated for average exposure to radiation. This is normally done by a lab specializing in the monitoring. Because of the simplicity of the system, its relative low cost and its sensitivity to nearly all forms of radiation, it is the primary means of radiation exposure monitoring of personnel.

Scintillation Counters

For accurate quantitative measurement of low energy beta emissions and for rapid measurement of gamma emissions, nothing surpasses the use of scintillation counters. Since they can range from low to high energy detection, they are also useful for the alpha emissions.

Scintillation counters are based on the use of light emitting substances, either in solution, or within a crystal. When a scintillant is placed in solution with a radioactive source (liquid scintillation counter), the radiation strikes the scintillant molecule, which will then fluoresce as it re-emits the energy. Thus the scintillant gives a flash of light for each radiation particle it encounters. The counter than converts light energy (either as counts of flashes, or as an integrated light intensity) to an electrical measure calibrated as either direct counts or counts per minute (CPM). If the efficiency of the system is known (the % of actual radioactive decays that result in a collision with a scintillant), then disintegrations per minute (DPM) can readily be calculated. DPM is an absolute value, whereas CPM is a function of the specific instrument used.

Low energy beta emissions can be detected with efficiencies of 40% or better with the inclusion of the scintillant directly into a cocktail solution. Alpha emissions can be detected with efficiencies in excess of 90%. Thus, with a liquid scintillation counter, very low doses of radiation can be detected. This makes it ideal for both sensitivity of detection and for safety.

If the system is modified such that the scintillant is a crystal placed outside of the sample chamber (vial) then the instrument becomes a gamma counter. Gamma emissions are capable of exiting the sample vial and entering into a fluorescent crystal. The light emitted from the crystal is then measured. Gamma counters are usually smaller than liquid scintillation counters, but are limited to use with gamma emittors. Modern scintillation counters usually combine the functional capabilities of both liquid scintillation and direct gamma counting.

Since all use of radioactive materials, and particularly the expensive counting devices are subject to local radiation safety regulations, the specific details for use must be left to the institutional discretion. Under no circumstances should radioactive materials be used without the express supervision of the radiation safety officer of the institution, following all specific institutional guidelines and manufacturer directions for the instrument used.

9. AUTORADIOGRAPHY

The process of localizing radioactive materials onto a cell is known as autoradiography. $_3$H (tritium) is used in cell analysis because it is a relatively weak Beta emitter (thus making it safer to handle) and more significantly, can be localized within cell organelles. $_{14}$C and $_{32}$P are also used, but are more radioactive, require significantly more precautions in handling and are inherently less capable of resolving intracellular details. They are used at the tissue or organ level of analysis.

Radioactive isotopes can be incorporated into cellular molecules. After the cell is labeled with radioactive molecules, it can be placed in contact with photographic film. Ionizing radiations are emitted during radioactive decay and silver ions in the photographic emulsion become reduced to metallic silver grains. The silver grains not only serve as a means of detecting radioactivity but, because of their number and distribution, provide information regarding the amount and cellular distribution of radioactive label.

The process of producing this picture is therefore called autoradiography and the picture is called an autoradiogram.

The number of silver grains produced depends on the type of photographic emulsion and the kind of ionizing particles emitted from the cell. Alpha particles produce straight, dense tracks a few micrometers in length. Gamma rays produce long random tracks of grains and are useless for autoradiograms. Beta particles or electrons produce single grains or tracks of grains. High energy Beta particles (such as those produced by 32P) may travel more than a millimeter before producing a grain. Low energy Beta particles ($_3$H and 14°C) produce silver grains within a few micrometers of the radioactive disintegration site and so provide very satisfactory resolution for autoradiography.

The site of synthesis of cellular molecules may be detected by feeding cells a radioactive precursor for a short period and then fixing the cells. During this pulse labelling, radioactivity is incorporated at the site of synthesis but does not have time to move from this site. The site of utilization of a particular molecule may be detected be chase labelling. Cells are exposed to a radioactive precursor, radioactivity is then washed or diluted away and the cells allowed to grow for a period of time. In this case, radioactivity is incorporated at the site of synthesis but then has time to move to a site of utilization in the cell.

$_3$H-thymidine can be used to locate sites of synthesis and utilization of DNA. Thymidine, the deoxyribose nucleoside of thymine, can be purchased with the tritium label attached to the methyl group of thymine. Thymidine is specifically incorporated into DNA in Tetrahymena. Some organisms can remove the methyl group from thymine, and incorporate the uracil product into RNA. Even in this case RNA would not be labelled because the tritium label would be removed with the methyl group. Methyl labelled thymidine, therefore, serves as a very specific label for DNA.

This is known as pulse labelling, after which the cells are washed free of the radioactive media. All remaining radioactivity would be due to the incorporation of the thymidine into the macromolecular structure of DNA. The cells will be fixed, covered with a photographic emulsion and allowed to develop.

During this time, the activity emanating from the $_3$H will expose the photographic emulsion, causing the presence of reduced silver grains immediately above the location of the radioactive source (DNA). Thus, it will be possible to localize the newly synthesized DNA, or that which was in the S phase of mitosis during the time period of the pulse labelling.

Rules for Safe Handling of Radioactive Isotopes

- All work with radioactive material must be done in a tray lined with absorbant paper.
- All glassware and equipment contacting radioactive material must be appropriately labelled and kept inside the tray. The only exception is that microscope slides of labelled cells may be removed from the tray after the drop of labelled cells has been applied to the slide and allowed to dry.
- Plastic gloves should be worn when handling radioactive material.
- All waste solutions containing radioisotopes, all contaminated gloves, paper, etc., must be placed in appropriate liquid or dry radioactive waste containers.

10. PHOTOGRAPHY

The use of photography within a cell biology laboratory allows for the capture of data and images for processing at a later time. It is a type of presentation, either through projection slides, or through illustrations.

Photomicrography

Photographically recording visual images observed through a light microscope is a useful means of obtaining a permanent record of activities. Using photomicrographs is the main means of recording electron microscope images.

Use of a camera on the microscope is straightforward. Merely center the object to be photographed, focus using the camera viewer and depress the camera shutter button. Equipping the shutter with a shutter release cable will help prevent vibrations. This assumes that you have the proper exposure.

Exposure and film type are the major problems of photomicrography. For most microscopes using a tungsten lamp source, there is very little light reaching the camera. Film that has a high enough exposure index (ASA speed) are too grainy to be used for effective work. In general, the faster the film the less inherent resolution the film will have. As in all things in photography, a compromise is called for.

The microscope projects an image of very low contrast, with low light intensity. A thick emulsion tends to lower the contrast even more. This results in photographs that are all gray, with no highlights (black and white). The tonal range is reduced significantly using general film for photomicrography.

Use Kodak Technical Pan Film at an ASA of 100 for photomicrography. This is a thin emulsion film with extremely high contrast. The contrast can even be controlled through the developing process and ranges from High (for photography of chromosomes), to moderate (used for general use) and low is not used in photomicrography. This same film can be used for copy work, since it reproduces images which are black and white.

Another means of increasing contrast is the use of colored filters within the microscope light path. Use a contrasting color to the object you wish to photograph. For example, chromosomes stained with aceto-orcein (dark red) can be contrast enhanced by the use of a green filter. Human chromosome spreads stained with Giemsa (blue) can be enhanced by the use of a red filter. This trick is also useful for routine viewing as well as photography.

The use of filters will increase the necessary exposure time. Technical Pan Film is also a relatively slow film. To establish the proper exposure, use the light meter built into the camera. If no light meter is available, you will have to shoot a roll of film and bracket several exposures to determine which is best. When using the built in meter, remember that all light meters are designed to give an image which is a medium gray. If you have a spot meter, be sure the spot is placed over an object which should be gray in the final image. If you have an averaging meter, be sure there is sufficient material in the viewfinder to give a proper average exposure. If you do not know whether you have a spot or averaging meter, find out. This is not trivial. Suppose you wish to photograph a chromosome spread. The chromosomes are typically less than 1-2% of the field of view. The meter will adjust the exposure such that the white field of light is exposed as gray, and your chromosomes will appear as darker gray on a gray field—in other words, extremely murky looking. Performing karyotype analysis on this type of image is difficult or impossible.

For 35 mm cameras, be sure to rewind the film when all exposures have been completed.

Processing

After exposure of the film, the film needs to be processed. Processing of black and white film has three steps. Develop the film, Stop it from developing, and fix the emulsion so that it is no longer light sensitive. OR You can send your film out for processing.

Macrophotography

Macrophotography is used to record things too large to be viewed in the microscope. This is an excellent means of making permanent records of electrophoresis gels, bands observed during ultracentrifugation, and whatever else you wish to capture on film.

Two changes are required from the use of photography through a microscope. The camera must be removed from the microscope and equipped with a lens, and secondly, the type of film used must be changed.

11. STATISTICS

Statistical manipulation is often necessary to order, define and/or organize raw data. A full analysis of statistics is beyond the scope of this work, but there are some standard analyses that anyone working in a cell biology laboratory should be aware of, and know how to perform. After data is collected, it must be ordered, or grouped according to the information which is to be sought. Data is collected in the forms:

Type of Data	*Type of Entry*
Nominal	yes or no
Ordinate	+, ++, +++
Numerical	0, 1, 1.3, etc.

When collected, the data may appear to be a mere collection of numbers, with little apparent trends. It is first necessary to order those numbers. One method is to count the times a number falls within a range increment. For example, in tossing a coin, one would count the number of Heads and Tails (eliminating the possibility of it landing on its edge). Coin flipping is nominal data, and thus would only have two alternatives. Should we flip the coin 100 times, we could count the number of times it lands Heads and the number of Tails. We would thus accumulate data relative to the categories available. A simple table of the grouping would be known as a frequency distribution, for example:

Coin Face	*Frequency*
Heads	45
Tails	55
Total	100

Similarly, if we examine the following numbers; 3,5,4,2,5,6,2,4,4, several things are apparent. First, the data needs to be grouped and the first task is to establish an increment for the categories. Let us group the data according to integers, with no rounding of decimals. We can construct a table which groups the data.

Integer	*Frequency*	*Total (Integer × Frequency)*
1	0	0
2	2	4
3	1	3
4	3	12
5	2	10
6	1	6
Totals	9	31

Mean, Median and Mode

From the data, we can now define and compute three important parameters of statistics.

Definition. The mean: The average of all the values obtained. It is computed by the sum of all of the values (Σx), divided by the number of values (n). The sum of all numbers is 31, while there are 9 values, thus, the mean is 3.44.

$$\overline{M} = \frac{\Sigma x}{n}$$

Definition. The median: The mid point in an arrangement of the categories by magnitude. Thus, the low for our data is 2, while the high is 6. The middle of this range is 4. The median is 4. It represents the middle of the possible range of categories.

Definition. The mode: The category that occurs with the highest frequency. For our data, the mode is equal to 4, since it occurs more often than any other value.

These values can now be used to characterize distribution patterns of data.

For our coin flipping, the likelihood of a Head or a Tail is equal. Another way of saying this is that there is equal probability of obtaining a Head or not obtaining a Head with each flip of the coin. When the situation exists that there is equal probability for an event as for the opposite event, the data will be graphed as a binomial distrution, and a Normal curve will result. If the coin is flipped ten times, the probability of one head and nine tails equals the probability of nine heads and one tail. The probability of two heads and eight tails equals the probability of eight heads and two tails and so on. However, the probability of the latter (two heads) is greater than the probability of the former (one head). The most likely arrangement is five heads and five tails.

When random data is arranged and displays a binomial distribution, a plot of frequency vs. occurency will result in a normal distribution curve. For an ideal set of data (i.e. no tricks, such as two headed coin, or gum on the edge of the coin), the data will be distributed in a bell shaped curve, where the median, mode and mean are equal.

This does not give an accurate indication of the deviation of the data, and in particular does not inform us of the degree of dispersion of the data about the mean. The measure of the dispersion of data is known as the Standard Deviation. It is given mathematically by the formula:

$$S = \text{sqrt}\frac{\Sigma(M - \overline{X})^2}{n-1}$$

This value gives a measure of the variability of the data, and in particular, how it varies from an ideal set of data generated by a random binomial distribution. In other words, how different is it from an ideal Normal Distribution. The more variable the data, the higher the value of the standard deviation.

Other measures of variability are the range, the Coefficient of Variation (Standard Deviation divided by the Mean and expressed as a Percent) and the Variance.

The variance is the deviation of several or all values from the mean and must be calculated relative to the total number of values. Variance can be calculated from the formula:

$$V = \frac{\Sigma(\overline{M} - X)^2}{n-1}$$

All of these calculated parameters are for a single set of data that conforms to a normal distribution. Unfortunately, biological data does not always conform in this way, and often sets of data must be compared. If the data does not fit a binomial distribution, often it fits a skewed plot known as a Poisson distribution. This distribution occurs when the probability of an event is so low, that the probability of its not occurring approaches 1. While this is a significant statistical event in biology, details of the Poisson distribution are left to texts on biological statistics.

Likewise, the proper handling of comparisons of multiple sets of data. Suffice it indicate that all statistics comparing multiple sets begin with calculation of the parameters detailed here, and for each set of data. For example, the standard error of the mean (also known simply as the standard error) is often used to measure distinctions among populations. It is defined as the standard deviation of a distribution of means. Thus, the mean for each population is computed and the collection of means are then used to calculate a standard deviation of those means.

Once all of these parameters are calculated, the general aim of statistical analysis is to estimate the significance of the data, and in particular the probability that the data represents effects of experimental treatment, or conversely, pure random distribution. Tests of significance will also be left to more extensive treatment in other volumes.

12. GRAPHS

Presentation of data in an orderly manner often calls for a graphic display. Now-a-days it is easier, with the advent of graphics programs for the computer, but still requires the application of basic techniques.

The first consideration for a graph, is whether the graph is needed, and if so, the type of graph to be used. For accuracy, a well constructed table of data usually gives more information than a graph. The values obtained and their variability are readily apparent in a table, and interpolation (reading the graph) is unnecessary. For visual impact, however, nothing is better than a graphic display.

There are a variety of graph types to be chosen from; *e.g.* line graphs, bar graphs and pie graphs. Each of these has its own characteristics and subdivisions. One also has to decide upon singular or multiple graphs, two-dimensional or three dimensional displays, presence or absence of error bars, and the aesthetics of the display. The latter include such details as legend bars, axes labels, titles and selection of the symbols to represent data, and patterns for bar graphs.

Selection of a good Graphic Program for computer use will make most of these choices available. Refer to Appendix D for a quick review of such programs.

The Basics

Perhaps the number one rule for graphic display has to do with the axes. Given a two-dimensional graph, with two values (x and y), which value is x and which is y? The answer is always the same - the KNOWN value is always the ordinate (x) value. The value that is MEASURED is the abscissa (y) value. For a standard curve of absorption in spectrophotometry, the known concentrations of the standards are placed on the x-axis, while the measured absorbance would be on the y. For measurements of the diameter of cells, the x-axis would be a micron scale, while the y axis would be the number of cells having a given diameter.

Unless you are specifically attempting to demonstrate an inverted function, the scales should always be arranged with the lowest value on the left of the x-axis, and the lowest value at the bottom of the y-value. The range of each scale should be determined by the lowest and higest value of your data, with the scale rounded to the nearest tenth, hundredth, thousandth, etc. That is, if the data ranges from 12 to 93, the scale should be from 10 to 100. It is not necessary to always range from 0, unless you wish to demonstrate the relationship of the data to this value (Spectrophotometric standard curve).

The number of integrals placed on the graph will be determined by the point you wish to make, but in general, one should use about ten divisions of the scale. For our range of 12 to 93, an appropriate scale would be from 0 to 100, with an integral of 10. Placing smaller integrals on the scale does not convey more information, but merely adds a lot of confusing marks to the graph. The user can estimate the values of 12 and 93 from such a scale without having every possible value ticked off on a scale.

Line Graph *vs.* Bar Graph or Pie Graph

If the presentation is to highlight various data as a percent of the total data, then a pie graph is ideal. Pie graphs might be used for example, to demonstrate the composition of the white cell differential count. They are the most often used graph type for business use, particularly in displaying budget details.

Pie Graphs are circular presentations which are drawn by summing your data and computing the percent of the total for each data entry. These percent values are then converted to portions of a circle (by multiplying the percent by 360°) and drawing the appropriate arc of a circle to represent the percent. By connecting the arc to the center point of the circle, the pie is divided into wedges, the size of which demonstrate the relative size of the data to the total. If one or more wedges are to be highlighted, that wedge can be drawn slightly out of the perimeter of the circle for what is referred to as an exploided view.

More typical of data presented in cell biology, however, are the line graph and the bar graph. There is no hard and fast rule for choosing between these graph types, except where the data is non-continuous. Then, a bar graph must be used. In general, line graphs are used to demonstrate data which is related on a continuous scale, whereas bar graphs are used to demonstrate discontinuous or interval data.

Suppose, for example, that you decide to count the number of T-lymphocytes in four slices of tissue, one each from the thymus, Payer's patches, a lymph node and a healing wound on the skin. Let's label each of these as T, P, L and S respectively. The numbers obtained per cubic centimeter of each tissue are T = 200, P = 150, L = 100 and S = 50. Note that there is a rather nice linear decrease in the numbers if T is placed on the left of an x-axis, and S to the right. A linear graph of this data would give a nice straight line, with a statistical regression fit and slope. But look at the data! There is no reason to place T (or P,L or S) to the right or left of any other point on the graph - the placement is totally arbitrary. A line graph for this data would be completely misleading since it would imply that there is a linear decrease from the thymus to a skin injury AND that there was some sort of quantitative relationship among the tissues. There is certainly a decrease, and a bar graph could demonstrate that fact, by arranging the tissue type on the x-axis in such a way to demonstrate that relationship - but there is no inherent quantitative relationship between the tissue types which would force one and only one graphic display. Certainly, the thymus is not four times some value of skin (although the numbers are).

However, were you to plot the number of lymphocytes with increasing distance from the point of a wound in the skin, an entirely different presentation would be called for. Distance is a continuous variable. We may choose to collect the data in 1 mm intervals, or 1 cm. The range is continuous from 0 to the limit of our measurements. That is we may wish to measure the value at 1 mm, or 1.2 mm or 1.23 mm or 1.23445 mm. The important point is that the 2 mm position is 2x the point at 1 mm. There is a linear relationship between the values to be placed on the x-axis. Therefore a linear graph would be appropriate, with the dots connected by a single line. If we choose to ignore the 1.2 and 1.23 and round these down to a value of 1, then a bar graph would be more appropriate. This latter technique (dividing the data in appropriate intervals and plotting as a bar graph) is known as a Histogram. This graph is very familiar to students since it is the graph used for the display of grade distributions.

Having decided that the data has been collected as a continuous series, and that the data will be plotted on a linear graph, there are still decisions to be made. Should the data be placed on the graph as individual points with no lines connecting them (a Scattergram)? Should a line be drawn between the points (known as a Dot-to-Dot)? Should the points be plotted, but curve smoothing be applied? If the latter, what type of smoothing?

There are many algorythms for curve fitting, with the two most commonly used being linear regression and polynomial regression. It is important to decide BEFORE graphing the data, which of these is appropriate.

Linear regression is used when there is good reason to suspect a linear relationship within the data (as for example in a spectrophotometric standard following the Beer-Lambert law). In general, the y value can be calculated from the equation for a straight line, $y = mx + b$, where m is the slope and b is the y-intercept.

Computer programs for this can be very misleading. Any set of data can be entered into a program to calculate and plot linear regression. It is important that there be a valid reason for supposing linearity before using this function, however. This is also true when using polynomial regressions. This type of regression calculates an ideal curve based on quadratic equations with increasing exponential values, that is $y = (mx + b)_n$, where n is greater than 1. The mathematics of this can become quite complex, but often the graphic displays begin to look better to the beginning student. It is important to note that use of polynomial regression must be warranted by the relationship within the data, not by the individual drawing the graph.

For single sets of data, that is the extent of the available options. For multiple sets the options increase. If the multiple sets are data collected pertaining to identical ordinate values, then error bars (standard deviation or standard error of the means) can be added to the graphics. Plots can be made where two lines are drawn, connecting the highest y values for each x, and a second connecting the lowest values (the Hi-Lo Graph). The area between the two lines presents a graphic depiction of variability at each ordinate value.

If the data collected involves two or more sets of data having a common x axis, but varying y axes (or values), then a multiple graph may be used. The rules for graphing apply to each set of data, with the following provision; Keep the number of data sets on any single graph to an absolute minimum. It is far better to have three graphs, each with 3 lines (or bars) than to have a single graph with 9 lines. A graph that contains an excess of information (such as 9 lines) is usually ignored by the viewer (as are tables with extensive lists of data). For this same reason, all unnecessary clutter should be removed from the graph; *e.g.* grid marks on the graph are rarely useful.

Finally, it is possible to plot two variables, y and z, against a common value, x. This is done with a 3D graphic program. The rules for designing a graph follow for this type of graph, and the use of these should clearly be left to computer graphics program. These graphs often look appealing with their hills and valleys, but rarely impart any more information than two separate 2D graphs. Perhaps the main reason is that people are familiar with two dimensional graphics, but have a more difficult time visually interpreting three dimensional graphs.

13. COMPUTERS

The advent of the personal computer has been one of the most significant factors impacting laboratory work in the past decades.

It doesn't make any difference which hardware system you use. The important point is that the computer is used as a tool to enhance learning in the laboratory.

Commercial Software

There are a number of types of programs available and useful to the cell biology laboratory

- Spread Sheets
- Graph Programs
- Equation Solvers
- Data Base Programs
- Word Processors
- Outline Programs
- Paint/Draw Programs
- Computer Assisted Instruction (CAI)
- Single Purpose Programs.

Spread Sheets

Without question, the single most useful program available for data collection and analysis is the spread sheet program. There are several available, each with its own merits. The leader in the PC dominated field is Lotus 1–2–3 or Excel. Integrated packages nearly always contain powerful spreadsheets.

A spreadsheet program is an electronic balance sheet divided in rows and columns. Pioneered by Visicalc, these programs may have had as much impact on computer use as the actual design of the hardware. Any data that can be tabulated in columns and rows can be added to this type of program. Functions are readily available for totals (sums), averages, means, maximum and minimum values, and full trignometric functions. The programs listed above also include capabilities of sorting data, searching through the data, and for automatic graphing of the data. Each allows the construction of blank masks that contain the instructions for coding input and output while allowing students ease of data entry.

Spreadsheet programs have become so powerful in their latest versions, that they can be used as word processors and for data base manipulations,

If you were to purchase only one software package for the cell biology laboratory, it should be a spreadsheet.

Graphics Programs

Separate programs for graphing data are useful in that they contain more options than those found within spreadsheets, and allow for more complex graphs. The better programs will also give regressional analyses, either linear or polynomial. Nearly all allow for data input from a spreadsheet or data base manager, in addition to direct entry.

Among the best in this area, are Sigma Plot, Energraphics, Harvard Graphics and Cricket Graph. Most of these programs are designed for business graphics, but can be used in the cell laboratory.

Data Base Managers

These programs are powerful for the long term storage and manipulation of data. They are more useful for research storage of data than for direct use in the undergraduate laboratory. Their strength lies in the ability to do with words what spreadsheets can do with numbers.

Data base managers are excellent means of filing references, sources and such things as equipment lists, chemical inventory, etc. They can also be used effectively for filing of nucleic acid sequences, but the data base must be created before it can be used. Whether or not to use a data base manager for this purpose depends on the availability of the data in an appropriate form for use within your program. If data base managers are used, however, a hard drive becomes nearly mandatory.

Word Processors

There are perhaps as many word processors available as pebbles on the shore. For many computer users, this function is synonymous with computers. For purposes of the cell biology laboratory, any one is sufficient. The programs are useful for writing lab reports, and marginally useful as alternatives to data base managers for such things as searching for nucleic acid codes.

Outline Programs

If you are heavily into writing reports and/or papers, you may want to tie together a good outlining program with a wordprocessor and top it off with RightWriter. The latter is a grammar and context checker that is quite enlightening when applied to literary creations.

Paint/Draw Programs

A useful adjunct to various graph programs are those listed as paint or draw. Paint programs are those which plot pixels on a graphic screen and allow simple drawing routines. Representative of these are PC-Paintbrush, PC-Paint, and PaintShow

Single Purpose Programs

These programs are sometimes available through commercial channels, but more often are written for specific purposes. It would be impossible to list all of the areas where these could be used in a cell biology laboratory, but a few should be mentioned.

- Resolution Calculations for Light and EM Microscopy
- Cell Morphometry-Area, Volume, Numbers
- Centrifugation, Conversion among Rotors
- Centrifugation, Sedimentation Coefficients, Molecular Weights
- Beer-Lambert Law
- Calculation of Molecular Weights from Electrophoresis Gels
- Ion Flux across Membranes

- Manometric Calculations
- H^+ Flux and Chemiosmosis
- Gene Mapping, Recombination
- Growth Curves
- Simulation of cAMP effects on Dictyostelium
- Radiation Dose Response.

Essentially, anything which requires repetitive calculations or data sorting is material for this section.

Included in this appendix are two programs. The first (CELLM) allows us to play the role of a cell membrane. The second, (BEER) allows rapid calculations of extinction coefficients using the Beer-Lambert law. Other programs are Quick Basic progarm for interconversion of centrifuge rotors and calculations of viscosity, sedimentation coefficients and clearing constants would require 12–15 pages if typed out. Programs can become very complex over time.

3 BIOCHEMISTRY

CARBOHYDRATES

EXERCISE 1. QUALITATIVE TESTS

Carbohydrates are a class of organic molecules with the general chemical formula Cn(H2O)n. These compounds are literally carbon hydrates. Only the monomeric form of these compounds, the monosaccharides, fit this description precisely. Two monosaccharides can be polymerized together through a glycosidic linkage to form a disaccharide. When a few monosaccharide molecules are polymerized together, the result is an oligosaccharide. A polysaccharide is an extensive polymer of carbohydrate monomers.

The monosaccharide glucose is our primary energy source. The function of the polysaccharides starch (plants) and glycogen (animals) is to store glucose in a readily accessible form, as well as to lower the osmotic potential of internal fluids. Some polysaccharides serve a structural role in living organisms. The glucose polymer cellulose is a major component of plant cell walls. Chitin, a polymer of N-acetylglucosamine, is a major structural component of the exoskeleton of insects and crustaceans. Hyaluronic acid and chondroitin sulfate occurs in the connective tissues of animals, especially in cartilage. Oligosaccharide side chains of glycoproteins may also serve as signals for intracellular sorting of the protein (i.e. mannose-6-phosphate signal designating lysosomal enzymes).

Several qualitative tests have been devised to detect members of this biologically significant class of compounds. These tests will utilize a test reagent that will yield a color change after reacting with specific functional groups of the compounds being tested. The following exercises are reactions that can detect the presence or absence of carbohydrates in test solutions. They range in specificity to the very general (i.e. Molisch test for carbohydrates) to the very specific (i.e. mucic acid test for galactose).

Exercise: You are given solutions containing: fructose, glucose, lactose, galactose, ribose, ribulose, sucrose, and starch. Devise a scheme by which you can systematically identify these compounds.

Procedure

Perform the following qualitative tests on 0.2 M solutions (unless otherwise stated) of starch, sucrose, glucose, lactose, galactose, ribose, and ribulose. Use the scheme you devised in the prelab section to identify an unknown solution. The unknown will be one of the above solutions or a mixture of two of the above solutions.

Test 1. Molisch Test for Carbohydrates

The Molisch test is a general test for the presence of carbohydrates. Molisch reagent is a solution of alpha-naphthol in 95% ethanol. This test is useful for identifying any compound which can be dehydrated to furfural or hydroxymethylfurfural in the presence of H2SO4. Furfural is derived from the dehydration of pentoses and pentosans, while hydroxymethylfurfural is produced from hexoses and hexosans. Oligosaccharides and polysaccharides are hydrolyzed to yield their repeating monomers by the acid. The alpha-naphthol reacts with the cyclic aldehydes to form purple colored condensation products. Although this test will detect compounds other than carbohydrates (i.e. glycoproteins), a negative result indicates the ABSENCE of carbohydrates.

Method: Add 2 drops of Molisch reagent to 2 ml of the sugar solution and mix thoroughly. Incline the tube, and GENTLY pour 5 ml of concentrated H2SO4 down the side of the test tube. A purple color at the interface of the sugar and acid indicates a positive test. Disregard a green color if it appears.

Test 2. Benedicts's Test for Reducing Sugars

Alkaline solutions of copper are reduced by sugars having a free aldehyde or ketone group, with the formation of colored cuprous oxide. Benedict's solution is composed of copper sulfate, sodium carbonate, and sodium citrate (pH 10.5). The citrate will form soluble complex ions with Cu++, preventing the precipitation of CuCO3 in alkaline solutions.

Method: Add 1 ml of the solution to be tested to 5 ml of Benedict's solution, and shake each tube. Place the tube in a boiling water bath and heat for 3 minutes. Remove the tubes from the heat and allow them to cool. Formation of a green, red, or yellow precipitate is a positive test for reducing sugars.

Test 3. Barfoed's Test for Monosaccharides

This reaction will detect reducing monosaccharides in the presence of disaccharides. This reagent uses copper ions to detect reducing sugars in an acidic solution. Barfoed's reagent is copper acetate in dilute acetic acid (pH 4.6). Look for the same color changes as in Benedict's test.

Method: Add 1 ml of the solution to be tested to 3 ml of freshly prepared Barfoed's reagent. Place test tubes into a boiling water bath and heat for 3 minutes. Remove the tubes from the bath and allow to cool. Formation of a green, red, or yellow precipitate is a positive test for reducing monosaccharides. Do not heat the tubes longer than 3 minutes, as a positive test can be obtained with disaccharides if they are heated long enough.

Test 4. Lasker and Enkelwitz Test for Ketoses

The Lasker and Enkelwitz test utilizes Benedict's solution, although the reaction is carried out at a much lower temperature. The color changes that are seen during this test are the same as with Benedict's solution. Use DILUTE sugar solutions with this test (0.02 M).

Method: Add 1 ml of the solution to be tested to 5 ml of Benedict's solution to a test tube and mix well. The test tube is heated in a 55oC water bath for 10–20 minutes.

Ketopentoses demonstrate a positive reaction within 10 minutes, while ketohexoses take about 20 minutes to react. Aldoses do not react positively with this test.

Test 5. Bial's Test for Pentoses

Bial's reagent uses orcinol, HCl, and $FeCl_3$. Orcinol forms colored condensation products with furfural generated by the dehydration of pentoses and pentosans. It is necessary to use DILUTE sugar solutions with this test (0.02M).

Method: Add 2 ml of the solution to be tested to 5 ml of Bial's reagent. Gently heat the tube to boiling. Allow the tube to cool. Formation of a green colored solution or precipitate denotes a positive reaction.

Test 6. Mucic Acid Test for Galactose

Oxidation of most monosaccharides by nitric acid yields soluble dicarboxylic acids. However, oxidation of galactose yields an insoluble mucic acid. Lactose will also yield a mucic acid, due to hydrolysis of the glycosidic linkage between its glucose and galactose subunits.

Method: Add 1 ml of concentrated nitric acid to 5 ml of the solution to be tested and mix well. Heat on a boiling water bath until the volume of the solution is reduced to about 1 ml. Remove the mixture from the water bath and let it cool at room temperature overnight. The presence of insoluble crystals in the bottom of the tube indicates the presence of mucic acid.

Test 7. Iodine Test for Starch and Glycogen

The use of Lugol's iodine reagent (IKI) is useful to distinguish starch and glycogen from other polysaccharides. Lugol's iodine yields a blue-black color in the presence of starch. Glycogen reacts with Lugol's reagent to give a brown-blue color. Other polysaccharides and monosaccharides yield no color change; the test solution remains the characteristic brown-yellow of the reagent. It is thought that starch and glycogen form helical coils. Iodine atoms can then fit into the helices to form a starch-iodine or glycogen-iodine complex. Starch in the form of amylose and amylopectin has less branches than glycogen. This means that the helices of starch are longer than glycogen, therefore binding more iodine atoms. The result is that the color produced by a starch-iodine complex is more intense than that obtained with a glycogen-iodine complex.

Method: Add 2–3 drops of Lugol's iodine solution to 5 ml of solution to be tested. Starch gives a blue-black color. A positive test for glycogen is a brown-blue color. A negative test is the brown-yellow color of the test reagent.

EXERCISE 2. QUALITATIVE ANALYSIS

This event is patterned after the middle school event, Science Crime Busters. The basis for doing well in this event is obviously practice. Be sure your students know that they are to be provided with the names (not the chemical formulas) of the 13 unknowns in this event.

The identification of the 13 unknowns is based on a flow chart approach that can be followed as presented here to separate and identify the unknowns.

NaCl, $NaHCO_3$, Na_2CO_3, $NaC_2H_3O_2$, NH_4Cl, $MgSO_4$, $Ca(NO_3)_2$, H_3BO_3, $CaCO_3$, $CaSO_4$, sucrose, cornstarch, fructose.

Add Water to Each

Soluble in Water	*Insoluble in Water*
NaCl, $NaHCO_3$, Na_2CO_3, $NaC_2H_3O_2$, NH_4Cl, $MgSO_4$, $Ca(NO_3)_2$, H_3BO_3, sucrose, fructose	$CaCO_3$, $CaSO_4$, cornstarch

Test for Insoluble Substances

1. $CaCO_3$ bubbles with vinegar-produces CO2 gas
2. Cornstarch turns purple with iodine, other two are brownish colored
3. $CaSO_4$ – the one left.

Test for Soluble Substances

1. NH_4Cl-acidic pH in solution-solid in a container will give small ammonia smell when opened sometimes.
2. Na_2CO_3, $NaC_2H_3O_2$ – both produce pink solutions with phenolphthalein but only Na_2CO_3 bubbles with vinegar, so test the two pink solutions with vinegar, one bubbles and the other does not.
3. $NaHCO_3$ bubbles with vinegar but does not produce pink solution with phenolphthalein, should produce slightly basic pH in water but not much.
4. $MgSO_4$, $Ca(NO_3)_2$

   $MgSO_4$ produces distinct precipitate with NaOH solution, $Ca(NO_3)_2$ produces milky appearance in solution with NaOH..."essentially lime water" but not a distinct precipitate like $MgSO_4$.
5. Fructose-produces a red precipitate with Benedict's test or copper(II) sulfate.
6. Have left at this point NaCl, sucrose and H3BO3 which is very soluble in alcohol but the other two are not. Sucrose dissolves readily in warm water while NaCl does not.

Practice with these solids and these tests will make these observations more recognizable.

The part of the test students sometimes has the most trouble with is the written part. Here are some helpful tips.

Benedict's Test Mechanism

Benedict's solution (copper(II) sulfate) works on sugars that are aldose (aldehyde) sugars in a basic solution. The Cu^{2+} ion in the solution causes it to be blue but it reacts with fructose or glucose to form Cu_2O which is a red precipitate.

Iodine Test for Starch

The alpha $-1 \rightarrow 4$ linkages between carbons in the starch cause the helical structure of the polysaccharide chain. The inner diameter of the helix is big enough for elementary iodine to become deposited, thus forming a blue complex (evidence for starch). When starch is mixed with iodine in water, an intensely colored starch/iodine complex is formed. Many of the details of the reaction are still unknown. But it seems that the iodine (in the form of I5- ions) gets stuck in the coils of beta amylose molecules (beta amylose is a soluble starch). The starch forces the iodine atoms into a linear arrangement in the central groove of the amylose coil. There is some transfer of charge between the starch and the iodine. That changes the way electrons are confined, and so, changes spacing of the energy levels. The iodine/starch complex has energy level spacings that are just so for absorbing visible light- giving the complex its intense blue color.

Tips for Writing Net Equations

1. Must note the state of the materials involved in the reaction originally to determine how the substance shows up in the net equation. Ex. solid sodium carbonate is reaction with a solution of hydrochloric acid would be different from solutions of hydrochloric acid and sodium carbonate are reacted.

   1st case: $Na_2CO_3(s) + 2H_3O+(aq) \longrightarrow 2Na+(aq) + CO_2(g) + 3H_2O(l)$

   2nd case: $CO32\text{-}(aq) + 2H_3O+(aq) \longrightarrow CO_2(g) + 3H_2O(l)$
2. Ionic solids in solution as well as strong acids and bases will be in "ion" form in the equations. Strong acids include HCl, HBr, HI, HNO3, HClO4 and H2SO4 and the strong bases include NaOH, KOH, $Ca(OH)_2$, etc.
3. Weak acids and bases (watch out for NH3) remain "intact" in net equations and precipitates are written as solids.

   Example. reaction of vinegar with a solution of sodium carbonate

   $2\ HC_2H_3O_2(aq) + CO_3^{2-}(aq) \longrightarrow CO_2(g) + H_2O(l) + 2\ C_2H_3O_2^-(aq)$
4. Be sure to balance net equations just like any other equation.

Qualitative Analysis

The following are the answers to this Sample Qualitative Analysis test.

1. $MgSO_4(s) + 2OH^-(aq) \longrightarrow Mg(OH)_2(s) + SO4^{2-}(aq)$
2. See information in tips above.
3. $CO_3^{2-}(aq) + 2H_3O^+(aq) \longrightarrow CO_2(g) + 3H_2O(l)$

   or $CO_3^{2-}(aq) + 2H^+(aq) \longrightarrow CO_2(g) + H_2O(l)$
4. This was a poorly worded question in that it was not stated to start with solid ammonium chloride or ammonium chloride already in solution. Students should ask if this is not stated.

   $NH_4^+(aq) + H_2O(l) \longrightarrow NH_3(aq) + H_3O+(aq)$

   or $NH4Cl(s) + H_2O(l) \longrightarrow NH_3(aq) + H_3O+(aq) + Cl^-(aq)$
5. $NaCl(s) + H_2O(l) \longrightarrow Na^+ (aq) + Cl^-(aq) + H_2O(l)$ or the water could be written over the arrow.

6. Carbonate is dibasic while bicarbonate is monobasic which is another way of stating that carbonate will produce more OH- in solution that bicarbonate.
7. Again, this question should have said to start with solid or aqueous sodium carbonate. Let's assume solid for this answer.

$$Na_2CO3(s) + 2H_2O(l) \longrightarrow 2NaOH(aq) + H_2CO_3(aq)$$

EXERCISE 3. TO ESTIMATE THE AMOUNT OF REDUCING SUGARS

To estimate the amount of reducing sugars in a grape, the method of test for reducing sugars must be known first. To test the presence of reducing sugars in a solution, Benedict's test can be carried out. In Benedict's test, equal volume of Benedict's solution should be added into the unknown solution. The mixture is then boiled. If reducing sugars are present, there should be some brick red precipitates. This method can be done quantitatively. The amount of precipitates can indicate the amount of reducing sugars. The more the precipitates are formed, the more the reducing sugars are present in the solution. So by comparing the amount of precipitates formed in the grape juice with the amount of precipitates formed in a standard sugar solution, one can estimate the amount of reducing sugars in a grape.

In this experiment, a standard glucose solution of concentration 0.01M was used. It was diluted into different concentration. The amount of reducing sugars was determined by Benedict's test and the results were plotted in a standard curve.

Procedure

1. Prepare 1 ml of glucose solutions of different concentrations from a stock 0.01M glucose solution by adding suitable amounts of distilled water. The dilution table was shown:

Molarity of glucose solution (M)	0.000	0.002	0.004	0.006	0.008	0.010
Volume of stock glucose solution added (ml)	0.0	0.2	0.4	0.6	0.8	1.0
Volume of distilled water added (ml)	1.0	0.8	0.6	0.4	0.2	0.0

2. Peel off the epidermis of the grape, put the grape into a beaker, grind the grape tissue with a glass rod to get as much juice as possible, pour the juice into a measuring cylinder and add distilled water into it up to 25 ml.
3. Pipette 1 ml of the prepared diluted juice into a test tube G. Pipette 1ml of the sugar solutions into test tube A to F.
4. Place 2 ml of Benedict's solution into each of the test tube A-G. Boil all tubes in a water bath for 5 minutes. Allow the precipitates to settle for 10 minutes. Estimate the amount of precipitates in each tube.

Discussion

1. The skin and seeds should be removed. They will affect the grinding procedure.

2. Excess Benedict's solution must be added so that all reducing sugars can be reacted.
3. Allow the precipitates to settle down before the estimation.
4. The amount of reducing sugar estimated must be in term of the amount of glucose.

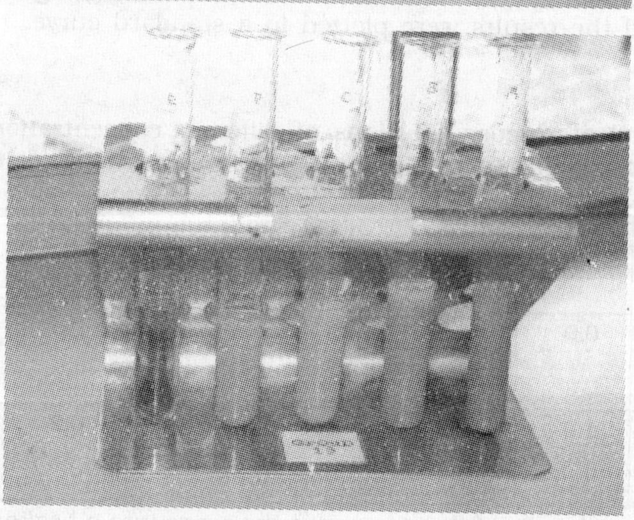

5. If the amount of reducing sugar in a grape is greater than the amount of reducing sugar in 0.01M glucose solution, we can dilute the grape juice by a dilution factor and do the comparison again.

EXERCISE 4. ESTIMATION OF REDUCING SUGAR BY SOMOGY'S METHOD

The reducing sugar selectivity reduces copper salt in alkaline medium. The oxidation of cuo by atmospheric O_2 is prevented by the addition of Na_2SO_4 on acidification of the reaction mixture with H_2SO_4. The KI and KIO_3 present in the reagent liberates iodine. In acidic medium cuprous oxide goes into the solution giving cuprous ions and these ions are oxidized by the liberated iodine. The excess of iodine is librated against N/200 $Na_2S_2O_3$ using starch as an indicator.

Reagents Required

1. *Copper reagent:*

   Dissolve the following substance in water and make up 1 liter with distilled water and filter.

   (*i*) Anhydrous sodium phosphate or disodium hydrogen phosphate 28 gms (or) 71 gms respectively.

   (*ii*) Sodium potassium tartorate-40 gm.

   (*iii*) 1n KOH 5.6 gm in 100 ml.

   (*iv*) 8% $CuSO_4$–8 gms of $CuSO_4$ in 100 ml.

   (*v*) Anhydrous sodium sulphate 180 mg.

   (*vi*) KI-8 gms.

   (*vii*) 1N KIO_3-0.892 gm in 50 ml.

2. In H_2SO_4—27.8 ml of H_2SO_4 in 1 litre.
3. Standard glucose solution, dissolve 100 mg in 100 ml of water.
4. 0.01N $K_2Cr_2O_7$ solution-dissolve 490 mg of $K_2Cr_2O_7$ in 100ml of water.
5. *1% starch indicator*: Make paste from 1 g soluble starch and 5ml distilled water, poured the suspension into 100ml boiling water boil for a minute cool and pressure by adding a drop of chloroform.

Procedure

Standardization of sodium thiosulphate solution: Take 10 ml of 0.01N $K_2Cr_2O_7$ in a conical flask to this add 1 gm of KI and half test tube of dilute H_2SO_4 is added. 1–2 drops of starch indicator of then this titrated against 0.01 N sodium thiosulphate solution. The end point is blue to colourless. From this 0.005 N sodium thiosulphate is prepared.

Pipette out 0.2 to 1 ml of standard glucose soln into different test tube and make up the volume to 2 ml with distilled water add 2 ml of Cu-reagent to each 10 min coal and add 2 ml of 1 N H_2SO_4 to each test tube and keep it aside for 2–3 min and then titrate the liberated iodine against sodium thiosulphate sodium using 1% starch as an indicator. Carryout a blank under the same condition using 2 ml distilled water instead of sugar sodium.

Result: The given unknown sample contains ———mg/% of sugar.

EXERCISE 5. ESTIMATION OF SUGAR BY FOLIN-WU METHOD

Principle

Glucose reduces the cupric ions present in the alkaline Copper-reagent to cuprous ions or the cupric sulphate is converted into cuprous oxide which reduces the phosphomolybdic acid to phosphomolybdous acid which is blue in colour when O.D is measured at 420 nm.

Reagents Required

1. *Alkaline copper-reagent:* Dissolve 40 gm of anhydrous sodium carbonate (Na_2CO_3) in about 400 ml of water and transfer it into 1 litre flask dissolve 7.5 gm of tartaric acid in this sodium and 4.5 gm of $CuSO_4$ which is dissolved in little water mix the soln and make it up to 1 litre.
2. *Phosphomolybdic acid reagent:* To 35 gm of phosphomolybdic acid add 5 gms sodium tungstate add 20 ml 10% NaOH and 50 ml distil water boiled vigorously for 20–40 min so as to remove the NH_3 present in molybdic acid cool. It and dilute to 350 ml and add 125 ml of o-phosphoric acid and make up to 500 ml.
3. *Standard glucose sodium:* Dissolve 100 mg of glucose in 100 ml distilled water and take 10 ml from this sodium and make it up to 100 ml.

Procedure

Pipette out standard glucose sodium 0–1 ml range and make up the sodium to 2 ml with distilled water. Add 2 ml of alkaline cu-reagent to all the test tube mix the content and keep them in boiling water bath for 8 min cool under running water without shaking and then add 2 ml of phosphomolybdic acid reagent to all test tube weight for 10 min and mix the contents and make up the volume to 25 ml with distilled water. Take the Optical density at 420 nm.

Report: 100 ml of unknown sample contains ——mg of sugar.

EXERCISE 6. ESTIMATION OF SUGAR BY HAGEDRON-JENSON METHOD

Principle

The standard sodium is treated with potassium ferricupnide. A part o ferricyanide is reduced by glucose to ferrocyanide. The remaining ferricyanide is determined from the amount of iodine liberated. The ferrocyanide froms a salt with zinc and is not deoxidized to ferricyanide by atmospheric oxygen.

The principle reactions are

$$2K_3\ [Fe(CN)_6] + 2KJ \longrightarrow 2K_4\ [Fe(CN)_6] + I_2$$

$$2K_4\ [Fe(CN)_6] + ZnSO_4 \longrightarrow K_2Zn\ [Fe(CN)_6] + 3K_2SO_4$$

The liberated iodine is estimated by titrating against sodium thiosulphate (0.005 N) using starch as indicator. The amount of thiosulphate required is noted down. The different between the blank and sample is measured. This gives the amount of thiosulphate required for sugar by plotting graph the conc. of sugar is unknown sample can be calculated.

Reagents Required

1. *Iodide sulphate chloride solution:* 25 g of $ZnSO_4$* 12.5 g of $NaCl_2$ is dissolved in 500 ml of water to this 12.5 g of KI is added on the day of experiment.
2. *3% acetic acid:* 3 ml of acetic acid made up to 100 ml.
3. *Potassium ferricyanide soln $K_3\ [F_3(CN)_6]$:* 0.82 gm of $K_3\ [F_3(CN)_6]$ and 5.3 g of anhydrous sodium carbonate is dissolved in water store it in a dark bottle.

4. Preparation of standard sugar soln = 100 mg of sugar in 100 ml is prepared and take 10 ml of stock and make it up to 100 ml
5. *0.1N $Na_2S_2O_3$ soln:* 12.4 g of sodium thiosulphate in 500ml of distilled water
6. *Starch indicator:* 1g of starch in 100 ml of H_2O and 5g of Nacl
7. *0.1N $K_2Cr_2O_7$:* 0.49 g in 100 ml of H_2O.

Procedure

Standardization of $Na_2S_2O_3$ soln. Take 10 ml 0.1N $K_2Cr_2O_7$ soln in a conical flask to this 1g of KI and half test tube. H_2SO_4 1–2 drops of starch indicator and then titrate it against 0.1 N $Na_2S_2O_3$ soln. The end point is blue to colourless seven clean and dry test tubes are taken and sugar soln ranging from 0.0, 0.4, 0.8 to 2 ml are added it is made up to 2 ml by adding distilled water. It gives 0–200 mg% of sugar soln add 3 ml of potassium ferricyanide to each of the conical flask boil the contents for 15 min and cool to the room temperature and then add 3 ml of iodine sulphate chlocide soln and shake well add 2 ml of 3% acetic acid just before tilt rating against 0.005 $Na_2S_2O_3$ using starch indicator. Titrate till blue colour disappear note down the reading substrate from the first value. The values are plotted against conc. of sugar soln from this graph conc. of unknown sugar is determined.

Report: The unknown soln contains ———mg of sugar.

EXERCISE 7. ESTIMATION OF REDUCING SUGAR'S BY DINITRO SALICYLIC ACID (DNS) METHOD

Principle

Reducing sugars have got the property to reduce many of the reagents one such reagent is 3.5 dinitro salicylic acid (DNS) 3,5 DNS in alkaline solution is reduced to 3 amino 5 nitro salicylic acid.

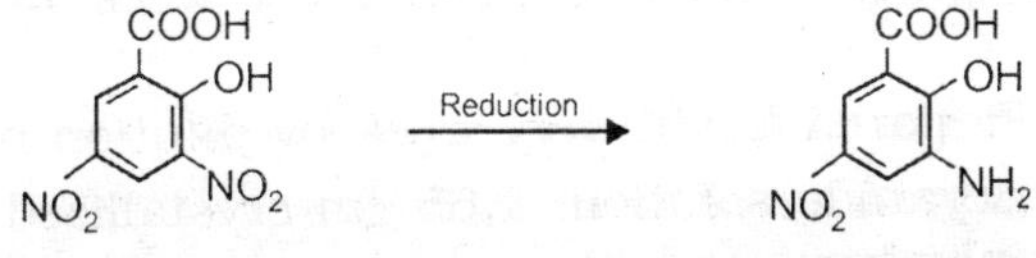

3, 5 dinitro salicylic acid (yellow) — 3 amino, 5-nitro salicylic acid (orange red)

Reagents Required

1. Sodium potassium tartrate:
   Dissolve 45 gm of sodium potassium tartrate in 75 ml of H_2O.
2. 3,5-DNS solution:
   Dissolve 1.5 gm of DNS reagent in 30 ml of 2m/lit NaOH.
3. 2 molar NaOH: 80 gm of NaOH dissolved in 1 litre.
4. DNS reagent:
   Prepare the fresh by mixing the reagent (1) and (2) make up to 150 ml with water.

5. Standard Sugar sodium:
   (i) Stock standard sugar sodium: 250 mg of glucose in water and make up the volume to 100 ml.
   (ii) Working standard sodium: Take 10 ml from this stock solution and make up the volume to 100 ml.

Procedure

Take seven clean dry test tubes. Pipette out std, sugar solution in range of 0 to 3ml in different test tube and make up the volume of all test tube to 3 ml with distilled water the concentration, is ranges from zero to 750 mg. Add 1ml DNS reagent to all the test tube and mix plug the test tube with cotton or marble and keep the test tube in a boiling water bath for 5 min take up the tubes and cool to room temperature and read extinction at 540 mm against the blank please note that all the tubes must be cooled to room temperature before reading since the absorbance is sensitive to temperature.

Prepared standard curves of the sugars provided and use them to estimate to concentration of the unknowns provided.

Result: The 100 ml of unknown solution contains ——— mgm of glucose.

EXERCISE 8. DETERMINATION OF BLOOD GLUCOSE BY HAGEDON-JENNSON METHOD

Principle

Blood proteins are precipitated with zinc hydroxide. The reducing sugars in the protein-free filtrate reduce potassium ferricyanide on heating. The amount of unreduced ferricyanide is determined iodimetrically.

Reagents

1. 0.45% zinc Sulphate: Prepared fresh every week by dilution of 45% stock solution.
2. 0.1 N NaOH: Prepared fresh every week by dilution of 2 N NaOH.
3. Potassium ferricyanide solution: 1.65 gm crystallized potassium ferricyanide and 10.6 gm anhydrous sodium carbonate were dissolved in 1 liter water are stored in a dark bottle, protected from light.
4. Iodide-Sulphate-Chloride Solution: Zinc Sulphate 10 gm and NaCl 50 gm were dissolved in 200 ml water. On the day of use 5 gm potassium iodide is added for 200 ml.
5. 3% acetic acid.
6. 0.005 N sodium thiosulphate: Prepared fresh daily by the diluting of 0.5 N $Na_2S_2O_3$ 0.5 N $Na_2S_2O_3$ was prepared by dissolving 70 gm of salt in 500 ml water. It was better if a solution of slightly higher normality was prepared since sodium thiosulphate decomposes soon. The solution should be protected from light and store in the cold the normality was cheeked daily with 0.005 N potassium iodate.

7. 0.005 N potassium iodate: That was a stable solution and should be made accurately. 0.3566 gm of the anhydrous salt was weighed accurately and dissolved in 2 liter H_2O. That was used to cheek $Na_2S_2O_3$ and potassium ferricyanide solution.
8. Starch indicator: 1gm soluble starch was dissolved in 100ml saturated NaCl solution.

Procedure

1. One ml 0.1N NaOH and 5 ml 0.45% Zinc sulphate were pipetted into a test tube 0.1ml blood was taken in a dry pipette and introduced into the gelatinous Zinc hydroxide in the test tube rinsing out the pipette twice with the mixture. The tube was kept in a boiling water both for 3min and cooled without disturbing the precipitate the mixture was filtered through a what man No.42 filter paper of lightly pressed moisture cotton.
2. The tube was washed twice with 3 ml portions of water and filter into. The same containers.
3. Two ml potassium ferricyanide was then added to the filtrate and heated in a boiling water both for 15min. after cooling 3 ml iodide-sulphate-chloride solution followed by 2 ml 3% acetic acid were added.
4. The liberated iodine was then titrated against 0.005 N $Na_2S_2O_3$ using 2–3 drops of 1% starch as indicator towards the end of the titration.
5. A blank was run through the entire procedure simultaneously. The blank should give a litre value of 1.97 to 2.00 ml.

Result: The concentration of blood glucose in given sample was ——— mg of glucose.

EXERCISE 9. TO DETERMINE BLOOD SUGAR BY NELSON AND SOMOGY'S METHOD

Principle

Blood proteins are precipitated by zinchydroxide. The filtrate is heated with alkaline copper reagent and the reduced cu formed is treated with arseno molybdate reagent resulting in the formation of violet colour, which is read in the photometer.

Reagents

1. 5% $ZnSO_4$ solution.
2. 0.34 N Barium hydroxide

   These two solution should be so adjusted that 5 ml $ZnSO_4$ requ 4.7–4.8 ml
   Ba $(OH)_2$ for complete neutralization.
3. Alkaline copper reagent:

   Solution A: 25 gm anhydrous $Na_2S_2O_3$, 25 gm Rochelle salt, 20 gm $NaHCO_3$ and 200 gm anhydrous Na_2SO_4 were dissolved in about 800 ml water and diluted to 1 litre. The solution was stored at room temperature and never bellow 20°C. It was filtered before use if any sediment was formed.

Solution B: 15% $CuSO_4$, $5H_2O$ containing one or two drops of concentration H_2SO_4. On the day of use 25 parts of solution A and part of solution B were mixed. That was the alkaline copper reagent.

4. Arsenomolybdate colour reagent: Ammonium molybdate 25 gm was dissolved in 450 ml H_2O 21 ml concentrated H_2SO_4 was added and mixed. Disodium ortho arsenate (Na_2H, ASO_4, $7H_2O$) 3 gm was dissolved in 25 ml H_2O and was added with stirring to the acidified molybdate solution. It was then placed in an incubator at 370C for 24–48 hrs and stored in a glass stoppered brown coloured bottle.
5. Standard glucose solution: [stock glucose solution]
   Exactly 0.1 gm of anhydrous pure glucose was dissolved in 10–15 ml 0.2% benzoic-acid and diluted to 100 ml with benzoid acid solution.

Three working standards were prepared by diluting 0.5, 1.0 and 2.00 ml of the stock solution to 100 ml with benzoic acid. These solution in benzoic acid kept indefinitely at room temp.

Working standard: 5 ml make up to 100 ml gives 50 mgm/m.

Procedure

Into a test tube containing 3.5 ml of water 0.1ml of blood was introduced through a clean dry micropipette and mixed well. To this tube 0.2 ml of 0.3 N BaCO + H_2 was added after the mixture turns brown 0.2 ml $ZnSO_4$ was added and mixed. After 10–15 min. The mixture was filtered through a what Mann No.1 filter paper.

Into 2 separate test tube 1ml aliquot of the filtrate was transferred and 1ml alk. cu reagent was then added. The tubes were covered with glass morbel and placed in a boiling water bath for 20min. The tubes were. The cooled under running water, 1ml arsonomolybdate reagent was added and the solution dilute to 25ml with H2O simultaneously standards in 0.2 to 1.0ml range were prepared and a reagent blank were similarly prepared.

The intensity of colour produced was reed at 680 mm. The colour was stable and reading may be taken at convince.

Report: The concentration of blood glucose in given sample is ——mgm/ml.

EXERCISE 10. DETERMINATION OF BLOOD GLUCOSE BY O-TALUIDINE METHOD

Principle

Protein in blood are precipitated with tri-chloro acetic acid as they interfere with estimation contents are filtrated obtained is known as protein free filtrate. It contains glucose whose concentrate is to be determined. Equal volumes of protein free filterate a std glucose solution are treated simultaneously with O-Toluidine reagent (in acetic acid) and kept in boiling water bath. A blue green N-glycosylamine derivative is formed. The intensity of blue green colour is proportional to the amount of glucose present. The O.D values of all three solution are read in a photoelectric colorimeter using a red filter (625 nm) and these O.D values the amount of glucose present in 100 ml blood is calculated.

Reagents

1. O-Toluidine reagent: To 5 gm thiourea, 90 ml O-Toluidine was added and dilute to 1 liter with glacial acetic acid stored in brown bottle and kept the reagent in a refrigerator.
2. 10% Tri-chloro Acetic Acid (TCA).
3. Glucose standard solution (0.1 mg/ml): 10 mg of glucose was dissolved in about 50 ml of distilled water in a 100 ml volumetric flask. To this 30 ml of 10% TCA was added and make up to 100 ml with distilled water.
4. Blank solution: 30ml of 10% TCA was diluted to 100ml.

Procedure

Preparation of protein free filtrate 3 ml of distilled water and 0.5ml of blood are taken in a dry test tube and they were mixed well. 1.5 ml of 10% TCA was added and thoroughly mixed and was allowed to stand for 10 min and filtered into dry test tube.

Development of colour: Standard glucose solution were taken in 6 test tubes in the range of 0.2–1 ml, 1 ml of protein free filtrate was taken in 7th test tube. To all these tubes 5 ml of 0 Toludine was added and mixed thoroughly.

The tubes were kept in boiling water bath for 10 min cooled and read the optical density at 620 mm.

Result: The concentration of blood glucose in given sample is ——— mg/ml

EXERCISE 11. ESTIMATION OF PROTEIN BY BIURET METHOD

Principle

This is the most commonly used method and this based on the fact that - CO - NH (peptide) group of protein from a purple complex with copper ions in an alkaline medium since all proteins contain the peptide bond the method is fairly specific and there is little interferences from other compounds some substance like urea and biuret interfere because they possess - CO – NH - group other interfering materials are reducing sugar like glucose which interact with Cu^{+3} ions (cupric) in the reagent.

Reagents Required

1. Biuret reagent:

   Dissolve 1.5 gm of $CuSO_4$ and 4.5 gm of Na-K tartarte in 250 ml 0.2 N NaOH solution to this solution is added 2.5 g of KI and volume made upto 500 ml with 0.2 N NaOH.
2. 0.2 N NaOH
3. Protein standard solution:

   Dissolve 500 mg of egg albumin in 50 ml of H_2O and make up the volume to 100 ml to get the final concentration 5 mg/ml.

Procedure

Pipette out standard protein solution into series of tubes 0.0, 0.2... 1 ml of protein sdn and make up the total volume to 4 ml 4 with the addition of H_2O. The blank tube will have only 4ml of water. Add 6ml of biuret reagent to each tube and mix well.

Keep the tubes at 37°C for 10 min during which a purple colour will develop. The O.D of each tube is measured at 52.0nm (green filter) using the reagent blank. Draw the graph to known the concentrate of a protein in unknown solution.

Result: The given sample contains ——mg of protein.

EXERCISE 12. ESTIMATION OF THE PROTEIN BY FC-METHOD

Principle

Protein reacts with the folin-ciocalton reagent (FC-reagent) to give a coloured complex. The colour so formed is due to the reaction of the alkaline copper sulphate with the protein and the reduction of phosphomolybdate by tyrosine and tryptophan present in the protein. The intensity of colour depends on the amount of these aromatic amino acids present in the protein and will thus vary for DLF protein.

Materials Required

1. Alkaline sodium carbonate solution: 20 gm/litre sodium carbonate (Na_2CO_3) in 0.1 m/litre NaOH.
2. Copper sulphate: Sodium potassium tartarate solution 5 gm/liter-hydrated copper sulphate ($CuSO_4$-$5H_2O$) in 10 gm/litre sodium potassium tartarate prepare freshly by meaning stalk solution.
3. Alkaline solution: Prepare on the day of use by mixing 50 ml of 1) and 1 ml of 2).
4. F.C. reagent: Dilute the commercial reagent with an equal volume of water on the day of use. This is a solution of sodium tangstate and sodium molybdate in phosphoric acid and HCL.
5. Standard protein solution: Weigh 100 mg egg albumin and dissolve it in approximate to 50 ml of water and make up the volume to 100ml.
6. Working protein solution: Pipette out 20 ml stock protein solution into other 100 ml std. volumetric flask and make up the volume to 100 ml.

Procedure

Pipette out protein solution in the range of 0–200 mg/1 ml into d/f test tubes make up the volume of all the test tubes to 1ml with water add 5 ml of alkaline solution to each test tube, mix thoroughly and allow to std room temp for 10 min add 0.5 ml of dilute FC reagent rapidly with immediately mixing. After 30 min read the extinction against the appropriate blank at 650 nm.

Estimate the protein concentrate of unknown solution after preparing a standard curve.

Result: The given unknown solution contains ————mg/ml of protein.

EXERCISE 13. PROTEIN ASSAY BY BRADFORD METHOD

Materials

1. Lyophilized bovine plasma gamma globulin or bovine serum albumin (BSA)
2. Coomasive Brilliant Blue
3. 0.15 M NaCl
4. Spectrophotometer and tubes
5. Micropipettes

Procedure (Standard Assay, 20–150 µg protein; 200–1500 µg/ml)

1. Prepare a series of protein standards using BSA diluted with 0.15 M NaCl to final concentrations of 0 (blank = NaCl only), 250, 500, 750 and 1500 µ g BSA/ml. Also prepare serial dilutions of the unknown sample to me measured.
2. Add 100 µ l of each of the above to a separate test tube (or spectrophotometer tube if using a Spec 20).
3. Add 5.0 ml of Coomasive Blue to each tube and mix by vortex, or inversion.
4. Adjust the spectrophotometer to a wavelength of 595 nm, and blank using the tube from step 3 which contains 0 BSA.
5. Wait 5 minutes and read each of the standards and each of the samples at 595 nm wavelength.
6. Plot the absorbance of the standards vs their concentration. Compute the extinction coefficient and calculate the concentrations of the unknown samples.

Procedure (Micro Assay, 1-10 µg protein)

1. Prepare standard concentrations of BSA of 1, 5, 7.5 and 10 µ g/ml. Prepare a blank of NaCl only. Prepare a series of sample dilutions.
2. Add 100 µ l of each of the above to separate tubes (use microcentrifuge tubes) and add 1.0 ml of Coomasie Blue to each tube.
3. Turn on and adjust a spectrophotometer to a wavelength of 595 nm, and blank the spectrophotometer using 1.5 ml cuvettes.
4. Wait 2 minutes and read the absorbance of each standard and sample at 595 nm.
5. Plot the absorbance of the standards v/s their concentration. Compute the extinction coefficient and calculate the concentrations of the unknown samples.

EXERCISE 14. ESTIMATON OF PROTEIN BY LOWRY PROTEIN ASSAY

Materials

1. 0.15% (w/v) sodium deoxycholate
2. 72% (w/v) Tri-chloroacetic acid (TCA)
3. Copper Tartrate/Carbonate (CTC)
4. 20% (v/v) Folin-Ciocalteu reagent

5. Bovine Serum Albumin (BSA)
6. Spectrophotometer and tubes
7. Micropipettes

Procedure

1. Prepare standard dilutions of BSA of 25, 50, 75 and 100 μ g/ml. Prepare appropriate serial dilutions of the sample to be measured.
2. Place 1.0 ml of each of the above into separate tubes. Add 100 μ l of sodium deoxycholate to each tube.
3. Wait 10 minutes and add 100 μ l of TCA to each tube.
4. Centrifuge each tube for 15 minutes at 3,000 xg and discard the supernatant.
5. Add 1.0 ml of water to each tube to dissolve the pellet. Add 1.0 ml of water to a new tube to be used as a blank.
6. Add 1.0 ml of CTC to each tube (including the blank), vortex and allow to set for 10 minutes.
7. Add 500 μ l Folin-Ciocalteu to each tube, vortex and allow to set for 30 minutes.
8. Turn on and zero a spectrophotometer to a wavelength of 750 nm. Use the blank from Step 7 to adjust for 100% T.
9. Read each of the standards and samples at 750 nm.
10. Plot the absorbance of the standards vs their concentration. Compute the extinction coefficient and calculate the concentrations of the unknown samples.

EXERCISE 15. BIURET PROTEIN ASSAY

The principle of the Biuret assay is similar to that of the Lowry, however it involves a single incubation of 20 min. There are very few interfering agents (ammonium salts being one such agent), and Layne (1957) reported fewer deviations than with the Lowry or ultraviolet absorption methods. However, the Biuret consumes much more material. The Biuret is a good general protein assay for batches of material for which yield is not a problem. The Bradford assay is faster and more sensitive.

Principle

Under alkaline conditions substances containing two or more peptide bonds form a purple complex with copper salts in the reagent.

Equipment

In addition to standard liquid handling supplies a visible light spectrophotometer is needed, with maximum transmission in the region of 450 nm. Glass or polystyrene (cheap) cuvettes may be used.

Materials

1. Biuret reagent
2. Bovine serum albumin (BSA)
3. Spectrophotometer and tubes

Procedure

1. Prepare standard dilutions of BSA containing 1, 2.5, 5.0, 7.5 and 10 mg/ml protein. Prepare serial dilutions of the unknown samples.
2. Add 1.0 ml of each of the standards, each sample, and 1.0 ml of distilled water to separate tubes. Add 4.0 ml of Biuret reagent to each tube. Mix by vortex.
3. Incubate all of the tubes at 37°C for 20 minutes.
4. Turn on and adjust a spectrophotometer to read at a wavelength of 540 nm.
5. Cool the tubes from Step 3, blank the spectrophotometer and read all of the standards and samples at 540 nm.
6. Plot the absorbance of the standards vs their concentration. Compute the extinction coefficient and calculate the concentrations of the unknown samples.

Analysis

Prepare a standard curve of absorbance versus micrograms protein (or *vice versa*), and determine amounts from the curve. Determine concentrations of original samples from the amount protein, volume/sample, and dilution factor, if any. If you are unfamiliar with how to obtain a protein concentration for a diluted sample from a standard curve.

Comments

The color is stable, but all readings should be taken within 10 min of each other. As with most assays, the Biuret can be scaled down for smaller cuvette sizes, consuming less protein. Proteins with an abnormally high or low percentage of amino acids with aromatic side groups will give high or low readings, respectively.

EXERCISE 16. ESTIMATION OF DNA BY DIPHENYLAMINE METHOD

Principle

When DNA is treated with diphenylamine under acidic condition a blue compound is formed with the sharp observation maximum at 595n. This reaction is given by 2 deoxypentose in general and is not specific for DNA. In acid sdn the straight from the deoxypentose is converted to the highly reactive b-hydroxy lerulin aldehyde which reacts with diphenylamine to give a blue complex. In DNA, only the deoxyribose of purine nudetide reacts so that the value obtained represents the ½ of the total deoxyribose are presented.

Materials

1. Commercial sample of DNA - 10 mg
2. Buffer saline 0.15 ml/lit NaCl, 0.15 ml/lit sodium citrate made up to 500 ml
3. Diphenyl amine reagent: dissolved in 10 gms of pure diphenyl amine in 1liter of glacial acetic acid and add 25 ml of concentrated H_2SO_4. This solution must be prepared fresh.
4. Boiling water bath.

Procedure

Dissolve 10 gms of nucleic acid in 50 ml of buffer saline remove 2 ml and add 4 ml of diphenly amine remove 2 ml and add 4 ml of DPA reagent heat on a boiling water bath for 10 nm cool and read the extension at 595 nm. Read the test and standard against nucleic acid and the commercial sample for DNA.

Result: Concentration of DNA present in the given sample——— mg/2 ml.

EXERCISE 17. ESTIMATION OF RNA BY ORCINOL METHOD

Principle

This is a general reaction for pentose and depends on the formation of furfural when pentose is heated with concentration HCL orcinol reacts in presence of furfusal in presence of ferric chloride as a catalyst purine to give green color only the purine nucleotide.

Materials

1. RNA commercial sample: 10 mg
2. RNA solution (0.2 mg/ml): Dissolve 10 mg of commercially available RNA is 50 ml of buffer saline and use the sample for standard preparation.
3. Orcinol reagent: Dissolve 10 gm of ferric chloride $(FeCl_3)_6$ H_2O in 1 litre of 1 gm of ferric chloride $(FeCl)_6$ H_2O in 1 litre of concentrated HCL and add 35 ml of 6% w/v orcinol in alcohol
4. Buffered saline.
5. Boiling water bath.

Procedure

Pipette out standard RNA solution in arrange of 0–2 ml into a series of test tube and make up the volume of each tube 2 ml with distil water and add 3 ml of orcinol reagent to each tube for standard solution .

For test solution : Take 2 ml of nucleic acid sodium (isolated from tissue source) add 3 ml of orcinol reagent to each tube, heat the tube on boiling water bath for 20 min, cool and take the optical density at 665 nm against orcinol blank.

Result: Concentration of RNA present in the given sample is ——— µgm/2 ml.

4 ENZYMOLOGY

ENZYMES

INTRODUCTION

Cells function by the action of enzymes. Life is a dynamic process that involves constant changes in chemical composition for these chemical enzymes are required. These changes are regulated by catalytic reactions, which are regulated by enzymes. What's why enzymes are called as Biological catalyst.

In this exercise, we will extract the enzyme tyrosinase and study its kinetic parameters. It is only one of thousands of enzymes working in concert within cells, but it is one but which readily demonstrates the main features of enzyme kinetics.

Since all enzymes are proteins, and proteins are differentially soluble in salt solutions, enzyme extraction procedures often begin with salt (typically, ammonium sulfate) precipitation. On the simplest level, proteins can be divided into albumins and globulins on the basis of their **solubility** in dilute salts. Albumins are considered to be soluble while globulins are insoluble. Solubility is relative, however, and as the salt concentration is increased, most proteins will **precipitate.**

Thus, if we homogenize a tissue in a solution that retains the enzyme in its soluble state, the enzyme can be subsequently separated from all insoluble proteins by centrifugation or filtration. The enzyme will be impure, since it will be in solution with many other proteins. If aliquots of a concentrated ammonium sulfate solution are then added serially, individual proteins will begin to precipitate according to their solubility. By careful manipulation of the salt concentrations, we can produce fractions which contain purer solutions of enzymes, or at least are enriched for a given enzyme. Fortunately, absolute purity of an enzyme extract is seldom required, but when it is, the fractions must be subjected to further procedures designed for purification (such as electrophoresis and/or column chromatography).

In order to determine the effectiveness of the purification, each step in the extraction procedure must be monitored for enzyme activity. That monitoring can be accomplished in many ways, but usually involves a measurement of the decrease in substrate, or the increase in product specific to the enzyme.

It is important to remember that enzymes act as catalysts to a reaction and that they affect only the reaction rate. The general formula for the action of an enzyme is given by the following:

$$E + S \underset{k_2}{\overset{k_1}{\rightleftarrows}} ES \longleftrightarrow EP \underset{k_4}{\overset{k_3}{\rightleftarrows}} E + P \qquad ...(4.1)$$

where E = concentration of the enzyme

S and P = concentrations of substrate and product, respectivelye

ES and EP = concentration of enzyme-substrate complex and enzyme-product complex

k_1–k_4 = rate constants for each step.

From equation 5.1, the rates(velocities) of each reaction can be given by:

$v_1 = k_1$ (E) (S); formation of enzyme-substrate complex

$v_2 = k_2$ (ES); reformation of free enzyme and substrate

$v_3 = k_3$ (ES); formation of product and free enzyme

$v_4 = k_4$ (E) (P); reformation of enzyme-product complex

In steady state equilibrium, $(v_1 - v_2) = (v_3 - v_4)$ and, if all product is either removed or does not recombine with the enzyme, then $k_4 = 0$, and k_1(E)(S) – k_2(ES) = k_3(ES)

This equation can then be rearranged to yield:

$$\frac{k_2 + k_3}{k_1} = \frac{(E)\ (S)}{(ES)} \qquad ...(4.2)$$

where the left side of this equation can be given as a single constant, known as K_m, the rate constant, or the Michaelis constant. Note that the units for this constant will be those of concentration.

One of the important concepts of metabolism is that enzymes from differing sources may have the same function (i.e. the same substrate and product), but possess significantly different K_m values. Since biological function is as dependent on the rate of a reaction as it is on the direction of a reaction, it becomes necessary to measure the K_m value for any enzyme under study.

Enzymes act as catalysts because of their three dimensional protein structure. This structure is controlled by many factors, but is particularly sensitive to changes in pH, salts and temperature. Small changes in the temperature of a reaction can significantly alter the reaction rate, and extremely high temperatures can irreversibly alter both the three dimensional structure of the enzyme and its activity. It may even render the enzyme non-functional; that is, to denature the enzyme. Salts can also cause denaturation, but the effects of ammonium sulfate are usually reversible. Heavy metal salts, by contrast, usually irreversibly alter the structure of the protein, and thus their routine use as fixatives in histological work.

Active Sites

An enzyme works by binding to a given substrate in such a geometrical fashion that the substrate is able to undergo its inherent reaction at a more rapid rate. This type of reaction is commonly referred to as the lock and key mode for enzyme action. It implies that there is a particular part of the enzyme structure, the active site, which specifically binds sterically to a substrate. The enzyme does not actually react with the substrate but merely brings the substrate into the proper alignment or configuration for it to react spontaneously or in conjunction with another substance. Since a reaction proceeds normally by a random kinetic action of molecules bumping into each other, any time molecules are aligned, they will react faster. Thus, for any given enzyme there will be a best fit configuration to the protein in order to align the substrate and to facilitate the reaction. When the enzyme is in its ideal configuration, the reaction will proceed at its maximum rate, and the overall rate of activity will be dependent upon substrate concentration.

Maximum reaction rate assumes that an optimal pH, salt environment and temperature have been established. demonstrates some typical effects of temperature and pH on the rate of an enzyme catalyzed reaction. Maximum rate further assumes the presence of any co-enzymes and/or cofactors that the enzyme requires. Co-enzymes are organic molecules which must bind to the protein portion of the enzyme in order to form the correct configuration for a reaction. Cofactors are inorganic molecules which do the same.

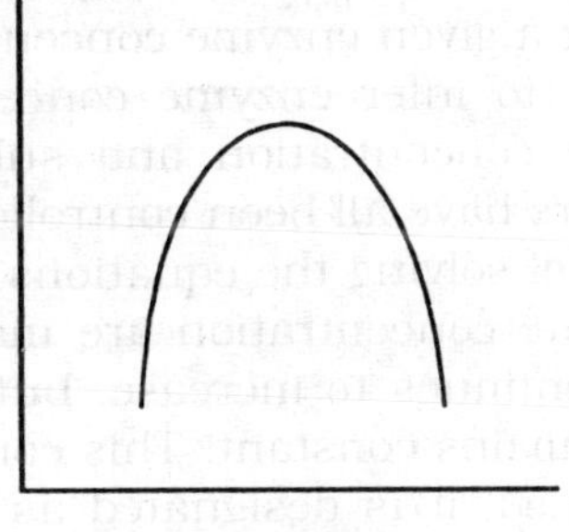

Fig. 1. For effect for pH and temperature.

Now, if we measure the concentration of an enzyme via its rate of activity (i.e. the velocity of the catalyzed reaction), we must control the reaction for the effects of temperature, pH, salt concentration, co-enzymes, cofactors, and substrate concentration. Each of these parameters affects the rate of an enzyme reaction. Thus, each must be carefully controlled if we attempt to study the affects of changes in the enzyme itself. For example, alterations in the rate of a reaction are directly dependent upon the concentration of functional enzyme molecules only when the enzyme is the limiting factor in the reaction. There must be sufficient substrate to saturate all enzyme molecules in order for this criterion to be met. If the substrate concentration is lowered to the point where it becomes rate limiting, it is impossible to accurately measure the enzyme concentration, because there will be two variables at work.

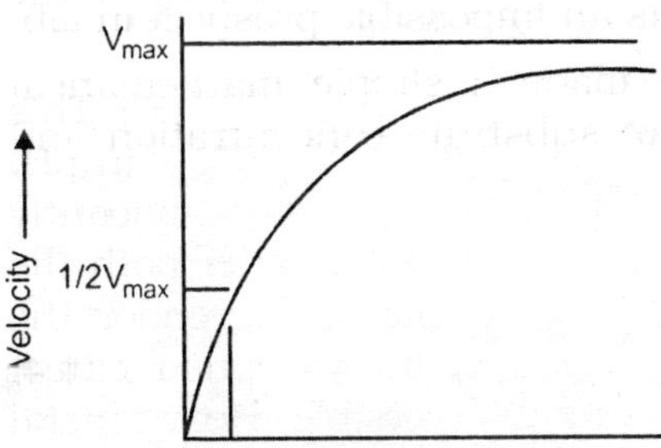

Fig. 2. Michaelis-Menten plots

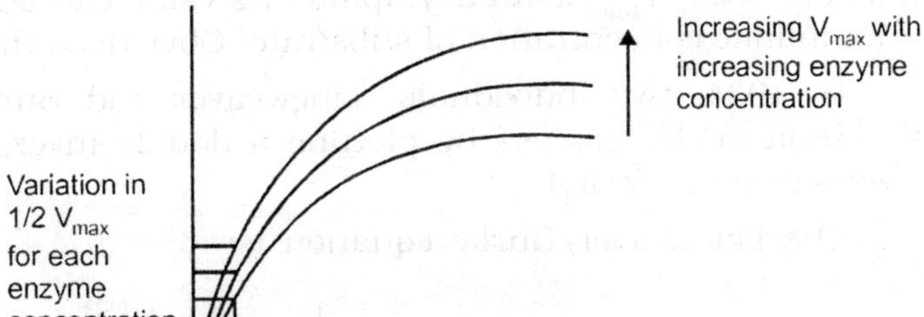

Fig. 3. Effect of increasing enzyme concentration

The relationship between substrate concentration and enzyme concentration was mathematically established by pioneering work of two biochemists, L. Michaelis and M.L. Menten in 1913. In recognition of their work, the plots of enzyme activity vs. substrate concentration are known as Michaelis-Menten plots. These are relatively simple plots in which the substrate concentration is on the x-axis, and the velocity of reaction is on the y-axis. The plot demonstrates that as the substrate increases, the velocity increases hyperbolically, and approaches a maximum rate known as V_{max}. This is dependent upon saturation of the enzyme. At V_{max}, all enzyme molecules are complexed with substrate, and thus any additional substrate added to the reaction has no effect on the rate of reaction.

However, this situation becomes more complex: as you change the enzyme concentration, V_{max} will also change. Thus, V_{max} is not a constant value, but is constant only for a given enzyme concentration. Consequently, the value of V_{max} can not be used directly to infer enzyme concentration. It is dependent upon at least two variables, enzyme concentration and substrate concentration (assuming temperature, pH and cofactors have all been controlled). What Michaelis and Menten discovered was a simple means of solving the equations for two variables. If multiple plots of enzyme activity vs. substrate concentration are made with increasing enzyme concentration, the value of V_{max} continues to increase, but the substrate concentration which corresponds to 1/2 V_{max} remains constant. This concentration is the **Michaelis constant** for an enzyme. As mentioned, it is designated as K_m and is operationally the concentration of substrate which will give exactly 1/2 V_{max} when reacted with an enzyme with maximum pH, temperature and cofactors.

According to the Michaelis-Menten equation:

$$V = \frac{V_{max}\,(S)}{K_m + (S)} \qquad \text{...(4.3)}$$

This equation is derived from the formula for a hyperbola ($c = xy$) where $K_m = (S)\,(V_{max}/v - 1)$

When $v = V_{max}/2$, $K_m = (S)\,(V_{max}/(V_{max}/2) - 1) = (S)$ confirming that the units of this constant are those of concentration.

A Michaelis-Menten plot thus can give us an easy way to measure the rate constant for a given enzyme. An immediate difficulty is apparent, however, when Michaelis-Menten plots are used. V_{max} is an asymptote. Its value can only be certain if the reaction is run at an infinite concentration of substrate. Obviously, this is an impossible prospect in lab.

In 1934, two individuals, Lineweaver and Burke made a simple mathematical alteration in the process by plotting a double inverse of substrate concentration and reaction rate (velocity).

The Lineweaver/Burke equation is:

$$\frac{1}{V} = \frac{K_m + (S)}{V_{max}\,(S)} \qquad \text{...(4.4)}$$

This equation fits the general form of a straight line, y = mx+b, where m is the slope of the line and b is the intercept. Thus, the Lineweaver/Burke Plot for an enzyme is more useful than Michaelis-Menten, since as velocity reaches infinity, $1/V_{max}$ approaches 0. Moreover, since the plot results in a straight line, the slope is equal to K_m/V_{max}, the y intercept equals $1/V_{max}$ (1/S = 0). Projection of the line back through the x axis yields the value $-1/K_m$ (when 1/V = 0). These values can easily be determined by using a linear regression plot and calculating the corresponding values for $x = 0$ and $y = 0$. The inverse of the intercept values will then yield V_{max} and K_m.

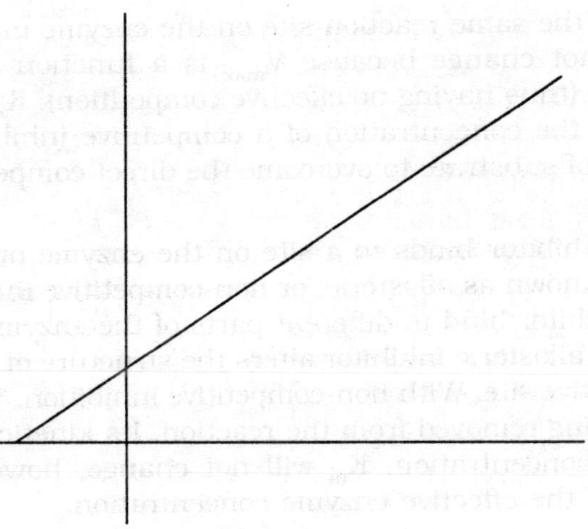

Fig. 4. Lineweaver -Burke plot of enzyme activity

Remember that the point of all of these calculations is to determine the true activity and thus the concentration of the enzyme. If the reaction conditions are adjusted so that the substrate concentration is at K_m, then alterations in the rate of reaction are linear and due to alterations in enzyme concentration. Kinetic analysis is the only means of accurately determining the concentration of active enzyme.

Specific Activity

This brings us to a definition for enzyme activity. **Specific activity** is defined in terms of enzyme units per mg enzyme protein. An enzyme unit is the amount of substrate converted to product per unit time under specific reaction conditions for pH and temperature.

As generally accepted, an enzyme unit is defined as that which catalyzes the transformation of 1 micromole of substrate per minute at 30°C and optimal chemical environment (pH and Substrate concentration). Specific Activity relates the enzyme units to the amount of protein in the sample.

While it is relatively easy to measure the protein content of a cell fraction, there may be a variable relationship between the protein content and a specific enzyme function. Remember that the initial extraction of an enzyme is accomplished by differential salt precipitation. Many proteins will precipitate together due to their solubility, but have no other common characteristics.

To determine both protein content and enzyme activity requires two different procedures. We can measure the amount of protein, or we can kinetically measure the enzyme activity. Combining the two will give us the specific activity.

Enzyme Inhibition

Finally, before studying a specific enzyme, let's examine the problem of **enzyme inhibition**. Remember that enzymes function by sterically binding to a substrate. If a molecule interferes with that binding, it will hinder or inhibit the activity of the enzyme.

If the inhibitor molecule binds to the same active site as the substrate, then the reaction is known as competitive inhibition because the two molecules, substrate and inhibitor, compete for the same reaction site on the enzyme molecule. With this type of inhibition, V_{max} will not change because V_{max} is a function of all enzyme molecules uniting with substrate (thus having no effective competition). K_m, on the other hand, will alter with changes in the concentration of a competitive inhibitor, because it requires larger concentrations of substrate to overcome the direct competition of the inhibitor for the active site.

If however, the inhibitor binds to a site on the enzyme other than the active site, then the inhibition is known as allosteric, or non-competitive inhibition. In this instance, the substrate and inhibitor bind to different parts of the enzyme molecule and thus are not in competition. An allosteric inhibitor alters the structure of the enzyme or physically blocks access to the active site. With non-competitive inhibition, V_{max} will change because, in effect, enzyme is being removed from the reaction. Its kinetic effects are equivalent to lowering the enzyme concentration. K_m will not change, however, since this value is constant regardless of the effective enzyme concentration.

Finally, there is a third class of inhibitor which can best be defined by its effect on V_{max} and K_m. Known as uncompetitive, it alters both of these values. It has effects on both the active site and allosteric sites.

Tyrosinase

This exercise involves the isolation (extraction) of the enzyme tyrosinase from potatoes and subsequent measurement of its activity. Tyrosinase is the common name for an enzyme that is formally termed Monophenol Monooxygenase and is listed as Enzyme #1.14.18.1 in the standard Enzyme Nomenclature. It is also known as phenolase, monophenol oxidase and as cresolase. It is, functionally, an oxygen oxidoreductase enzyme.

This nomenclature points out another difficulty of working with enzymes. Their names are derived from the known activities. Enzymes isolated from different sources and measured for their catalytic activity with varying substrates can turn out to be the same protein. Thus, the enzyme tyrosinase, discovered in animal systems, was named for its action on the amino acid tyrosine, and specifically for its ability to form dopaquinone, an intermediate metabolite in the production of melanin.

The same enzyme isolated from plant materials had been examined for its ability to oxidize phenolic residues, and thus the names phenolase, monophenol oxidase and cresolase. Since it has been extensively studied in melanin production, we will continue to use the common name of tyrosinase.

The enzyme tyrosinase is fairly ubiquitous; that is, it is found in nearly all cells. In research, it has been purified from the fungus N. crassa by freezing kilogram quantities of the fungal mycelia in liquid nitrogen, homogenizing the frozen tissue with a French Press, precipitating the proteins in ammonium sulfate, and purifying the enzyme chromatographically on Sephadex and Celite columns. A fairly complex undertaking.

Tyrosinase has been extracted from hamster melanomas by modifications of this technique and with the addition of acetone extractions as well as DEAE-cellulose chromatography and alumina treatments. Tyrosinase has also been separated from many plant tissues utilizing a far simpler technique based principally on ammonium sulfate precipitation of proteins.

The catalytic action of this enzyme is the conversion of tyrosine + O_2 to yield dihydroxyphenylalanine (DOPA), which is then converted to dopaquinone + H_2O.

Dopaquinone in turn can be readily converted to dopachrome, an orange to red pigment (found in human red hair), which can then be converted to the black/brown melanin pigments (found in virtually all human pigments).

$$\text{Tyrosine} + 1/2\ O_2 \longrightarrow \text{DOPA}$$

$$2\ \text{DOPA} + O_2 \longrightarrow 2\ \text{Dopaquinone} + 2\ H_2O$$

$$\text{Dopaquinone} \longrightarrow \text{Leukodopachrome}$$

$$\text{Leukodopachrome} + \text{Dopaquinone} \longrightarrow \text{Dopachrome} + \text{DOPA}$$

The enzyme catalyzes the first two of these reactions, namely the conversion of tyrosine and the conversion of DOPA. The formation of dopachrome from dopaquinone is spontaneous.

We can now monitor the activity of the enzyme by analyzing the disappearance of tyrosine and/or DOPA as substrates, the appearance of leukodopachrome or dopachrome as products, or by monitoring the use of oxygen. Physiologists and chemists have long preferred the manometric determination of gaseous oxygen exchange, but far simpler is the determination of dopachrome, a natural pigment with an absorbance maximum at 475 nm. This absorbance allows us to use standard spectrophotometric analysis by analyzing the formation rate of dopachrome from the substrate DOPA.

The summary reaction for tyrosinase activity, as used in this exercise is

$$\text{DOPA} + 1/2\ O_2 \longrightarrow \text{Dopachrome}$$

EXERCISE 1. TO DEMONSTRATE THE PRESENCE OF CATALASE IN PIG'S LIVER

Background Information

Catalase is an enzyme found in many animal tissues such as meat, liver, and many plant tissues such as stems of seedlings. Catalase is found in the organelle peroxisomes and microbodies in a cell. In industry, it is used to generate oxygen to convert latex to foam rubber. The presence of catalase is important because it can decompose hydrogen peroxide which is a by-product of certain cell oxidations and is also very toxic. Catalase can eliminate the hydrogen peroxide immediately. It is a fastest-acting enzyme know. Its activity can be demonstrated by dropping a piece of fresh liver into hydrogen peoxide, when rapid evolution of oxygen is observed.

Principles

The evolution of gas should be observed when catalase is placed into hydrogen peroxide. The hydrogen peroxide is decomposed into water

1. Grind 10g of fresh pig's liver in 10cm3 of distilled water in a mortar with a pestle.
2. Filter the liver extract.
3. Dilute the liver filtrate by 100% with distilled water.
4. Add a drop of the filtrate to 5cm3 of hydrogen peroxide.
5. Observe the evolution of gas.

6. Repeat step (4) with distilled water instead of hydrogen peroxide. Then repeat again with distilled water instead of liver filtrate. These are the controls.
7. If you want to test any other tissues, repeat the procedure with any other tissues.

Precautions

1. Hydrogen peroxide solution is corrosive. It may hurt your skin. Wash your fingers if you can get contact with the solution.
2. The extract must be filtrated so that the residues will not affect the result.

EXERCISE 2. TO DETERMINE THE OPTIMUM pH FOR TRYPSIN

Introduction

Trypsin is a kind of protease. This enzyme is present in small intestine and can break down protein into amino acid. Different enzymes may have different optimum pH. In the optimum pH, the enzymes work best. The activity is the highest. In lower pH or higher pH, the excessive hydrogen ions or hydroxide ions may break the ionic bonds. This changes the shape of the enzyme. The shape of active site also alters. This lowers the catalytic activity.

Photographic film has a protein called gelatin which coats on the surface of it. If it is removed, a whitish stain will appear.

Principles

In this experiment, we can put a photographic film strips into a trypsin solution. The protease will digest the gelatin coating and make it whitish. Different buffer solutions will be added to the solutions as well to change the surrounding pH. The degree of digestion of the gelatin reflects the activity of the enzyme. Therefore, the optimum pH can be estimated.

Procedure

1. Set up the water bath and adjust the temperature to 37 degrees Celsius.
2. Pipette 1mL buffer solution of pH 1.0 into a test tube.
3. Add a strip of photographic film into the solution and put it into the water bath for 5 minutes.
4. Pipette 1mL of trypsin solution into another test tube. Put it into the water bath for 5 minutes.
5. Pour the 1mL of trypsin solution into test tube with film.
6. Record the time taken for complete disappearance of gelatin on the film.
7. Repeat the experiment with buffer solution of pH around 2.0, 4.0, 6.0, 8.0, 10.0.

Precautions

1. The temperature must be kept constant since temperature is a factor affecting the enzymatic activity.
2. The gelatin coating is easily scratched and damaged and it should be handled with care.

3. The enzyme, and the film have to be put into the water bath for 5 minutes before mixing. This allows both substrates and enzymes reach the optimum temperature before mixing.
4. The tubes must be mixed from time to time.
5. If the gelatin coat remains for over 1 hour. We can assume the time to be infinity.

EXERCISE 3. EXTRACTION OF TYROSINASE

Materials

- Potatoes
- Paring Knife
- Blender
- 0.1 M NaF
- Rubber Gloves
- Saturated Ammonium Sulfate (4.1 M, 25°C)
- Volumetric Cylinders (50 ml, 100 ml, 250 ml)
- Cheesecloth
- Beakers (100 ml, 250 ml)
- Chilled Centrifuge Tubes (30–50 ml)
- Refrigerated Centrifuge
- 0.1 M Citrate Buffer, pH 4.8
- Glass Stirring Rod.

Procedure

1. Peel a small potato and cut into pieces about 1 inch square.
2. Add 100 grams of the potato to a blender, along with 100 ml of sodium fluoride (NaF). Homogenize for about one minute at high speed.
   Caution: Sodium Fluoride is a poison! Wear rubber gloves while handling, and wipe up any spills immediately.
3. Pour the homogenate (mixture) through several layers of cheese cloth and into a beaker.
4. Measure the volume of the homogenate and add an equal volume of saturated ammonium sulfate. That is, if the fluid volume of your homogenate is 150 ml, add 150 ml of ammonium sulfate. This will cause a floculent white precipitate to appear as many of the previously soluble potato proteins become insoluble. The enzyme tyrosinase is one of these proteins and thus will be found in the subsequent precipitate.
5. Divide the ammonium sulfate-treated homogenate into chilled centrifuge tubes and centrifuge at 1,500 xg for 5 minutes at 4°C.
6. Collect the centrifuge tubes, and carefully pour off and discard the fluid (supernatant). **Save the pellets.** Combine all of the pellets into a 100 ml beaker.

7. Add 60 ml of citrate buffer, pH 4.8, to the pooled pellet and stir the contents well. Use a glass rod to break up the pellet. Continue to stir for 2 minutes while keeping the solution cool.
8. Again divide the solution into centrifuge tubes and recentrifuge at 300 xg for 5 minutes at 4°C.
9. Collect and save the supernatant. **This is your enzyme extract!** Place it in an erlenmeyer flask, label it as Enzyme Extract and place it in an ice bucket. The enzyme tyrosinase is insoluble in 50% ammonium sulfate, but is soluble in the citrate buffer. Keep this extract chilled for the duration of the laboratory. Tyrosinase is stable for about an hour under the conditions of this exercise. If not used within this period, you will need to extract more enzyme from a fresh potato.

EXERCISE 4. PREPARATION OF STANDARD CURVE

Materials

- 8 mM DOPA
- Enzyme Extract-From Expt. No. 5.1 Test tubes
- 5 ml Pipette
- 0.1M Citrate Buffer, pH 4.8
- Spectrophotometer and Cuvettes.

Procedure

1. Begin by preparing a standard solution of the orange colored dopachrome from L-DOPA. To 10 ml of 8 mM DOPA, add 0.5 ml of your enzyme extract and allow the solution to sit for 15 minutes at room temperature. During this period, all of the DOPA will be converted to dopachrome, and your solution will now contain 8 mM dopachrome. Dopachrome is somewhat unstable in the presence of light and should be stored in an amber bottle or out of the light.
2. Prepare a 1:1 series of dilutions of the 8 mM Dopachrome to yield the concentrations in the following table: Add 3.0 ml of each indicated concentration to tubes #1–8.

Tube #	*Final Concentration of Dopachrome (mM)*
1	0
2	0.125
3	0.25
4	0.5
5	1.0
6	2.0
7	4.0
8	8.0

3. With these dilutions, you have prepared tubes containing concentrations from 0 to 8 mM dopachrome (tubes 1–8). Tube 1 contains no dopachrome and is used for blanking the spectrophotometer.
4. The units of concentration are millimolar (mM). A 1.0 mM solution contains .001 moles per liter or .000001 moles per ml. Thus, with a volume of 3.0 ml, there are .000003 moles of dopachrome, or 3 micromoles. Correspondingly, tubes 2–8 contain 1 to 24 micromoles of dopachrome. For the remainder of this exercise, be sure to distinguish between concentration (mM) and total amount of substance present (micromoles).
5. Turn on your spectrophotometer and set a wavelength of 475. Use tube #1 from the above dilutions as a blank and adjust the spectrophotometer for 0 and 100% T. Read the absorbance (or read and convert transmittance) of each of the solutions in tubes 2–8 and complete the following table:

Tube	*Concentration of Dopachrome (mM)*	*Absorbance*	*A/C*
#1	0	0	—
#2	0.125		
#3	0.25		
#4	0.5		
#5	1.0		
#6	2.0		
#7	4.5		
#8	8.0		

6. Calculate the values for the last column of the table. This column represents the simplest calculation of the extinction coefficient for dopachrome absorbance. Average the values in this column and enter the number at the bottom of the column. This is the average extinction coefficient and can be used in subsequent determinations of dopachrome concentrations according to the Beer-Lambert law.

   You can more accurately determine the extinction coefficient by performing a linear regression analysis of your data, and computing the slope and y intercept. The slope of the linear regression will represent the extinction coefficient for your sample.
7. Plot a scattergram of the absorbance value against the concentration of dopachrome. The known concentration of dopachrome should be the x axis, while absorbance should be the y axis.
8. Plot the computed slope and intercept of the linear regression as a straight line overlaying your scattergram. The equation for a straight line is $y = mx + b$, where m is the slope and b the intercept.

Notes

Since tyrosinase catalyzes the conversion of L-DOPA to dopachrome, this exercise measures the conversion of colorless DOPA to the dark orange dopachrome. Substrate and product are in a 1:1 ratio for this reaction, thus the amount of product formed equals the amount of substrate used. The optical density of dopachrome at 475 nm is directly proportional to the intensity of orange color formation in solution (Beer-Lambert Law).

EXERCISE 5. ENZYME CONCENTRATION

Materials

- Enzyme Extract
- 0.1 M Citrate Buffer, pH 4.8
- 10 ml Pipette
- 8 mM DOPA
- Spectrophotometer and Cuvettes
- Ice Bath

Procedure

1. To determine the kinetic effects of the enzyme reaction, first determine an appropriate dilution of your enzyme extract. This will give a rate of reaction of 5–10 micromoles of DOPA converted per minute. Prepare a serial dilution of your enzyme extract. Place 9.0 ml of citrate buffer into each of three test-tubes. Label the tubes 1/10, 1/100 and 1/1000.
2. Pipette 1.0 ml of your enzyme extract into the first of these tubes (the one labeled 1/10) and mix by inversion.
3. Pipette 1.0 ml of the 1/10 dilution into the second tube (labeled as 1/100) and mix by inversion.
4. Pipette 1.0 ml of the 1/100 dilution into the third tube (labeled as 1/1000) and mix by inversion.
5. Place all of the dilutions in the ice bath until ready to use.
6. If not already done, turn on a spectrophotometer, adjust to 475 nm and blank with a tube containing 2.5 ml of citrate buffer and 0.5 ml of enzyme extract.
7. Add 2.5 ml of 8 mM DOPA to each of 4 cuvettes or testtubes. Note that each tube contains .0025 X .008 moles or 20 micromoles of DOPA.
8. Add 0.5 ml of undiluted enzyme extract to one of the tubes containing the 8 mM DOPA. Mix by inversion, place into the spectrophotometer and immediately begin timing the reaction. Carefully measure the time required for the conversion of 8 micromoles of DOPA. Note that since the cuvette will contain a volume of 3.0 ml, the concentration when 8 micromoles are converted will be 8/3.0 or 2.67 mM dopachrome. Use the data from the standard curve (Exercise 5.2) to determine the absorbance equal to 2.67 mM dopachrome. This absorbance value will be the end point for the reaction.

The Absorbance equal to 3.33 mM dopachrome = ______

9. As the reaction takes place within the spectrophotometer, the absorbance will increase as dopachrome is formed. When the absorbance reaches the value above, note the elapsed time from the mixing of the enzyme extract with the 10 mM DOPA. Express the time as a decimal rather than minutes, seconds.

The time should be between three and five minutes. If the end point is reached before three minutes, repeat Step 8, but using the next dilution of enzyme (i.e. the 1/10 after the undiluted, the 1/100 after the 1/10 and the 1/1000 after the 1/100).

The rate of activity = ______________ micromoles/minute/0.5 ml of diluted extract. The dilution factor (inverse of dilution, 1,10,100 or 1000) is _____.
The activity of the undiluted enzyme is ___________ micromoles/minute/0.5 ml or ____________ micromoles/minute/1.0 ml of extract.

10. For the enzyme dilution which reaches the end point between 3 and 5 minutes, calculate the velocity of reaction. Divide the amount of product formed (10 micromoles) by the time required to reach the end point.

EXERCISE 6. EFFECTS OF pH

Materials

- 8 mM DOPA in citrate buffer adjusted to pH values of 3.6, 4.2, 4.8, 5.4, 6.0, 6.6, 7.2, 7.8
- Enzyme Extract
- Spectrophotometer and Cuvettes
- Stopwatch

Procedure

1. Set up a series of test tubes each containing 2.5 ml of 8 mM DOPA, but adjusted to the following pH values: 3.6, 4.2, 4.8, 5.4, 6.0, 6.6, 7.2, and 7.8
2. Begin with the tube containing DOPA at pH 3.6, add 0.5 ml of the diluted enzyme extract which will convert 10 micromoles of DOPA in 3–5 minutes as determined in expt. no. 5.3. Start timing the reaction, mix by inversion and insert into the spectrophotometer. Note the time for conversion of 10 micromoles of DOPA.
3. Repeat Step 2 for each of the indicated pH values. Complete the following table:

pH	*Time(Minutes)*	*Micromoles of Dopachrome*	*Velocity (Micromoles/Minute)*
3.6		10	
4.2		10	
4.8		10	
5.4		10	
6.0		10	
6.6		10	
7.2		10	
7.8		10	

4. Plot pH (x-axis) versus Reaction Velocity (y-axis).

EXERCISE 7. EFFECTS OF TEMPERATURE

Materials

- Enzyme Extract
- 8 mM DOPA pH 6.6
- Incubators or Water baths adjusted to 10, 15, 20, 25, 30, 35 and 40°C
- Spectrophotometer and Cuvettes
- Stopwatch.

Procedure

1. Set up a series of test tubes each containing 2.5 ml of 8 mM DOPA buffered to a pH = 6.6. Place one tube in an ice bath or incubator adjusted to the following temperature; 10, 15, 20, 25, 30, 35 and 40°C
2. Add 0.5 ml of an appropriately diluted enzyme extract (to yield 10 micromoles dopachrome in 3–5 minutes) to each of a second series of tubes. Place one each in the corresponding temperature baths. Allow all of the tubes to temperature equilibrate for 5 minutes. **Do not mix the tubes.**
3. Beginning with the 10°C tube, and with the spectrophotometer adjusted to 475 nm and properly blanked, pour the enzyme (0.5 ml at 10°C) into the tube containing the DOPA and begin timing the reaction. Mix thoroughly. Note the time to reach the end point equivalent to the conversion of 10 micromoles of substrate.
4. Repeat Step 3 for each of the listed temperatures, complete the following and plot the data.

Temperature (°C)	*Time (Minutes)*	*Micromoles of Dopachrome*	*Velocity (Micromoles/Minute)*
10		10	
15		10	
20		10	
25		10	
30		10	
35		10	
40		10	

EXERCISE 8. COMPUTER SIMULATION OF ENZYME ACTIVITY

You will note that for the determination of pH and temperature effects, we established conditions that scanned a wide range of the variables. To refine these procedures, we would add more values, for example, temperatures from 30–40°C at steps of 1.0°C or 36–37°C with a step of 0.1°C. This is time consuming.

To establish the principle, however, let's shift to computer simulation of enzyme activity. Computer simulations will also assist in the rapid accumulation of data for kinetic analysis.

There are several excellent commercial programs available for enzyme kinetic analysis, and use of these must be left to the descretion of the instructor. The author has had excellent results with a rather old BASIC program, ENZKIN, which is still available from Conduit.

This is a simple, straight-forward and inexpensive program which is available for Apple Computers and for main frames. It can readily be converted to run on other machines with some simple source code changes. Unfortunately, Conduit may not continue supplying this program in the future.

If you do not wish to convert programs, or if you prefer a more sophisticated program is desired, the author recommends ENZPACK Ver. 2.0. In addition to simulation of enzyme kinetics, this program allows entry of data with subsequent graphing and analysis. It also allows for high order kinetic analysis in addition to Lineweaver-Burk plots.

EXERCISE 9. KINETIC ANALYSIS

Materials

- 8 mM DOPA pH 6.6
- Enzyme Extract, diluted to yield 10 micomoles of dopachrome in 3–5 minutes (Expt. No. 5.3)
- Spectrophotometer and Cuvettes
- Stopwatch

Procedure

1. Prepare a reaction blank in a clean cuvette to contain 2.5 ml of citrate buffer and 0.5 ml of enzyme extract. Use this blank to adjust your spectrophotometer for 100% transmittance. Remove the blank and save.
2. Add 2.5 ml of 8 mM DOPA, pH 6.6 to a clean cuvette.
3. Add 0.5 ml of appropriately diluted enzyme extract. Shake well and immediately insert the tube into the spectrophotometer. Record the absorbance or transmittance as quickly as possible. Designate this reading as time 0.
4. At 30 second intervals read and record the transmittance until a transmittance value of 10% (Absorbance = 1.0) is reached. Complete the following table:

Time (Minutes)	*Absorbance*	*Concentration of Dopachrome(mM)*	*Micromoles of Dopachrome*
0.5			
1.0			
2.0			
2.5			
3.0			
3.5			
4.0			
4.5			
5.0			

5. Plot time in minutes (x-axis) versus the amount of dopachrome formed (y-axis).

EXERCISE 10. DETERMINATION OF K_m AND V_{max}

Materials

- Enzyme Extract
- 8 mM L-DOPA adjusted to pH 6.6
- Spectrophotometer and Cuvettes
- Stopwatch

Procedure

1. Dilute the DOPA standard (8 mM) to obtain each of the following concentrations of L-DOPA: 0.5 mM, 1 mM, 2 mM 4 mM, and 8 mM.
2. Repeat exercise 7 for each of the substrate concentrations listed, substituting the change in concentration where appropriate.
3. Plot each set of data and from the data calculate the time required to convert 10 micromoles of DOPA to dopachrome. Compute the velocity of enzyme reaction for each substrate concentration. Fill in the following table:

Substrate (DOPA) Concentration (mM)	*Velocity Micromoles/Minute*	*1/s*	*1/υ*
0.5		2.00	
1.0		1.00	
2.0		0.50	
4.0		0.25	
8.0		0.125	

4. Plot the rate of DOPA conversion (v) against substrate concentration in the appropriate place below. This is a Michaelis-Menten plot.
5. Plot a double reciprocal of the values plotted in step 4; that is, 1/s versus 1/v. This is a Lineweaver-Burke plot.
6. Perform a linear regression analysis on the second plot and compute the slope and both y and x intercepts.

   Note that the x intercept is $-1/K_m$, the negative inverse of which is the Michaelis-Menten Constant. The y intercept is $1/V_{max}$ and the slope equals K_m/V_{max}.

EXERCISE 11. ADDITION OF ENZYME INHIBITORS

Materials

- Enzyme Extract
- 8 mM L-DOPA
- 8 mM Benzoic Acid
- 8 mM KCN
- 0.1 M Citrate Buffer, pH 6.6.

Procedure

1. Tyrosinase is inhibited by compounds that complex with copper, as well as by benzoic acid and cyanide. To determine the inhibitory effects of benzoic acid and cyanide, set up a series of tubes as indicated.
2. Using one tube at a time, add 0.5 ml of the enzyme dilution previously calculated to yield 10 micromoles of dopachrome in 2–3 minutes. For each tube, measure the time required to convert 10 micromoles of DOPA to dopachrome. Enter those times in the table below. Compute the reaction velocity for each substrate concentration:

Tube #	*Time (Minutes)*	*Final [DOPA] nM*	*Velocity Micromoles/Minute*
1 Benzoic Acid		6.67	
2 Inhibited		6.00	
3 Series		5.33	
4		4.67	
5		4.00	
6		3.33	
7		2.67	
8		2.00	
9		1.33	
10		0.67	
11		0	
12 KCN		6.00	
13 Inhibited		5.33	
14 SERIES		4.67	
15		4.00	
16		3.33	
17		2.67	
18		2.00	
19		1.33	
20		0.67	
21		0	

3. Calculate the values for 1/s and 1/v for each of the corresponding s and v in the table below. Plot 1/v *vs* 1/s for the presence of benzoic acid and a second plot for the presence of KCN. Compute the values of V_{max} and K_m for the presence of each inhibitor. Determine whether these inhibitors are competitive, non-competitive or uncompetitive.

Benzoic Acid Inhibition			
Tube #	*8 mM DOPA*	*8 mM Benzoic Acid*	*Buffer*
1	2.0	0.5	0
2	1.8	0.5	0.2
3	1.6	0.5	0.4
4	1.4	0.5	0.6
5	1.2	0.5	0.8
6	1.0	0.5	1.0
7	0.8	0.5	1.2
8	0.6	0.5	1.4
9	0.4	0.5	1.6
10	0.2	0.5	1.8
11	0	0.5	2.0

KCN Inhibition			
Tube #	*8 mM DOPA*	*8 mM KCN*	*Buffer*
12	1.8	0.5	0.2
13	1.6	0.5	0.4
14	1.4	0.5	0.6
15	1.2	0.5	0.8
16	1.0	0.5	1.0
17	0.8	0.5	1.2
18	0.6	0.5	1.4
19	0.4	0.5	1.6
20	0.2	0.5	1.8
21	0	0.5	2.0

EXERCISE 12. PROTEIN CONCENTRATION/ENZYME ACTIVITY

Materials

- Commercially pure tyrosinase
- UV Spectrophotometer or Materials for Lowry or Bradford Protein determination
- L-DOPA
- 0.1 M Citrate buffer, pH 6.6

Procedure

1. Prepare a solution of 0.7 micrograms of commercially pure tyrosinase diluted to 4 ml. with 0.1 M Citrate Buffer, pH 6.6.
2. Measure the OD_{280} of your sample and prepare a dilution of your enzyme extract to a final concentration of 0.7 micrograms in 4 ml of citrate buffer.
3. Place both enzyme samples in a water bath at 30° C for 5 minutes to temperature equilibrate.

4. Turn on the spectrophotometer, set the wavelength to 475 nm and blank the instrument using citrate buffer as the blank.
5. Select the commercial preparation and add exactly 1.0 ml of L-DOPA (4 mg/ml in citrate buffer) and immediately read the absorbance at 475 nm.
6. Replace the tube in the water bath and wait exactly 5 minutes. Read the OD_{475} immediately.
7. The molar absorbance coefficient for dopachrome is 3.7 x 10_4. Use this value to compute the specific activity of the commercial enzyme preparation. Check this activity against that listed with the enzyme preparation.
8. Repeat steps 5 and 6 with your extracted enzyme preparation. Compute the specific activity (enzyme units of activity/ mg protein) of your enzyme preparation. ***Enzyme Unit:*** *The absorbance reading under the conditions specified in this exercise is proportional to the enzyme concentration, where 1 unit of enzyme activity yield a 0.81 OD change in readings.*

Notes

The protcin content can be measured by the Lowry or Biuret procedures found in the Appendix G, or more simply by a single spectrophotometric measure of thc absorbance of the sample at 280 nm. Without going into mathematical detail, a 1% pure solution of tyrosinase has an OD_{280} equal to 15.6/cm. The Beer-Lambert law can thus be used to determine protein content in a non-destructive manner.

EXERCISE 13. TO STUDY THE ACTION AND ACTIVITY OF AMYLASE ON STARCH DIGESTION

Principle

In order to find out the action of amylase on starch, amylase solution can be added to starch solution. After some times, the presence of starch in the mixture can be tested by iodine solution. The positive test for the starch is dark blue colouration. If there is no starch, iodine solution remains brown. If there is less starch or no starch in the mixture, it can be concluded that amylase can change starch into another chemicals. This can show amylase can have some action on starch. To the study the activity of amylase on starch, the amount of starch in the mixture can be estimated by observing the colour change in the iodine solution at a regular time interval. The colour changes from dark blue to brwon indicates the disappcarance of starch in the mixture. If the colour changes very rapidly, it shows that the activity of amylase is high. If the change is slow, it shows the acitivity of amylase is slow.

Procedure

1. Two drops of iodine solution were put into each depression in a spotting tile.
2. 5 ml of 1% starch solution was added into test tube A with a 5 ml pipette.
3. 0.5 ml of 1% amylase solution was added into the same tube with a 1ml pipette.
4. The solution was mixed thoroughly.
5. At a half-minute interval, two drops of mixture were taken out from test tube A and transferred to each depression of the spotting tile.

6. The iodine solution and the mixture was mixed into the depression by a glass rod.
7. The colour of the iodine solution was observed and recorded.
8. Step 5 to 7 were repeated until the colour of the iodine solution remained brown.

Explanation of Results

Starch molecules can combine with the active sites of the amylase. Amylase can then speed up the digestion of starch into maltose. When amylase was mixed with starch, starch is digested. Since less starch is present in the mixture, if we test the presence of the starch by iodine solution, the colour of the iodine is less dark blue. When there is a complete digestion of starch, the colour of the iodine solution will remain brown. The reaction rate is fast at first since the amount of starch in the mixture is large. However, the reaction rate is slower after 1.5 minutes, since the concentration of substrate decreases.

Precautions

1. The experiment should be carried out at the same temperature.
2. The solutions must be well mixed since the starch and amylase are not completely dissolved.
3. Adding too much iodine solution and mixture into the depressions must be avoided. This may cause the over-flow of the solution from the depressions. Mixing of the solutions in the depressions is not easily done.

Sources of Errors

1. The volume of each drop of solutions may not be the same.
2. The amylase or starch solution left on the inside surface of the test tube may cause an inaccuracy in the amount of solutions added.
3. The amylase and starch are not completely soluble.

Improvement

1. The colour change is not easy to observe. If a colorimeter and data-logging device are used. The results can be quantitized and the results can be easily compared.
2. A shorter time-interval can be used. This makes the result more clearly observed. Of course, this may have difficulty when the process is done by one person.

EXERCISE 14. DETERMINATION OF EFFECT OF PHON THE ACTIVITY OF HUMAN SALIVARY α-AMYLASE

Principle

α-amylase catalyses the hydrolysis of α 174 linkage of starch and produces reducing sugars. Maltose reduces 3.5 DNA into 3 amino –5 nitro salicylic acid a orange red colour complex. Read the O.D at 540 nm.

Enzyme are active over a limited pH range only and a plot of activity against pH usually gives a bell shaped curve. The pH value of max activity is known as optimum pH and this is characteristic of the enzyme. The variation of activity with pH is due to the change in the state of ionization of the enzyme protein 4 other compounds of the reaction mixture.

Reagents

1. DNS reagent
2. 1% starch: 1gm of starch dissolved in buffer of different pH
3. Crude solution
4. Buffer solution
   (i) Acetate buffer pH 4.0, 5.0 and 5.6
   (ii) Acetate acid: 11.5 ml/lit (0.2M)
      (i) Sodium acetate: 16.4 gm/lit (0.2m)
      (ii) Phosphate buffer – pH, 6.0, 7.0 and 8.0
   (a) Na_2HPO_4, $2H_2O$ – 35.61 gm/lit
   (b) Na_2HPO_4, H_2O – 27.67 gm/lit

Pm	*A sdn (m) mone basic*	*B sdn (ml) dibasic*	*Distilled H_2O (ml)*	*Total value in ml*
Acetate buffer				
4.0	41.0	9.0	50	100
5.0	14.8	35.2	50	100
5.6	13.7	36.3	50	100
Phosphate buffer				
6.0	87.7	12.3	100	200
7.0	39.0	61.0	100	200
8.0	5.3	94.7	100	200

Procedure

Take seven test tube add 0.5 ml of any pH buffer to the first tube and 0.5ml of substrate of any pH, which serves as a blank. Then add 0.5 ml of enzyme to each tube except for blank. Add 0.5 ml of substrate of different pH 4.0, 5.0, 5.6, 6.0, 7.0 and 8.0 to the respective test tube incubate all the test tubes exactly for 20 at room temperature add 1 ml of DNS reagent to each test tube and keep all tube in boiling water bath for exactly 5 ml. Then add 10 ml of distilled H_2O to each tube and read O.D at 540 nm.

Plot the graph by taking different pH along the x-axis and activity on y-axis.

Result: The optimum pH of Human Salivary x-amylase is ______ μm/ml/min.

EXERCISE 15. TO DETERMINE THE EFFECT OF TEMPERATURE ON THE ACTIVITY OF HUMAN SALIVARY α-AMYLASE

Principle

α-amylase catalyes the hydrolysis of α 1,4 linkage of starch and produce reducing sugars. The liberated reducing sugars acid a orange red colour complex. Read the O.D at 540 nm.

The effect of temperature on enzyme catalyzed reaction indicates the structure changes in enzyme. Molecules must posses a certain energy of attraction before they can react and enzyme functions as catalyst lowering the energy of attraction. If an enzyme catalyzed reaction can be determined by measuring the minimum velocity at different temperature.

The energy is more active at d/f temperature below this temperature the enzyme activity decreases and above this temperature the enzyme become denatured or lose its structure required for catalytic activity.

Reagents

1. DNS reagent
2. 1% starch
3. Phosphate buffer
4. Crude enzyme (1 : 20 dil)

Procedure

Take 6 clean and dry test tube pipette out 0.5 ml of enzyme to each it except for 1st tube which serves as a blank. Add 0.5 ml of substrate to each tube incubate each it for 20 at respective temperature (13°C, 25°C, 36°C, 50°C and 60°C) and 1 ml of DNS reagent to each test tube and keep all test tube in boiling water bath for exactly 5 min and add 10 ml of distill water to each tube and read the O.D at 540 nm.

Plot the graph by taking temperature on x-axis and activity in y-axis.

Result: The Optimum temperature of the humans salivary α-amylases is 36°C.

EXERCISE 16. CONSTRUCTION OF MALTOSE CALIBRATION CURVE

Principle

Maltose is a reducing disaccharide. Maltose reduces the alkaline solution of 3.5 dinitro salicylic acid (DNS) which is pale yellow in colour into orange red colour complex of 3 amino – 5-nitro salicylic acid. The optical density is measured at 540 nm. The intensity of the colour depends on the concentration of maltose.

Reagent

1. 3,5-dinitrosalicylic acid (DNS):

   Dissolve 10 gms of DNS in 200 ml of 2N NaOH to this add 500 ml of distill H_2O and 300 ml of sodium pot tartrate and make up the volume to 100 ml with dist H_2O.
2. Standard maltose solution: 1mg/ml in dist. H_2O.
3. Distilled water.

Procedure

Pipette out standard maltose solution ranging from 0.0 to 20 ml into test tubes and make up the volume to 2.0 ml with distilled water. The first tubes with 2.0 ml of distilled water serves as blank.

Add 1ml of alkaline DNS-reagent to each tube keep all the tubes in boiling water bath for exactly 5 min and cool to room temperature. Add 4 ml of distilled water to each tubes and read the O.D at 540 nm.

Plot the graph by taking case of maltose along x-axis and optical density along y-axis.

Result: Maltose calibration curve is constructed which is straight line passing through origin.

5 ELECTROPHORESIS

ELECTROPHORESIS

Electrophoresis is defined as the separation (migration) of charged particles through a solution or gel under influence of an electrical field.

The rate of movement of particle depends on the following factors.

1. The charge of the particle
2. Applied electric field
3. Temperature
4. Nature of the suspended medium.

What is gel electrophoresis?

Gel electrophoresis is a method that separates macromolecules-either nucleic acids or proteins on the basis of size, electric charge, and other physical properties. A gel is a colloid in a solid form. The term electrophoresis describes the migration of charged particle under the influence of an electric field. Electro refers to the energy of electricity. Phoresis, from the Greek verb phoros, means "to carry across." Thus, gel electrophoresis refers to the technique in which molecules are forced across a span of gel, motivated by an electrical current. Activated electrodes at either end of the gel provide the driving force. A molecule's properties determine how rapidly an electric field can move the molecule through a gelatinous medium.

Many important biological molecules such as amino acids, peptides, proteins, nucleotides, and nucleic acids, posses ionisable groups and, therefore, at any given pH, exist in solution as electrically charged species either as cations (+) or anions (–). Depending on the nature of the net charge, the charged particles will migrate either to the cathode or to the anode.

How does this technique work?

Gel electrophoresis is a technique used for the separation of nucleic acids and proteins. Separation of large (macro) molecules depends upon two forces: charge and mass. When a biological sample, such as proteins or DNA, is mixed in a buffer solution and applied to a gel, these two forces act together. The electrical current from one electrode repels the molecules while the other electrode simultaneously attracts the molecules. The frictional force of the gel material acts as a "molecular sieve," separating the molecules by size. During electrophoresis, macromolecules are forced to move through the pores when the electrical current is applied. Their rate of migration through the electric field depends on the strength of the field, size and shape of the molecules, relative hydrophobicity of the samples, and on the ionic strength and temperature of the buffer in which the molecules are moving. After staining, the separated macromolecules in each lane can be seen in a series of bands spread from one end of the gel to the other.

Agarose

There are two basic types of materials used to make gels: agarose and polyacrylamide. Agarose is a natural colloid extracted from sea weed. It is very fragile and easily destroyed by handling. Agarose gels have very large "pore" size and are used primarily to separate very large molecules with a molecular mass greater than 200 kdal. Agarose gels can be processed faster than polyacrylamide gels, but their resolution is inferior. That is, the bands formed in the agarose gels are fuzzy and spread far apart. This is a result of pore size and it cannot be controlled.

Agarose is a linear polysaccharide (average molecular mas about 12,000) made up of the basic repeat unit agarobiose, which comprises alternating units of galactose and 3,6-anhydrogalactose. Agarose is usually used at concentrations between 1% and 3%.

Agarose gels are formed by suspending dry agarose in aqueous buffer, then boiling the mixture until a clear solution forms. This is poured and allowed to cool to room temperature to form a rigid gel.

Polyacrylamide

There are two basic types of materials used to make gels: agarose and polyacrylamide. The polyacrylamide gel electrophoresis (PAGE) technique was introduced by Raymond and Weintraub (1959). Polyacrylamide is the same material that is used for skin electrodes and in soft contact lenses. Polyacrylamide gel may be prepared so as to provide a wide variety of electrophoretic conditions. The pore size of the gel may be varied to produce different molecular seiving effects for separating proteins of different sizes. In this way, the percentage of polyacrylmide can be controlled in a given gel. By controlling the percentage (from 3% to 30%), precise pore sizes can be obtained, usually from 5 to 2,000 kdal. This is the ideal range for gene sequencing, protein, polypeptide, and enzyme analysis. Polyacrylamide gels can be cast in a single percentage or with varying gradients. Gradient gels provide continuous decrease in pore size from the top to the bottom of the gel, resulting in thin bands. Because of this banding effect, detailed genetic and molecular analysis can be performed on gradient polyacrylamide gels. Polyacrylamide gels offer greater flexibility and more sharply defined banding than agarose gels.

Mobility of a molecule = (applied voltage) × (net charge of the molecule)/ friction of the molecule (in the electrical field)

$$\upsilon \text{ (velocity)} = E \text{ (voltage)} \times q \text{ (charge)}/f \text{ (frictional coefficient)}.$$

Polyacrylamide Gel Electrophoresis

Polyacrylamide is the solid support for electrophoresis when polypeptides, RNA or DNA fragments are analyzed. Acrylamide plus N,N'-methylene-bis-acrylamide in a given percentage and ratio are polymerized in the presence of ammonium persulfate and TEMED (N,N,N',N'-tetra-methyl-ethylene-diamine) as catalysts.

Safety and Practical Points

1. Acrylamide and bis-acrylamide are toxic as long as they are not polymerized.
2. Buffer (usually Tris) and other ingredients (detergents) are mixed with acrylamide before polymerization.

3. Degassing of acrylamide solution is necessary before pouring the gel because O_2 is a strong inhibitor of the polymerization reaction.

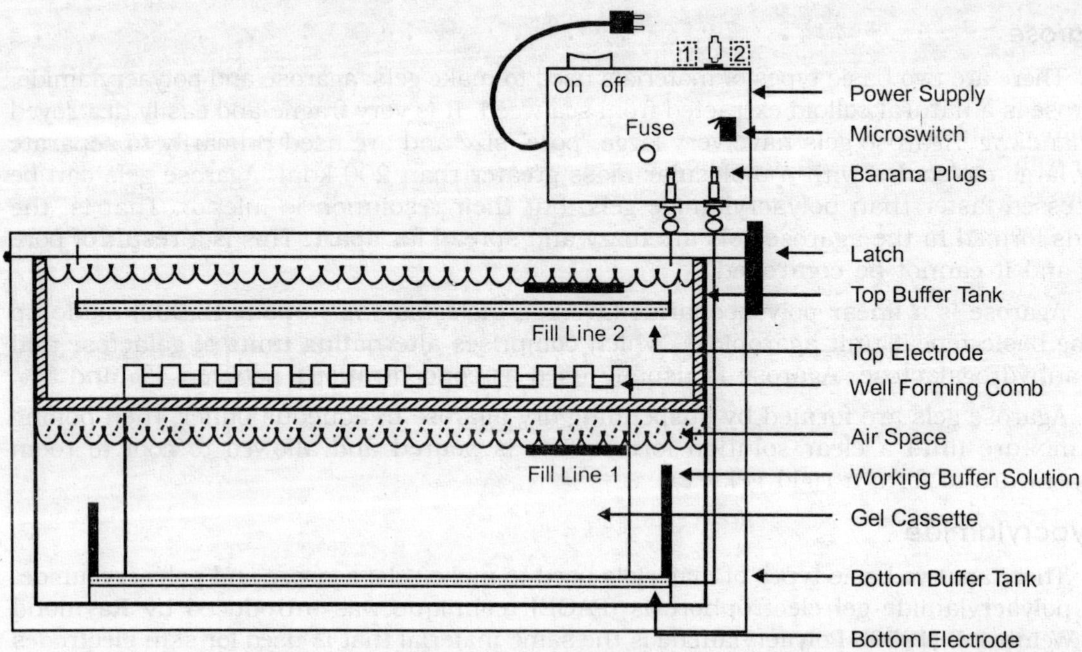

Fig. 1. Electrophoresis

Polyacrylamide Gel Electrophoresis of Proteins

1. Under non-denaturing conditions.
2. Under denaturing conditions.
3. Isoelectric focusing.

These techniques used to analyze certain properties of a protein such as: isoelectric point, composition of a protein fraction or complex, purity of a protein fraction, size of a protein.

We will concentrate on denaturing polyacrylamide gel electrophoresis in the presence of sodium dodecylsulfate (SDS-PAGE) and a reducing agent (DTT, or dithioerithritol, DTE).

The protein is denatured by boiling in "sample buffer" which contains:

1. Buffer pH 6.8 (Tris-HCl).
2. SDS to give 1%.
3. Glycerol to give 5%.
4. DTT or DTE to give 50 mM.
5. Bromophenol blue (tracking dye).

Discontinuous Polyacrylamide Gel Electrophoresis

This type of polyacrylamide gel consists of two parts:

(1) The larger running (resolving) gel

(2) The shorter upper stacking gel.

The running gel has a higher percentage (usually 10–15%) of acrylamide and a Tris-HCl buffer of pH 8.8.

The stacking gel usually contains 5% acrylamide and a Tris-HCl buffer of pH 6.8.

The buffer used in SDS-PAGE is Tris-glycine with a pH of about 8.3.

Note. In the figure below (from Switzer-Garrity,) all buffers are Tris buffers (either with glycine or with HCl). The pH of the buffer in which the sample is prepared is 6.8 — not what the figure says!

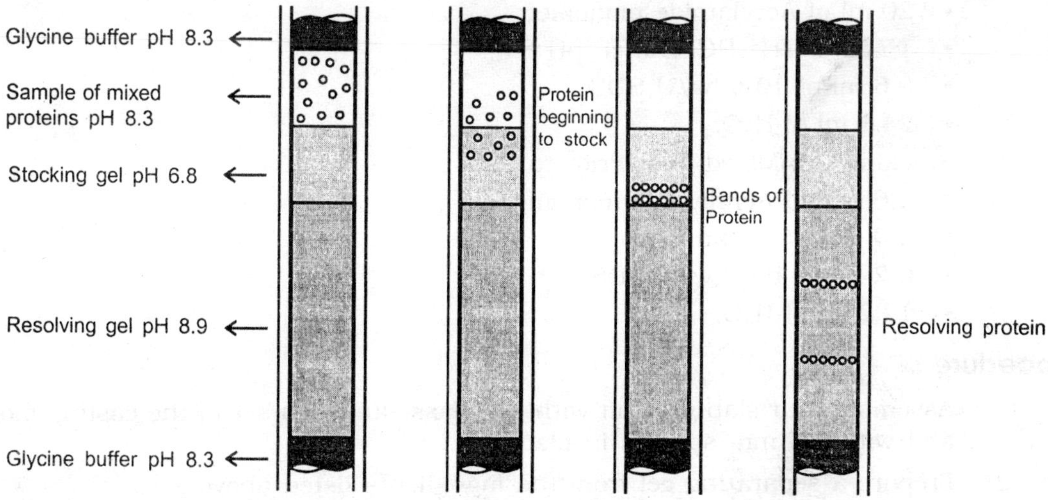

Fig. 2. Protein passage through a disc-gel Electrophoresis System.

Determination of the Molecular Weight of a Polypeptide by SDS-PAGE

As all polypeptides are wrapped with SDS and thus are strongly negatively charged, they migrate through the running gel according to their size (small polypeptides migrate faster than large ones!).

There is a linear relationship between the log of the molecular weight of the polypeptide and its migration during SDS-PAGE.

Standard polypeptides have to be run on the same gel and a curve of their migration vs. the log of their molecular weight has to be generated.

EXERCISE 1. PREPARATION OF SDS-POLYACRYLAMIDE GELS

Materials

- Casting gel unit for electrophoresis
- Siliconized pasteur pipettes
- Syringes equipped with blunt, stub nosed, needles
- Vacuum chamber for degassing gels

- Micropipettes (10–300 µl)
- Stock 30%T:0.8%C Acrylamide monomer
- 1.5 M Tris-HCl buffer, pH 8.8
- 10% (w/v) SDS
- 10% (w/v) Ammonium persulfate.

TEMED acrylamide is a powerful neurotoxin. Do not breathe powder or otherwise come in contact with the monomer. Wear gloves at all times.

- Separation Gel Mixed Just Prior to Use
    - 20 ml of Acrylamide monomer
    - 15 ml of Tris-HCl Buffer, pH 8.8
    - 0.6 ml of 10% (w/v) SDS
    - 24.1 ml of H_2O.
- Stacking Gel Mixed Just Prior to Use
    - 2.66 ml of Acrylamide monomer
    - 5.0 ml of Tris buffer, pH 8.8
    - 0.2 ml of 10% (w/v) SDS
    - 12.2 ml of H_2O.

Procedure

1. Assemble your slab gel unit with the glass sandwich set in the casting mode, and with 1.5 mm spacers in place.
2. Prepare a separating gel from the ingredients listed above.
3. Add the separating gel to a side arm flask, stopper the flask and attach to a vacuum pump equipped with a cold trap. Turn on the vacuum and degas the solution for approximately 10 minutes. During this period, gently swirl the solution in the flask.
4. Turn off the vacuum, open the flask and add 200 µl of ammonium persulfate and 20 µl of TEMED to the solution.
5. Add the stopper to the flask and degas for an additional 2 minutes while gently swirling the solution to mix the two accelerators. Use this solution within a few minutes of mixing, or it will gel in the flask.
6. Transfer the degassed acrylamide solution to the casting chamber with a pasteur pipette. Gently fill the center of the glass chamber with the solution by allowing the solution to run down the side of one of the spacers. Be careful not to introduce air bubbles during this step.
7. Adjust the level of the gel in the chamber by inserting a syringe equipped with a 22 gauge needle into the chamber and removing excess gel.
8. Immediately water layer the gels to prevent formation of a curved meniscus. Using a second syringe and needle, add approximately 0.5 ml of water to the chamber by placing the tip of the needle at an angle to a spacer and gently allowing the water to flow down the edge of the spacer and over the gel. Add an additional 0.5 ml of water to the chamber by layering it against the spacer on the opposite side of the chamber. Done appropriately, the water will form

a layer over the gel, and a clear line of demaration will be observed as the gel polymerizes.

9. After 30 minutes, the gel should be polymerized. If degassing was insufficient, or the ammonium persulfate not fresh, the polymerization may take an hour or more. When the gel is polymerized, lift the gel in its casting chamber and tilt to decant the water layer.
10. Prepare a stacking gel from the listed ingredients.
11. Degas the stacking gel as in step 3.
12. Add 75 μl ammonium persulfate and 10 μl TEMED to the stacking gel and degas for an additional 2 minutes.
13. Add approximately 1 ml of stacking gel to the gel chamber and gently rock back and forth to wash the surface of the separating gel. Pour off the still liquid stacking gel and dispose of properly. Remember that liquid acrylamide is extremely hazardous!
14. Add fresh stacking gel to nearly fill the chamber, but allow room for the insertion of a teflon comb used to form sample wells. Carefully insert a teflon comb into the chamber. Adjust the volume of the stacking gel as needed to completely fill the spaces in the comb.

    Be careful not to trap any air bubbles beneath the combs. Oxygen inhibits polymerization, and will subsequently result in poor protein separations.
15. Allow the gels to polymerize for at least 30 minutes prior to use.

EXERCISE 2. SEPARATION OF PROTEIN STANDARDS: SDS-PAGE

Materials

- 10% SDS-polyacrylamide gel
- Protein standards
- 2X-SDS Sample Buffer
- 1X-SDS Electrophoresis Running Buffer (Tris-Glycine + SDS)
- 0.001% (w/v) Bromophenol Blue
- Micropipettes with flat tips for electrophoresis wells.

Procedure

1. Remove the teflon combs from prepared gels by gently lifting the combs from the chamber. Rinse the wells (formed by the removal of the combs) with distilled water and drain off the water.
2. Fill the wells and the chamber with running buffer.
3. Prepare aliquots of a known protein standard by mixing equal parts of the protein standard with 2X sample buffer.
4. Using a micropipette, add the sample to the bottom of a well, blue to a separate well.
5. Remove the gel from its casting stand and assemble it into the appropriate slab unit for running the electrophoresis. Be sure to follow the manufacturer's directions for assembly.

6. Pour a sufficient quantity of running buffer into both the lower and upper chambers of the electrophoresis apparatus until the bottom of the gel is immersed in buffer, and the top is covered, while the electrodes reach into the buffer of the upper chamber. Be careful not to disturb the samples in the wells when adding buffer to the upper chamber.
7. Assemble the top of the electrophoresis apparatus and connect the system to an appropriate power source. Be sure that the cathode (+) is connected to the upper buffer chamber.
8. Turn on the power supply and run the gel at 20 mA constant current per 1.5 mm gel.

   For example, if two gels are run, each having 1.5 mm spacers, the current should be adjusted to 40 mA. One gel with 1.5 mm spacers should be run at 20 mA, while a gel with 0.75 mm spacers should be run at 10 mA.
9. When the tracking dye reaches the separating gel layer, increase the current to 30 mA per 1.5 mm gel.
10. Continue applying the current until the tracking dye reaches the bottom of the separating gel layer (approximately 4 hours).
11. Turn off and disconnect the power supply. Disassemble the gel apparatus and remove the glass sandwich containing the gel. Place the sandwich flat on paper towels and carefully remove the clamps from the sandwich.
12. Working on one side of the sandwich, carefully slide one of the spacers out from between the two glass plates. Using the spacer or a plastic wedge as a lever, gently pry the glass plates apart without damaging the gel contained within.
13. Lift the bottom glass plate with the gel and transfer the gel to an appropriate container filled with buffer, stain or preservative.

    The gel may at this point be used for Coomasie Blue staining, silver staining, enzyme detection, Western blots or for more advanced procedures, such as electroblotting or electroelution.

    If prestained protein standards were used, the gels may be scanned directly for analysis. Place the gel into 50% methanol and gently rock the container for about 30 minutes prior to scanning. This can be accomplished by placing the gels into a flat dish and gently lifting the edge of the disk once every 30 seconds. There are commercially available rocker units for this purpose. If the gel is to be dried, use a commercial gel dryer, (SE 1160 Slab Gel Dryer). Follow the manufacturer's directions demonstrates a dried and stained gel containing a series of proteins of known molecular weights.
14. Plot the relative mobility of each protein against the log of its molecular weight.

    Relative mobility is the term used for the ratio of the distance the protein has moved from its point of origin (the beginning of the separating gel) relative to the distance the tracking dye has moved (the gel front). The ratio is abbreviated as Rf. Molecular weight is expressed in daltons, presents a plot of the relative molecular weight of protein standards against the log of their molecular weight.

EXERCISE 3. COOMASIE BLUE STAINING OF PROTEIN GELS

Materials

- Protein gel from exercise 2
- 0.25% (w/v) Coomasie Brilliant Blue R 250 in methanol-water-glacial acetic acid (5/5/1), filtered immediately before use.
- 7% (v/v) acetic acid
- Commercial destaining unit (Optional).

Procedure

1. Place a gel prepared as in exercise 2 in at least 10 volumes of Coomasie Blue staining solution for 2–4 hours. Rock gently to distribute the dye evenly over the gel.
2. At the conclusion of the staining, wash the gels with several changes of water.
3. Place the gels into a solution of 7% acetic acid for at least 1 hour.
4. If the background is still deeply stained at the end of the hour, move the gels to fresh 7% acetic acid as often as necessary.

   If a commercial destainer is available, this will decrease the time required for stain removal. Follow the manufacturer's directions for use of the destainer.
5. Place the gels into containers filled with 7% acetic acid as a final fixative.
6. Photograph the gels or analyze the gels spectrophotometrically.

Notes

Coomasie Brilliant Blue R 250 is the most commonly used staining procedure for the detection of proteins. It is the method of choice if SDS is used in the electrophoresis of proteins, and is sensitive for a range of 0.5 to 20 micrograms of protein. Within this range, it also follows the Beer-Lambert law and thus can be quantitative as well as qualitative. The major drawback is the length of time for the procedure and a requirement for destaining. Overstaining results in a significant retention of stain within the gel, and thus a high background stain, which might obliterate the bands. The length of time for staining must be carefully monitored, and can range from 20 minutes to several hours. If maximum sensitivity is desired, one should try 2 hours for a 5% gel and 4 hours for a 10% gel. Destaining must be monitored visually and adjusted accordingly.

EXERCISE 4. SILVER STAINING OF GELS

Materials

- Protein gel from exercise 2
- 45% (v/v) Methanol + 12% (w/v) acetic acid
- 5% (v/v) Methanol + 7% (w/v) acetic acid
- 10% Glutaraldehyde
- 0.01M Dithiothreitol
- Silver nitrate solution
- Sodium citrate/formaldehyde
- Kodak Farmer's Reducer or Kodak Rapid Fixer.

Procedure

1. Fix gels by gently rocking them in a solution of 45% methanol/12% acetic acid until the gels are completely submerged. Fix for 30 minutes at room temperature.
2. Remove the fixative and wash 2X for 15' each with 5% ethanol/7% acetic acid. (Gels thicker than 1 mm require longer washing.)
3. Soak the gels for 30 minutes in 10% glutaraldehyde.
4. Wash 3X with deionized water, 10 minutes each.
5. Place in dithiothreitol for 30 minutes.
6. Place in silver nitrate solution for 30 minutes.
7. Wash for 1 minute with deionized water.

   Dispose of used silver nitrate solution immediately with continuous flushing. This solution is potentially explosive when crystals form upon drying.
8. Place in sodium citrate/formaldehyde solution for 1 minute.
9. Replace the sodium carbonate/formaldehyde solution with a fresh batch, place gels on a light box and observe the development of the bands. Continue to rock gently as the gel develops.
10. When the desired degree of banding is observed (and before the entire gel turns black), withdraw the citrate/formaldehyde solution and immediately add 1% glacial acetic acid for 5 minutes.
11. Replace the glacial acetic acid with Farmer's reducer or Kodak Rapid Fixer for 1 minute. Remove Farmer's reducer and wash with several changes of deionized water.
12. Photograph or scan the gel with a densitometer, demonstrates a typical silver stained protein gel.
13. For storage soak the gel in 3% glycerol for 5 minutes and dry between dialysis membranes under reduced pressure at 80–82° C for 3 hours. Alternatively, place the wet gel into a plastic container (a storage bag will do) and store at room temperature. If desired, the gels may be dried between Whatman 3MM filter paper for autoradiography, or dried using a commercial gel dryer.

EXERCISE 5. DOCUMENTATION

Materials

- Polaroid camera (Fotodyne Foto/Phoresis I or equivalent) or
- 35 mm camera equipped with macro lens
- Stained gel.

Procedure

1. Photograph the gels.
2. Use the photographs or negatives to measure the distance from the point of protein application (or for two gel systems, the line separating the stacking

and separating gels) to the final location of the tracking dye near the bottom of the gel.

3. Measure the distance from the point of origin to the center of each band appearing on the gel.
4. Divide each of the values obtained in Step 3 by that obtained in Step 2 to obtain the relative mobility (the Rf value) for each band.
5. Using either the graph of Rf values and molecular weights from exercise 2, compute the molecular weights of each band.

Optional

Scan the negative with a densitometer and compute Rf values based on the distances from the point of origin to the peak tracing for each protein band. Integration of the area of each peak will yield quantitative data as well as the molecular weight.

EXERCISE 6. WESTERN BLOTS

Materials

- Blot Cell
- BA 83 0.2 μm pore nitrocellulose sheets
- Buffer, PBS-Tween 20
- Antigenic proteins, antibodies, and horseradish peroxidase labelled antiglobulins.

Procedure

1. Run an electrophoretic separation of known antigenic proteins according to the procedures in exercise 1 and 2.
2. Draw a line 0.5 cm from the top edge of an 8 × 10 cm nitrocellulose sheet and soak it in blot buffer for about 5 minutes.

   Nitrocellulose is both fragile and flammable and easily contaminated during handling. Wear gloves which are prewashed.

   When soaking the microcellulose wet first one side and then turn the sheet over and wet the other, to prevent trapping air within the filter.
3. Place 200 ml of blot buffer into a tray and add a piece of filter paper slightly larger than the electrophoretic gel from Step 1.
4. Remove the gel from the electrophoresis chamber after the proteins have been separated, and place the gel into the tray containing the filter paper. Do not allow the gel to fall onto the paper, but place it next to the paper in the tray.
5. Gently slide the gel onto the top of the filter paper. Keep the stacking gel off of the paper until the last moment, since it tends to stick and make repositioning difficult.
6. Holding the gel and the filter paper together, carefully remove them from the tray of blot buffer and transfer the paper and gel to a pad of the blot cell with the gel facing up.

7. Transfer the nitrocellulose sheet (ink side down) onto the top of the gel and line up the line drawn on the sheet with the top of the stacking gel.

   Once the gel and nitrocellulose touch they can not be separates.

8. Roll a glass rod across the surface of the nitrocellulose to remove any air bubbles and insure good contact between the gel and nitrocellulose.
9. Lay another sheet of wet filter paper on top of the nitrocellulose creating a sandwich of paper-gel-nitrocellulose-paper, all lying on the pad of the blot cell.
10. Add a second pad to the top of the sandwich and place the entire group inside of the support frame of the blot cell, and assemble the blot cell so that the nitrocellulose side of the sandwich is toward the positive terminal.
11. Check that the buffer levels are adequate and that the cooling water bath is adjusted to at least 5° C. Subject the gel to electrophoresis for 30 minutes with the electrodes in the high field-intensity position. Follow the manufacturer directions during this phase. Failure to closely monitor the electrophoresis buffer or temperature can result in a fire. Use a circulating cold bath appropriate to the apparatus and hold the voltage to a constant 100 vdc.
12. Upon completion of the electrophoresis (timed according to manufacturer's directions), turn off the power and disassemble the apparatus. Remove the blot pads from the sandwich and remove the filter paper from the nitrocellulose side.
13. Place the sandwich, nitrocellulose side down, onto a glass plate and remove the other filter paper.
14. Use a ball point pen to outline the edges of the separating gel onto the nitrocellulose, including the location of the wells. Carefully lift the gel away from the nitrocellulose and mark the locations of the pre-stained molecular weight standards as the gel is peeled away. Peel the gel from the separating gel side, not the stacking gel.
15. Wash the blot (the nitrocellulose sheet) at least four times with 100 ml of PBS-Tween 20 for five minutes each on a rocking platform.
16. Cut the blot into 0.5 cm strips.
17. Inactivate sera containing positive and negative antibody controls to the antigens under examination by treating at 56°C for 30 minutes. Make dilutions of 1 : 100 and 1 : 1000 of the controls with PBS-Tween 20.
18. Place 3 ml of the diluted sera or controls onto a strip from Step 16 and incubate for 1 hour at room temperature while continuously rocking the sample.
19. Wash the strips four times for 5 minutes each with 10 ml quantities of PBS-Tween 20. The first wash should be done at 50° C but the last three may be done at room temperature.
20. Add 3 ml of horseradish peroxidase-labelled antiglobulin, optimally diluted in PBS-Tween and incubate at room temperature for 1 hour with continuous agitation.
21. Wash the strips four times for 5 minutes each with PBS-Tween 20 and one more time with PBS only.

22. Remove the PBS and add 5 ml of substrate solution. Positive reaction bands usually appear within 10 minutes. Stop the reaction by washing with water.

Notes

One of the more difficult tasks of electrophoretic separations is the identification of specific bands or spots within a developed gel. As observed with LDH isozymes, one method of doing this is to react the bands with an enzyme substrate that can be detected calorimetrically.

As a rule, however, most peptides are denatured during electrophoresis, and of course, nucleic acids have no enzyme activity. The methods employed for identifying non-enzymatic proteins and nucleic acids have been termed Western for immunoblotting of proteins, Southern for techniques using DNA probes Northern when using RNA probes. The probes are radioactive complimentary strands of nucleic acid. The first of these techniques was the Southern, named for the developer of the procedure, Edward Southern. Northern and then Western blots were named by analogy.

Blotting techniques first develop a primary gel: protein on acrylamide; or DNA/RNA on agarose. The gel patterns are then transferred to nitrocellulose membrane filters and immobilized within the nitrocellulose membrane. This process of transfer to an immobilizing substrate is where the term blotting originated. The process is widely used in today's laboratories because the immobilization allows for extensive biochemical and immunological binding assays that range from simple chemical composition to affinity purification of monospecific antibodies and cell-protein ligand interactions.

In practice, the electrophoresis gel is sandwiched between two layers of filters, two foam pads (for support) and two layers of a stainless steel mesh. This entire apparatus can be submerged in a buffer and transfer allowed to occur by diffusion (yielding two blots, one on each filter), or can be arranged in an electro-convective system so that transfer occurs in a second electrophoretic field.

Once the transfer has occurred, the blots can be probed with any number of specific or non-specific entities. DNA can be probed, for example, with cDNA or even a specific messenger RNA to identify the presence of the gene for that message.

SODIUM DODECYL SULPHATE POLY ACRYLAMIOE GEL ELECTROPHORESIS (SDS-PAGE)

INTRODUCTION

The analytical electrophoresis of proteins is carried out in polyacrylamide gels under conditions that ensure dissociation of the proteins into their individual polypeptide subunits and that minimize aggregation. Most commonly, the strongly anionic detergent Sodium Dodecyl Sulphate (SDS) is used in combination with a reducing agent and heat to dissociate the proteins before they are loaded on the gel. The denatured polypeptides bind 50S and become negatively charged. Because the amount of SDS bound is almost always proportional to the molecular weight of the polypeptide and is independent of its sequence, 50S-polypeptide complexes migrate through poly-acryl amide gels in accordance with the size of polypeptide. At saturation, approximately 1.4 g of detergent is bound per gram of polypeptide. By using markers of known molecular weight, it is therefore possible to estimate the molecular weigh of the polypeptide chains.

SDS-poly-acrylamide gel electrophoresis is carried out with a discontinuous buffer system ill which the buffer in the reservoirs is of a different pH and ionic strength from the buffer used to cast the gel. The 50S-polypeptide complexes in the sample that is applied to the gel are swept along by a moving boundary created when an electric current is passed between the electrodes. After migrating through a stacking gel of high porosity, the complexes are deposited in a very thin zone (or stack) on the surface of the resolving gel. The ability of the discontinuous buffer systems to concentrate all of the complexes in the sample into a very small volume greatly increases the resolution of SDS-poly-acrylamide gels.

The sample and the stacking gel contain Tris-CI (pH 6.8), the upper and lower buffer reservoirs contain Tris-glycine (pH 8.3), and the resolving gel contains Tris-CI (pH 8.8). AI-components of the system contain 0.1% 50S. The chloride ions in the sample and stacking gel form the leading edge of the moving boundary, and the trailing edges of the moving boundary is a zone of, lower conductivity and steeper voltage gradient which sweeps the polypeptides from the sample. There, the higher pH of the resolving gel favors the ionization of glycine, and the resulting glycine ions migrate through the stacked polypeptides and travel through the resolving gel immediately behind the chloride ions. Freed from the moving boundary, the 50S-poly-acrylamide complexes move through the resolving gel in a zone of uniform voltage and pH and are separated according to size by sieving.

Polyacrylamide gels are composed of chains of polymerized acryl amide that are cross-linked by a bi-functional agent such as N, N'-Methylene-bisacrylamide. The effective range of separation of SOS-poly-acrylamide gels depends on the concentration of poly-acrylamide used to cast the gel and on the amount of cross-linking.

Effective Range of Separation of SDS-Polyacrylamide Gels

Acrylamide concentration	*Linear range of separation*
15	12–43
10	16–68
7.5	36–94
5.0	57–212

Molar ratio of bis-acrylamide : acrylamide is 1 : 29.

Cross-links formed from bisacrylamide add rigidity and tensile strength to the gel and form pores through which the 50S-polypeptide complexes must pass.

The sieving properties of the gel are determined by the size of the pores, which is a function of the absolute concentrations of acrylamide and bisacrylamide used to cast the gel.

PREPARATION OF POLYACRYLAMIDE GELS

Role of Reagents Involved

Reagents. Acrylamide and N, N' -Methylene bisacrylamide A stock solution containing 29% (w/v) acrylamide and 1% (w/v) N, N' Methylene-bisacrylamide should be prepared in deionized, warm water (to assist the dissolution of the bisacrylamide. Check that the pH of the solution is 7.0 or less, and store the solution in dark bottles at room temperature. Fresh soluLions should be prepared every few months.

Sodium dodecyl sulfate (SDS). A 10% stock solution should be prepared in deionized water and stored at room temperature.

Tris buffers for the preparations of resolving and stacking gels. It is essential that these buffers be prepared with Tris base. After the Tris base has been dissolved in deionized water, the pH of the solution should be adjusted with HCI.

TEMED (N, N, N', N'-tetramethylethylenediamine) TEMEO accelerates the polymerization of acrylamide and bisacrylamide by catalyzing the formulation of free radicals from ammonium persulfate.

Ammonium persulfate : Ammonium persulfate provides the free radicals that drive polymerization of acrylamide and bisacrylamide. A small amount of a 10% (w Iv) stock solution should be prepared in deionized water and stored at 4C. Ammonium persulfate decomposes slowly, and fresh solutions should be prepared weekly.

Tris-glycine electrophoresis buffer : This buffer contains 25 mM Tris base, 250 mM glycine (electrophoresis grade) (pH 8.3), 0.1% 50S. A 5X stock can be made by dissolving 15.1 g of Tris base and 94 g of glycine in 900 ml of deionized water. Then 50 ml of a 10% (w/v) stock solution of electrophoresis-gralic 50S is adjusted to 1000 ml with water.

Casting of 50S-Polyacrylamide Gels

Assemble the glass plates according to the apparatus manufacturer's instructions. Determine the volume of the gel mold (this information is usually provided by the manufacturer). In an Erlenmeyer flask, prepare the appropriate volume of solution containing the desired concentration of acrylamide for the resolving gel. Polymerization will begin as soon as the TEMED has been added. Without delay, swirl the mixture rapidly.

Pour the acrylamide solution into the gap between the glass plates. Leave sufficient space for the stacking gel (the length to the teeth of the comb plus 1 cm). Using a Pasteur pipette, carefully overlay the acrylamide solution with 0.1 % 505 (for gels containing 8% acrylamide) or isobutanol (for gels containing ~ 10% acrylamide). Place the gel in a vertical position at room temperature.

Large-scale isolation method. After polymerization is complete (30 min), pour off the overlay and wash the top of the gel several times with deionized water to remove any un-polymerized acrylamide. Drain as much fluid as possible from the top of the gel, and then remove any remaining water with the edge of the paper towel.

Prepare the stacking gel as follows: In a disposable plastic tubes, prepare the appropriate volume of solution containing the desired concentration of acrylamide. Polymerization will begin as soon as the TEMEO has been added. Without delay, swirl the mixture rapidly and proceed, to the next step.

Pour the stacking gel solution directly onto the surface of the polymerized resolving gel. Immediately insert a clean Teflon comb into the stacking gel solution, being careful to avoid trapping air bubbles. Add more stacking gel solution to fill the paces of the comb completely. Place the gel in a vertical position at room temperature.

While the stacking gel is polymerizing, prepare the samples by heating them to 100 degrees for 3 minutes in 1 × 50S gel loading buffer to denature the proteins.

1 × 50S gel loading buffer	
50 mM Tris-CI (pH 6.8)	1.2 ml
100 mM dithiothreitol/p mercaptoethanol	0.95 ml

2% 50S (electrophoresis grade)	2 ml
0.1 % bormophenol blue	0.5 ml
10% glycerol	1 ml

1 X SDS gel loading buffer lacking dithiothreitol/p mercaptoethanol can be stored at room temperature. Dithiothreitol/p mercaptoethanol should then be added, just before the buffer is used, from a 1M stock.

After polymerization is complete (30 minutes), remove the Teflon comb carefully. Wash the wells immediately with deionized water to remove any unpolymerized acrylamide. Mount the gel in the electrophoresis apparatus. Add Tris-glycine electrophoresis buffer to the top and bottom reservoirs. Remove the bubbles that become trapped at the bottom of the gel between the glass plates. This is best done with a bent hypodermic needle attached to a syringe.

Load up to 15 III of each of the samples in the predetermined order into the bottom of the wells. Load an equal volume of l X SDS gel loading buffer into any wells that are unused. Attach the electrophoresis apparatus to an electric power supply (the positive electrode should be connected to the bottom buffer reservoir). Apply a voltage of 8 V/cm to the gel. After the dye front has moved into the resolving gel, increase the voltage to 50 V/cm and run the gel until the bromophenol blue reaches the bottom of the resolving gel (about 4 hours). Then turn off the power supply.

Remove the glass plates from the electrophoresis apparatus and place them on a paper towel. Using a spatula, dry the plates apart. Mark the orientation of the gel by cutting a comer from the bottom of the gel that is closest to the left most well (slot 1). Important Do not cut the comer from gels that are to be used for western blotting.

The gel can now be fixed, stained with Coomassie-Brilliant Blue, fluorographed or autoradiographed or used to establish a western blot.

Staining of SDS-Polyacrylamide Gels

Polypeptides separated by SDS-poly-acrylamide gels can be simultaneously fixed with methanol: glacial acetic acid and stained with Coomassie-Brilliant Blue R250, a triphenylmethane textile dye also known as Acid Blue 83. The gel is immersed for several hours in a concentrated methanol/acetic acid solution of the dye, and excess dye is then allowed to diffuse from the gel during a prolonged period of destaining.

Dissolve 0.15 g of Coomassie-Brilliant Blue R250 in 90 ml of methanol: water (1:1 v/v) and 10 ml of glacial acetic acid. Filter the solution through a Whatman No.1 filter to remove any particulate matter.

Immerse the gel in at least 5 volumes of staining solution and place on a slowly rotating platform for a maximum of 4 hours at room temperature.

Remove the stain and save it for future use. Destain the gel by soaking it in the methanol/acetic acid solution (step 1) without the dye on a slowly rocking platform for 4-8 hours, changing the destaining solution three or four times.

The more thoroughly the gel is destained, the smaller the amount of protein that can be detected by staining with Coomassie-Brilliant Blue.

After destaining, gels may be stored indefinitely in water containing 20% glycerol.

AGAROSE GEL ELECTROPHORESIS

Introduction

Electrophoresis through agarose or polyacrylamide gels is the standard method used to separate, identify and purify DNA fragments. The technique is simple, rapid to perform and capable of resolving fragments of DNA that cannot be separated adequately by other procedures, such as density gradient centrifugation. Furthermore, the location of DNA within the gel can be determined directly by staining with low concentrations of the fluorescent intercalating dye ethidium bromide; bands containing as little as 1-10 mg of DNA can be detected by direct examination of the gel in ultraviolet light. If necessary, these bands of DNA can be recovered from the gel and used for a variety of cloning purposes.

This method whereby charged molecules in solution, chiefly proteins and nucleic acids, migrates in response to electric field. Their rate of migration, or mobility, through the electric field depends on the strength of the field, on the net charge, size and shape of the molecules and also on the ionic strength, viscosity and temperature of the medium in which the molecules are moving.

Movement of the DNA in the gel depends on its molecular weight, conformation, and concentration of the agarose, voltage applied and strength of the electrophoresis buffer.

Requirements

(*a*) Submarine gel apparatus including glass plate, comb and surround.

(*b*) Ethidium bromide: 10 mg/ml.

(*c*) Agarose

(*d*) T AE buffer: 0.04 M tris-acetate, 0.001 M EDTA, pH 8.0

(*e*) Ethanol

(*f*) 6 × Gel loading buffer.

Preparation and Examination of Agarose Gels

Seal the edges of the clean, dry, glass plate (or the open ends of the plastic tray supplied with the electrophoresis apparatus) with autoclave tape so as to form a mould. Set the mold on the horizontal section of the bench (check with a level).

1. Prepare sufficient electrophoresis buffer (usually 1 × T AE or 0.5 × TBE) to fill the electrophoresis tank and to prepare the gel. Add the correct amount of powdered agarose to measured quantity of electrophoresis buffer in an Erlenmeyer flask or a glass bottle with a loose-fitting cap. The buffer should not occupy more than 50% of the volume of the flask or bottle.
2. Heat the slurry in a boiling-water bath or a microwave oven until the agarose dissolves.
3. Cool the solution to 60 degrees and if desired add. Ethidium bromide (from a stock solution of 1.0 mg/ml in water) to a final concentration of 0.5 mg/ml and mix thoroughly.

    (*a*) Treat the solution with 100 mg of powdered activated charcoal for each 100 ml solution.

    (*b*) Store the solution for 1 hour at room temperature with intermittent shaking.

(*c*) Filter the solution through a Whatman No. 1 filter and discard the filtrate.

(*d*) Seal the filter and activated charcoal in a plastic bag and dispose the bag in safe place.

AGAROSE GEL ELECTROPHORESIS

Materials

- Agarose solution in TBE or TAE (generally 0.7-1%)
- 1X TBE or TAE (same buffer as in agarose)
- Gel loading dye.

Procedure

1. To prepare 100 ml of a 0.8% agarose solution, measure 0.8 g agarose into a glass beaker or flask and add 100 ml 1X TBE or TAE. 10 mg/ml ethidium bromide.
2. Stir on a hot plate until agarose is dissolved and solution is clear. Allow solution to cool to about 55°C before pouring. (Ethidium bromide)
3. Can be added at this point to a concentration of 0.5 μg/ml.
4. Prepare gel tray by sealing ends with tape or other custom-made dam.
5. Place comb in gel tray about 1 inch from one end of the tray and position the comb vertically such that the teeth are about 1–2 mm above the surface of the tray.
6. Pour 50°C gel solution into tray to a depth of about 5 mm. Allow gel to solidify about 20 minutes at room temperature.
7. To run, gently remove the comb, place tray in electrophoresis chamber, and cover (just until wells are submerged) with electrophoresis buffer (the same buffer used to prepare the agarose).
8. Excess agarose can be stored at room temperature and remelted in a microwave.
9. To prepare samples for electrophoresis, add 1 μl of 6X gel loading dye for every 5 μl of DNA solution. Mix well. Load 5–12 μl of DNA per well (for minigel).
10. Electrophoresis at 50–150 volts until dye markers have migrated an appropriate distance, depending on the size of DNA to be visualized.

If the gel was not stained with ethidium during the run, stain the gel in 0.5 μg/ml ethidium bromide until the DNA has taken up the dye and is visible under short wave UV light, if the DNA will not be used further, or with a hand-held long-wave UV light if the DNA is to be cut out and purified.

6 MICROBIOLOGY

INTRODUCTION

(a) *Laboratory Rules.* For the safety and convenience of everyone working in the laboratory, it is important that the following laboratory rules be observed at all times:

1. Place only those materials needed for the day's laboratory exercise on the benchtops.
2. Since some of the microorganisms used in this class are pathogenic or potentially pathogenic (opportunistic), it is essential to always follow proper aseptic technique in handling and transferring all organisms.
3. No eating, drinking, or any other hand-to-mouth activity while in the lab. If you need a short break, wash your hands with disinfectant soap and leave the room.
4. Using a wax marker, properly label all inoculated culture tubes or petri plates with the name or the initials of the microorganism you are growing, your initials or a group symbol, and any other pertinent information.

   Place all inoculated material only on your assigned incubator shelf. Culture tubes should be stored upright in plastic beakers, while petri plates should be stacked and incubated upside-down.
5. Always clean the oil off of the oil immersion lens of the microscope with a piece of lens paper at the completion of each microscopy lab.
6. Disinfect the benchtop with isopropyl alcohol before and after each lab period. Be sure your Bunsen burner is turned off before you spray any alcohol.
7. Always wash your hands with disinfectant soap.

(b) *General Directions*

1. Familiarize yourself in advance with the procedure of the experiment to be performed.
2. Disinfect the working table with isopropyl alcohol before and after each lab.
3. The first part of each lab period will be used to complete and record the results of previous experiments. We will always go over these results as a class. You may wish to purchase a set of coloured pencils to aid you in recording your results in the lab manual.
4. Wash your hands with disinfectant soap before leaving the lab.

(c) *Binomial Nomenclature.* Microorganisms are given specific scientific names based on the binomial (two names) system of nomenclature. The first name is referred to as the genus and the second name is termed the species. The names usually come from Latin or Greek and describe some characteristic of the organism.

To correctly write the scientific name of a microorganism, the first letter of the genus should be capitalized while the species name should be in lower case letters. Both the genus and species names are italicized or underlined. Several examples are given below:

Bacillus subtilus

Bacillus: L. dim. noun *Bacillum*, a small rod

subtilus: L. adj. *subtilus*, slender.

Escherichia coli

Escherichia: after discoverer, Prof. Escherich

coli: L. gen. noun *coli*, of the colon.

Staphylococcus aureus

Staphylococcus: Gr. noun *Staphyle*, a bunch of grapes; Gr. noun *coccus*, berry
aureus: L. adj. *aureus*, golden

(d) *Metric Length and Fluid Volume.* The study of microorganisms necessitates an understanding of the metric system of length. The basic unit of length is the meter (m), which is approximately 39.37 inches. The basic unit for fluid volume is the litre (l), which is approximately 1.06 quarts. The prefix placed in front of the basic unit indicates a certain fraction or multiple of that unit. The most common prefixes we will be using are:

- centi = 10^{-2} or 1/100
  centimeter (cm) = 10^{-2} m or 1/100 m
- milli = 10^{-3} or 1/1000
  millimeter (mm) = 10^{-3} m or 1/1000 m
  millilitre (ml) = 10^{-3} l or 1/1000 l
- micro = 10^{-6} or 1/1,000,000
  micrometer (μm) = 10^{-6} m or 1/1,000,000 m
  microlitre (μl) = 10^{-6} l or 1/1,000,000 l
- nano = 10^{-9} or 1/1,000,000,000
  nanometer (nm) = 10^{-9} m or 1/1,000,000,000 m.

In microbiology, we deal with extremely small units of metric length (micrometer, nanometer). The main unit of length is the micrometer (μm) which is 10^{-6} (1/1,000,000) of a meter or approximately 1/25,400 of an inch.

The average size of a rod-shaped (cylindrical) bacterium is 0.5-1.0 μm wide by 1.0-4.0 μm long. An average coccus-shaped (spherical) bacterium is about 0.5-1.0 μm in diameter. A volume of one cubic inch is sufficient to contain approximately nine trillion average-sized bacteria. It would take over 18,000,000 average-sized cocci lined up edge-to-edge to span the diameter of a dime.

In several labs we will be using pipettes to measure fluid volume in ml.

1. THE MICROSCOPY

Introduction

The microscope has been a valuable tool in the development of scientific theory and study of cells, microbes etc. the modern compound microscope that the device was used in Biology. A compound microscope is composed of two elements; a primary magnifying lens and a secondary lens system. Light is caused to pass through an object and is then focused by the primary and secondary lens. If the beam of light is replaced by an electron beam, the microscope becomes a transmission electron microscope. If light is bounced off of the object instead of passing through, the light microscope becomes a dissecting scope. If electrons are bounced off of the object in a scanned pattern, the instrument becomes a scanning electron microscope.

The function of any microscope is to enhance **resolution.** The microscope is used to create an enlarged view of an object such that we can observe details not otherwise possible with the human eye. Because of the enlargement, resolution is often confused with **magnification,** which refers to the size of an image. In general, the greater the magnification, the greater the resolution, but this is not always true. There are several practical limitations of lens design which can result in increased magnification without increased resolution.

If an image of a cell is magnified from 10X to 45X, the image gets larger, but not necessarily any clearer. The image on the left is magnified with no increase in resolution. The image on the right is magnified the same, but with increasing resolution. Note that by the time the image is magnified 10X (from 10X to 100X), the image on the left is completely unusable. The image on the right, however, presents more detailed information. Without resolution, no matter how much the image is magnified, the amount of observable detail is fixed, and regardless of how much you increase the size of the image, no more detail can be seen. At this point, you will have reached the limit of resolution or the resolving power of the lens. This property of the lens is fixed by the design and construction of the lens. To change the resolution, a different lens is often the only answer.

The reason for a dichotomy between magnification and resolution is the ability of the human eye to see two objects. It is necessary that two objects be about 0.1 mm apart when held 10" from the face in order for us to detect them as two objects. If they are closer than 0.1 mm, we will perceive them as a single object. If two objects are 0.01 mm apart, we can not detect them unless we magnify an image of them by 10X. What has happened is that we have effectively altered our resolution ability from 0.1 mm to 0.01 mm through the use of a magnifying lens. We would say that our limit of resolution has changed from 0.1 mm to 0.01 mm, or inversely, our resolving power (resolution) has increased by a factor of 10.

How to Use Microscope

1. *Moving and transporting the microscope.* Grasp the arm of the microscope with one hand and support the base of the microscope with the other. Handle the microscope gently.
2. Before you plug in the microscope, turn the voltage control dial on the right side of the base of the microscope to 1. Now plug in the microscope and use the on/off switch in the front of the microscope on the base to turn it on.

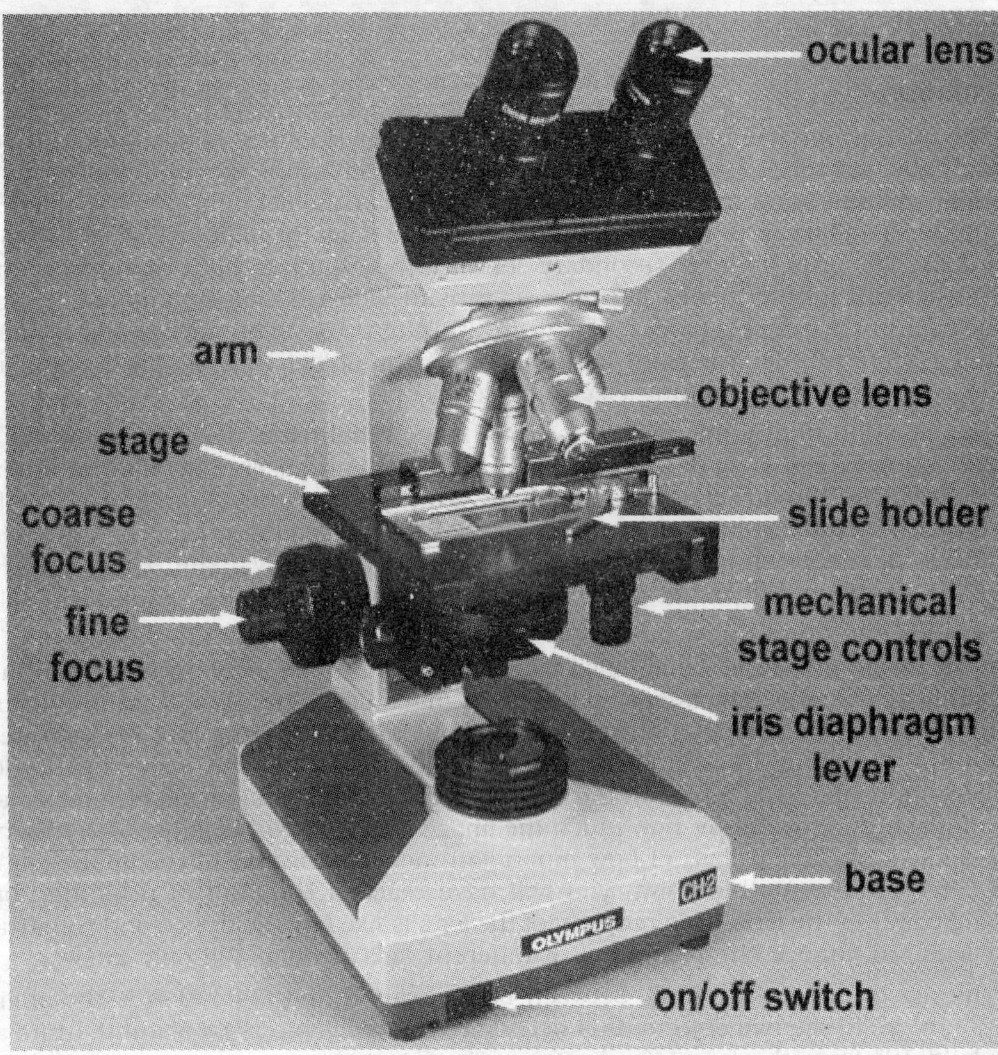

Fig. 1. Olympus Microscope

Make sure the entire cord is on the bench top and not hanging down where it could be caught by a leg. **Adjust the voltage control dial to 10.**

3. *Adjusting the eyepieces.* These microscopes are binocular, that is, they have 2 ocular lenses (eyepieces). To adjust them, first find the proper distance between your eyes and the eyepieces by closing one eye and slowly moving your head toward that eyepiece until you see the complete field of view-about 1 inch away. Keep your head steady and both eyes in the same plane. Now open the other eye and gradually increase the distance between the eyepieces until it matches the distance between your eyes. At the correct distance you will see one circular field of view with both eyes.
4. *Positioning the slide.* Place the slide specimen-side-up on the stage so that the specimen lies over the opening for the light in the middle of the stage. Secure the slide between not under-the slide holder arms of the mechanical stage.

The slide can now be moved from place to place using the 2 control knobs located under the stage on the right of the microscope.

5. *Adjusting the illumination.* (*a*) Adjust the *voltage* by turning the *voltage control dial* located in the rear right-hand side of the microscope base. For oil immersion microscopy (1000X) set the light on 9 or 10. At lower magnifications less light will be needed.

   (*b*) Adjust the *amount of light coming through the condenser* using the *iris diaphragm lever* located under the stage in the front of the microscope. Light adjustment using the iris diaphragm lever is critical to obtaining proper contrast. For oil immersion microscopy (1000X), the iris diaphragm lever should be set almost all the way open (to your left for maximum light). For low powers such as 100X the iris diaphragm lever should be set mostly closed (to your right for minimum light).

   (*c*) The *condenser height control* (the single knob under the stage on the left-hand side of the microscope should be set so the condenser is all the way up.

6. *Obtaining different magnifications.* The final magnification is a product of the 2 lenses being used. The *eyepiece or ocular lens* magnifies *10X*. The *objective lenses* are mounted on a turret near the stage. The small *yellow-striped lens* magnifies *10X;* the *blue-striped lens* magnifies *40X*, and the *white-striped oil immersion lens* magnifies *100X*. Final magnifications are as follows:

Ocular lens	X	*Objective lens*	=	*Total magnification*
10X	X	10X (yellow)	=	100X
10X	X	40X (blue)	=	400X
10X	X	100X (white)	=	1000X

7. *Focusing from lower power to higher power*

   (*a*) Rotate the *yellow-striped 10X objective* until it locks into place (total magnification of 100X).

   (*b*) Turn the *coarse focus control* (larger knob) all the way *away from you* until it stops.

   (*c*) Look through the eyepieces and turn the *coarse focus control* (larger knob) *towards you* slowly until the specimen comes into focus.

   (*d*) Get the specimen into sharp focus using the *fine focus control* (smaller knob) and adjust the light for optimum contrast using the iris diaphragm lever.

   (*e*) If higher magnification is desired, simply rotate the *blue-striped 40X objective* into place (total magnification of 400X) and the specimen should still be in focus. (Minor adjustments in fine focus and light contrast may be needed.)

   (*f*) For maximum magnification *(1000X or oil immersion)*, rotate the blue-striped 40X objective slightly out of position and place a *drop of immersion oil* on the slide. Now rotate the *white-striped 100X oil immersion objective* into place. Again, the specimen should remain in focus, although minor adjustments in fine focus and light contrast may be needed.

8. *Cleaning the microscope.* Clean the exterior lenses of the eyepiece and objective before and after each lab using *lens paper* only. (Paper towel or kim-wipes may scratch the lens.) *Remove any immersion oil from the oil immersion lens before putting the microscope away.*
9. *Reason for using immersion oil.* Normally, when light waves travel from one medium into another, they bend. Therefore, as the light travels from the glass slide to the air, the light waves bend and are scattered similar to the "bent pencil" effect when a pencil is placed in a glass of water. The microscope magnifies this distortion effect. Also, if high magnification is to be used, more light is needed.

   Immersion oil has the same refractive index as glass and, therefore, provides an optically homogeneous path between the slide and the lens of the objective. Light waves thus travel from the glass slide, into glass-like oil, into the glass lens without being scattered or distorting the image. In other words, the immersion oil "traps" the light and prevents the distortion effect that is seen as a result of the bending of the light waves.

Bright Field, Dark Field, Phase Contrast

All microscopes actually allow visualization of objects through minute shifts in the *wavelength phase* as the light passes through the object. Further image forming can be had through the use of colour, or through a complete negative image of the object. If the normal phase shift is increased (usually by 1/4 wavelength), then the microscope becomes a *phase contrast microscope.* Phase contrast microscopes can be designed to have medium phase or dark phase renditions, by altering the degree of additional shift to the wavelength from 1/4 to 1/2 wavelengths, respectively.

If the beam of light is shifted in phase by a variable amount, the system becomes a *differential interference contrast microscope.*

If the light image is reversed, then the microscope becomes a *dark field microscope.* All standard bright field microscopes can be readily converted to dark field by inserting a round opaque disk beneath the condenser. Dark field microscopy was first utilized to examine trans-filterable infectious agents, later to be termed viruses, and to determine that they were particulate in nature. Small objects, even those below the limits of resolution, can be detected easily with dark field, as the object appears to emit light on a dark field. (Look at the sky for a comparison. It is fairly easy to see stars in a dark sky, but impossible during the day. The same is true for dark field vs bright field microscopy).

Finally, if the normal light microscope is functionally turned upside down, the microscope becomes an *inverted microscope.* This is particularly useful in tissue culture since it allows observation of cells through the bottom of a culture vessel, without opening the container, and without the air interface normally present between the objective and the surface of the culture. By adding phase contrast optics to the inverted microscope, it is possible to monitor tissue cultures directly, without the aid of stains or other enhancements.

The Electron Microscope

The transmission electron microscope (TEM) is having resolving power (3-10 Å). The scanning electron microscope (SEM) is becoming increasingly popular with cell biologists because of its remarkable ability for quantifiable mapping of surface detail, along with improved resolution (30-100 Å) and its ability to show 3D structure.

The transmission electron microscope is identical in concept to the modern binocular light microscope. It is composed of a light source (in this case an electron source), a substage condenser to focus the electrons on the specimen, and an objective and ocular lens system. In the electron microscope, the ocular lens is replaced with a projection lens, since it projects an image onto a fluorescent screen or a photographic plate. Since the electrons do not pass through glass, they are focused by electro-magnetic fields. Instead of rotating a nose-piece with different fixed lenses, the EM merely changes the current and voltage applied to the electromagnetic lenses.

The size of an electron microscope is dependent upon two factors. Primary is the need for a good vacuum through which the electrons must pass (it takes less than 1 cm of air to completely stop an electron beam). Peripheral pumps and elaborate valves controls are needed to create the vacuum. A substantial electrical potential (voltage) is also needed to accelerate the electrons out of the source. The source is usually a tungsten filament, very much like a light bulb, but with 40-150 Killivolts of accelerating voltage applied to an anode to accelerate the electrons down the microscope column. Modern electronics have produced transformers which are reasonably small but capable of generating 60,000 volts.

Another characteristic of electron microscopes is that they are usually designed upside-down, similar to an inverted light microscope. The electron source is on top, and the electrons travel down the tube, opposite to light rays traveling up a microscope tube. This is merely a design feature that allows the operator and technicians ease of access to its various components. The newer electron microscope is beginning to look like a desk with a TV monitor on it.

Until recently, the major advantage of an electron microscope also has been its major disadvantage. In theory, the transmission electron microscope should be capable of giving a resolution of several angstroms. This would give excellent molecular resolution of cell organelles. However, as the resolution increases, the field of view decreases and it becomes increasingly difficult to view the molecular detail within the cell. Electron microscopes designed to yield high resolution have to be compromised to view larger objects. Cell structures fall within the size range that was most problematic for viewing. For example, if we wished to resolve the architecture of an entire eucaryotic chromosome, not just the chromosome, but the cell itself was too large to be seen effectively in an electron microscope. Zooming in on paired chromosomes was impossible. Modern electron microscope design allows for this zooming, and the observation of whole tissues while retaining macromolecular resolution.

The Scanning Electron Microscope

The scanning electron microscope works by bouncing electrons off of the surface and forming an image from the reflected electrons. Actually, the electrons reaching the specimen (the 1° electrons) are normally not used (although they can form a transmitted image, similar to standard TEM), but they incite a second group of electrons (the 2° electrons) to be given off from the very surface of the object. Thus, if a beam of primary electrons is scanned across an object in a raster pattern (similar to a television scan), the object will give off secondary electrons in the same scanned pattern. These electrons are gathered by a positively charged detector, which is scanned in synchrony with the emission beam scan. Thus, the name scanning electron microscope, with the image formed by the collection of secondary electrons.

It is possible to focus the primary electrons in exactly the same manner as a TEM. Since the primary electrons can be focused independently of the secondary electrons, two images can be produced simultaneously. Thus, an image of a sectioned material can be superimposed on an image of its surface. The instrument then becomes a STEM, or Scanning-Transmission Electron Microscope. It has the same capabilities of a TEM, with the added benefits of an SEM.

SEM allows a good deal of analytical data to be collected in addition to the formed image. As the primary electrons bombard the surface of an object, they interact with the atoms of the surface to yield even more particles and radiations other than secondary electrons. Among these radiations are Auger electrons, and characteristic X-rays. The X-rays have unique, discreet energy values, characteristic of the atomic structure of the atom from which they emanated. If one collects these X-rays and analyzes their inherent energy, the process becomes Energy Dispersive X-ray Analysis. Combining the scan information from secondary and Auger electrons, together with the qualitative and quantitative X-ray information allows the complete molecular mapping of an object's surface.

Finally, the scanning microscope has one further advantage that is useful in cell structure analysis. As the electron beam scans the surface of an object, it can be designed to etch the surface. That is, it can be made to blow apart the outermost atomic layer. As with the emission of characteristic X-rays, the particles can be collected and analyzed with each pass of the electron beam. Thus, the outer layer can be analyzed on the first scan, and subsequently lower layers analyzed with each additional scan. Electrons are relatively small, and the etching can be enhanced by bombarding the surface with ions rather than electrons (the equivalent of bombarding with bowling balls rather than BB's). The resultant Secondary Emmisions-Ion Scanning data can finally be analyzed and the three-dimensional bit-mapped atomic image of an object reconstructed.

EXERCISE 1. THE BRIGHT FIELD MICROSCOPE

Materials

- Binocular Microscope
- Microscope slide.

Procedure

1. Pick up a microscope from the cabinet by placing one hand under the base and the other on the arm of the microscope. Most microscope damage is due to careless transport.

2. Place the microscope in front of you, unwind the power cord and plug it in. The microscope is normally provided in its storage position, that is, with its eyepieces pointed back over the arm. This takes less room in a cabinet, but is not the position for which it was designed to be used. If your instructor approves, slightly loosen the screw holding the binocular head and rotate the entire binocular head 180°. Carefully (and gently) tighten the screw to prevent the head from falling off.

You will notice that all parts of the microscope are now conveniently located for your use, with an uninterrupted view of the stage, and substage. The focus controls are conveniently at arms-length.

Note the magnification power and the numerical aperture of the lenses which are on your microscope's nose-piece. These values are stamped or painted onto the barrels of the objectives. Record the magnification power and numerical aperture of each lens in the space provided below:

Magnification (x#)	*Numerical Aperture (NA)*
Enter the magnification of the oculars and whether they are normal or widefield ________.	Enter the numerical aperture of the condenser ________.
Your maximum resolution will depend upon the highest effective numerical aperture of the system. The highest value is normally given by the 100X, or oil immersion lens.	The numerical aperture for an air interphase = 1.0.
Indicate the numerical aperture of the 100X lens ________.	The numerical aperture for oil interphase = 1.3–1.5.
Indicate the numerical aperture of the condenser ________.	The maximum effective numerical aperture is the lowest of those listed. It depends on the angle and thus on maximum positioning of the condenser. Using the lowest NA value from above as the working numerical aperture, calculate the limit of resolution for your microscope, assuming violet light with a wavelength of 400 pm.
Limit of resolution = ________ μ.	From equation the limit of resolution = 0.61 × λ/NA, and therefore, the calculated value for your microscope is:

3. Obtain a prepared microscope slide. Place the slide on the stage and ensure that it is locked in place with the slide holder.

Rotate the condenser focusing knob to move the condenser to its highest position of travel. Although there is an ideal location for the condenser, the correct position of the condenser will vary slightly for each objective. Unless directed otherwise, it will not be necessary to move the condenser during any of the intended uses in this course.

If, however, you wish to find the ideal location, focus the microscope on any portion of a slide, and then simply close down the condenser aperture and move the **condenser** until you have a sharply focused view of the condenser aperture (usually with a slight blue hazy edge). If you do this, you can then open the aperture until it just fills the field of view (different for each objective). This is the correct location and use of the condenser and aperture and the condenser should not be moved from this position. Never use the condenser aperture for control of light intensity. Control of light intensity is the purpose of the variable rheostat (dimmer switch, or voltage regulator) on the light source.

4. Turn on the microscope by rotating the dimmer switch and adjust the light intensity to a comfortable level. Be sure that the condenser aperture is open if you have not set it as directed in the previous paragraph (slide the condenser diaphragm lever back and forth to check).

5. Looking down into the microscope, adjust the eyepieces to your interpupillary distance and diopter. The Nikon microscope is equipped with a knob between the eye tube extensions for this adjustment. Many microscopes simply require pushing the eye tubes together or apart directly. Move the eye tubes back or forth until you see one uniform field of view.

The first time you use the microscope, adjust the eyepieces for your personal comfort. Note that modern microscopes have HK (high eye point) eyepieces and consequently you need not remove eyeglasses if you are wearing them. Quite the contrary, they should be worn to prevent eyestrain while you constantly shift from looking through the microscope to reading the lab manual.

- Begin by focusing the microscope on any object within the field of view.
- Find a suitably contrasty location in the center of the field of view and close your left eye. Using the coarse and fine adjustments, focus until you obtain a sharp image with your right eye only.
- Now close your right eye and adjust the focus of the left eyepiece by rotating the diopter adjusting ring located on the left eyepiece. Do not readjust the focus of the left eye with the coarse or fine adjustments of the microscope-use the adjustment ring on the eye tube.

All subsequent uses of the same microscope will involve use of the coarse and fine focus adjustments, without reference to the procedures in step 2. That is, step 2 need only be performed once at the beginning of your lab. It may, of course, be checked periodically if desired, and will need to be readjusted if someone else uses your microscope.

Optional

Familiarize yourself with the operation of any tension adjustment options or pre-set devices that may be attached to the microscope.

Coarse Adjustment Tension: The coarse adjustment may be eased or tightened by the adjusting ring. If the rotation of the coarse focus knob is too loose, turn the adjusting ring counterclockwise. Too much tension may be adjusted by turning clockwise. Avoid excessive rotation as it will place undo stress on the internal gears. Adjust the tension so that the stage will remain stationary after focusing but can be moved with relative ease by turning the coarse adjustment knob. Some microscopes require turning the two coarse adjustment knobs in opposite directions, while others require the use of a screwdriver. Be sure to check with your instructor or the manufacturer's directions before adjusting this feature.

EXERECISE 2. INTRODUCTION TO THE MICROSCOPE AND COMPARISON OF SIZES AND SHAPES OF MICROORGANISMS

A. Bacterial Shapes and Arrangements

Bacteria are unicellular prokaryotic microorganisms. There are three common shapes of bacteria: the coccus, the bacillus, and the spiral. Bacteria divide by **binary fission**, a process by which one bacterium splits into two.

1. Coccus

A coccus-shaped bacterium is usually spherical, although some appear oval, elongated, or flattened on one side. Cocci are approximately 0.5 micrometer (μm) in diameter and may be seen, based on their planes of division and tendency to remain attached after replication, in one of the following arrangements:

(*a*) Division in one plane produces either a *diplococcus* or *streptococcus* arrangement.

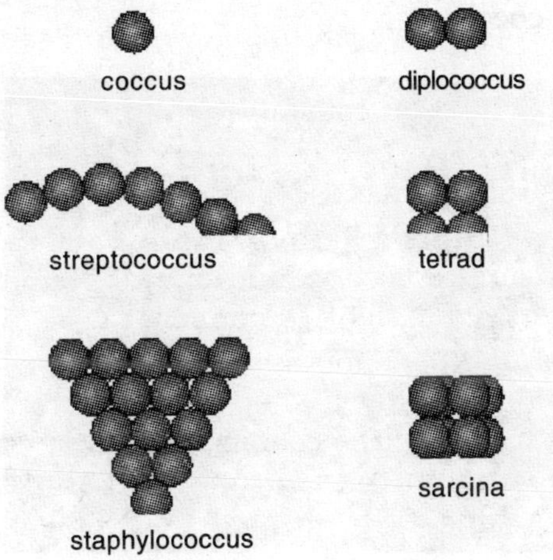

Fig. 2. Arrangements of Cocci.

Note: Cocii in pairs arrows.

(*i*) *Diplococcus:* pair of cocci

Fig. 3. Diplococcus arrangement.

Note: Gram-positive (purple) cocci in chains (arrows).

(*ii*) *Streptococcus:* chain of cocci

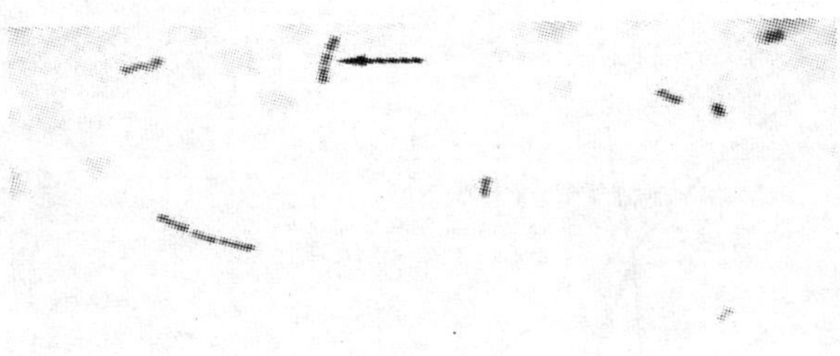

Fig. 4. Streptococcus pyogenes.

(*b*) Division in **two planes** produces a **tetrad** arrangement.

Note: A tetrad appears as a square of four cocci (arrows).

A tetrad: square of 4 cocci

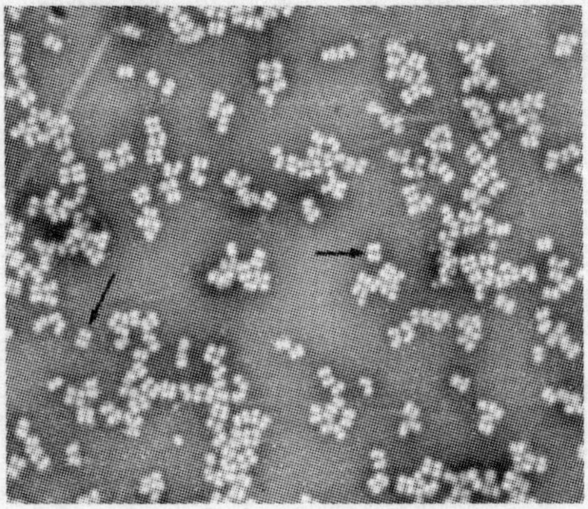

Fig. 5. Tetrad arrangement.

(c) Division in **three planes** produces a **sarcina** arrangement. sarcina: cube of 8 **cocci.**

(d) Division in **random planes** produces a **staphylococcus** arrangement. staphylococcus: cocci in irregular, often grape-like clusters.

As you observe these different cocci, keep in mind that the procedures used in slide preparation may cause some arrangements to break apart or clump together. The correct form, however, should predominate. Also remember that each coccus in an arrangement represents a complete, single, one-celled organism.

2. Bacillus (rod)

A bacillus or rod is a hotdog-shaped bacterium having one of the following arrangements:

(a) bacillus: a single bacillus.

(b) streptobacillus: bacilli in chains-Streptobacillus Arrangement.

Note: bacilli in chains.

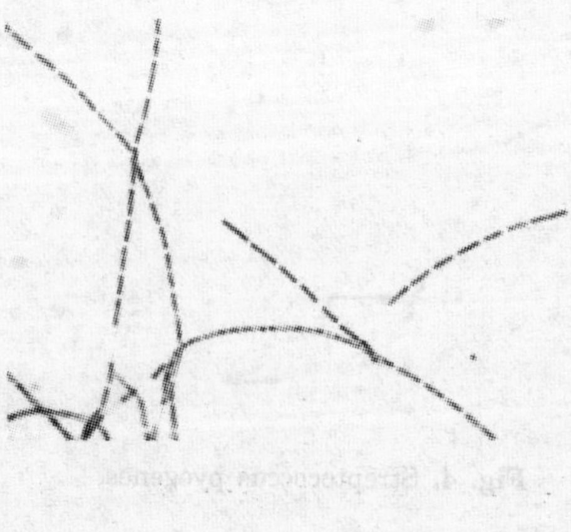

(c) *Coccobacillus:* oval and similar to a coccus.

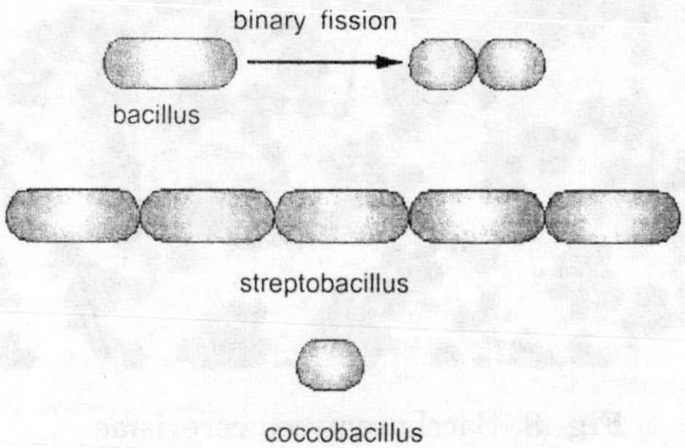

Fig. 6. Arrangements of Bacilli.

A single bacillus is typically 0.5-1.0 μm wide and from 1-4 μm long. Small bacilli or bacilli that have just divided by binary fission may at first glance be confused for cocci so they must be observed carefully. You will, however, be able to see bacilli that have not divided and are definitely rod-shaped as well as bacilli in the process of dividing.

3. Spiral

Spiral-shaped bacteria occur in one of three forms.

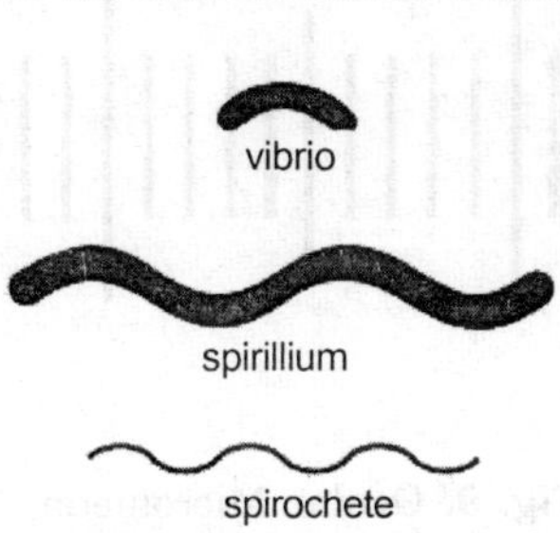

Fig. 7. Spiral forms.

(a) *vibrio:* an incomplete spiral or comma-shaped.

(b) *spirillum:* a thick, rigid spiral.

(c) *spirochete:* a thin, flexible spiral.

The spirals you will observe range from 5-40 μm long but some are over 100 μm in length. The spirochetes are the thinnest of the bacteria, often having a width of only 0.25-0.5 μm.

B. Yeasts

Yeasts, such as the common baker's yeast *Saccharomyces cerevisiae* are **unicellular fungi.** They usually appear spherical and have a diameter of 3-5 μm. Yeasts commonly reproduce asexually by a process called budding. Unlike bacteria, which are prokaryotic, yeast are eukaryotic.

Note: budding yeast (arrows).

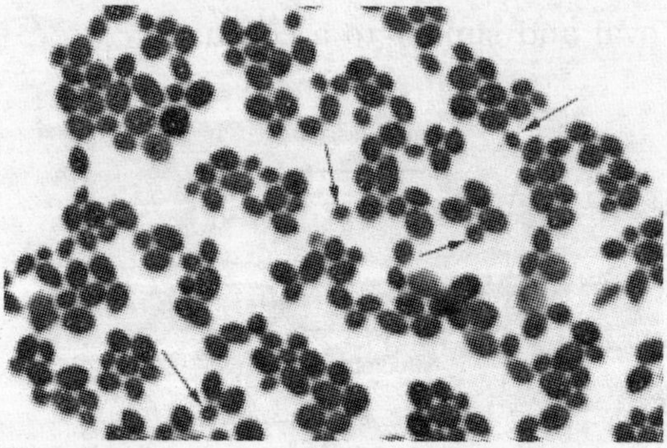

Fig. 8. Saccharomyces cerevisiae.

C. Measurement of Microorganisms

The ocular micrometers provided are calibrated so that when using 1000X oil immersion microscopy, the distance between any two lines on the scale represents a length of approximately one micrometer. Remember this does not hold true when using other magnifications.

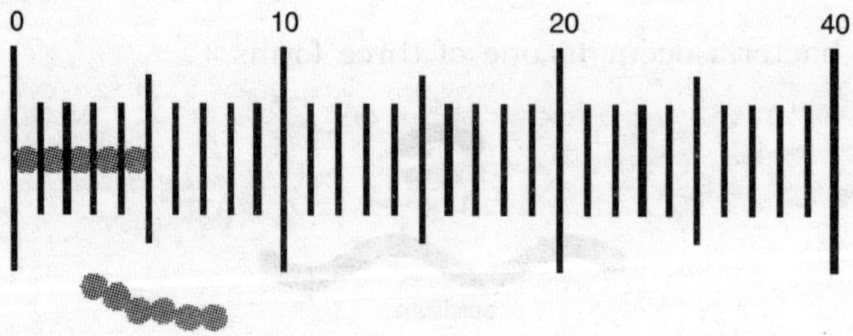

Fig. 9. Ocular Micrometer.

The approximate size of a microorganism can be determined using an ocular micrometer, an eyepiece that contains a scale which will appear superimposed upon the focused specimen.

For a comparison of the size of a virus, a bacterium, and a human cell.

D. Focusing with Oil Immersion

1. Before you plug in the microscope, turn the voltage control dial on the right-hand side of the base of the microscope to 1. Now plug in the microscope and turn it on.
2. Place the slide in the slide holder, center the slide using the two mechanical stage control knobs under the stage on the right-hand side of the microscope, and place a rounded drop of immersion oil on the area to be observed.
3. Rotate the white-striped 100X oil immersion objective until it is locked into place. This will give a total magnification of 1000X.
4. Turn the voltage control dial on the right-hand side of the base of the microscope to 9 or 10. Make sure the iris diaphragm lever in front under the stage is almost wide open, (toward the left side of the stage), and the knob under the

stage on the left-hand side of the stage controlling the height of the condenser is turned so the condenser is all the way up.

5. Watching the slide and objective lens carefully from the front of the microscope, lower the oil immersion objective into the oil by raising the stage until the lens just touches the slide. Do this by turning the coarse focus (larger knob) away from you until the spring-loaded objective lens just begins to spring upward.
6. While looking through the eyepieces, turn the fine focus (smaller knob) towards you at a slow steady speed until the specimen comes into focus. (If the specimen does not come into focus within a few complete turns of the fine focus control and the lens is starting to come out of the oil, you missed the specimen when it went through focus. Simply reverse direction and start turning the fine focus away from you.)
7. Using the iris diaphragm lever, adjust the light to obtain optimum contrast.
8. When finished, wipe the oil off of the oil immersion objective with lens paper, turn the voltage control dial back to 1, turn off the microscope, unplug the power cord, and wrap the cord around the base of the microscope.

An alternate focusing technique is to first focus on the slide with the yellow-striped 10X objective by using only the coarse focus control and then without moving the stage, add immersion oil, rotate the white-striped 100X oil immersion objective into place, and adjust the fine focus and the light as needed. This procedure is discussed in the introduction to the lab manual.

Specimens

Prepared slides of the following bacteria:

- *Staphylococcus aureus*
- *Escherichia coli*
- *Borrelia recurrentis* or *Borrelia burgdorferi*
- *Spirillum* species.

Demonstration slides of the following bacteria:

Micrococcus luteus

Neisseria gonorrhoeae

Streptococcus species

Bacillus megaterium

Broth culture of *Saccharomyces cerevisiae*

Human hair.

Procedure

1. Using **oil immersion microscopy** (1000X), observe and measure the bacteria listed below:

Tips for Microscopic Observations

Remember that in the process of making the slide, some of the coccal arrangements will clump together and others will break apart. Move the slide around until you see an area representing the true arrangement of each organism. Also remember that small bacilli (such as *Escherichia coli*) that have just divided by binary fission will look similar to cocci. Look carefully for bacilli that are not dividing and are definitely rod-shaped as well as bacilli in the process of dividing to confirm the true shape. Also, bacilli do not divide so as to form clusters. Any such clusters you see are artifacts from preparing the slide. Finally, you will have to look carefully to see the spirochetes as they are the thinnest of the bacteria. When seen microscopically, spirochetes resemble extremely thin, wavy pencil lines.

(a) *Staphylococcus aureus: Staphylococcus* species, as the genus name implies, have a staphylococcus arrangement (cocci in irregular, often grape-like clusters). Measure the diameter of a single coccus.

(b) *Escherichia coli: Escherichia coli* is a small bacillus. Estimate the length and width of a typical rod.

(c) *Borrelia recurrentis: Borrelia* species are spirochetes (thin, flexible spirals). You are examining blood infected with *Borrelia recurrentis*. Measure the length and width of a typical spirochete and the diameter of a red blood cell.

(d) *Spirillum: Spirillum* species appear as thick, rigid spirals. Measure the length and width of a typical spirillum.

When finished, remove the oil from the prepared slides using paper towel and return them to their proper tray.

2. Observe the demonstration slides of the following bacteria:

   (a) *Micrococcus luteus: Micrococcus luteus* can appear as tetrads, cubes of 8, or in irregular clusters. This strain usually exhibits a tetrad or a sarcina arrangement. Measure the diameter of a single coccus.

   (b) *Neisseria gonorrhoeae: Neisseria* species usually have a diplococcus arrangement. Measure the diameter of a single coccus.

   (c) *Streptococcus pyogenes: Streptococcus* species, as the genus name implies, usually have a streptococcus arrangement (cocci in chains). Measure the diameter of a single coccus.

   (d) *Bacillus megaterium: Bacillus megaterium* appears as large bacilli in chains (a streptobacillus). Measure the length and width of a single bacillus.

3. Prepare a wet mount of baker's yeast (Saccharomyces cerevisiae) by putting a *small drop* of the yeast culture on a microscope slide and placing a cover slip over the drop. Using your iris diaphragm lever, *reduce the light* for improved contrast by moving the lever almost all the way to the right and observe using *oil immersion microscopy.* Measure the diameter of a typical yeast.

   When finished, wash the slide and use it again for step 4. Discard the coverslip in the biowaste disposal container at the front of the room and under the hood.

4. Remove a small piece of a hair from your head and place it in a small drop of water on a slide. Place a cover slip over the drop and observe using oil immersion microscopy. Measure the diameter of your hair and compare this with the size of each of the bacteria and the yeast observed in steps 1-3. Discard the slide and coverslip in the biowaste disposal containers at the front of the room and under the hood.

5. At the completion of the lab, remove the oil from the oil immersion objective using lens paper and put your microscope away.

Results

1. Make drawings of several of the bacteria from each of the four prepared slides and indicate their approximate size in micrometers.

diameter = _____ micrometers arrangement =	length = _____ micrometers width = _____ micrometers
length = _____ micrometers width = _____ micrometers diameter of RBC = _____ micrometers	length = _____ micrometers width = _____ micrometers

2. Make drawings of several of the bacteria from each of the four demonstration slides and indicate their approximate size in micrometers.

diameter = _____ micrometers arrangement =	diameter = _____micrometers arrangement =
diameter = _____ micrometers arrangement =	length = _____ micrometers width = _____ micrometers arrangement =

3. Make a drawing of several yeast cells and indicate their size in micrometers.
   Saccharomyces cerevisiae
   diameter = _____ micrometers.
4. Make a drawing indicating the size of the bacteria and yeast observed above relative to the diameter of your hair.
   diameter = _____ micrometers.

PERFORMANCE OBJECTIVES

Discussion

1. State three basic shapes of bacteria.
2. State and describe five different arrangements of cocci.
3. State and describe three different arrangements of bacilli.
4. State and describe three different spiral forms.
5. Describe the appearance of a typical yeast.

Results

1. When given an oil immersion microscope, a prepared slide of a microorganism, and an ocular micrometer, determine the size of that organism in micrometers.
2. Using a microscope, identify different bacterial shapes and arrangements.
3. Differentiate a yeast from a coccus-shaped bacterium by its size.
4. Compare the size of the microorganisms observed in lab with the diameter of a hair when using oil immersion microscopy.

EXERCISE 3. CELL SIZE MEASUREMENTS: OCULAR AND STAGE MICROMETERS

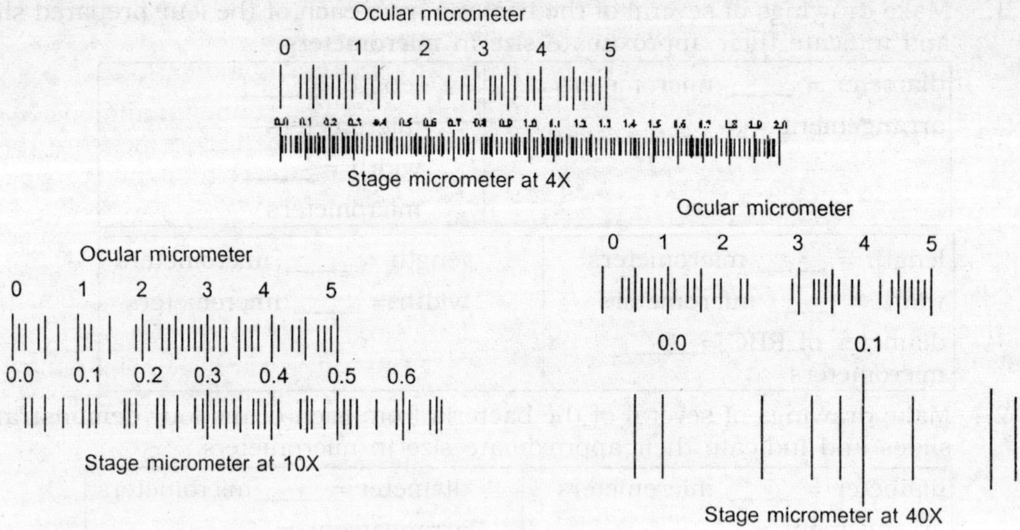

Fig. 10. Micrometery.

Materials

- Microscope
- Ocular micrometer
- Stage micrometer
- Millimeter ruler
- Prepared slide.

Procedure

1. Place a stage micrometer on the microscope stage, and using the lowest magnification (4X), focus on the grid of the stage micrometer.
2. Rotate the ocular micrometer by turning the appropriate eyepiece. Move the stage until you superimpose the lines of the ocular micrometer upon those of the stage micrometer. With the lines of the two micrometers coinciding at one end of the field, count the spaces of each micrometer to a point at which the lines of the micrometers coincide again.
3. Since each division of the stage micrometer measures 10 micrometers, and since you know how many ocular divisions are equivalent to one stage division, you can now calculate the number of micrometers in each space of the ocular scale.
4. Repeat for 10X and 40X, and 100X. Record your calculations below.

Microscope #	
Value for each ocular unit at 4X	
Value for each ocular unit at 10X	
Value for each ocular unit at 40/45X	
Value for each ocular unit at 100X	

Using the stage micrometer, determine the smallest length (in microns) which can be resolved with each objective. This is the **measured** limit of resolution for each lens.

Compare this value to the theoretical limit of resolution calculated on the basis of the numerical aperture of the lens and a wavelength of 450 nm (blue light).

Using the calculated values for your ocular micrometer, determine the dimensions of the letter 'e' found on your microscope slide. Use a millimeter ruler to measure the letter 'e' directly and compare with the calculated values obtained through the microscope.

Notes

To measure an object seen in a microscope, an ocular micrometer serves as a scale or rule. This is simply a disc of glass upon which equally spaced divisions are etched. The rule may be divided into 50 subdivisions, or more rarely 100 subdivisions. To use the ocular micrometer, calibrate it against a fixed and known ruler, the stage micrometer. Stage micrometers also come in varying lengths, but most are 2 mm long and subdivided into 0.01 mm (10 micrometer) lengths. Each objective will need to be calibrated independently. To use, simply superimpose the ocular micrometer onto the stage micrometer and note the relationship of the length of the ocular to the stage micrometer. Note that at different magnifications, the stage micrometer changes, but the ocular micrometer is fixed in dimension. In reality, the stage micrometer is also fixed, and what is changing is the power of the magnification of the objective.

EXERCISE 4. MEASURING DEPTH

Materials

- Microscope
- Prepared slide with three coloured, crossed threads.

Procedure

1. Place a slide containing three coloured and crossed threads on the microscope stage.
2. Determine the width (diameter) of the threads using the procedures from Expt. No. 3.
3. Locate a spot where all three threads cross each other at the same point. Use the fine focus control to focus first on the lowermost thread, then the middle thread and finally, the uppermost thread. List the order of the threads from the top to the bottom, by indicating their colour.
4. Focus on the top of the uppermost thread. Note the scale markings on the fine focus knob and record the calibrated reading directly from the fine focus control. Carefully rotate the fine focus and stop when the microscope is just focused on the upper edge of the next thread. Record the reading from the focus control below. The difference between the two readings is the depth (thickness) of the upper thread.

Position	*Colour*	*Diameter*	*Depth*
Top			
Middle			
Bottom			

EXERCISE 5. MEASURING AREA

Materials

- Microscope
- Square ocular grid
- Prepared slide of blood smear.

Procedure

1. Obtain an ocular grid etched with a square, and insert it into an ocular of the microscope (or use a microscope previously setup by the instructor).
2. Calibrate the ocular grid in a manner similar to that outlined in Exercise 1.3, and determine the area of each marked grid section. Draw the grid in the space below and add all pertinent dimensions.
3. Place a prepared slide of a blood smear on the microscope and focus on the slide using the 40X objective. Count the number of cells within the four margins of the grid area.

   Count only cells which touch the top and left margins of the grid. Do not count any cell which touches the right or bottom margin of the grid. Record the number in the box on the next page.
4. Select four additional random fields of view with approximately the same density of cells and count the number of cells per grid. Record the numbers in the space provided.
5. Average the results and, based on the known dimensions of your grid, calculate the number of cells per mm^2 for the blood smear.

   Area of the grid (40X) ____________________

Cell Count	*Number of cells/mm²*
Average number of cells per mm^2 ____________	

EXERCISE 6. CELL COUNT BY HEMOCYTOMETER OR MEASURING VOLUME

Materials

- Microscope
- Hemocytometer and coverslip
- Suspension of yeast.

Procedure

1. Make a serial dilution series of the yeast suspension, from 1/10 to 1/1000.
2. Obtain a hemocytometer and place it on the desk before you. Place a clean coverslip over the center chamber.
3. Starting with the 1/10 dilution, use a pasteur pipette to transfer a small aliquot of the dilution to the hemocytometer. Place the tip of the pipette into the V shaped groove of the hemocytometer and allow the cell suspension to flow into the chamber of the hemocytometer by capillary action until the chamber is filled. Do not overfill the chamber.

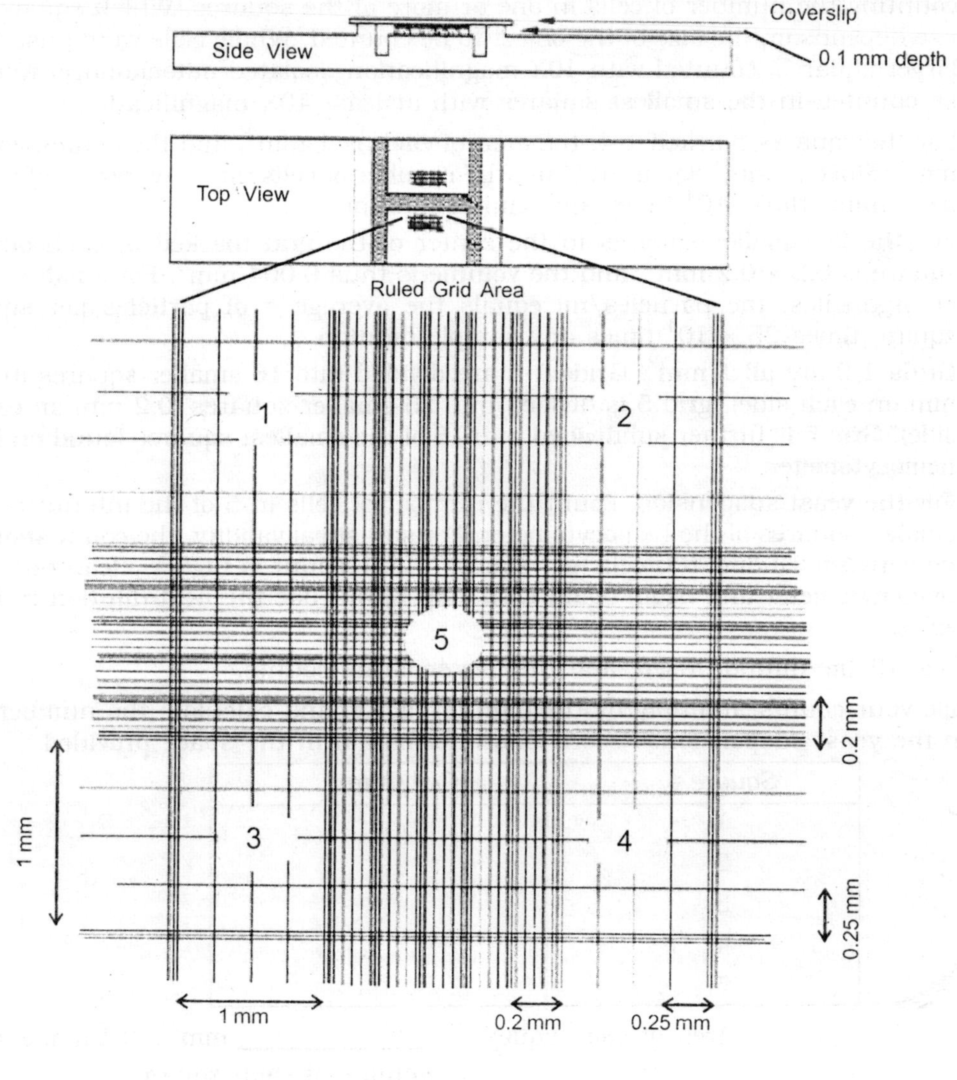

4. Add a similar sample of diluted yeast to the opposite side of the chamber and allow the cells to settle for about 1 minute before counting.
5. Refer to the diagram of the hemocytometer grid in figure and note the following.

6. The coverslip is 0.1 mm above the grid, and the lines etched on the grid are at preset dimensions.
7. The four outer squares, marked 1-4, each cover a volume of 10^{-4} ml.
8. The inner square, marked as 5, also covers a volume of 10^{-4} ml, but is further subdivided into 25 smaller squares. The volume over each of the 25 smaller squares is 4.0×10^{-6} ml.
9. Each of the 25 smaller squares is further divided into 16 squares, which are the smallest gradations on the hemocytometer. The volume over these smallest squares is $.25 \times 1^{-6}$ ml.
10. Given these volumes, the number of cells in a sample can be determined by counting the number of cells in one or more of the squares. Which square to use depends on the size of the object to be counted. Whole cells would use the larger squares, counted with 10X magnification. Isolated mitochondria would be counted in the smallest squares with at least 40X magnification.
11. For the squares marked 1-4, the area of each is 1 mm^2, and the volume is .1 mm^3. Since .1 mm^3 equals 10^{-4} ml, the number of cells/ml = Average # of cells per 1 mm_2 times 10^4 times any sample dilution.
12. For the 25 smaller squares in the center of the grid marked 5, each small square is 0.2×0.2 mm^2, and the volume is thus 0.004 mm^3. For small cells, or organelles, the particles/ml equals the average # of particles per small square times 25×10^4 times any sample dilution.
13. Grids 1-5 are all 1 mm^2. Grids 1-4 are divided into 16 smaller squares (0.25 mm on each side), grid 5 is divided into 25 smaller squares (0.2 mm on each side). Grid 5 is further subdivided into 16 of the smallest squares found on the hemocytometer.
14. For the yeast suspension, count the number of cells in 5 of the intermediate, smaller squares of the hemocytometer. For statistical validity, the count should be between 10 and 100 cells per square. If the count is higher, clean out the hemocytometer and begin again with step 3, but use the next dilution in the series.

    Record the dilution used, and the five separate counts.

Average your counts, multiply by the dilution factor, and calculate the number of cells/ml in the yeast suspension. Record this information in the space provided.

Square #	*Cell Count*
1	
2	
3	
4	
5	

Area of each square = ____________ $mm^2 \times 0.1$ mm depth

= volume of each square.

Volume of each square = ____________ mm^3

Average number of cells per mm^3 = ____________

Number of cells per cm^3 (1000 × above) = ____________

Note: Number of cells per cm^3 is also number per ml.

Number of cells per ml ________ × dilution factor (200) = ________ cells per ml of whole blood.

EXERCISE 7. MEASUREMENT OF CELL ORGANELLES

Materials

- Prepared Slides of the following:
- Paramecium
- Euglena
- Spirogyra
- Onion Root Tip
- Trachea
- Mitochondria.
- Dicot Leaf x.s.
- Microscope.

Procedure

1. Each of the slides presents a representative cell, with various organelles of particular interest. Observe each slide, and make a drawing of each cell and its representative organelle(s).
2. Measure the appropriate dimensions of all structures.

EXERCISE 8. USE OF DARKFIELD ILLUMINATION

Materials

- Microscope with dark field stop
- Suspension of *Amoeba proteus*
- Transfer pipette
- Slides, coverslips.

Procedure

1. Make a simple wet mount of the Amoeba by placing a drop of the culture on a slide and placing a cover slip on the slide.
2. Observe and measure the Amoeba using normal bright field optics.
3. Place a dark field stop in the filter holder below the microscope's substage condenser and and continue to monitor the movements of the amoeba.
4. Note the differences in observable structure possible with dark field microscopy when compared to bright field. Draw and label the amoeba viewed in both ways.

EXERCISE 9. THE PHASE CONTRAST MICROSCOPE

Materials

- Phase contrast microscope
- Telescopic ocular for centering phase rings
- Culture of *Amoeba proteus*
- Transfer pipettes, slides, coverslips
- Prepared, prestained slide of *Amoeba proteus*.

Procedure

1. Establish Koehler illumination on the microscope. If the instructor approves, center the phase annular ring and its corresponding phase plate.
2. Place a slide on the stage of the microscope, move the condenser to its highest position and focus at 10X magnification.
3. Open the condenser diaphragm to its maximum setting, and close the field diaphragm completely.
4. Using the condenser movement control, move the condenser until a sharp image of the field diaphragm is observed. To determine that the focus is indeed the field diaphragm, slightly open and close the field diaphragm to see if its movement can be detected in the field of view. When focused, there will be a slight blue haze on the edge of the diaphragm.
5. Open the field diaphragm until it nearly fills the field, but can be still seen. Center the field diaphragm in the field of view using the centering screws on the substage condenser. Open the field diaphragm to completely fill the field of view.
6. Remove one of the oculars from its tube and while peering down the tube, open the condenser diaphragm until it just fills the field of view at the bottom of the tube. Replace the ocular in the tube.
7. Completing steps a-d establishes Koehler illumination, where the field diaphragm is superimposed onto the object and centers the major optical components of the microscope.
8. To check on the phase annulus and its corresponding phase plate, remove an ocular and replace it with a telescopic ocular designed to focus on the rear lens of a phase objective. Match the phase objective with its corresponding setting on the phase condenser and visually verify that the phase annulus (a clear ring) is perfectly matched to the phase plate (a darker ring). If it is not, ask the instructor for assistance in centering the phase annulus. This is most often accomplished by adjusting a second set of centering screws attached to the phase condenser. Replace the normal ocular before using the phase contrast optics. Return the phase condenser setting to the normal bright field position.
9. Make a simple wet mount of the amoeba and observe under bright field microscopy at 10X.
10. Locate an active amoeba and center it in the field of view. Rotate the condenser phase ring to match the 10X phase with the 10X objective.
11. Observe the difference in the appearance of the amoeba between normal bright field and phase contrast.

12. Draw the amoeba viewed under phase contrast. Label organelles which are more clearly visible with phase contrast than with bright field microscopy.
13. Return the phase control on the condenser to the normal bright field setting, switch to a higher magnification (20X or 40X) and observe the amoeba at the higher magnifications with and without phase enhancement.
14. Compare the view of the amoeba under phase contrast, normal unstained bright field and darkfield with the view of the prestained commercial preparation of *Amoeba proteus*. List the organelles and/or structures which are more clearly demonstrated by each optical technique.
15. Drawing of amoeba observed with phase contrast optics.

EXERCISE 10. THE INVERTED PHASE MICROSCOPE

Materials

Inverted phase microscope. Monolayer cultured cells or a suspension of protozoa in a plastic disposable petri plate.

Procedure

1. Using as a guide, familiarize yourself with the components of the inverted microscope. Note that the operation of the microscope is the same as in expt. No. 1, the only exception being that the components are inverted. The phase condenser may be an annular ring or a phase slider.
2. Place either a culture flask containing a monolayer of cells or a plastic petri plate containing a suspension of protozoa on the stage of the microscope.
3. Observe the culture with normal bright field and with phase contrast. Note any movements of the cells and over a period of time, note any changes in the nuclear material of the monolayer cells.
4. Draw and label your observations in the space provided.

Drawing of cells observed with inverted phase microscope.

2. ASEPTIC TECHNIQUE AND TRANSFER OF MICROORGANISMS

INTRODUCTION

In natural environments, microorganisms usually exist as mixed populations. However, if we are to study, characterize, and identify microorganisms, we must have the organisms in the form of a pure culture. A pure culture is one in which all organisms are descendants of the same organism. Techniques for obtaining pure cultures from a mixed population will be described in exercise 3.

In working with microorganisms we must also have a sterile nutrient-containing-medium in which to grow the organisms. Anything in or on which we grow microorganisms is termed a medium. A sterile medium is one which is free of all life forms. It is usually sterilized by heating it to a temperature at which all contaminating microorganisms are destroyed.

Finally, in working with microorganisms, we must have a method of transferring growing organisms (called the **inoculum**) from a pure culture to a sterile medium without introducing any unwanted outside contaminants. This method of preventing unwanted microorganisms from gaining access is termed aseptic technique.

A. Aseptic Technique

The procedure for aseptically transferring microorganisms is as follows :

1. *Sterilize the inoculating loop.* The inoculating loop is sterilized by passing it at an angle through the flame of a gas burner until the entire length of the wire becomes orange from the heat. In this way all contaminants on the wire are incinerated. **Never lay the loop down once it is sterilized** or it may again become contaminated. Allow the loop to cool a few seconds to avoid killing the inoculum.

2. *Remove the inoculum.*

(*a*) *Removing inoculum from a broth culture* (organisms growing in a liquid medium):

1. Hold the culture tube in one hand and in your other hand, hold the sterilized inoculating loop as if it were a pencil.
2. Remove the cap of the pure culture tube with the little finger of your loop hand. *Never lay the cap down* or it may become contaminated.
3. Very briefly flame the lip of the culture tube. This creates a convection current which forces air out of the tube and preventing airborne contaminants from entering the tube. The heat of the gas burner also causes the air around your work area to rise, and this also reduces the chance of airborne microorganisms contaminating your cultures.
4. Keeping the culture tube at an angle, insert the inoculating loop and remove a loopful of inoculum.
5. Again flame the lip of the culture tube.
6. Replace the cap.

(*b*) *Removing inoculum from a plate culture* (organisms growing on an agar surface in a petri plate):

1. Sterilize the inoculating loop in the flame of a gas burner.
2. Lift the lid of the culture plate slightly and stab the loop into the agar away from any growth to cool the loop.
3. Scrape off a **small amount** of the organisms and close the lid.

3. *Transfer the Inoculum to the Sterile Medium.*

(*a*) *Transferring the inoculum into a broth tube:*

1. Pick up the sterile broth tube and remove the cap with the little finger of your loop hand. **Do not set the cap down.**
2. Briefly flame the lip of the broth tube.
3. Place the loopful of inoculum into the broth, and withdraw the loop. Do not lay the loop down.
4. Again flame the lip of the tube.
5. Replace the cap.

6. Resterilize the loop by placing it in the flame until it is orange. Now you may lay the loop down until it is needed again.

(*b*) *Transferring the inoculum into a petri plate:*

1. Lift the edge of the lid just enough to insert the loop.
2. Streak the loop across the surface of the agar medium. These streaking patterns allow you to obtain single isolated bacterial colonies originating from a single bacterium or arrangement of bacteria.
3. In order to avoid digging into the agar as you streak the loop over the top of the agar you must keep the loop parallel to the agar surface. Always start streaking at the "12:00 position" of the plate and streak side-to-side as you pull the loop toward you. Each time you flame and cool the loop between sectors, rotate the plate counterclockwise so you are always working in the "12:00 position" of the plate. This keeps the inoculating loop parallel with the agar surface and helps prevent the loop from digging into the agar.
4. Remove the loop and close the lid.
5. Resterilize the inoculating.

In the future, every procedure in the lab will be done using similar aseptic technique.

B. Forms of Culture Media

1. Broth tubes are tubes containing a liquid medium. A typical nutrient containing broth medium such as Trypticase Soy broth contains substrates for microbial growth such as pancreatic digest of casein, papaic digest of soybean meal, sodium chloride, and water. After incubation, growth (development of many cells from a few cells) may be observed as one or a combination of three forms:

   (*a*) *Pellicle:* A mass of organisms is floating on top of the broth.

Turbidity: The organisms appear as a general cloudiness throughout the broth.

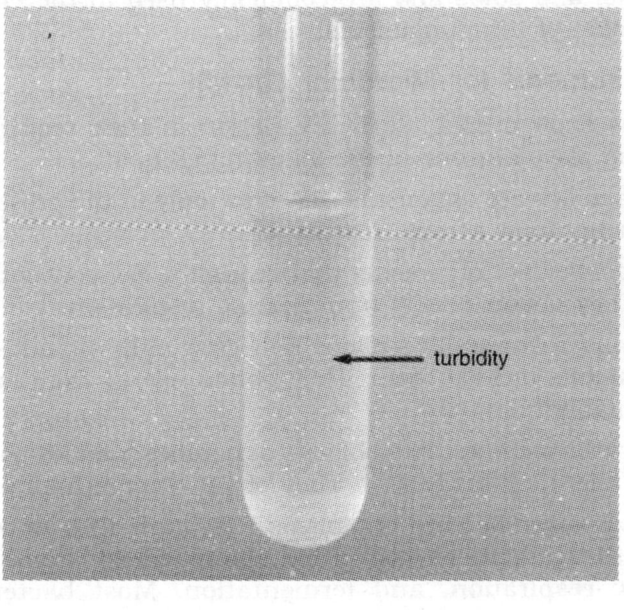

Broth culture showing turbidity

(b) *Sediment:* A mass of organisms appears as a deposit at the bottom of the tube Broth.

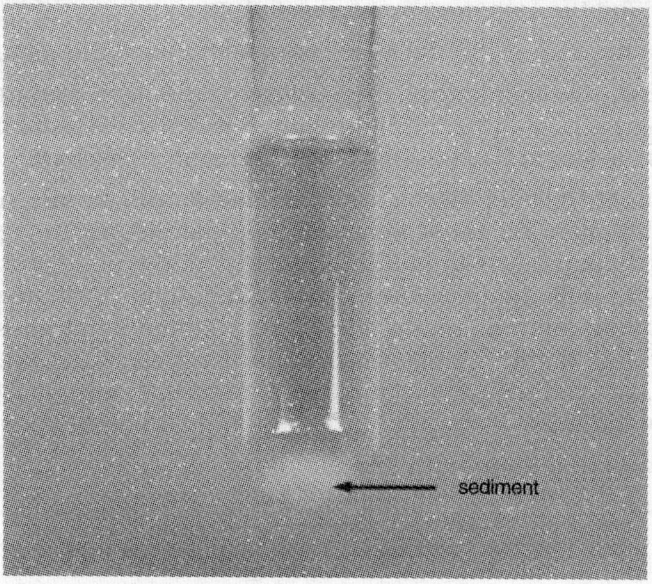

Culture showing sediment

2. Slant tubes are tubes containing a nutrient medium plus a solidifying agent, agar-agar. The medium has been allowed to solidify at an angle in order to get a flat inoculating surface.
3. Stab tubes (deeps) are tubes of hardened agar medium which are inoculated by "stabbing" the inoculum into the agar.
4. Agar plates are sterile petri plates that are aseptically filled with a melted sterile agar medium and allowed to solidify. Plates are much less confining than slants and stabs and are commonly used in the culturing, separating, and counting of microorganisms.

C. Oxygen Requirements for Microbial Growth

Microorganisms show a great deal of variation in their requirements for gaseous oxygen. Most can be placed in one of the following groups:

1. *Obligate aerobes* are organisms that grow **only** in the presence of oxygen. They obtain energy from aerobic respiration.
2. *Microaerophiles* are organisms that require a low concentration of oxygen for growth. They obtain energy from aerobic respiration.
3. *Obligate anaerobes* are organisms that grow **only** without oxygen and, in fact, oxygen inhibits or kills them. They obtain energy from anaerobic respiration or fermentation.
4. *Aerotolerant anaerobes,* like obligate anaerobes, cannot use oxygen for growth but they tolerate it fairly well. They obtain energy from fermentation.
5. *Facultative anaerobes* are organisms that grow with or without oxygen, but generally better with oxygen. They obtain energy from aerobic respiration, anaerobic respiration, and fermentation. Most bacteria are facultative anaerobes.

D. Temperature Requirements

Microorganisms are divided into groups on the basis of their preferred range of temperature:

1. *Psychrophiles* are cold-loving bacteria. Their optimum growth temperature is between –5°C and 15°C. They are usually found in the Arctic and Antarctic regions and in streams fed by glaciers.
2. *Mesophiles* are bacteria that grow best at moderate temperatures. Their optimum growth temperature is between 25°C and 45°C. Most bacteria are mesophilic and include common soil bacteria and bacteria that live in and on the body.
3. *Thermophiles* are heat-loving bacteria. Their optimum growth temperature is between 45°C and 70°C and are commonly found in hot springs and in compost heaps.
4. *Hyperthermophiles* are bacteria that grow at very high temperatures. Their optimum growth temperature is between 70°C and 110°C. They are usually members of the *Archae* and are found growing near hydrothermal vents at great depths in the ocean.

E. Colony Morphology and Pigmentation

A colony is a visible mass of microorganisms growing on an agar surface and usually originating from a single organism or arrangement of organisms. Different microorganisms will frequently produce colonies which differ in their morphological appearance (form, elevation, margin, surface, optical characteristics, and pigmentation). Probably the most visual characteristic is **pigmentation** (colour). Some microorganisms produce pigment during growth and are said to be **chromogenic**. Often, however, formation of pigment depends on environmental factors such as temperature, nutrients, pH and moisture. For example, *Serratia marcescens* produces a deep red pigment at 25°C, but does not produce pigment at 37°C.

Pigments can be divided into two basic types: water insoluble and water soluble. If the pigment is **water insoluble,** as in the case of most chromogenic bacteria, it does not diffuse out of the organism. As a result, the colonies are pigmented but the agar remains the normal colour. If the pigment is **water soluble** as in the case of *Pseudomonas aeruginosa*, it will diffuse out of the organism into the surrounding medium. Both the colonies and the agar will appear pigmented.

Below is a list of several common chromogenic bacteria:

- *Staphylococcus aureus* - gold; water insoluble
- *Micrococcus luteus* - yellow; water insoluble
- *Micrococcus roseus* - pink; water insoluble
- *Mycobacterium phlei* - orange; water insoluble
- *Serratia marcescens* - orange/red; water insoluble.

Media. Trypticase Soy Broth tubes (4), Trypticase Soy Agar slant tubes (4), Trypticase Soy Agar stab tubes (4), and Trypticase Soy Agar plates (7).

Organisms. Trypticase Soy Broth cultures of *Bacillus subtilis, Escherichia coli* and *Micrococcus luteus*, and Trypticase Soy Agar plate cultures of *Mycobacterium phlei.*

Procedure

1. Aseptically inoculate one Trypticase Soy Broth tube, one Trypticase Soy Agar slant tube, one Trypticase Soy Agar stab tube, and one Trypticase Soy Agar plate with *B. subtilis*.

   Remember to label all tubes with a wax marker. When streaking the agar plates. This procedure is termed **streaking for isolation** and has a diluting effect. The friction of the loop against the agar causes organisms to fall off the loop. Near the end of the streaking pattern, individual organisms become separated far enough apart on the agar surface to give rise to **isolated single colonies** after incubation.
2. Aseptically inoculate one Trypticase Soy Broth tube, one Trypticase Soy Agar slant tube, one Trypticase Soy Agar stab tube, and one Trypticase Soy Agar plate with *E. coli*.
3. Aseptically inoculate one Trypticase Soy Broth tube, one Trypticase Soy Agar slant tube, one Trypticase Soy Agar stab tube, and one Trypticase Soy Agar plate with *M. luteus*.
4. Aseptically inoculate one Trypticase Soy Broth tube, one Trypticase Soy Agar slant tube, one Trypticase Soy Agar stab tube, and one Trypticase Soy Agar plate with *M. phlei*.
5. Incubate all the tubes and plates inoculated with *B. subtilis, E. coli, M. luteus,* and *M. phlei* at 37°C. Place the tubes in a plastic beaker to keep them upright. Incubate the plates upside down (lid on the bottom) to prevent condensing water from falling down on the growing colonies and causing them to run together.
6. In order to illustrate that microorganisms are all around us and to demonstrate the necessity for proper aseptic technique, **contaminate** three Trypticase Soy Agar plates as follows:

   (*a*) Remove the lid from the first agar plate and place the exposed agar portion in or out of the building for the duration of today's lab. Replace the lid, label the plate "air", and incubate it **at room temperature. Do this plate first.**

   (*b*) Using a wax marker, divide a second petri plate in half. You and your partner both moisten a sterile cotton swab in sterile water. Rub your swab over some surface in the building or on yourself. Use this swab to inoculate your half of the second agar plate. Label the plate and incubate **at room temperature.**

   (*c*) With a wax marker, divide a third petri plate in half. Rub your fingers over the surface of your half of the third agar plate. Label and incubate **at 37°C. Do this plate last.**

Results

1. Draw and describe the growth seen in each of the four broth cultures.

Bacillus subtilis growth =	*Escherichia coli growth* =
Micrococcus luteus growth =	*Mycobacterium phlei growth* =

2. Observe the growth in the slant cultures and stab cultures for pigmentation and purity.
3. Observe the results of the three "contamination" plates and note the differences in colony appearances.
4. Observe the demonstration plates of chromogenic bacteria and state the color and water solubility of each pigment.

PERFORMANCE OBJECTIVES

Discussion

1. Define the following terms: pure culture, sterile medium, inoculum, aseptic technique, and colony.
2. State and define the three types of growth that may be seen in a broth culture.
3. Define the following terms: obligate aerobe, microaerophile, obligate anaerobe, aerotolerant anaerobe, and facultative anaerobe.
4. Define the following terms: psychrophile, mesophile, thermophile, and hyperthermophile.
5. Define the following terms: chromogenic, water soluble pigment, and water insoluble pigment.

Procedure

1. Using an inoculating loop, demonstrate how to aseptically remove some inoculum from either a broth tube, slant tube, stab tube, or petri plate, and inoculate a sterile broth tube, slant tube, stab tube, or petri plate without introducing outside contamination.
2. Label all tubes and plates and place them on the proper shelf in the incubator.
3. Dispose of all materials when the experiment is completed, being sure to remove all markings from the glassware. Place all tubes and plates in the designated areas.

Results

1. Recognize and identify the following types of growth in a broth culture: pellicle, turbidity, sediment, and any combination of these.
2. State the colour and water solubility of pigment seen on a plate culture of a chromogenic bacterium.

3. CONTROL OF MICROORGANISMS BY USING PHYSICAL AGENTS

A. INTRODUCTION TO THE CONTROL OF MICROORGANISMS

The next three labs deal with the inhibition, destruction, and removal of microorganisms. Control of microorganisms is essential in order to prevent the transmission of diseases and infection, stop decomposition and spoilage, and prevent unwanted microbial contamination.

Microorganisms are controlled by means of physical agents and chemical agents. Physical agents include such methods of control as high or low temperature, desiccation, osmotic pressure, radiation, and filtration. Control by chemical agents refers to the use of disinfectants, antiseptics, antibiotics, and chemotherapeutic antimicrobial chemicals.

Basic terms used in discussing the control of microorganisms include:

1. *Sterilization:* The process of destroying all forms of life. A sterile object is one free of all life forms, including endospores.
2. *Disinfection:* The reduction or elimination of pathogenic microorganisms in or on materials, so they are no longer a health hazard.
3. *Disinfectant:* Chemical agents used to disinfect inanimate objects but generally to toxic to use on human tissues.
4. *Antiseptic:* Chemical agents that disinfect, but are not harmful to human tissues.
5. *Antibiotic:* A metabolic product produced by one microorganism that inhibits or kills other microorganisms.
6. *Chemotherapeutic antimicrobial chemical:* Synthetic chemicals that can be used therapeutically.
7. *Cidal:* Kills microorganisms.
8. *Static:* Inhibits the growth of microorganisms.

These three labs will demonstrate the control of microorganisms with physical agents, disinfectants and antiseptics, and antimicrobial chemotherapeutic agents. Keep in mind that when evaluating or choosing a method of controlling microorganisms, you must consider the following factors which may influence antimicrobial activity:

1. the concentration and kind of a chemical agent used;
2. the intensity and nature of a physical agent used;
3. the length of exposure to the agent;
4. the temperature at which the agent is used;
5. the number of microorganisms present;
6. the organism itself; and
7. the nature of the material bearing the microorganism.

B. TEMPERATURE

Microorganisms have a minimum, an optimum, and a maximum temperature for growth. Temperatures *below the minimum* usually have a *static* action on microorganisms. They inhibit microbial growth by slowing down metabolism but do not necessarily kill the organism. Temperatures *above the maximum* usually have a *cidal* action, since they denature microbial enzymes and other proteins. Temperature is a very common and effective way of controlling microorganisms.

1. High Temperature

Vegetative microorganisms can generally be killed at temperatures from 50°C to 70°C with moist heat. Bacterial *endospores*, however, are very resistant to heat and extended exposure to much higher temperature is necessary for their destruction. High temperature may be applied as either moist heat or dry heat.

(a) Moist heat. Moist heat is generally more effective than dry heat for killing microorganisms because of its ability to *penetrate* microbial cells. Moist heat kills microorganisms by *denaturing their proteins* (causes proteins and enzymes to lose their three-dimensional functional shape). It also may *melt lipids* in cytoplasmic membranes.

1. *Autoclaving.* Autoclaving employs steam under pressure. Water normally boils at 100°C; however, when put under pressure, water boils at a higher temperature. During autoclaving, the materials to be sterilized are placed under 15 pounds per square inch of pressure in a pressure-cooker type of apparatus. Under 15 pounds of pressure, the boiling point of water is raised to 121°C, a temperature sufficient to kill bacterial endospores.

   The time the material is left in the autoclave varies with the nature and amount of material being sterilized. Given sufficient time (generally 15-45 minutes), autoclaving is *cidal* for both vegetative organisms and endospores, and is the most common method of sterilization for materials not damaged by heat.

2. *Boiling water.* Boiling water (100°C) will generally kill vegetative cells after about 10 minutes of exposure. However, certain viruses, such as the hepatitis viruses, may survive exposure to boiling water for up to 30 minutes, and endospores of certain *Clostridium* and *Bacillus* species may survive even hours of boiling.

(b) Dry heat. Dry heat kills microorganisms through a process of protein oxidation rather than protein coagulation. Examples of dry heat include:

1. *Hot air sterilization.* Microbiological ovens employ very high dry temperatures: 171°C for 1 hour; 160°C for 2 hours or longer; or 121°C for 16 hours or longer depending on the volume. They are generally used only for sterilizing glassware, metal instruments, and other inert materials like oils and powders that are not damaged by excessive temperature.

2. *Incineration.* Incinerators are used to destroy disposable or expendable materials by burning. We also sterilize our inoculating loops by incineration.

(c) Pasteurization. Pasteurization is the mild heating of milk and other materials to kill particular spoilage organisms or pathogens. It does not, however, kill all organisms. Milk is usually pasteurized by heating to 71.6°C for at least 15 seconds in the flash method or 62.9°C for 30 minutes in the holding method.

2. Low Temperature

Low temperature inhibits microbial growth by slowing down microbial metabolism. Examples include refrigeration and freezing. Refrigeration at 5°C slows the growth of microorganisms and keeps food fresh for a few days. Freezing at –10°C stops microbial growth, but generally does not kill microorganisms, and keeps food fresh for several months.

C. DESICCATION

Desiccation, or drying, generally has a **static** effect on microorganisms. Lack of water inhibits the action of microbial enzymes. Dehydrated and freeze-dried foods, for example, do not require refrigeration because the absence of water inhibits microbial growth.

D. OSMOTIC PRESSURE

Microorganisms, in their natural environments, are constantly faced with alterations in osmotic pressure. Water tends to flow through semipermeable membranes, such as the cytoplasmic membrane of microorganisms, towards the side with a higher concentration of dissolved materials **(solute).** In other words, water moves from greater water (lower solute) concentration to lesser water (greater solute) concentration.

When the concentration of dissolved materials or solute is higher inside the cell than it is outside, the cell is said to be in a hypotonic environment and water will flow into the cell. The rigid cell walls of bacteria and fungi, however, prevent bursting or plasmoptysis. If the concentration of solute is the same both inside and outside the cell, the cell is said to be in an isotonic environment. Water flows equally in and out of the cell. Hypotonic and isotonic environments are not usually harmful to microorganisms. However, if the concentration of dissolved materials or solute is higher outside of the cell than inside, then the cell is in a hypertonic environment. Under this condition, water flows out of the cell, resulting in shrinkage of the cytoplasmic membrane or plasmolysis. Under such conditions, the cell becomes dehydrated and its growth is inhibited.

The canning of jams or preserves with a high sugar concentration inhibits bacterial growth through hypertonicity. The same effect is obtained by salt-curing meats or placing foods in a salt brine. This static action of osmotic pressure thus prevents bacterial decomposition of the food. Molds, on the other hand, are more tolerant of hypertonicity. Foods, such as those mentioned above, tend to become overgrown with molds unless they are first sealed to exclude oxygen. (Molds are aerobic.)

E. RADIATION

1. Ultraviolet Radiation

The ultraviolet portion of the light spectrum includes all radiations with wavelengths from 100 nm to 400 nm. It has low wave length and low energy. The microbicidal activity of ultraviolet (UV) light depends on the length of exposure: the longer the exposure the greater the cidal activity. It also depends on the wavelength of UV used. The most cidal wavelengths of UV light lie in the 260 nm-270 nm range where it is absorbed by nucleic acid.

In terms of its mode of action, UV light is absorbed by microbial DNA and causes adjacent thymine bases on the same DNA strand to covalently bond together, forming what are called thymine-thymine dimers. As the DNA replicates, nucleotides do not complementary base pair with the thymine dimers and this terminates the replication of that DNA strand. However, most of the damage from UV radiation actually comes from the cell trying to repair the damage to the DNA by a process called SOS repair. In very heavily damaged DNA containing large numbers of thymine dimers, a process called SOS repair is activated as kind of a last ditch effort to repair the DNA. In this process, a gene product of the SOS system binds to DNA polymerase allowing it to synthesize new DNA across the damaged DNA. However, this altered DNA polymerase loses its proofreading ability resulting in the synthesis of DNA that itself now contains many misincorporated bases. In other words, UV radiation causes mutation and can lead to faulty protein synthesis. With sufficient mutation, bacterial metabolism is blocked and the organism dies. Agents such as UV radiation that cause high rates of mutation are called mutagens.

The effect of this inproper base pairing may be reversed to some extent by exposing the bacteria to strong visible light immediately after exposure to the UV light. The visible light activates an enzyme that breaks the bond that joins the thymine bases, thus enabling correct complementary base pairing to again take place. This process is called **photoreactivation.**

UV lights are frequently used to reduce the microbial populations in hospital operating rooms and sinks, aseptic filling rooms of pharmaceutical companies, in microbiological hoods, and in the processing equipment used by the food and dairy industries.

An important consideration when using UV light is that it has very poor penetrating power. Only microorganisms on the surface of a material that are exposed directly to the radiation are susceptible to destruction. UV light can also damage the eyes, cause burns, and cause mutation in cells of the skin.

2. Ionizing Radiation

Ionizing radiation, such as X-rays and gamma rays, has much more energy and penetrating power than ultraviolet radiation. It ionizes water and other molecules to form radicals (molecular fragments with unpaired electrons) that can break DNA strands. It is often used to sterilize pharmaceuticals and disposable medical supplies such as syringes, surgical gloves, catheters, sutures, and petri plates. It can also be used to retard spoilage in seafoods, meats, poultry, and fruits.

F. FILTRATION

Microbiological membrane filters provide a useful way of sterilizing materials such as vaccines, antibiotic solutions, animal sera, enzyme solutions, vitamin solutions, and other solutions that may be damaged or denatured by high temperatures or chemical agents. The filters contain pores small enough to prevent the passage of microbes but large enough to allow the organism-free fluid to pass through. The liquid is then collected in a sterile flask. Filters with a pore diameter from 25 nm to 0.45 μm are usually used in this procedure. Filters can also be used to remove microorganisms from water and air for microbiological testing.

PROCEDURE

A. Osmotic Pressure Procedure

Media. 2 plates of Trypticase Soy agar, 2 plates of 5% glucose agar, 2 plates of 10% glucose agar, 2 plates of 25% glucose agar, 2 plates of 5% NaCl agar, 2 plates of 10% NaCl agar, and 2 plates of 15% NaCl agar.

Organisms. Trypticase Soy broth cultures of *Escherichia coli* and *Staphylococcus aureus*; a spore suspension of the mold *Aspergillus niger.*

1. Divide one plate of each of the following media in half. Using your inoculating loop, streak one half of each plate with *E. coli* and the other half with *S. aureus.* Incubate at 37°C until the next lab period.

(*a*) Trypticase Soy agar (control)
(*b*) Trypticase Soy agar with 5% glucose
(*c*) Trypticase Soy agar with 10% glucose
(*d*) Trypticase Soy agar with 25% glucose
(*e*) Trypticase Soy agar with 5% NaCl
(*f*) Trypticase Soy agar with 10% NaCl
(*g*) Trypticase Soy agar with 15% NaCl.

2. Using a sterile swab, streak one plate of each of the following media with a spore suspension of the mold *A. niger*. Incubate at room temperature for 1 week.

(*a*) Trypticase Soy agar (control)
(*b*) Trypticase Soy agar with 5% glucose
(*c*) Trypticase Soy agar with 10% glucose
(*d*) Trypticase Soy agar with 25% glucose
(*e*) Trypticase Soy agar with 5% NaCl
(*f*) Trypticase Soy agar with 10% NaCl
(*g*) Trypticase Soy agar with 15% NaCl.

B. Ultraviolet Radiation Procedure

Media. 5 plates of Trypticase Soy agar.

Organisms. Trypticase Soy broth culture of *Serratia marcescens*

1. Using *sterile swabs*, streak all 5 Trypticase Soy agar plates with *S. marcescens* as follows:
   (*a*) Dip the swab into the culture.
   (*b*) *Remove all of the excess liquid by pressing the swab against the side of the tube.*
   (*c*) Streak the plate so as to cover the entire agar surface with organisms.
2. Expose 3 of the plates to UV light as follows:
   (a) Remove the lid of each plate and place a piece of cardboard with the letter "V" cut out of it over the top of the agar.
   (b) Expose the first plate to UV light for 3 seconds, the second plate for 10 seconds, and the third plate for 60 seconds.
   (c) Replace the lids and incubate at room temperature until the next lab period.
3. Leaving the lid on, lay the cardboard with the letter "V" cut out over the fourth plate and expose to UV light for 60 seconds. Incubate at room temperature with the other plates.
4. Use the fifth plate as a nonirradiated control and incubate at room temperature with the other plates.

Note. Do not look directly at the UV light as it may harm the eyes.

C. Filtration Procedure

Medium. 2 plates of Trypticase Soy agar.

Organism. Trypticase Soy broth cultures of *Micrococcus luteus.*

1. Using alcohol-flamed forceps, aseptically place a sterile membrane filter into a sterile filtration device.
2. Pour the culture of *M. luteus* into the top of the filter set-up.
3. Vacuum until all the liquid passes through the filter into the sterile flask.
4. With alcohol-flamed forceps, remove the filter and place it organism-side-up on the surface of a Trypticase Soy agar plate.
5. Using a sterile swab, streak the surface of another Trypticase Soy agar plate with the filtrate from the flask.
6. Incubate the plates at 37°C until the next lab period.

RESULTS

A. Osmotic Pressure

Observe the 2 sets of plates from the osmotic pressure experiment and record the results below:

Plate	*Escherichia coli*	*Staphylococcus aureus*	*Aspergillus niger*
Control			
5% NaCl			
10% NaCl			
15% NaCl			
5% glucose			
10% glucose			
25% glucose			

+ = scant growth
++ = moderate growth
+++ = abundant growth
– = no growth.

Escherichia coli and Staphlococcus aureus	*Aspergillus niger*
control	control
5% NaCl	5% NaCl
10% NaCl	10% NaCl
15% NaCl	15% NaCl
5% glucose	5% glucose
10% glucose	10% glucose
25% glucose	25% glucose

B. Ultraviolet Radiation

1. Make drawings of the 5 plates from the ultraviolet light experiment.
2. Observe the plates exposed to UV light for any nonpigmented colonies. Aseptically pick-off one of these nonpigmented colonies and streak it in a plate of Trypticase Soy agar. Incubate at room temperature until the next lab period.

3. After incubation, observe the plate you streaked with the nonpigmented colony. Does the organism still lack chromogenicity?

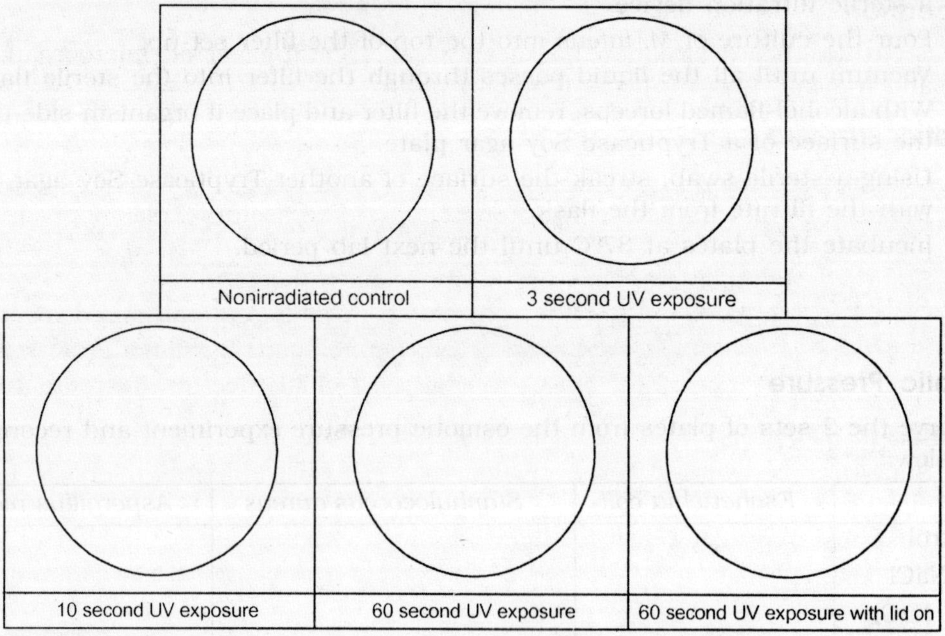

C. Filtration

Observe the 2 filtration plates and describe the results below:

plate containing the filter (growth or no growth)	
plate sttealed eith filinte (growth or no growth)	

PERFORMANCE OBJECTIVES

A. Introduction to the Control of Microorganism

1. Define the following terms: sterilization, disinfection, static, and cidal.

B. Temperature

1. State whether moist or dry heat is more effective in controlling microorganisms, and indicate why?
2. State specifically how moist heat kills microorganisms.
3. State two methods of applying moist heat.
4. Briefly describe the process of autoclaving (pressure, time, and temperature).
5. State whether or not boiling is an effective means of sterilization, and indicate why?
6. State specifically how dry heat kills microorganisms.
7. State two methods of applying dry heat.
8. Define pasteurization.

9. State whether low temperature has a static or cidal effect on microorganisms, and indicate why?

C. Desiccation

1. State whether desiccation has a static or a cidal effect on microorganisms, and indicate how it affects the cell.

D. Osmotic Pressure

1. Describe osmosis in terms of water flow through a semipermeable membrane.
2. Define the following terms: hypotonic, hypertonic, isotonic, plasmoptysis, and plasmolysis.
3. State why hypotonicity does not normally harm bacteria.
4. Describe how bacterial growth is inhibited in jams and salt-cured meats.
5. State why jams still must be sealed even though bacteria will not grow in them.
6. State whether hypertonicity has a static or a cidal effect on microorganisms.

E. Radiation

1. State how the wavelength and the length of exposure influence the bacteriocidal effect of UV light.
2. Describe specifically how UV light kills microorganisms.
3. State why UV light is only useful as a means of controlling surface contaminants and give several practical applications.
4. Describe how ionizing radiation kills microorganisms and state several common applications.

F. Filtration

1. State the concept behind sterilizing solutions with micropore membrane filters.
2. State why filters are preferred over autoclaving for such materials as vaccines, antibiotic solutions, sera, and enzyme solutions.

4. CONTROL OF MICROORGANISMS BY USING DISINFECTANTS AND ANTISEPTICS

INTRODUCTION

A. Disinfectants and Antiseptics

Disinfection is the reduction or elimination of **pathogenic** microorganisms in or on materials so that they are less of a health hazard. The term **disinfectant** is generally used for chemical agents employed to disinfect inanimate objects, whereas the term **antiseptic** is used to indicate a nontoxic disinfectant suitable for use on animal tissue. Because disinfectants and antiseptics often work slowly on some viruses (such as the hepatitis viruses), *Mycobacterium tuberculosis*, and especially bacterial **endospores**, they are usually **unreliable for sterilization** (the destruction of **all** life forms).

There are a number of factors which influence the antimicrobial action of disinfectants and antiseptics, including:

1. The concentration of the chemical agent.
2. The temperature at which the agent is being used. Generally, the lower the temperature, the lower the effectiveness.

3. The kinds of microorganisms present (endospore producers, *Mycobacterium tuberculosis*, etc.).
4. The number of microorganisms present. The more organisms present, the harder it is to disinfect.
5. The nature of the material bearing the microorganisms. Organic material such as dirt and excreta interferes with some agents.

The best results are generally obtained when the initial microbial numbers are low and when the surface to be disinfected is clean and free of possible interfering substances.

There are two common antimicrobial modes of action for disinfectants and antiseptics:

1. They may damage the lipids and/or proteins of the semipermeable cytoplasmic membrane of microorganisms resulting in leakage of cellular materials needed to sustain life.
2. They may denature microbial enzymes and other proteins, usually by disrupting the hydrogen and disulfide bonds that give the protein its three-dimensional functional shape. This blocks metabolism.

A large number of such chemical agents are in common use. Some of the more common groups are listed below:

1. Phenol and phenol derivatives

Phenol (5-10%) was the first disinfectant commonly used. However, because of its toxicity and odor, phenol derivatives are now generally used. These include orthophenylphenol, hexachlorophene, triclosan, hexylresorcinol, and chlorhexidine. Orthophenylphenol is the agent in Lysol®, O-syl®, Staphene®, and Amphyl®. Hexachlorophene in a 3% solution is combined with detergent and is found in PhisoHex®. Triclosan is a phenolic antiseptic very common in antimicrobial soaps and other products. Hexylresorcinol is in throat lozenges and ST-37. A 4% solution of chlorhexidine in isopropyl alcohol and combined with detergent (Hibiclens®) is a common handwashing agent and surgical handscrub. These agents kill most bacteria, most fungi, and some viruses, but are usually ineffective against endospores. They alter membrane permeability and denature proteins.

2. Soaps and detergents

Soaps are only mildly microbicidal. Their use aids in the mechanical removal of microorganisms by breaking up the oily film on the skin (emulsification) and reducing the surface tension of water so it spreads and penetrates more readily. Many cosmetic soaps also contain added disinfectants to increase antimicrobial activity.

Detergents may be anionic or cationic. Anionic (negatively charged) detergents, such as laundry powders, mechanically remove microorganisms and other materials but are not very microbicidal. Cationic (positively charged) detergents alter membrane permeability and denature proteins. They are effective against many vegetative bacteria, some fungi, and some viruses. However, endospores, *Mycobacterium tuberculosis*, and *Pseudomonas* species are usually resistant. They are also inactivated by soaps and organic materials like excreta. Cationic detergents include the quaternary ammonium compounds (zephiran, diaprene, roccal, ceepryn, and phemerol).

3. Alcohols

70% solutions of ethyl or isopropyl alcohol are effective in killing vegetative bacteria, enveloped viruses, and fungi. However, they are usually ineffective against endospores and non-enveloped viruses. Once they evaporate, their cidal activity will cease. Alcohols denature membranes and are often combined with other disinfectants, such as iodine, mercurials, and cationic detergents for increased effectiveness.

4. Acids and alkalies

Acids and alkalies alter membrane permeability and denature proteins and other molecules. Salts of organic acids, such as calcium propionate, potassium sorbate, and methylparaben, are commonly used as food preservatives. Undecylenic acid (Desenex®) is used for dermatophyte infections of the skin. An example of an alkali is lye (sodium hydroxide).

5. Heavy metals

Heavy metals, such as mercury, silver, and copper, denature proteins. Mercury compounds (mercurochrome, metaphen, merthiolate) are only bacteriostatic and are not effective against endospores. Silver nitrate (1%) is sometimes put in the eyes of newborns to prevent gonococcal ophthalmia. Copper sulfate is used to combat fungal diseases of plants and is also a common algicide. Selinium sulfide kills fungi and their spores.

6. Chlorine

Chlorine gas reacts with water to form hypochlorite ions, which in turn denature microbial enzymes. Chlorine is used in the chlorination of drinking water, swimming pools, and sewage. Sodium hypochlorite is the active agent in household bleach. Calcium hypochlorite, sodium hypochlorite, and chloramines (chlorine plus ammonia) are used to sanitize glassware, eating utensils, dairy and food processing equipment, and hemodialysis systems.

7. Iodine and iodophores

Iodine also denatures microbial proteins and is usually dissolved in an alcohol solution to produce a tincture. Iodophores are a combination of iodine and an anionic detergent (such as polyvinylpyrrolidone) which reduces surface tension and slowly releases the iodine. Iodophores are less irritating than iodine and do not stain. They are generally effective against vegetative bacteria, *Mycobacterium tuberculosis*, fungi, some viruses, and some endospores. Examples include Wescodyne®, Ioprep®, Ioclide®, Betadine®, and Isodine®.

8. Aldehydes

Aldehydes, such as formaldehyde and glutaraldehyde, denature microbial proteins. Formalin (37% aqueous solution of formaldehyde gas) is extremely active and kills most forms of microbial life. It is used in embalming, preserving biological specimens, and in preparing vaccines. Alkaline glutaraldehyde (Cidex®), acid glutaraldehyde (Sonacide®), and glutaraldehyde phenate solutions (Sporocidin®) kill vegetative bacteria in 10-30 minutes and endospores in about 4 hours. A 10 hour exposure to a 2% glutaraldehyde solution can be used for cold sterilization of materials.

9. Ethylene oxide gas

Ethylene oxide is one of the very few chemicals that can be relied upon for sterilization (after 4-12 hours exposure). Since it is explosive, it is usually mixed with inert gases such as freon or carbon dioxide. Gaseous chemosterilizers, using ethylene oxide, are commonly used to sterilize heat-sensitive items such as plastic syringes, petri plates, textiles, sutures, artificial heart valves, heart-lung machines, and mattresses. Ethylene oxide has very high penetrating power and denatures microbial proteins. Vapors are toxic to the skin, eyes, and mucous membranes and are also carcinogenic.

B. Evaluation of Disinfectants and Antiseptics

It is possible to evaluate disinfectants and antiseptics using either in vitro or in vivo tests. An in vitro test is one done under artificial, controlled laboratory conditions. An in vivo test is one done under the actual conditions of normal use.

A common in vitro test is to compare the antimicrobial activity of the disinfectant being tested with that of phenol. The resulting value is called a phenol coefficient and has some value in comparing the strength of disinfectants under standard conditions. Phenol coefficients may be misleading, however, because as mentioned earlier, the killing rate varies greatly with the conditions under which the chemical agents are used. The concentration of the agent, the temperature at which it is being used, the length of exposure to the agent, the number and kinds of microorganisms present, and the nature of the material bearing the microorganisms all influence the antimicrobial activity of a disinfectant. If a disinfectant is being evaluated for possible use in a given in vivo situation, it must be evaluated under the same conditions in which it will actually be used.

Today we will do a test to see how thermometers might carry microorganisms if not properly disinfected or cleaned.

C. Effectiveness of Hand Washing

There are 2 categories of microorganisms, or flora, normally found on the hands. Resident flora are the normal flora of the skin. Transient flora are the microorganisms you pick up from what you have been handling. It is routine practice to wash the hands prior to and after examining a patient and to do a complete regimented surgical scrub prior to going into the operating room. This is done in order to remove the potentially harmful transient flora, reduce the number of resident flora, and disinfect the skin.

Actual sterilization of the hands is not possible since microorganisms live not only on the surface of the skin but also in deeper skin layers, in ducts of sweat glands, and around hair follicles. These normal flora are mainly nonpathogenic staphylococci (Lab 15) and diphtheroid bacilli.

Today we will qualitatively evaluate the effectiveness of the length of washing time on the removal of microorganisms from the hands.

PROCEDURES

A. Evaluations of Disinfectants and Antiseptics

Materials

- 8 sterile glass rods ("thermometers")
  12 plates of Trypticase Soy agar (TSA); 3 each
  16 tubes of sterile water; 4 each
  4 tubes of a particular disinfectant; 1 each
  1 bottle of dishwashing detergent per class.

Organisms

- Trypticase Soy broth culture of *Escherichia coli*
  Trypticase Soy broth culture of *Bacillus subtilis*
  Oral sample (your mouth)
  Stool specimen.

Disinfectants

4 tubes of one of the following disinfectants per group of 4:

- 70% isopropyl alcohol
  3% hydrogen peroxide
  brand "X" mouthwash
  brand "Y" mouthwash.

Procedure for Evaluation of Disinfectants and Antiseptics

Each group of 4 will test one particular disinfectant against each of the 4 organisms or samples. One person will test the normal flora of his or her own mouth, one will test a stool specimen, one will test *E. coli*, and one will test *B. subtilis*.

1. Take 2 plates of TSA and, using your wax marker, divide each plate in half. Label the 4 halves as follows: control, 5 seconds, 30 seconds, and 3 minutes. Also place the name of the disinfectant your group is testing, the name of the specimen being tested, and your group name or symbol on each plate. Take the third TSA plate and label it "soap and water."
2. Holding the sterile glass rod by one tip only, place it in your mouth, in the stool specimen, in the *E. coli*, or in the *B. subtilis* for 3 minutes.
3. After 2 minutes, place the rod in your first tube of sterile water to rinse it briefly.
4. Remove the rod from the water, let the excess water drip off, and streak the tip of the rod on the control sector of the TSA plate. Be careful that the inoculum does not enter the other sector of the plate.
5. Place the rod in the mouth (use a new sterile glass rod for the mouth), the stool, the *E. coli*, or the *B. subtilis* a second time for 3 minutes. Then place it in your tube of disinfectant for 5 seconds. Remove the rod from the disinfectant and rinse it briefly in your second tube of sterile water. Streak the tip of the rod on the 5 second sector of the TSA plate.
6. Place the rod in the specimen a third time (use a new sterile rod for the mouth) for 3 minutes. Then place it in your disinfectant tube for 30 seconds. Rinse it briefly in your third tube of sterile water, and streak the tip on the 30 second sector of the TSA plate.
7. Place the rod a fourth time in the specimen (use a new sterile rod for the mouth) for 3 minutes. Then place it in your tube of disinfectant for 3 minutes. Rinse it briefly in your fourth tube of sterile water, and streak the tip on the 3 minute sector of the TSA plate.
8. Place the rod a final time in the specimen (use a new sterile rod for the mouth) for 3 minutes. Squeeze a small amount of dishwashing detergent on the rod and clean the rod using a wet paper towel. Rinse the rod under running water and streak the tip of the rod on the TSA plate labelled "soap and water."
9. Incubate the TSA plates at 37°C until the next lab period.

B. Effectiveness of Hand Washing

Materials

- 2 plates of Trypticase Soy agar (TSA)
  sterile scrub brush
  soap.

Procedure for Effectiveness fo Hand Washing

1. Using your wax marker, divide each TSA plate in half and label the halves 1 through 4.
2. Rub your fingers over sector 1 prior to washing your hands.
3. Using a scrub brush, soap, and water, scrub your hands for 2 minutes. Rub your damp fingers over sector 2.
4. Again scrub your hands with soap and water for 2 minutes and rub your fingers over sector 3.
5. Again scrub your hands with soap and water for 2 minutes and rub your fingers over sector 4.
6. Incubate the TSA plates at 37°C until the next lab period.

RESULTS

A. Evaluation of Disinfectants and Antiseptics

Record your group's results and the results of the other groups who used different disinfectants below:

Organism	*time*	*70% isopropyl alcohol*	*3% hydrogen peroxide*	*mouthwash "A"*	*mouthwash "B"*
Escherichia coli	control				
	5 seconds				
	30 seconds				
	3 minutes				
	soap and water				
Bacillus subtilis	control				
	5 seconds				
	30 seconds				
	3 minutes				
	soap and water				
stool	control				
	5 seconds				
	30 seconds				
	3 minutes				
	soap and water				
mouth	control				
	5 seconds				
	30 seconds				
	3 minutes				
	soap and water				

– = no growth

\+ = some growth

++ = abundant growth

B. Effectiveness of Hand Washing

Record your results of the 2 TSA "hand washing" plates:

sector	*growth*
sector 1 (no washing)	
sector 2 (2 min. of washing)	
sector 3 (4 min. of washing	
sector 4 (6 min. of washing)	

– = no growth

\+ = some growth

++ = abundant growth

PERFORMANCE OBJECTIVES

A. Disinfectants and Antisptics

1. Define the following terms: disinfection, disinfectant, antiseptic.
2. State why chemical agents are usually unreliable for sterilization.
3. List five factors which may influence the antimicrobial action of disinfectants and antiseptics.
4. Describe two modes of action of disinfectants and antiseptics (i.e., how they harm the microorganisms).
5. Name two chemical agent which are reliable for sterilization.

B. Evaluation of Disinfectants and Antiseptics

1. State why the results of an in vitro test to evaluate chemical agents may not necessarily apply to in vivo situations.

C. Evaluation of Hand Washing

1. Define transient flora and resident flora and compare the two groups in terms of easc of removal.

5. CONTROL OF MICROORGANISMS BY USING ANTIMICROBIAL CHEMOTHERAPY

INTRODUCTION

A. Antimicrobial Chemotherapeutic Agents

Antimicrobial chemotherapy is the use of chemicals to inhibit or kill microorganisms in or on the host. Chemotherapy is based on selective toxicity. This means that the agent used must inhibit or kill the microorganism in question without seriously harming the host.

In order to be selectively toxic, a chemotherapeutic agent must interact with some microbial function or microbial structure that is either not present or is substantially different from that of the host. For example, in treating infections caused by prokaryotic bacteria, the agent may inhibit peptidoglycan synthesis or alter bacterial (prokaryotic) ribosomes. Human cells do not contain peptidoglycan and possess eukaryotic ribosomes. Therefore, the drug shows little if any effect on the host (selective toxicity). Eukaryotic microorganisms, on the other hand, have structures and functions more closely related to those of the host. As a result, the variety of agents selectively effective against eukaryotic microorganisms such as fungi and protozoans is small when compared to the number available against prokaryotes. Also keep in mind that viruses are not cells and, therefore, lack the structures and functions altered by antibiotics so antibiotics are not effective against viruses.

Based on their origin, there are 2 general classes of antimicrobial chemotherapeutic agents:

1. *Antibiotics:* substances produced as metabolic products of one microorganism which inhibit or kill other microorganisms.
2. *Antimicrobial chemotherapeutic chemicals:* chemicals synthesized in the laboratory which can be used therapeutically on microorganisms.

Today the distinction between the 2 classes is not as clear, since many antibiotics are extensively modified in the laboratory (semisynthetic) or even synthesized without the help of microorganisms.

Most of the major groups of antibiotics were discovered prior to 1955, and most antibiotic advances since then have come about by modifying the older forms. In fact, only 3 major groups of microorganisms have yielded useful antibiotics: the actinomycetes (filamentous, branching soil bacteria such as *Streptomyces*), bacteria of the genus *Bacillus*, and the saprophytic molds *Penicillium* and *Cephalosporium*.

To produce antibiotics, manufacturers inoculate large quantities of medium with carefully selected strains of the appropriate species of antibiotic-producing microorganism. After incubation, the drug is extracted from the medium and purified. Its activity is standardized and it is put into a form suitable for administration.

Some antimicrobial agents are *cidal* in action: they kill microorganisms (e.g., penicillins, cephalosporins, streptomycin, neomycin). Others are *static* in action: they inhibit microbial growth long enough for the body's own defenses to remove the organisms (e.g., tetracyclines, erythromycin, sulfonamides).

Antimicrobial agents also vary in their spectrum. Drugs that are effective against a variety of both gram-positive and gram-negative bacteria are said to be broad spectrum (e.g., tetracycline, streptomycin, cephalosporins, ampicillin, sulfonamides). Those effective against just gram-positive bacteria, just gram negative bacteria, or only a few species are termed narrow spectrum (e.g., penicillin G, erythromycin, clindamycin, gentamycin).

If a choice is available, a narrow spectrum is preferable since it will cause less destruction to the body's normal flora. In fact, indiscriminate use of broad spectrum antibiotics can lead to superinfection by opportunistic microorganisms, such as *Candida* (yeast infections) and *Clostridium difficile* (antibiotic-associated ulcerative colitis), when the body's normal flora is destroyed. Other dangers from indiscriminate use of antimicrobial chemotherapeutic agents include drug toxicity, allergic reactions to the drug, and selection for resistant strains of microorganisms.

Below are examples of commonly used antimicrobial chemotherapeutic agents arranged according to their mode of action:

1. Antimicrobial agents that inhibit peptidoglycan synthesis

Inhibition of peptidoglycan synthesis in actively-dividing bacteria results in osmotic lysis. (A list of common antimicrobial chemotherapeutic agents listed by both their generic and brand names and arranged by their mode of action).

(a) *Penicillins* (produced by the mold *Penicillium*). There are several classes of penicillins:

1. *Natural penicillins* are highly effective against gram-positive bacteria (and a very few gram-negative bacteria) but are inactivated by the bacterial enzyme penicillinase. Examples include *penicillin G, F, X, K, O, and V.*
2. *Semisynthetic penicillins* are effective against gram-positive bacteria but are not inactivated by penicillinase. Examples include *methicillin, dicloxacillin, and nafcillin.*
3. *Semisynthetic broad spectrum penicillins* are effective against a variety of gram-positive and gram-negative bacteria but are inactivated by penicillinase. Examples include *ampicillin, carbenicillin, oxacillin, azlocillin, mezlocillin, and piperacillin.*
4. *Semisynthetic broad spectrum penicillins combined with beta lactamase inhibitors such as clavulanic acid and sulbactam.* Although the clavulanic acid and sulbactam have no antimicrobial action of their own, they inhibits penicillinase thus protecting the penicillin from degradation. Examples include *amoxicillin plus clavulanic acid, ticarcillin plus clavulanic acid, and ampicillin plus sulbactam.*

(b) *Cephalosporins* (produced by the mold *Cephalosporium*). Cephalosporins are effective against a variety of gram-positive and gram-negative bacteria and are resistant to penicillinase (although some can be inactivated by other beta-lactamase enzymes similar to penicillinase). Four "generations" of cephalosporins have been developed over the years in an attempt to counter bacterial resistance.

1. First generation cephalosporins include *cephalothin, cephapirin, and cephalexin.*
2. Second generation cephalosporins include *cefamandole, cefaclor, cefazolin, cefuroxime, and cefoxitin.*
3. Third generation cephalosporins include *cefotaxime, cefsulodin, cefetamet, cefixime, ceftriaxone, cefoperazone, ceftazidine, and moxalactam.*

(c) *Carbapenems.* Carbapenems consist of a broad spectrum beta lactam antibiotic to inhibit peptidoglycan synthesis combined with cilastatin sodium, an agent which prevents degradation of the antibiotic in the kidneys. An example is *imipenem.*

(d) *Monobactems.* Monobactems are broad spectrum beta lactam antibiotics resistant to beta lactamase. An example is *aztreonam.*

(e) *Carbacephem.* A synthetic cephalosporins. An example is *loracarbef.*

(f) *Vancomycin* (produced by the bacterium *Streptomyces*). Vancomycin and teichoplanin are glycopeptides that are effective against gram-positive bacteria.

(g) *Bacitracin* (produced by the bacterium *Bacillus*). Bacitracin is used topically against gram-positive bacteria.

2. Antimicrobial agents that alter the cytoplasmic membrane

Alteration of the cytoplasmic membrane of microorganisms results in leakage of cellular materials. (A list of common antimicrobial chemotherapeutic agents listed by both their generic and brand names and arranged by their mode of action.)

(a) *Polymyxin B* (produced by the bacterium *Bacillus*): Polymyxin B is used in severe *Pseudomonas* infections.

(b) *Amphotericin B* (produced by the bacterium *Streptomyces*): Amphotericin B is used for systemic fungal infections.

(c) *Nystatin* (produced by the bacterium *Streptomyces*): Nystatin is used mainly for *Candida* yeast infections.

(d) *Imidazoles* (produced by the bacterium *Streptomyces*): The imidazoles are antifungal antibiotics used for yeast infections, dermatophytic infections, and systemic fungal infections. Examples include clotrimazole, miconazole, ketoconazole, itraconazole, and fluconazole.

3. Antimicrobial agents that inhibit protein synthesis

(A list of common antimicrobial chemotherapeutic agents listed by both their generic and brand names and arranged by their mode of action.)

These agents prevent bacteria from synthesizing structural proteins and enzymes.

(a) Agents that block transcription (prevent the synthesis of mRNA off of DNA).

1. *Rifampins* (produced by the bacterium *Streptomyces*): Rifampins are effective against some gram-positive and gram-negative bacteria and *Mycobacterium tuberculosis*. They inhibit the enzyme RNA polymerase.

(b) Agents that block translation (alter bacterial ribosomes to prevent mRNA from being translated into proteins).

1. Agents such as the aminoglycosides (produced by the bacterium *Streptomyces)* that bind irreversibly to the 30s ribosomal subunit and prevent the 50S ribosomal subunit from attaching to the translation initiation complex. Aminoglycosides also cause a misreading of the mRNA. Examples include streptomycin, kanamycin, tobramycin, and amikacin. Most are effective against gram-positive and gram-negative bacteria.
2. Agents that bind reversibly to the 30S ribosomal subunit in such a way that anticodons of charged tRNAs cannot align properly with the codons of the mRNA. Examples include tetracycline, minocycline, and doxycycline, produced by the bacterium *Streptomyces*. They are effective against a variety of gram-positive and gram-negative bacteria.
3. Agents that bind reversibly to the 50s ribosomal subunit and block peptide bond formation during protein synthesis. Examples include lincomycin and clindamycin, produced by the bacterium *Streptomyces*. Most are used against gram-positive bacteria.

4. Agents that bind reversibly to the 50s ribosomal subunit and block translation by inhibiting elongation of the protein by the enzyme peptidyltransferase that forms peptide bonds between the amino acids, by preventing the ribosome from translocating down the mRNA, or both. Macrolides such as erythromycin, roxithromycin, clarithromycin, and azithromycin are examples and are used against gram-positive bacteria and some gram-negative bacteria.
5. The **oxazolidinones** (linezolid) bind to the 50S ribosomal subunit appear to interfere with the initiation of translation.
6. The **streptogramins** (a combination of quinupristin and dalfopristin) bind to different sites on the 50S ribosomal subunit and work synergistically to inhibit translocation.

4. Antimicrobial agents that interfere with DNA synthesis

(A list of common antimicrobial chemotherapeutic agents listed by both their generic and brand names and arranged by their mode of action.)

(*a*) *Fluoroquinolones* (synthetic chemicals): The fluoroquinolones inhibiting one or more of a group of enzymes called topoisomerase, enzymes needed for bacterial nucleic acid synthesis. For example, DNA gyrase (topoisomerase II) breaks and rejoins the strands of bacterial DNA to relieve the stress of the unwinding of DNA that occurs during DNA replication and transcription. Fluoroquinolones are broad spectrum and examples include norfloxacin, ciprofloxacin, enoxacin, levofloxacin, and trovafloxacin.

(*b*) *Sulfonamides and trimethoprim* (synthetic chemicals): Co-trimoxazole is a combination of sulfamethoxazole and trimethoprim. Both of these drugs block enzymes in the bacteria pathway required for the synthesis of tetrahydrofolic acid, a cofactor needed for bacteria to make the nucleotide bases thymine, guanine, uracil, and adenine.

(*c*) *Metronidazole* is a drug that is activated by the microbial proteins flavodoxin and feredoxin found in microaerophilc and anaerobic bacteria and certain protozoans. Once activated, the metronidazole puts nicks in the microbial DNA strands.

B. Microbial Resistance to Antimicrobial Chemotherapeutic Agents

A common problem in antimicrobial chemotherapy is the development of resistant strains of bacteria. Most bacteria become resistant to antimicrobial agents by one or more of the following mechanisms:

1. Producing enzymes which detoxify or inactivate the antibiotic, *e.g.*, penicillinase and other beta-lactamases.
2. Altering the target site in the bacterium to reduce or block binding of the antibiotic, *e.g.*, producing a slightly altered ribosomal subunit that still functions but to which the drug can't bind.
3. Preventing transport of the antimicrobial agent into the bacterium, *e.g.*, producing an altered cytoplasmic membrane or outer membrane.
4. Developing an alternate metabolic pathway to by-pass the metabolic step being blocked by the antimicrobial agent, *e.g.*, overcoming drugs that resemble substrates and tie-up bacterial enzymes.

5. Increasing the production of a certain bacterial enzyme, *e.g.*, overcoming drugs that resemble substrates and tie-up bacterial enzymes.

These changes in the bacterium that enable it to resist the antimicrobial agent occur naturally as a result of mutation or genetic recombination of the DNA in the nucleoid, or as a result of obtaining plasmids from other bacteria. Exposure to the antimicrobial agent then selects for these resistant strains of organism.

The spread of antibiotic resistance in pathogenic bacteria is due to both direct selection and indirect selection. Direct selection refers to the selection of antibiotic resistant pathogens at the site of infection. Indirect selection is the selection of antibiotic-resistant normal floras within an individual anytime an antibiotic is given. At a later date, these resistant normal floras may transfer resistance genes to pathogens that enter the body. In addition, these resistant normal flora may be transmitted from person to person through such means as the fecal-oral route or through respiratory secretions.

As an example, many gram-negative bacteria possess R (resistance) plasmids which have genes coding for multiple antibiotic resistance through the mechanisms stated above, as well as transfer genes coding for a sex pilus. Such an organism can conjugate with other bacteria and transfer an R plasmid to them. *Escherichia coli, Proteus, Serratia, Salmonella, Shigella,* and *Pseudomonas* are examples of bacteria which frequently have R plasmids. Because of the problem of antibiotic resistance, antibiotic susceptibility testing is usually done in the clinical laboratory to determine which antimicrobial chemotherapeutic agents will most likely be effective on a particular strain of microorganism. This is discussed in the next section.

To illustrate how plasmids carrying genes coding for antibiotic resistance can be picked up by antibiotic-sensitive bacteria, in today's lab we will use plasmid DNA to transform an Escherichia coli sensitive to the antibiotic ampicillin into one that is resistant to the drug.

The *E. coli* will be rendered more "competent" to take up plasmid DNA (pAMP), which contains a gene coding for ampicillin resistance, by treating them with a solution of calcium chloride, cold incubation, and a brief heat shock. They will then be plated on two types of media: Lauria-Bertani agar (LB) and Lauria-Bertani agar with ampicillin (LB/amp). Only *E. coli* that have picked up a plasmid coding for ampicillin resistance will be able to form colonies on the LB/amp agar.

C. Antibiotic Susceptibility Testing

For some microorganisms, susceptibility to chemotherapeutic agents is predictable. However, for many microorganisms (*Pseudomonas, Staphylococcus aureus,* and gram-negative enteric bacilli such as *Escherichia coli, Serratia, Proteus,* etc.) there is no reliable way of predicting which antimicrobial agent will be effective in a given case. This is especially true with the emergence of many antibiotic-resistant strains of bacteria. Because of this, antibiotic susceptibility testing is often essential in order to determine which antimicrobial agent to use against a specific strain of bacterium.

Several tests may be used to tell a physician which antimicrobial agent is most likely to combat a specific pathogen:

1. Tube dilution tests

In this test, a series of culture tubes are prepared, each containing a liquid medium and a different concentration of a chemotherapeutic agent. The tubes are then inoculated with the test organism and incubated for 16-20 hours at 35C. After incubation, the tubes are examined for turbidity (growth). The lowest concentration of chemotherapeutic agent capable of preventing growth of the test organism is the **minimum inhibitory concentration (MIC).**

Subculturing of tubes showing no turbidity into tubes containing medium but no chemotherapeutic agent can determine the **minimum bactericidal concentration (MBC).** MBC is the lowest concentration of the chemotherapeutic agent that results in no growth (turbidity) of the subcultures. These tests, however, are rather time consuming and expensive to perform.

2. The agar diffusion test (Bauer-Kirby test)

A procedure commonly used in clinical labs to determine antimicrobial susceptibility is the Bauer-Kirby disc diffusion method. In this test, the in vitro response of bacteria to a standardized antibiotic-containing disc has been correlated with the clinical response of patients given that drug.

In the development of this method, a single high-potency disc of each chosen chemotherapeutic agent was used. Zones of growth inhibition surrounding each type of disc were correlated with the minimum inhibitory concentrations of each antimicrobial agent (as determined by the tube dilution test). The MIC for each agent was then compared to the usually-attained blood level in the patient with adequate dosage. Categories of "Resistant," "Intermediate," and "Sensitive" were then established.

The basic steps for the Bauer-Kirby method of antimicrobial susceptibility testing are given below:

(*a*) Prepare a **standard turbidity inoculum** of the test bacterium so that a certain density of bacteria will be put on the plate.

(*b*) **Inoculate a 150mm Mueller-Hinton agar plate** with the standardized inoculum so as to cover the entire agar surface with bacteria.

(*c*) Place **standardized antibiotic-containing discs** on the plate.

(*d*) **Incubate** the plate at 35°C for 18-20 hours.

(*e*) Measure the **diameter** of any resulting **zones of inhibition** in millimeters (mm).

(*f*) Determine if the bacterium is **susceptible, moderately susceptible, intermediate, or resistant** to each antimicrobial agent.

The term intermediate generally means that the result is inconclusive for that drug-organism combination. The term moderately susceptible is usually applied to those situations where a drug may be used for infections in a particular body site, e.g., cystitis because the drug becomes highly concentrated in the urine.

3. Automated tests

Computerized automated tests have been developed for antimicrobial susceptibility testing. These tests measure the inhibitory effect of the antimicrobial agents in a liquid medium by using light scattering to determine growth of the test organism. Results can be obtained within a few hours. Labs performing very large numbers of susceptibility tests frequently use the automated methods but the equipment is quite expensive.

PROCEDURES

A. Microbial Resistance to Antimicrobial Chemotherapeutic Agents

Materials

Plasmid DNA (pAMP) on ice, calcium chloride solution on ice, 2 sterile culture tubes, 1 tube of LB broth, 2 plates of LB agar, 2 plates of LB agar with ampicillin (LB/amp), sterile 1 ml transfer pipettes, sterile plastic inoculating loops, bent glass rod, turntable, alcohol, beaker of ice, water bath at 42°C.

Organism

LB agar culture of *Escherichia coli*

Microbial Resistance Procedure

1. Label one LB agar plate "Transformed Bacteria, Positive control" and the other LB agar plate "Wild-Type Bacteria, Positive Control."

   Label one LB/amp agar plate "Transformed Bacteria, Experiment" and the other LB/amp agar plate "Wild-Type Bacteria, Negative Control."

2. Label one sterile culture tube "(+)AMP" and the other "(–)AMP." Using a sterile 1ml transfer pipette, add 250 µl of ice cold calcium chloride to each tube. Place both tubes on ice.

   Using a sterile plastic inoculating loop, transfer 1-2 large colonies of *E. coli* into the (+)AMP tube and vigorously tap against the wall of the tube to dislodge all the bacteria. Immediately suspend the cells by repeatedly pipetting in and out with a sterile transfer pipette until no visible clumps of bacteria remain. Return tube to the ice.

3. Repeat step 3 this time using the (–)AMP tube and return to the ice.
4. Using a sterile plastic inoculating loop, add one loopful of pDNA (plasmid DNA) solution to the (+)AMP tube and swish loop to mix the DNA. Return to the ice.
5. Incubate both tubes on ice for 15 minutes.
6. After 15 minutes, "heat-shock" both tube of bacteria by immersing them in a 42°C water bath for 90 seconds. Return both tubes to the ice for 1 minute or more.
7. Using a sterile 1ml transfer pipette, add 250 µl of LB broth to each tube. Tap tubes with your fingers to mix. Set tubes in a test tube rack at room temperature.
8. Using a sterile 1ml transfer pipette, add 100 µl of *E. coli* suspension from the (–)AMP tube onto the LB/amp agar plate labelled "Wild-Type Bacteria, Negative Control." Add another 100 l of *E. coli* from the (–)AMP to the LB agar plate labelled "Wild-Type Bacteria, Positive Control."
9. Using a bent glass rod dipped in alcohol and flamed, spread the bacteria thoroughly over both agar plates. Make sure you reflame the glass rod between plates.
10. Using a sterile 1ml transfer pipette, add 100 µl of *E. coli* suspension from the (+)AMP tube onto the LB/amp agar plate labelled "Transformed Bacteria, Experiment." Add another 100 l of *E. coli* from the (+)AMP to the LB agar plate labelled "Transformed Bacteria, Positive Control."

11. Immediately spread as in step 10.
12. Incubate all plates at **37°C.**

B. Antibiotic Susceptibility Testing

Materials

- 150mm Mueller-Hinton agar plates (3).

  Sterile swabs (3).

  An antibiotic disc dispenser containing discs of antibiotics commonly effective against gram-positive bacteria, and one containing discs of antibiotics commonly effective against gram-negative bacteria.

Organisms

- Trypticase Soy broth cultures of *Staphylococcus aureus* (gram-positive),
- *Escherichia coli* (gram-negative), and *Pseudomonas aeruginosa* (gram-negative).

Antibiotic Susceptibility Testing Procedure

1. Take **3 Mueller-Hinton agar plates.** Label one *S. aureus*, one *E. coli*, and one *P. aeruginosa.*
2. Using your wax marker, divide each plate into **thirds** to guide your streaking.

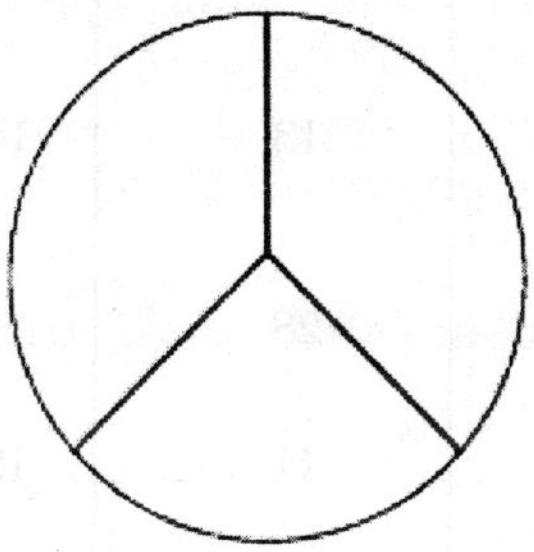

3. Dip a sterile swab into the previously-standardized tube of S. aureus. Squeeze the swab against the inner wall of the tube to remove excess liquid.
4. Streak the swab perpendicular to each of the 3 lines drawn on the plate overlapping the streaks to assure complete coverage of the entire agar surface with inoculum.

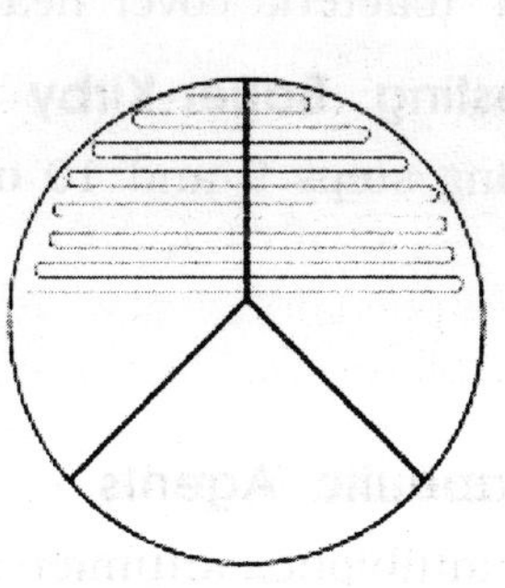

5. Repeat steps 3 and 4 for the *E. coli* and *P. aeruginosa* plates.
6. Using the appropriate antibiotic disc dispenser, place gram-positive antibiotic-containing discs on the plate of S. aureus and gram-negative antibiotic-containing discs on the plates of E. coli and P. aeruginosa.
7. Make sure that one of each of the antibiotic-containing discs in the dispenser is on the plate and touch each disc lightly with sterile forceps to make sure it adheres to the agar surface.
8. Incubate the 3 plates upside-down at 37°C until the next lab period.
9. Using a metric ruler, measure the diameter of the zone of inhibition around each disc on each plate in mm by placing the ruler on the bottom of the plate.
10. Determine whether each organism is susceptible, moderately susceptible, intermediate, or resistant to each chemotherapeutic agent using the standardized table and record your results.

Table 2

Zone Size Interpretive Chart for Bauer-Kirby Test

antimicrobial agent	*disc code*	*R = mm or less*	*I = mm*	*MS = mm*	*S = mm or more*
amikacin	AN-30	15	15-16	-	16
amoxicillin/ clavulanic acid-staphylococci	AmC-30	19	-	-	20
amoxicillin/ clavulanic acid-other organisms	AmC-30	13	14-17	-	18
ampicillin-staphylococci	AM-10	28	-	-	29
ampicillin-G-enterics	AM-10	11	12-13	-	14

RESULTS

A. Microbial Resistance to Antimicrobial Chemotherapeutic Agents

Count the number of colonies on each plate. If the growth is too dense to count individual colonies, record "lawn" (bacteria cover nearly the entire agar surface).

B. Antibiotic Susceptibility Testing: Bauer-Kirby Method

Interpret the results following steps 9 and 10 of the procedure and record your results in the tables below.

PERFORMANCE OBJECTIVES

A. Antimicrobial Chemotherapeutic Agents

1. Define the following: antibiotic, antimicrobial chemotherapeutic chemical, narrow spectrum antibiotic, broad spectrum antibiotic.
2. Discuss the meaning of selective toxicity in terms of antimicrobial chemotherapy.

3. List four genera of microorganisms that produce useful antibiotics.
4. Describe four different major modes of action of antimicrobial chemotherapeutic chemicals and give three examples of drugs fitting each mode of action.

B. Microbial Resistance to Antimicrobial Agents

Discussion

1. State five mechanisms by which microorganisms may resist antimicrobial chemotherapeutic agents.
2. Briefly describe R plasmids and name four bacteria that commonly possess these plasmids.

Results

1. Interpret the results of the *Escherichia coli* plasmid transformation experiment.

C. Antibiotic Susceptibility Testing

Discussion

1. State why antimicrobial susceptibility testing is often essential in choosing the proper chemotherapeutic agent to use in treating an infection.
2. State what is meant by MIC.

Results

1. Interpret the results of a Bauer-Kirby antimicrobial susceptibility test when given a Mueller-Hinton agar plate, a metric ruler, and a standardized zone-size interpretation table.

6. ISOLATION OF PURE CULTURES FROM A MIXED POPULATION

INTRODUCTION

The microorganisms exist in nature as mixed populations. However, to study microorganisms in the laboratory we must have them in the form of a pure culture, that is, one in which all organisms are descendants of the same organism.

Two major steps are involved in obtaining pure cultures from a mixed population:

1. First, the mixture must be diluted until the various individual microorganisms become separated far enough apart on an agar surface that after incubation they form visible colonies isolated from the colonies of other microorganisms. This plate is called an isolation plate.

2. Then, an isolated colony can be aseptically **"picked off"** the isolation plate.

Before removing bacteria from the petri plate, first cool the loop by sticking it into the agar away from any growth.

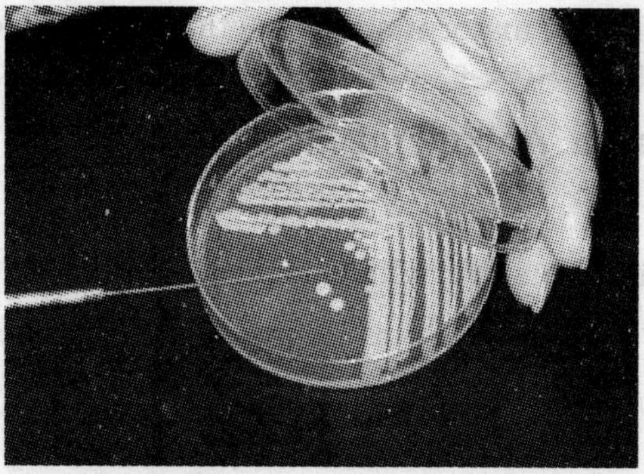

Fig. 11. Picking a single colony off of a petri plate in order to obtain a pure culture and transferred to new sterile medium.

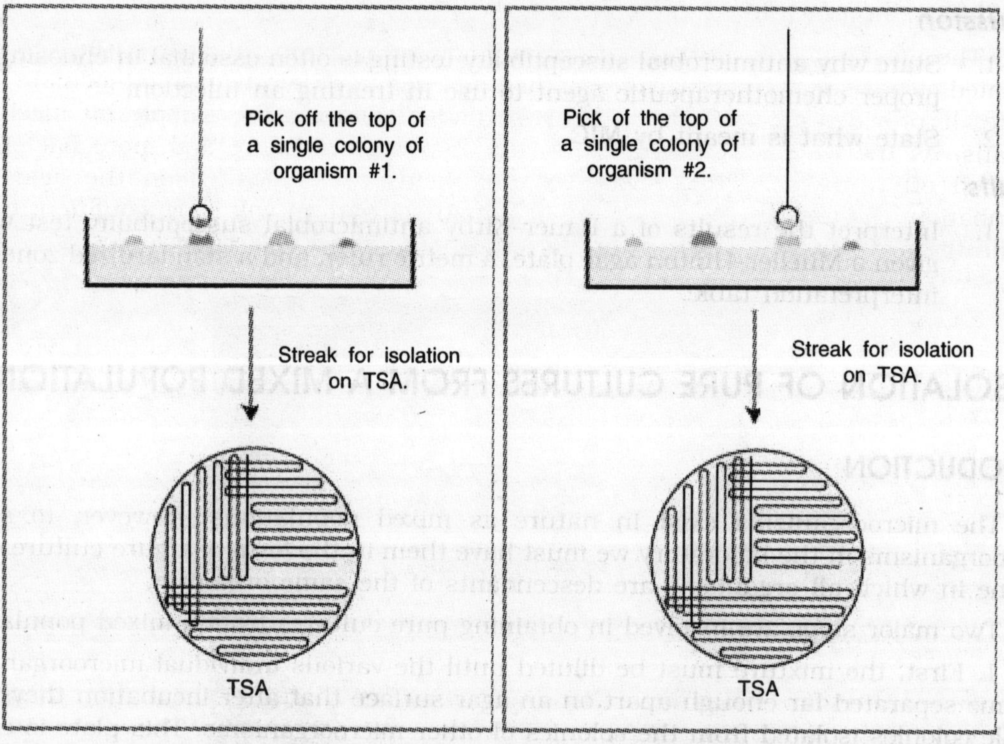

Fig. 12. Obtaining pure cultures from an isolation plate.

After incubation, all organisms in the new culture will be descendants of the same organism, that is, a pure culture.

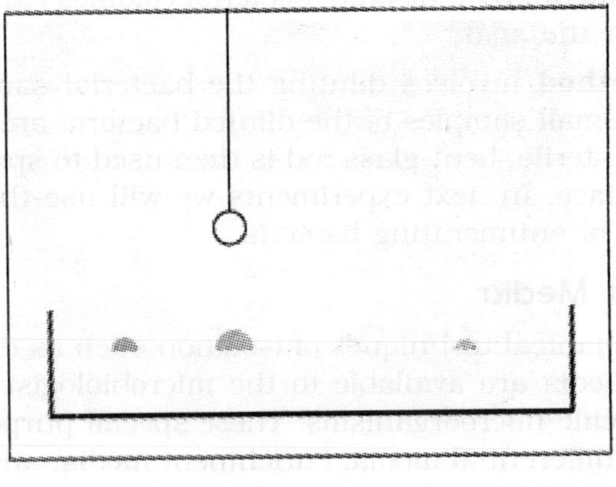

Fig. 13. Picking off a single colony from a plate culture.

A. Streak Plate Method of Isolation

The most common way of separating bacterial cells on the agar surface to obtain isolated colonies is the streak plate method we used in Experiment 2 to inoculate a petri plate. It provides a simple and rapid method of diluting the sample by mechanical means. As the loop is streaked across the agar surface, more and more bacteria are rubbed off until individual separated organisms are deposited on the agar. After incubation, the area at the beginning of the streak pattern will show confluent growth, while the area near the end of the pattern should show discrete colonies.

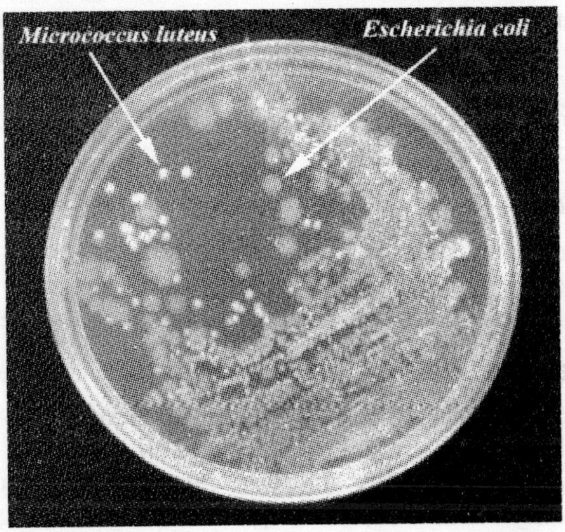

Micrococcus luteus produces a yellow, water insoluble pigment. *Escherichia coli* is nonchromogenic.
Fig. 14. Single isolated colonies of *micrococcus luteus* and *escherichia coli* growing on TSA.

B. The Pour Plate and Spin Plate Methods of Isolation

Another method of separating bacteria is the pour plate method. With the **pour plate method,** the bacteria are mixed with melted agar until evenly distributed and separated throughout the liquid. The melted agar is then poured into an empty plate and allowed to solidify. After incubation, discrete bacterial colonies can then be found growing both on the agar and in the agar.

The **spin plate method** involves diluting the bacterial sample in tubes of sterile water, saline, or broth. Small samples of the diluted bacteria are then pipetted onto the surface of agar plates. A sterile, bent-glass rod is then used to spread the bacteria evenly over the entire agar surface. In next experiments we will use this technique as part of the plate count method of enumerating bacteria.

C. Use of Specialized Media

To supplement mechanical techniques of isolation such as the streak plate method, many special-purpose media are available to the microbiologist to aid in the isolation and identification of specific microorganisms. These special purpose media fall into four groups: selective media, differential media, enrichment media, and combination selective and differential media.

1. Selective media

A selective medium has agents added which will inhibit the growth of one group of organisms while permitting the growth of another. For example, Columbia CNA agar has the antibiotics colistin and nalidixic acid added which inhibit the growth of gram-negative bacteria but not the growth of gram-positives. It is, therefore, said to be selective for gram-positive organisms, and would be useful in separating a mixture of gram-positive and gram-negative bacteria.

2. Differential media

A differential medium contains additives that cause an observable colour change in the medium when a particular chemical reaction occurs. They are useful in differentiating bacteria according to some biochemical characteristic. In other words, they indicate whether or not a certain organism can carry out a specific biochemical reaction during its normal metabolism. Many such media will be used in future labs to aid in the identification of microorganisms.

3. Enrichment media

An enrichment medium contains additives that enhance the growth of certain organisms. This is useful when the organism you wish to culture is present in relatively small numbers compared to the other organisms growing in the mixture.

4. Combination selective and differential media

A combination selective and differential medium permits the growth of one group of organisms while inhibiting the growth of another. In addition, it differentiates those organisms that grow based on whether they can carry out particular chemical reactions. For example, Eosin Methylene Blue (EMB) agar is selective for gram-negative bacteria. The dyes eosin Y and methylene blue found in the medium inhibit the growth of gram-positive bacteria but not the growth of gram-negatives. In addition, it is useful in differentiating the various gram-negative enteric bacilli belonging to the bacterial family Enterobacteriaceae. The appearance of typical members of this bacterial family on EMB agar is as follows:

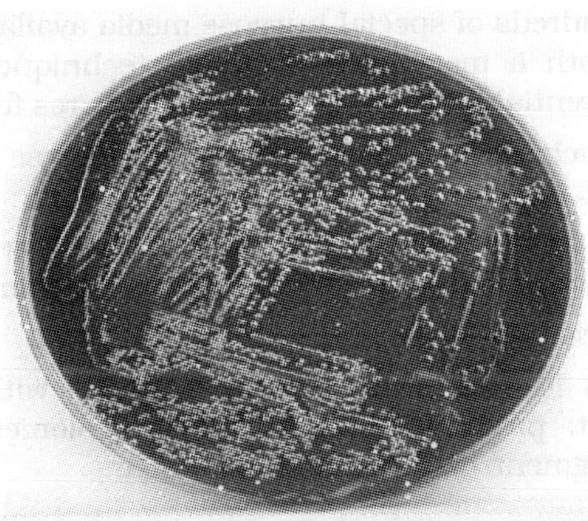

Only gram-negative bacteria grow on EMB agar. (Gram-positive bacteria are inhibited by the dyes eosin and methylene blue added to the agar.) Based on its rate of lactose fermentation, *Escherichia coli* produces dark, blue-black colonies with a metallic green sheen on EMB agar.

- large, blue-black colonics with a green metallic sheen

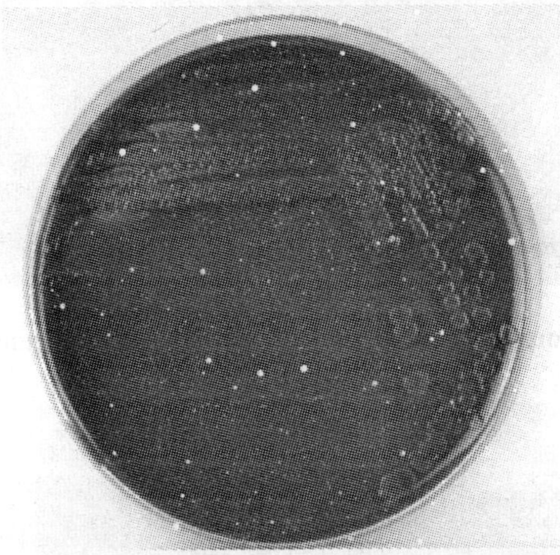

Enterobacter aerogenes growing on EMB Agar

Only gram-negative bacteria grow on EMB agar. (Gram-positive bacteria are inhibited by the dyes eosin and methylene blue added to the agar.) Based on its rate of lactose fermentation, *Enterobacter* produces large, mucoid, pink to purple colonies with no metallic green sheen on EMB agar.

- and *Klebsiella*: large, mucoid, pink to purple colonies with no metallic sheen
- *Salmonella* and *Shigella* and *Proteus*: large, colourless colonies
- *Shigella*: colourless to pink colonies.

The colour changes in the colonies are a result of bacterial fermentation of the sugar lactose while colourless colonies indicate lactose nonfermenters.

There are literally hundreds of special-purpose media available to the microbiologist. Today we will combine both a mechanical isolation technique (the streak plate) with selective and selective-differential media to obtain pure cultures from a mixture of bacteria.

Media. One plate of each of the following media: Trypticase Soy agar, Columbia CNA agar, and EMB agar.

Organisms. A broth culture containing a mixture of one of the following gram-positive bacteria and one of the following gram-negative bacteria:

- Possible gram-positive bacteria:
  - o Micrococcus luteus. A gram-positive coccus with a tetrad or a sarcina arrangement; produces circular, convex colonies with a yellow, water-insoluble pigment on Trypticase Soy agar.

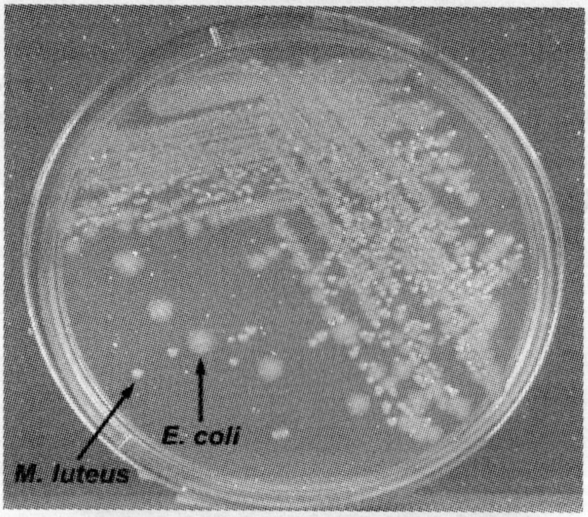

Isolation Plate: Mixture of *Escherichia coli* and *Micrococcus luteus* grown on TSA

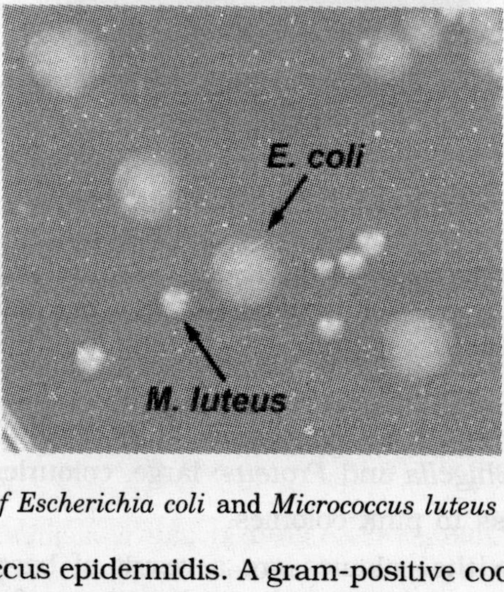

Clos up view of Escherichia coli and *Micrococcus luteus* grown on TSA

  - o Staphylococcus epidermidis. A gram-positive coccus with a staphylococcus arrangement; produces circular, convex, nonpigmented colonies on Trypticase Soy agar.

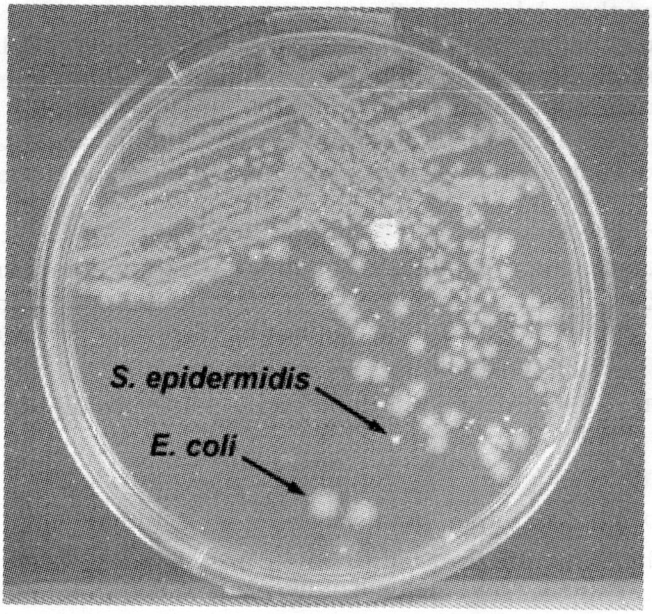

Isolation Plate: Mixture of *Escherichia coli* and *Staphylococcus epidermidis* grown on TSA

- Possible gram-negative bacteria:
  - *Escherichia coli*. A gram-negative bacillus; produces irregular, raised, nonpigmented colonies on Trypticase Soy agar.

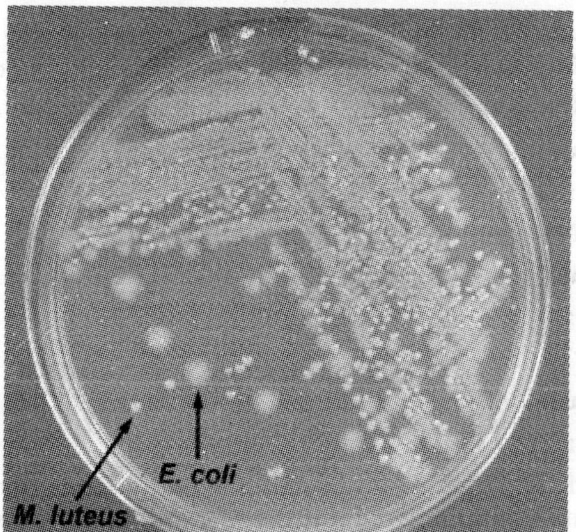

Isolation Plate: Mixture of *Escherichia coli* and *Micrococcus luteus* grown on TSA

  - *Enterobacter aerogenes*. A gram-negative bacillus; produces irregular raised, nonpigmented, possibly mucoid colonies on Trypticase Soy agar.

During the next three labs you will attempt to obtain pure cultures of each organism in your mixture and determine which two bacteria you have. Today you will try to separate the bacteria in the mixture in order to obtain isolated colonies; next lab you will identify the two bacteria in your mixture and pick off single isolated colonies of each of the two bacteria in order to get a pure culture of each. The following lab you will prepare microscopy slides of each of the two pure cultures to determine if they are indeed pure.

Procedure

1. First attempt to obtain isolated colonies of the two organisms in your mixture by using mechanical methods on an all-purpose growth medium, Trypticase Soy agar. Streak the mixture on a plate of Trypticase Soy agar using one of the two streaking patterns illustrated in Lab 2.
2. Streak the same mixture for isolation on a plate of Columbia CNA agar (selective for gram-positive bacteria).
    - *Micrococcus leteus* growing on Columbia CNA agar.
    - *Staphylococcus epidermidis* growing on Columbia CNA agar.
3. Streak the same mixture for isolation on a plate of EMB agar (selective for gram-negative bacteria and differential for certain members of the bacterial family *Enterobacteriaceae*).
    - *Escherichia* coli growing on EMB agar.
    - Enterobacter aerogenes growing on EMB agar.
4. Incubate the three plates at 37°C until the next lab period.

Results

1. Observe isolated colonies on the plates of Trypticase Soy agar, Columbia CNA agar, and EMB agar. Record your observations and conclusions.

    Trypticase Soy agar

    Observations:

    Conclusions:

    Columbia CNA agar

    Observations:

    Conclusions:

    EMB agar

    Observations:

    Conclusions:

2. Using any of the three plates, pick off a single isolated colony of each of the two organisms in your original mixture and aseptically transfer them to separate plates of Trypticase Soy agar. When picking off single colonies, remove the top portion of the colony without touching the agar surface itself to avoid picking up any inhibited bacteria from the surface of the agar. Use your regular plate streaking pattern to inoculate these plates and incubate at 37°C until the next lab period. These will be your pure cultures for Lab 5 (Direct and Indirect stains).

Discussion

1. Given a mixture of two bacteria and plates of Trypticase Soy agar, describe the steps you would take in order to eventually obtain pure cultures of each organism.
2. Define: selective medium, differential medium, enrichment medium, and combination selective-differential medium.
3. State the usefulness of Columbia CNA agar and EMB agar.
4. Describe how each of the following would appear when grown on EMB agar:
   (a) *Escherichia coli*
   (b) *Enterobacter aerogenes*
   (c) *Salmonella.*

Procedure

1. Using the streak plate method of isolation, obtain isolated colonies from a mixture of microorganisms.
2. Pick off isolated colonies of microorganisms growing on a streak plate and aseptically transfer them to sterile media to obtain pure cultures.

Results

1. When given a plate of Columbia CNA agar or EMB agar showing discrete colonies, correctly interpret the results.

7. BACTERIAL STAINING

INTRODUCTION

Visualization of microorganisms in the living state is most difficult, not only because they are minute, but also because they are transparent and practically colourless when suspended in an aqueous medium.

To study their properties and divide microorganisms into specific groups for diagnostic purposes, biological stains and staining procedures in conjunction with light microscopy have become major tools in microbiology.

Chemically a stain may be defined as an organic compound containing a benzene ring pus a chromophore and auxochrome.

Stains are of two types:

1. Acidic stains *e.g.* Picric acid
2. Basic stains *e.g.* Methylene blue.

Types of staining techniques:

1. *Simple staining* (use of single stain)

For visualization of morphological shape (cocci, bacilli, and spirilli) and arrangement (chains, clusters, Pairs, and tetrads).

2. *Differential staining.* (use of two contrasing stains)

It is divided in two groups, *i.e.*

(*a*) separation into groups, Gram stain and Acid-fast stain.

(*b*) Visualization of structures, Flagella stain, apsule stain, spore stain, Nuclear stain.

STAINING MICROBES

- Microbes are invisible to the naked eye and difficult to see and identify even when using a microscope. Staining microbes makes them easier to observe and reveals the presence microscopic structures because charged portions of the stain bind to specific macromolecules within the structures. A simple stain visualizes the microorganisms, a differential stain displays the chemical differences in cellular structures, including the cell wall and cell membrane because the macromolecules within the structure bind to different components of the stain.
- An example of this differential staining is seen in staining used for blood smears.
- Staining white blood cells with a differential stain displays the difference between the five white blood cell types; basophils, eosinophils, neutrophils, monocytes and lymphocytes. The intracellular granules of basophils stain dark blue because of their affinity to basic portion of the stain Basophil means basic loving. On the other hand, the eosinophil (acid loving) stains red as a result of the intracellular granule's affinity to the acidic portion of the stain. Treatment of microbial diseases depends upon the correct identification of microorganisms and relies upon the ability to identify specific internal structures. Bacterial cells are commonly stained with a differential stain called the **Gram stain** and protozoal cells with the **Trichrome stain,** in order to reveal the internal structural differences in order to identify the organisms by present. Properly preparing slides for staining is important to ensuring good results. Remember, you cannot see the material you are working with so you must develop good technique based upon principles.
- Always start with clean slides using lens paper to clean them. Slides can be made from direct clinical material (a wound, sputum, knee fluid, the throat etc.), broth cultures and from solid media cultures. The first principle is that some fluid is needed to emulsify the material if it is dry, however too much fluid may make the microbes hard to find. Slides from clinical cultures are usually placed directly on the slide without the addition of water, as are slides from broth cultures. Slides from solid media require water to emulsify and separate the individual bacterial cells for better observation, but a very single drop of water is usually adequate. Next the material must be attached to the slide so they don't wash off with the staining process. The second principle involves fixing the slide using either a chemical fixative or heat. In this lab a heat-fixing tray will be used.
- This lab will use the principles and techniques above to make and stain bacterial slides using a differential staining technique called the Gram stain. Initially 3 stock cultures (known types) of bacteria will be stained, and then the 3 isolated unknown microbes from the environmental cultures will be

stained and examined. The environmental culture will contain a variety of bacteria and possibly some fungi. Bacterial cells can be observed for shape (rod, coccus, or spirilla) and arrangement (in chains, clusters, etc.). Arrangements of cells are best observed from clinical and broth cultures because the emulsification process disrupts the natural arrangement from colonies "picked" from solid media.

Materials

Three isolation plates

Original environmental broth culture original TSA plate

Stock Cultures

(1) *S. epidermidis*,

(2) *E. coli*,

(3) *Bacillus sp.*

Slides, transfer loops.

Reagents

- Gram stain reagents
- Crystal violet

Iodine

Decolourizer-Acetone-Ethanol

Safranin.

PREPARING THE SMEARS

1. Collect the broth and pure subcultures you made from example. 2. Observe the colony morphology. Ask the instructor to critique your isolation technique. This will be very important in later labs. Practice the isolation technique on a new plate if you need some more experience.
2. *Preparing a smear.* There are 3 steps in preparing a smear for staining. Remember to use aseptic technique and flame the loop before and after each use.
   - Preparation of the slide—Clean and dry the slide thoroughly to remove oils.
   - Preparation of the smear—From the broth culture use the loop to spread one or two drops of specimen in the center of the slide spreading it until it is approximately the size of a nickel. When making a smear from solid media cultures, start by putting a very small drop of water in the center of the slide and then mix a loopful of bacteria from the surface of solid media in the water, spreading it out to the size of a nickel.
   - Fixation—The point of fixation is to attach the organisms and cells to the slide without disrupting them. In this class we will use an electric fixing tray that will dry and fix the smears in one step. Slides must be completely dry and fixed before staining, or they will wash off.

Note: Smears made from a broth look shiny even when they are dry.

STAINING THE SMEARS

1. In the beginning it is wise to make a single broth slide and a single solid medium slide and then stain and observe these before making the other slides. This allows you to alter your technique if the results are not optimal.
2. Begin with the known culture smears (*S. epidermidis, E. coli, or Bacillus sp.*). Place the smear on the staining rack over the sink.
3. Cover the smear area with the crystal violet stain and leave it for 15 seconds and then rinse the slide with a gentle stream of water.
4. Apply Gram's iodine covering the smear completely for 15 seconds and then rinse.
5. Using the Gram's decolorizer, apply it a drop at a time to the smear area until no more colour leaves the area. Quickly rinse with water to stop the decolorizing process.
6. Apply Safranin to the smear for 15 seconds, then rinse with water.
7. Allow the smear to air dry or place it on the drying and fixation tray.

A. Simple Staining

To perform the simple staining procedure to compare morphological shapes and arrangements of bacterial cells

Principles

In simple staining, the bacterial smear is stained with single reagent. Basic stains with a positively charged chromogen are preferred, because bacterial nucleic acids and certain cell we all components carry a negative charges that strongly attracts and binds to the cationic chromogen. The propose of simple staining is to elucidate the morphology and arrangement of bacterial cells.

The most commonly used basic stains are methylene blue, crystal violet, and carbol fuchsin.

Materials required

- *Cultures:* 24 hrs nutrient agar slant culture of *E.coli* and *Bacillus cereus*, and a 24 hour nutrient broth culture of *S. aureus*.
- *Reagents:* Methylene blue, crystal violet, and carbol fuchsin.

Procedure

1. Prepare separate bacterial smears of the organisms, all smears must be heat fixed prior to staining.
2. Place a slide on the staining tray and flood the seamer with one of the indicated stains, using the appropriate exposure time for each carbol fuchsin, 15 to 30 seconds. Crystal violet, 20 to 60 seconds and methylene blue, 1 to 2 minute.
3. Wash the smear with tap water to remove excess stain. During this step, hold the slide parallel to the stream of water, in this way you can reduce the loss of organisms from the preparation.
4. Using bibulous paper, blot dry but do not wipe the slide.
5. Repeat this procedure with the remaining two organisms, using a different stain for each.
6. Examine all stained slides under oil immersion.

B. Negative Staining

Principle

Negative Staining requires the use of an acidic stain such as nigrosin. The acidic stain with its negatively charged chromogen, will not penetrate the cells because of the negative charge on the surface of bacterial. Therefore the unstained cells are easily discernible against the coloured background.

The pratical application of the negative staining is two fold: first, since heat fixation is not required and the cells are not subjected to the distorting effects of the chemicals and geat, their natural size and shape can be seen.

Second, it is possible to observe bacteria that are difficult to stain, such as some spirille.

Materials Required

Cultures: 24 hours agar slant cultures of *Micrococcus luteus*, *Bacillus cereus*, and *Aquaspirillum itersonii.*

Procedure

1. Place a small drop of nigrosin close to one end of a clean slide.
2. Using sterile technique, place a loopful of inoculum from the *Micrococcus luteus* culture in the drope of nigrosin and mix.
3. With the edge of a second slide held at a 30° angle and placed in front of the bacterial suspension, push the mixture to form a thin smear.
4. Air dry. Donot heat fix the slide.
5. Repeat steps 1 to 4 for slide preparations of *Bacillus cerus* and *Aquasprillum itersonii.*
6. Examine the slide under oil immersion.

8. DIRECT STAIN AND INDIRECT STAIN

A. INTRODUCTION TO STAINING

Bacterial morphology (form and structure) may be examined in two ways:

- by observing living unstained organisms (wet mount), or
- by observing killed stained organisms.

Since bacteria are almost colourless and therefore show little contrast with the broth in which they are suspended, they are difficult to observe when unstained. Staining microorganisms enables one to:

- see greater contrast between the organism and the background
- differentiate various morphological types (by shape, arrangement, gram reaction, etc.)
- observe certain structures (flagella, capsules, endospores, etc.).

Before staining bacteria, you must first understand how to "fix" the organisms to the glass slide. If the preparation is not fixed, the organisms will be washed off the slide during staining. A simple method is that of air drying and heat fixing. The organisms are heat fixed by passing an air-dried smear of the organisms through the flame of a gas burner. The heat coagulates the organisms' proteins causing the bacteria to stick to the slide.

The procedure for heat fixation is as follows:

1. If the culture is taken from an agar medium:
   (*a*) Using the dropper bottle of distilled water found in your staining rack, place 1/2 a drop of water on a clean slide by touching the dropper to the slide.
   (*b*) Aseptically remove a small amount of the culture from the agar surface and just touch it several times to the drop of water until it just turns cloudy.
   (*c*) Burn the remaining bacteria off of the loop. (If too much culture is added to the water, you will not see stained individual bacteria.)
   (*d*) Using the loop, spread the suspension over the entire slide to form a thin film.
   (*e*) Allow this thin suspension to completely air dry.
   (*f*) Pass the slide (film-side up) through the flame of the bunsen burner 3 or 4 times to heat-fix.
   Caution: Too much heat might distort the organism and, in the case of the gram stain, may cause gram-positive organisms to stain gram-negatively. The slide should feel very warm but not too hot to hold.
2. If the organism is taken from a broth culture:
   (*a*) Aseptically place 2 or 3 loops of the culture on a clean slide. Do not use water.
   (*b*) Using the loop, spread the suspension over the entire slide to form a thin film.
   (*c*) Allow this thin suspension to completely air dry.
   (*d*) Pass the slide (film-side up) through the flame of the bunsen burner 3 or 4 times to heat-fix.

In order to understand how staining works, it will be helpful to know a little about the physical and chemical nature of stains. Stains are generally salts in which one of the ions is coloured. (A salt is a compound composed of a positively charged ion and a negatively charged ion.) For example, the dye methylene blue is actually the salt methylene blue chloride which will dissociate in water into a positively charged methylene blue ion which is blue in colour and a negatively charged chloride ion which is colourless.

Dyes or stains may be divided into two groups: basic and acidic. If the colour portion of the dye resides in the positive ion, as in the above case, it is called a basic dye (examples: methylene blue, crystal violet, safranin). If the colour portion is in the negatively charged ion, it is called an acidic dye (examples: nigrosin, congo red).

Because of their chemical nature, the cytoplasm of all bacterial cells have a slight negative charge when growing in a medium of near neutral pH. Therefore, when using a basic dye, the positively charged colour portion of the stain combines with the negatively charged bacterial cytoplasm (opposite charges attract) and the organism becomes directly stained.

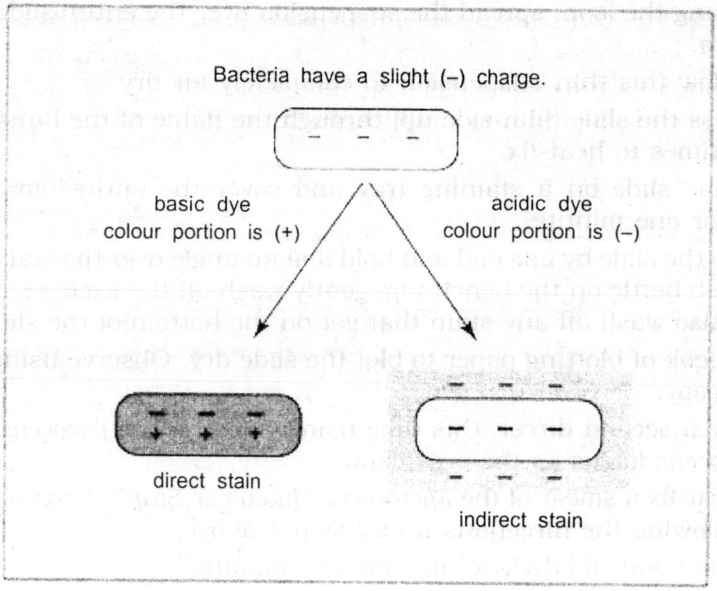

Fig. 15. Direct staining and indirect staining.

An **acidic dye,** due to its chemical nature, reacts differently. Since the colour portion of the dye is on the negative ion, it will not readily combine with the negatively charged bacterial cytoplasm (like charges repel). Instead, it forms a deposit around the organism, leaving the organism itself colourless. Since the organism is seen indirectly, this type of staining is called indirect or negative, and is used to get a more accurate view of bacterial size, shapes, and arrangements.

Try both direct and indirect stains of several microorganisms.

B. DIRECT STAIN USING A BASIC DYE

In direct staining the positively charged colour portion of the basic dye combines with the negatively charged bacterium and the organism becomes directly stained.

Organisms

Your pure cultures of *Staphylococcus epidermidis* (coccus with staphylococcus arrangement) or *Micrococcus luteus* (coccus with a tetrad or a sarcina arrangement) and *Escherichia coli* (small bacillus) or *Enterobacter aerogenes* (small bacillus).

Procedure

1. Heat-fix a smear of either *Escherichia coli* or *Enterobacter aerogenes* as follows:
   (*a*) Using the dropper bottle of distilled water found in your staining rack, place a small drop of water on a clean slide by touching the dropper to the slide.
   (*b*) Aseptically remove a small amount of the culture from the agar surface and touch it several times to the drop of water until it just turns cloudy.
   (*c*) Burn the remaining bacteria off of the loop. (If too much culture is added to the water, you will not see stained individual bacteria.)

(*d*) Using the loop, spread the suspension over the entire slide to form a thin film.

(*e*) Allow this thin suspension to completely air dry.

(*f*) Pass the slide (film-side up) through the flame of the bunsen burner 3 or 4 times to heat-fix.

2. Place the slide on a staining tray and cover the entire film with **safranin.** Stain for one minute.
3. Pick up the slide by one end and hold it at an angle over the staining tray. Using the wash bottle on the bench top, gently wash off the excess safranin from the slide. Also wash off any stain that got on the bottom of the slide as well.
4. Use a book of blotting paper to blot the slide dry. Observe using oil immersion microscopy.
5. Prepare a second direct, this time using either *Staphylococcus epidermidis* or *Micrococcus luteus* as the organism.
   (*a*) Heat-fix a smear of the *Micrococcus luteus* or *Staphylococcus epidermidis* by following the directions under step 1 above.
   (*b*) Stain with methylene blue for one minute.
   (*c*) Wash off the excess methylene blue with water.
   (*d*) Blot dry and observe using oil immersion microscopy.
6. Prepare a third slide of the normal flora and cells of your mouth.
   (a) Using a sterile cotton swab, vigorously scrape the inside of your mouth and gums.
   (*b*) Rub the swab over the slide (do not use water), air dry, and heat-fix.
   (c) Stain with crystal violet for 30 seconds.
   (*d*) Wash off the excess crystal violet with water.
   (*e*) Blot dry and observe. Find epithelial cells using your 10X objective, center them in the field, and switch to oil immersion to observe your normal flora bacteria on and around your epithelial cells.

C. INDIRECT STAIN USING AN ACIDIC DYE

In negative staining, the negatively charged colour portion of the acidic dye is repelled by the negatively charged bacterial cell. Therefore the background will be stained and the cell will remain colourless.

Organism

Your pure culture of *Staphylococcus epidermidis* or *Micrococcus luteus*.

Procedure

1. Place a small drop of nigrosin on a clean slide.
2. Aseptically add a small amount of *Staphylococcus epidermidis* or *Micrococcus luteus* to the dye and mix gently with the loop.
3. Using the edge of another slide, spread the mixture with varying pressure across the slide so that there are alternating light and dark areas. Make sure the dye is not to thick or you will not see the bacteria!
4. Let the film of dyed bacteria air dry completely on the slide. Do not heat fix and do not wash off the dye.

5. Observe using oil immersion microscopy. Find an area that has neither too much nor too little dye (an area that appears **light purple** where the light comes through the slide). If the dye is too thick, not enough light will pass through; if the dye is too thin, the background will be too light for sufficient contrast.

Results

Make drawings of your three direct stain preparations and your indirect stain preparation.

PERFORMANCE OBJECTIVES

A. Introduction to Staining

1. Describe the procedure for heat fixation.
2. Define the following: acidic dye, basic dye, direct stain, and indirect stain.
3. State in chemical and physical terms the principle behind direct staining and the principle behind indirect staining.

B. Direct Staining

Procedure

1. Transfer a small number of bacteria from an agar surface or a broth culture to a glass slide and heat-fix the preparation.
2. Prepare a direct stain when given all the necessary materials.

Results

1. Recognize a direct stain preparation when it is observed through a microscope and state the shape and arrangement of the organism.

C. Indirect Staining

Procedure

1. Perform an indirect stain when given all the necessary materials.
2. State why the dye is not washed off when doing an indirect stain.

Results

1. Recognize an indirect stain preparation when it is observed through a microscope and state the shape and arrangement of the organism.

9. GRAM STAIN AND CAPSULE STAIN

A. THE GRAM STAIN

Introduction

The gram stain is the most widely used staining procedure in bacteriology. It is called a **differential stain** since it differentiates between gram-positive and gram-negative bacteria. Bacteria which stain purple with the gram staining procedure are termed gram-positive; those which stain pink are said to be gram-negative. The terms positive and negative have nothing to do with electrical charge, but simply designate two distinct morphological groups of bacteria.

Gram-positive and gram-negative bacteria stain differently because of fundamental differences in the structure of their cell walls. The bacterial cell wall serves to give the organism its size and shape as well as to prevent osmotic lysis. The material in the bacterial cell wall which confers rigidity is peptidoglycan.

In electron micrographs, the gram-positive cell wall appears as a broad, dense wall 20-80 nm thick and consisting of numerous interconnecting layers of peptidoglycan. Chemically, 60 to 90% of the gram-positive cell wall is peptidoglycan. Interwoven in the cell wall of gram-positive are teichoic acids. Teichoic acids, which extend through and beyond the rest of the cell wall, are composed of polymers of glycerol, phosphates, and the sugar alcohol ribitol. Some have a lipid attached (lipoteichoic acid). The outer surface of the peptidoglycan is studded with proteins that differ with the strain and species of the bacterium.

The gram-negative cell wall, on the other hand, contains only 2-3 layers of peptidoglycan and is surrounded by an outer membrane composed of phospholipids, lipopolysaccharide, lipoprotein, and proteins. Only 10%-20% of the gram-negative cell wall is peptidoglycan. The phospholipids are located mainly in the inner layer of the outer membrane, as are the lipoproteins that connect the outer membrane to the peptidoglycan. The lipopolysaccharides, located in the outer layer of the outer membrane, consist of a lipid portion called lipid A embedded in the membrane and a polysaccharide portion extending outward from the bacterial surface. The outer membrane also contains a number of proteins that differ with the strain and species of the bacterium.

The gram staining procedure involves four basic steps:

1. The bacteria are first stained with the basic dye **crystal violet.** Both gram-positive and gram-negative bacteria become directly stained and appear purple after this step.
2. The bacteria are then treated with gram's iodine solution. This allows the stain to be retained better by forming an insoluble crystal violet-iodine complex. Both gram-positive and gram-negative bacteria remain purple after this step.
3. Gram's decolourizer, a mixture of ethyl alcohol and acetone, is then added. This is the differential step. Gram-positive bacteria retain the crystal violet-iodine complex while gram-negative are decolourized.
4. Finally, the counterstain safranin (also a basic dye) is applied. Since the gram-positive bacteria are already stained purple, they are not affected by the counterstain. Gram-negative bacteria, which are now colourless, become directly stained by the safranin. Thus, garm-positive appear purple and gram-negative appear pink.

With the current theory behind gram staining, it is thought that in gram-positive bacteria, the crystal violet and iodine combine to form a larger molecule that precipitates out within the cell. The alcohol/acetone mixture then causes dehydration of the multilayered peptidoglycan, thus decreasing the space between the molecules and causing the cell wall to trap the crystal violet-iodine complex within the cell. In the case of gram-negative bacteria, the alcohol/acetone mixture, being a lipid solvent, dissolves the outer membrane of the cell wall and may also damage the cytoplasmic membrane to which the peptidoglycan is attached. The single thin layer of peptidoglycan is unable to retain the crystal violet-iodine complex and the cell is decolourized.

It is important to note that gram-positivity (the ability to retain the purple crystal violet-iodine complex) is not an all-or-nothing phenomenon but a matter of degree. There are several factors which could result in a gram-positive organism staining gram-negatively:

1. *The method and techniques used.* Overheating during heat fixation, over decolourization with alcohol, and even too much washing with water between steps may result in gram-positive bacteria losing the crystal violet-iodine complex.
2. *The age of the culture.* Cultures more than 24 hours old may lose their ability to retain the crystal violet-iodine complex.
3. *The organism itself.* Some gram-positive bacteria are more able to retain the crystal violet-iodine complex than others.

Therefore, one must use very precise techniques in gram staining and interpret the results with discretion.

Organisms

Trypticase Soy agar plate cultures of *Escherichia coli* (a small, gram-negative rod) and *Staphylococcus epidermidis* (a gram-positive coccus in irregular, often grape-like clusters).

Procedure

1. Heat-fix a smear of a mixture of *Escherichia coli* and *Staphylococcus epidermidis* as follows:
   (*a*) Using the dropper bottle of distilled water found in your staining rack, place a small drop of water on a clean slide by touching the dropper to the slide.
   (*b*) Aseptically remove a small amount of Staphylococcus epidermidis from the agar surface and mix it generously with the water. Flame the loop and let it cool. Now aseptically remove a small amount of Escherichia coli and sparingly add it to the water. Flame the loop and let it cool.
   (*c*) Using the loop, spread the mixture over the entire slide to form a thin film.
   (*d*) Allow this thin suspension to completely air dry.
   (*e*) Pass the slide (film-side up) through the flame of the bunsen burner 3 or 4 times to heat-fix.
2. Stain with Hucker's crystal violet for one minute. Gently wash with water. Shake off the excess water but do not blot dry between steps.
3. Stain with gram's iodine solution for one minute and gently wash with water.
4. Decolourize by adding gram's decolourizer drop by drop until the purple stops flowing. Wash immediately with water.
5. Stain with safranin for one minute and wash with water.
6. Blot dry and observe using oil immersion microscopy.

B. THE CAPSULE STAIN

Introduction

Many bacteria secrete a slimy, viscous covering called a capsule or glycocalyx. This is usually composed of polysaccharide, polypeptide, or both.

The ability to produce a capsule is an inherited property of the organism, but the capsule is not an absolutely essential cellular component. Capsules are often produced only under specific growth conditions. Even though not essential for life, capsules probably help bacteria to survive in nature. Capsules help many pathogenic and normal flora bacteria to initially **resist phagocytosis** by the host's phagocytic cells. In soil and water, capsules help prevent bacteria from being engulfed by protozoans. Capsules also help many bacteria to **adhere to surfaces** and thus resist flushing.

Organisms

Skim Milk broth culture of *Enterobacter aerogenes*. The skim milk supplies essential nutrients for capsule production and also provides a slightly stainable background.

Procedure

1. Stir up the Skim Milk broth culture with your loop and place 2-3 loops of *Enterobacter aerogenes* on a slide.
2. Using your loop, spread it out over the entire slide to form a thin film.
3. Let it completely air dry. Do not heat fix. Capsules stick well to glass, and heat may destroy the capsule.
4. Stain with crystal violet for one minute.
5. Wash off the excess stain with copper sulfate solution. Do not use water!
6. Blot dry and observe using oil immersion microscopy. The organism and the milk dried on the slide will pick up the purple dye while the capsule will remain colourless.
7. Observe the demonstration capsule stain of *Streptococcus pneumoniae* (the pneumococcus), an encapsulated bacterium often with a diplococcus arrangement.

RESULTS

A. The Gram Stain

Make drawings of each bacterium on your gram stain preparation.

colour = gram reaction = shape =	colour = gram reaction = shape = arrangement =

B. The Capsule Stain

Make a drawing of your capsule stain preparation of *Enterobacter aerogenes* and the demonstration capsule stain of *Streptococcus pneumoniae.*

PERFORMANCE OBJECTIVES

A. The Gram Stain

Discussion

1. State why the gram stain is said to be a differential stain.
2. Describe the differences between a gram-positive and a gram-negative cell wall.
3. Describe a theory as to why gram-positive bacteria retain the crystal violet-iodine complex while gram-negatives become decolourized.
4. Describe three conditions that may result in a gram-positive organism staining gram-negatively.

Procedure

1. State the procedure for the gram stain.
2. Perform a gram stain when given all the necessary materials.

Results

1. Determine if a bacterium is gram-positive or gram-negative when microscopically viewing a gram stain preparation and state the shape and arrangement of the organism.

B. The Capsule Stain

Discussion

1. State the chemical nature and major functions of bacterial capsules.

Results

1. Recognize capsules as the structures observed when microscopically viewing a capsule stain preparation.

10. ENDOSPORE STAIN AND BACTERIAL MOTILITY

A. ENDOSPORE STAIN

Introduction

A few genera of bacteria, such as *Bacillus* and *Clostridium* have the ability to produce **resistant survival forms** termed **endospores.** Unlike the reproductive spores of fungi and plants, these endospores are resistant to heat, drying, radiation, and various chemical disinfectants.

Endospore formation (sporulation) occurs through a complex series of events. One is produced within each vegetative bacterium. Once the endospore is formed, the vegetative portion of the bacterium is degraded and the dormant endospore is released.

First the DNA replicates and a cytoplasmic membrane septum forms at one end of the cell. A second layer of cytoplasmic membrane then forms around one of the DNA molecules (the one that will become part of the endospore) to form a **forespore.** Both of these membrane layers then synthesize peptidoglycan in the space between them to form the first protective coat, the **cortex.** Calcium dipocolinate is also incorporated into the forming endospore. A **spore coat** composed of a keratin-like protein then forms around the cortex. Sometimes an outer membrane composed of lipid and protein and called an exosporium is also seen.

Finally, the remainder of the bacterium is degraded and the endospore is released. Sporulation generally takes around 15 hours.

The endospore is able to survive for long periods of time until environmental conditions again become favorable for growth. The endospore then germinates, producing a single vegetative bacterium.

Bacterial **endospores** are resistant to antibiotics, most disinfectants, and physical agents such as radiation, boiling, and drying. The impermeability of the spore coat is thought to be responsible for the endospore's resistance to chemicals. The heat resistance of endospores is due to a variety of factors:

- Calcium-dipicolinate, abundant within the endospore, may stabilize and protect the endospore's DNA.
- Specialized DNA-binding proteins saturate the endospore's DNA and protect it from heat, drying, chemicals, and radiation.
- The cortex may osmotically remove water from the interior of the endospore and the dehydration that results is thought to be very important in the endospore's resistance to heat and radiation.
- Finally, DNA repair enzymes contained within the endospore are able to repair damaged DNA during germination.

Due to the resistant nature of the endospore coats, endospores are difficult to stain. Strong dyes and vigorous staining conditions such as heat are needed. Once stained, however, endospores are equally hard to decolourize. Since few bacterial genera produce endospores, the endospore stain is a good diagnostic test for species of *Bacillus* and *Clostridium.*

Organisms

Trypticase Soy agar plate cultures of *Bacillus megaterium.*

Procedure

1. Heat-fix a smear of *Bacillus megaterium* as follows:
   (*a*) Using the dropper bottle of distilled water found in your staining rack, place a small drop of water on a clean slide by touching the dropper to the slide.
   (*b*) Aseptically remove a small amount of the culture from the edge of the growth (see Fig. 14) on the agar surface and generously mix it with the drop of water until the water turns cloudy.
   (*c*) Burn the remaining bacteria off of the loop.
   (*d*) Using the loop, spread the suspension over the entire slide to form a thin film.
   (*e*) Allow this thin suspension to completely air dry.
   (*f*) Pass the slide (film-side up) through the flame of the bunsen burner 3 or 4 times to heat-fix.

2. Place a piece of blotting paper over the smear and saturate with malachite green.
3. Let the malachite green sit on the slide for one minute and proceed to the next step.
4. Holding the slide with forceps, carefully heat the slide in the flame of a bunsen burner until the stain just begins to steam. Remove the slide from the heat until steaming stops; then gently reheat. Continue steaming the smear in this manner for five minutes. As the malachite green evaporates, continually add more. Do not let the paper dry out.
5. After five minutes of steaming, wash the excess stain and blotting paper off the slide with water. Don't forget to wash of any dye that got onto the bottom of the slide.
6. Blot the slide dry.
7. Now flood the smear with safranin and stain for one minute.
8. Wash off the excess safranin with water, blot dry, and observe using oil immersion microscopy. With this endospore staining procedure, endospores will stain green while vegetative bacteria will stain red.
9. Observe the demonstration slide of Bacillus anthracis. With this staining procedure, the vegetative bacteria stain blue and the endospores are colourless. Note the long chains of rod-shaped, endospore-containing bacteria.
10. Observe the demonstration slide of *Clostridium tetani.* With this staining procedure, the vegetative bacteria stain blue and the endospores are colourless. Note the "tennis racquet" appearance of the endospore-containing *Clostridium.*

B. BACTERIAL MOTILITY

Introduction

Many bacteria are capable of motility (the ability to move under their own power). Most motile bacteria propel themselves by special organelles termed **flagella.** The bacterial flagellum, is a noncontractile, semi-rigid, helical tube composed of protein and anchors to the bacterial cytoplasmic membrane and cell well by means of disk-like structures. The rotation of the inner disk causes the flagellum to act much like a propeller.

Bacterial motility constitutes unicellular behavior. In other words, motile bacteria are capable of a behavior called **taxis.** Taxis is a motile response to an environmental stimulus and functions to keep bacteria in an optimum environment.

The arrangement of the flagella about the bacterium is of use in classification and identification. The following flagellar arrangements may be found:

1. *monotrichous* a single flagellum at one pole.
2. *amphitrichous* single flagella at both poles.
3. *lophotrichous* two or more flagella at one or both poles of the cell.
4. *peritrichous* completely surrounded by flagella.

One group of bacteria, the **spirochetes,** has internally-located axial filaments or endoflagella. Axial filaments wrap around the spirochete towards the middle from both ends. They are located above the peptidoglycan cell wall but underneath the outer membrane or sheath.

To detect bacterial motility, we can use any of the following three methods: 1) direct observation by means of special-purpose microscopes (phase-contrast and dark-field), 2) motility media, and indirectly, 3) flagella staining.

(i) Direct observation of motility using special-purpose microscopes

(a) *Phase-contrast microscopy.* A phase-contrast microscope uses special phase-contrast objectives and a condenser assembly to control illumination and give an optical effect of direct staining. The special optics convert slight variations in specimen thickness into corresponding visible variation in brightness. Thus, the bacterium and its structures appear darker than the background.

(b) *Dark-field microscopy.* A dark-field microscope uses a special condenser to direct light away from the objective lens. However, bacteria (or other objects) lying in the transparent medium will scatter light so that it enters the objective. This gives the optical effect of an indirect stain. The organism will appear bright against the dark background. Dark field microscopy is especially valuable in observing the very thin spirochetes.

(ii) Motility test medium

Semi-solid Motility Test medium may also be used to detect motility. The agar concentration (0.3%) is sufficient to form a soft gel without hindering motility. When a nonmotile organism is stabbed into Motility Test medium, growth occurs only along the line of inoculation. Growth along the stab line is very sharp and defined. When motile organisms are stabbed into the soft agar, they swim away from the stab line. Growth occurs throughout the tube rather than being concentrated along the line of inoculation. Growth along the stab line appears much more cloud-like as it moves away from the stab. A tetrazolium salt (TTC) is incorporated into the medium. Bacterial metabolism reduces the TTC producing formazan which is red in colour. The more bacteria present at any location, the darker red the growth appears.

(iii) Flagella staining

If we assume that bacterial flagella confer motility, flagella staining can then be used indirectly to denote bacterial motility. Since flagella are very thin (20-28 nm in diameter), they are below the resolution limits of a normal light microscope and cannot be seen unless one first treats them with special dyes and mordants which build up as layers of precipitate along the length of the flagella, making them microscopically visible. This is a delicate staining procedure and will not be attempted here. We will, however look at several demonstration flagella stains.

Organisms

Trypticase Soy broth cultures of *Pseudomonas aeruginosa* and *Staphylococcus aureus*. *Caution: handle these organisms as pathogens.*

Medium. Motility Test medium (2 tubes).

Procedure

1. Observe the phase-contrast video microscopy demonstration of motile *Pseudomonas aeruginosa.*
2. Observe the dark-field microscopy demonstration of motile *Pseudomonas aeruginosa.*

3. Take 2 tubes of Motility Test medium per pair. Stab one with *Pseudomonas aeruginosa* and the other with *Staphylococcus aureus*. Incubate at 37°C until the next lab period.
4. Observe the flagella stain demonstrations of *Pseudomonas aeruginosa* (monotrichous) *Proteus vulgaris* (peritrichous) and *Spirillum undula* (lophotrichous) as well as the dark-field photomicrograph of the spirochete. When observing flagella stain slides, keep in mind that flagella often break off during the staining procedure so you have to look carefully to observe the true flagellar arrangement.

RESULTS

A. Endospore Stain

Make drawings of the various endospore stain preparations.

B. Bacterial Motility

1. Observe the phase contrast and dark-field microscopy demonstrations of bacterial motility.
2. Observe the two tubes of Motility Test medium.
3. Make drawings of the flagella stain demonstrations.

PERFORMANCE OBJECTIVES

A. Endospore Stain

Discussion

1. Name two endospore-producing genera of bacteria.
2. State the function of bacterial endospores.

Results

1. Recognize endospores as the structures observed in an endospore stain preparation.
2. Identify a bacterium as an endospore-containing *Clostridium* by its "tennis racquet" appearance.

B. Bacterial Motility

Discussion

1. Define the following flagellar arrangements: monotrichous, lophotrichous, amphitrichous, peritrichous, and axial filaments.
2. State the chemical nature and function of bacterial flagella.
3. Describe three methods of testing for bacterial motility and indicate how to interpret the results.

Results

1. Recognize bacterial motility when using phase-contrast or dark-field microscopy.
2. Interpret the results of Motility Test Medium.
3. Recognize monotrichous, lophotrichous, amphitrichous, and peritrichous flagellar arrangements.

11. ENUMERATION OF MICROORGANISMS

A. INTRODUCTION (PLATE COUNT)

The laboratory microbiologist often has to determine the number of bacteria in a given sample as well as having to compare the amount of bacterial growth under various conditions. Enumeration of microorganisms is especially important in dairy microbiology, food microbiology, and water microbiology.

Since the enumeration of microorganisms involves the use of extremely small dilutions and extremely large numbers of cells, scientific notation is routinely used in calculations.

The number of bacteria in a given sample is usually too great to be counted directly. However, if the sample is serially diluted and then plated out on an agar surface in such a manner that single isolated bacteria form visible isolated colonies.

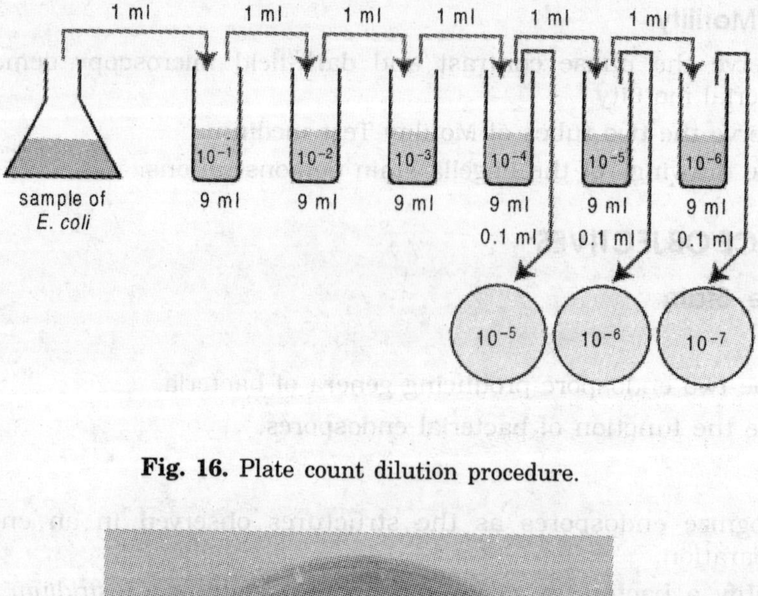

Fig. 16. Plate count dilution procedure.

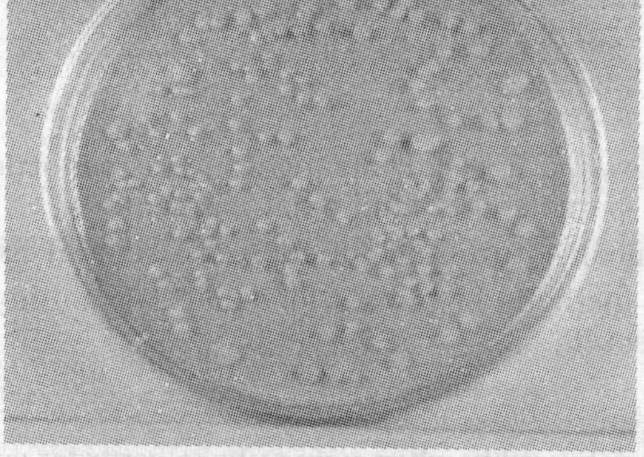

Fig. 17. Single isolated colonies obtained during the plate count

The number of colonies can be used as a measure of the number of viable (living) cells in that known dilution. However, keep in mind that if the organism normally forms multiple cell arrangements, such as chains, the colony-forming unit may consist of a chain of bacteria rather than a single bacterium. In addition, some of the bacteria may be clumped together. Therefore, when doing the plate count technique, we generally say we are determining the number of Colony-Forming Units (CFUs) in that known dilution. By extrapolation, this number can in turn be used to calculate the number of CFUs in the original sample.

Normally, the bacterial sample is diluted by factors of 10 and plated on agar. After incubation, the number of colonies on a dilution plate showing between 30 and 300 colonies is determined. A plate having 30-300 colonies is chosen because this range is considered statistically significant. If there are less than 30 colonies on the plate, small errors in dilution technique or the presence of a few contaminants will have a drastic effect on the final count. Likewise, if there are more than 300 colonies on the plate, there will be poor isolation and colonies will have grown together.

Generally, one wants to determine the number of CFUs per millilitre (ml) of sample. To find this, the number of colonies (on a plate having 30-300 colonies) is multiplied by the number of times the original ml of bacteria was diluted (the dilution factor of the plate counted). For example, if a plate containing a 1/1,000,000 dilution of the original ml of sample shows 150 colonies, then 150 represents 1/1,000,000 the number of CFUs present in the original ml. Therefore the number of CFUs per ml in the original sample is found by multiplying 150 × 1,000,000 as shown in the formula below:

number of CFUs per ml of sample = number of colonies (30-300 plate) × the dilution factor of the plate counted

In the case of the example above, 150 × 1,000,000 = 150,000,000 CFUs per ml.

For a more accurate count it is advisable to plate each dilution in duplicate or triplicate and then find an average count.

B. DIRECT MICROSCOPIC METHOD (TOTAL CELL COUNT)

In the direct microscopic count, a counting chamber consisting of a ruled slide and a coverslip is employed. It is constructed in such a manner that a known volume is delimited by the coverslip, slide, and ruled lines. The number of bacteria in a small known volume is directly counted microscopically and the number of bacteria in the larger original sample is determined by extrapolation.

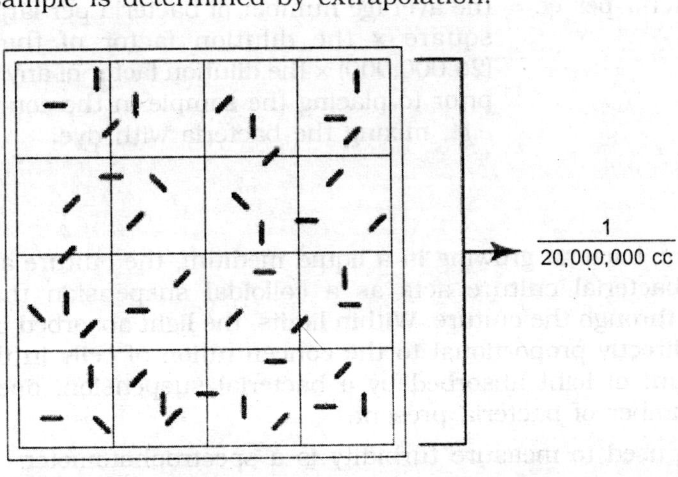

The Petroff-Hausser counting chamber

Fig. 18. Large Double Lined Square of a Petroff-Hausser Counter.

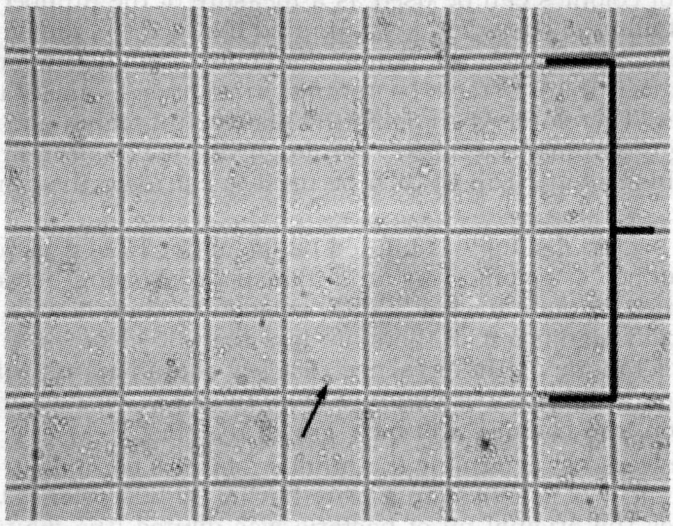

Fig. 19. Petroff-Hausser Counter as seen through a Microscope.

The double-lined "square" holding 1/20,000,000 cc is shown by the bracket. The arrow shows a bacterium.

The square holds a volume of 1/20,000,000 of a cubic centimeter. Using a microscope, the bacteria in the square are counted. For example, has squares 1/20 of a millimeter (mm) by 1/20 of a mm and is 1/50 of a mm deep. The volume of one square therefore is 1/20,000 of a cubic mm or 1/20,000,000 of a cubic centimeter (cc). The normal procedure is to count the number of bacteria in five large double-lined squares and divide by five to get the average number of bacteria per large square. This number is then multiplied by 20,000,000, since the square holds a volume of 1/20,000,000 cc, to find the total number of organisms per cc in the original sample.

If the bacteria are diluted (such as by mixing with dye) before being placed in the counting chamber, then this dilution must also be considered in the final calculations.

The formula used for the direct microscopic count is:

number of bacteria per cc = the average number of bacteria per large double-lined square × the dilution factor of the large square (20,000,000) × the dilution factor of any dilutions made prior to placing the sample in the counting chamber, *e.g.*, mixing the bacteria with dye.

C. TURBIDITY

When we mix the bacteria growing in a liquid medium, the culture appears **turbid.** This is because a bacterial culture acts as a colloidal suspension that blocks and reflects light passing through the culture. Within limits, the light absorbed by the bacterial suspension will be directly proportional to the concentration of cells in the culture. By measuring the amount of light absorbed by a bacterial suspension, one can estimate and compare the number of bacteria present.

The instrument used to measure turbidity is a spectrophotometer.

It consists of a light source, a filter which allows only a single wavelength of light to pass through, the sample tube containing the bacterial suspension, and a photocell that compares the amount of light coming through the tube with the total light entering the tube.

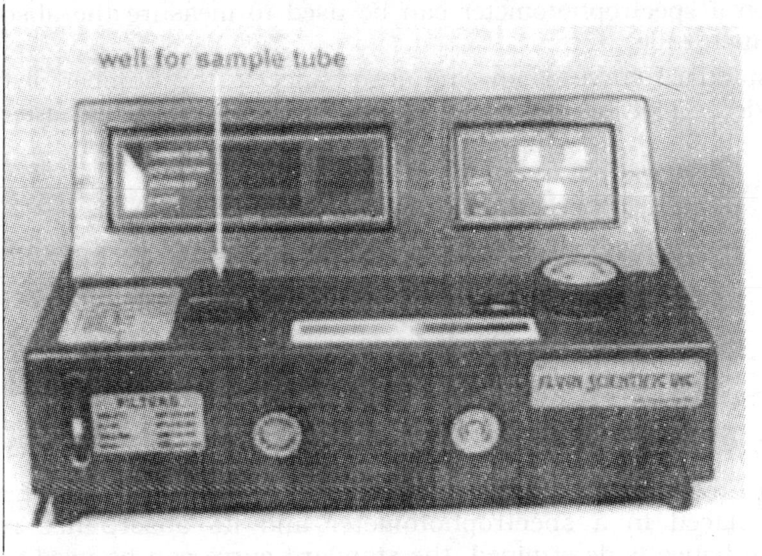

Fig. 20. A Spectrophotometer.

The ability of the culture to block the light can be expressed as either percent of light transmitted through the tube or the amount of light absorbed in the tube.

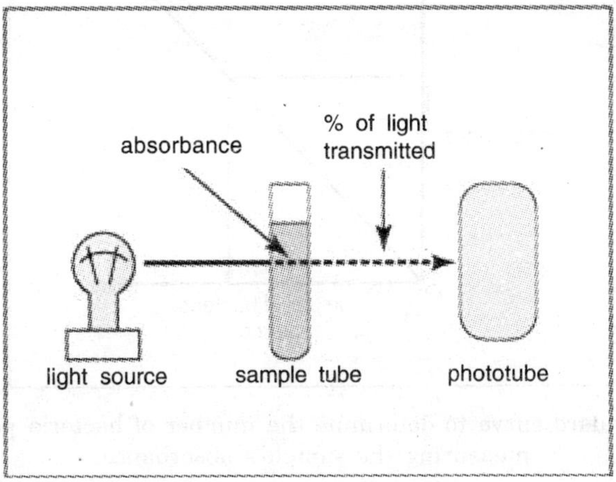

Fig. 21. A Spectrophotometer.

The percent of light transmitted is inversely proportional to the bacterial concentration. (The greater the percent transmittance, the lower the number of bacteria.) The absorbance (or optical density) is directly proportional to the cell concentration. (The greater the absorbance, the greater the number of bacteria.)

Turbidimetric measurement is often correlated with some other method of cell count, such as the direct microscopic method or the plate count. In this way, turbidity can be used as an indirect measurement of the cell count. For example:

1. Several dilutions can be made of a bacterial stock.
2. A Petroff-Hausser counter can then be used to perform a direct microscopic count on each dilution.
3. Then a spectrophotometer can be used to measure the absorbance of each dilution tube.
4. A standard curve comparing absorbance to the number of bacteria can be made by plotting absorbance versus the number of bacteria per cc.

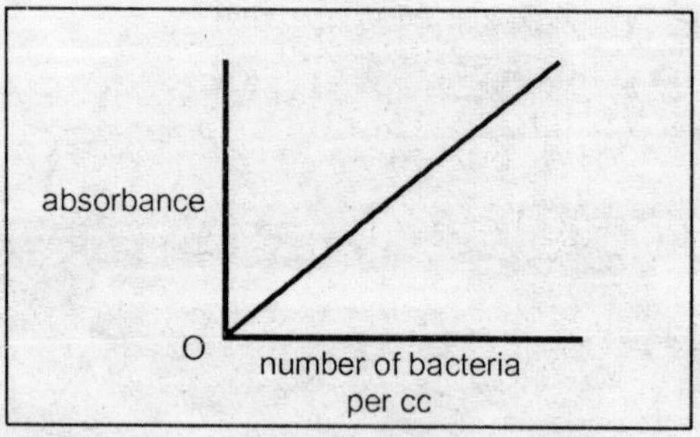

Fig. 22. A standard curve plotting the number of bacteria per cc versus absorbance.

5. Once the standard curve is completed, any dilution tube of that organism can be placed in a spectrophotometer and its absorbance read. Once the absorbance is determined, the standard curve can be used to determine the corresponding number of bacteria per cc.

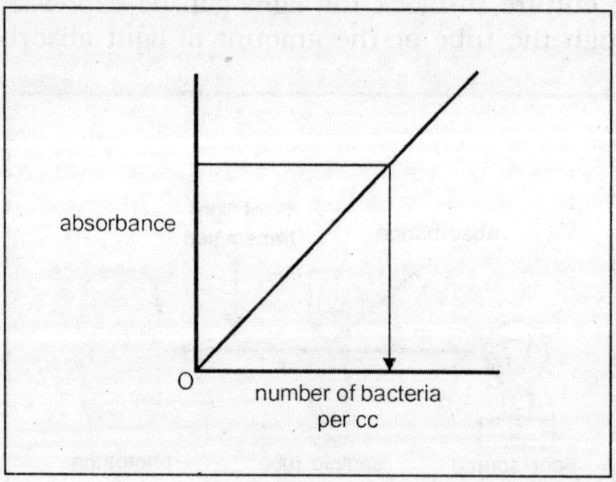

Fig. 23. Using a standard curve to determine the number of bacteria per cc in a sample by measuring the sample's absorbance.

Materials. 6 tubes each containing 9.0 ml sterile saline, 3 plates of Trypticase Soy agar, 2 sterile 1.0 ml pipettes, pipette filler, turntable, bent glass rod, dish of alcohol.

Organism. Trypticase Soy broth culture of *Escherichia coli.*

PROCEDURE

A. Plate Count

1. Take 6 dilution tubes, each containing 9.0 ml of sterile saline. Aseptically dilute 1.0 ml of a sample of *E. coli* as shown in and described below.

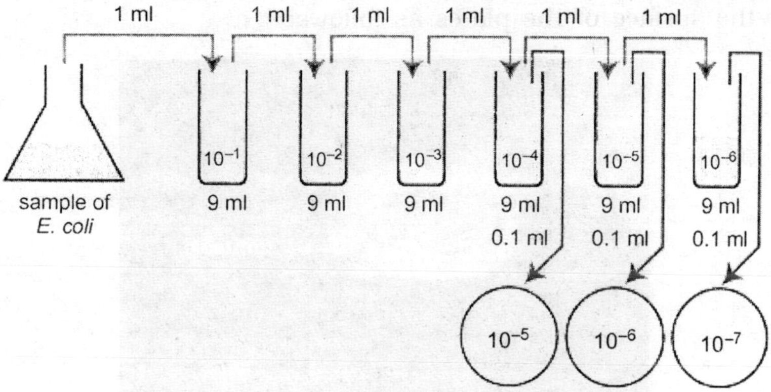

Fig. 24. Plate count dilution procedure.

(*a*) Remove a sterile 1.0 ml pipette from the bag. Do not touch the portion of the pipette tips that will go into the tubes and do not lay the pipette down. From the tip of the pipette to the "0" line is 1 ml; each numbered division (0.1, 0.2, etc.) represents 0.1 ml.

(*b*) Insert the cotton-tipped end of the pipette into a blue 2 ml pipette filler.

(*c*) Flame the sample flask, insert the pipette to the bottom of the flask, and withdraw 1.0 ml (up to the "0" line of the sample by turning the filler knob towards you. Draw the sample up slowly so that it isn't accidentally drawn into the filler itself. Reflame and cap the sample.

(*d*) Flame the first dilution tube and dispense the 1.0 ml of sample into the tube by turning the filler knob away from you. Draw the liquid up and down in the pipette several times to rinse the pipette and help mix. Reflame and cap the tube.

(*e*) Mix the tube thoroughly by either holding the tube in one hand and vigorously tapping the bottom with the other hand or by using a vortex mixer. This is to assure an even distribution of the bacteria throughout the liquid.

(*f*) Using the same procedure, aseptically withdraw 1.0 ml from the first dilution tube and dispense into the second dilution tube. Continue doing this from tube to tube as shown in until the dilution is completed. Discard the pipette in the biowaste disposal containers at the front of the room and under the hood.

These pipetting and mixing techniques will be demonstrated by your instructor.

2. Using a new 1.0 ml pipette, aseptically transfer 0.1 ml from each of the last three dilution tubes onto the surface of the corresponding plates of trypticase soy agar as shown in figure. Note that since only 0.1 ml of the bacterial dilution (rather than the desired 1.0 ml) is placed on the plate, the bacterial dilution on the plate is 1/10 the dilution of the tube from which it came.

Using a turntable and sterile bent glass rod immediately spread the solution over the surface of the plates as follows:

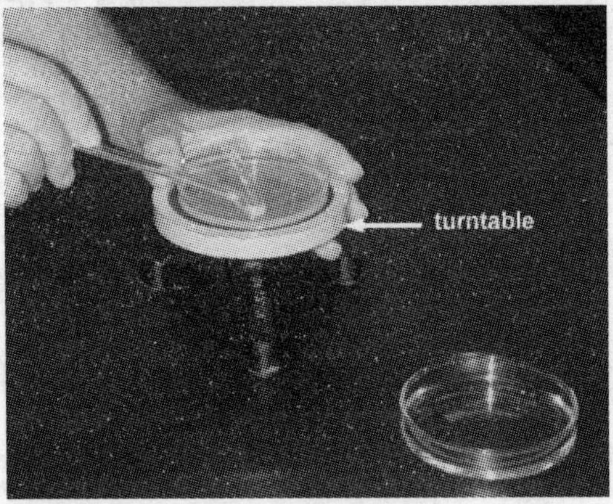

Fig. 25. Using a bent glass rod and a turntable to spread a bacterial sample

(*a*) Place the plate containing the 0.1 ml of dilution on a turntable.

(*b*) Sterilize the glass rod by dipping the bent portion in a dish of alcohol and igniting the alcohol with the flame from your burner. Let the flame burn out.

(*c*) Place the bent portion of the glass rod on the agar surface and spin the turntable for about 30 seconds to distribute the 0.1 ml of dilution evenly over the entire agar surface.

(*d*) Replace the lid and resterilize the glass rod with alcohol and flaming.

(*e*) Repeat for each plate.

(*f*) Discard the pipette in the biowaste disposal containers at the front of the room and under the hood.

3. Incubate the 3 agar plates upside down **at 37°C** until the next lab period. Place the used dilution tubes in the disposal baskets in the hood.

B. Direct Microscopic Method

1. Pipette 1.0 ml of the sample of *E. coli* into a tube containing 1.0 ml of the dye methylene blue. This gives a 1/2 dilution of the sample.
2. Using a Pasteur pipette, fill the chamber of a Petroff-Hausser counting chamber with this 1/2 dilution.
3. Place a coverslip over the chamber and focus on the squares using 400X (40X objective).
4. Count the number of bacteria in 5 large double-lined squares. For those organisms on the lines, count those on the left and upper lines but not those on the right and lower lines. Divide this total number by 5 to find the average number of bacteria per large square.
5. Calculate the number of bacteria per cc as follows:

Number of bacteria per cc = the average number of bacteria per large square × the dilution factor of the large square (20,000,000) × the dilution factor of any dilutions made prior to placing the sample in the counting chamber, such as mixing it with dye (2 in this case).

C. Turbidity

Your instructor will set up a spectrophotometer demonstration illustrating that as the number of bacteria in a broth culture increases, the absorbance increases (or the percent light transmitted decreases).

RESULTS

A. Plate Count

1. Choose a plate that appears to have between 30 and 300 colonies.
   - Sample 1/100,000 dilution plate
   - Sample 1/1,000,000 dilution plate
   - Sample 1/10,000,000 dilution plate.
2. Count the exact number of colonies on that plate using the colony counter (as demonstrated by your instructor).
3. Calculate the number of CFUs per ml of original sample as follows:

   number of CFUs per ml of sample = number of colonies (30-300 plate) × the dilution factor of the plate counted

   ____________ = number of colonies

   ____________ = dilution factor of plate counted

   ____________ = number of CFUs per ml.
4. Record your results on the blackboard.

B. Direct Microscopic Method

Observe the demonstration of the Petroff-Hausser counting chamber.

C. Turbidity

Observe your instructor's demonstration of the spectrophotometer.

PERFORMANCE OBJECTIVES

Discussion

1. State the formula for determining the number of CFUs per ml of sample when using the plate count technique.
2. When given a diagram of a plate count dilution and the number of colonies on the resulting plates, choose the correct plate for counting, determine the dilution factor of that plate, and calculate the number of CFUs per ml in the original sample.
   - Plate count practice problems

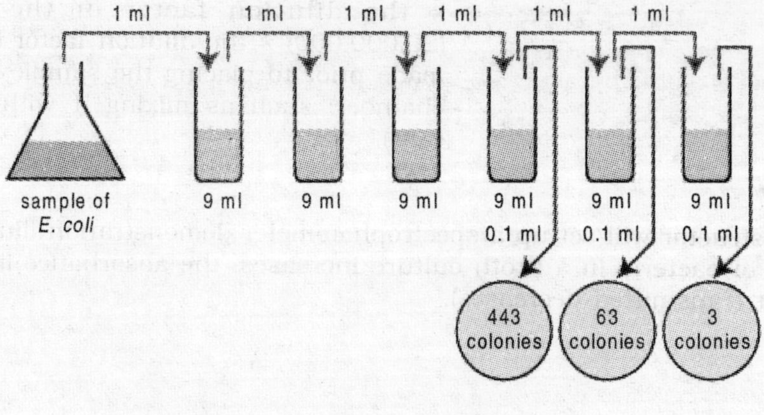

Plate Count : Practice problem #1.

A sample of E.coli is diluted according to the above diagram. The number of colonies that grew is indicated on the petri plates. How many CFUs are there per ml in the original sample?

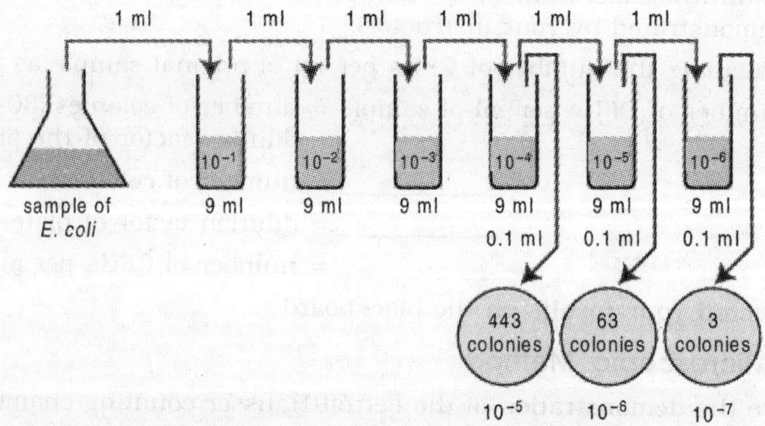

Plate Count : Answer to practice problem #1.

The correct dilutions are shown on the tubes and plates above.

The formula to be used is:

the number of CFUs per ml of sample = the number of colonies (30-300) × the dilution factor of the plate counted.

- First choose the correct plate to count, that is, one having between 30 and 300 colonies.
  - The correct plate is the one having 63 colonies on the 1/1,000,000 or 10^{-6} dilution.
  - Multiply 63 by the dilution factor of that plate.

- o Since the dilution factor is 1/1,000,000 or 10^{-6}, the dilution factor or inverse is 1,000,000 or 10^6.
- o 63 × 1,000,000 = 63,000,000 CFUs per ml (6.3×10^7 in scientific notation.)

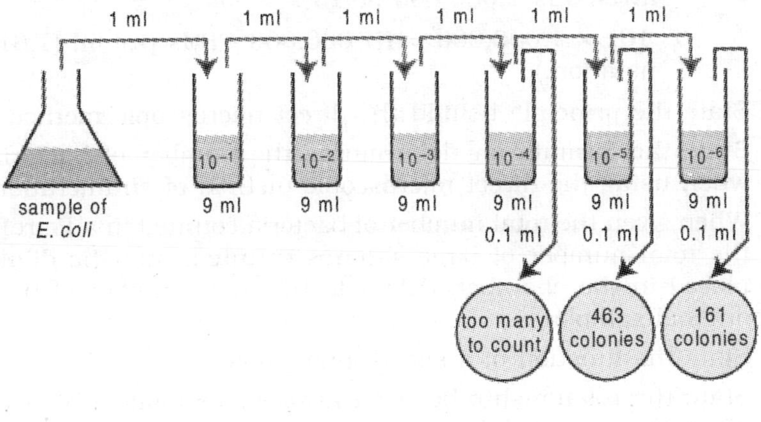

Plate Count : Practice problem #2.

A sample of E.coli is diluted according to the above diagram. The number of colonies that grew is indicated on the petri plates. How many CFUs are there per ml in the original sample?

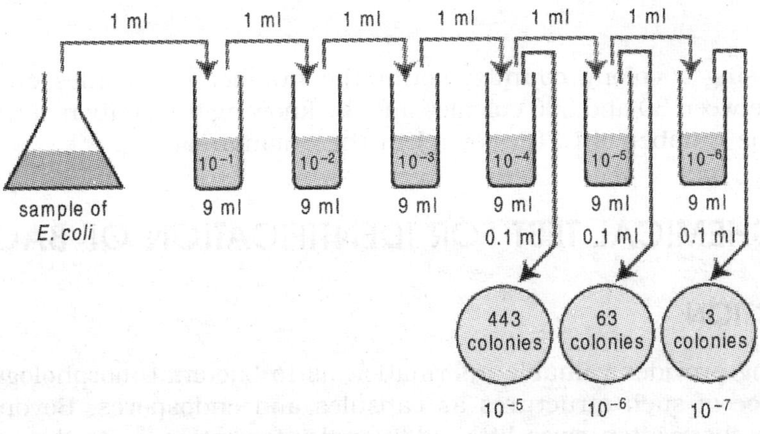

Plate Count : Answer to practice problem #2.

The correct dilutions are shown on the tubes and plates above.

The formula to be used is:

the number of CFUs per ml of sample = the number of colonies (30-300) × the dilution factor of the plate counted.

- First choose the correct plate to count, that is, one having between 30 and 300 colonies.

- o The correct plate is the one having 161 colonies on the 1/1,000,000 or 10^{-6} dilution.
- Multiply 161 by the dilution factor of that plate.
  - o Since the dilution factor is 1/1,000,000 or 10^{-6}, the dilution factor or inverse is 1,000,000 or 10^{6}.
  - o 161 × 1,000,000 =161,000,000 CFUs per ml (1.61 × 10^{8} in scientific notation.)

3. State the principle behind the direct microscopic method of enumeration.
4. State the formula for determining the number of bacteria per cc of sample when using the direct microscopic method of enumeration.
5. When given the total number of bacteria counted in a Petroff-Hausser chamber, the total number of large squares counted, and the dilution of the bacteria placed in the chamber, calculate the total number of bacteria per cc in the original sample.
6. State the function of a spectrophotometer.
7. State the relationship between absorbance (optical density) and the number of bacteria in a broth sample.
8. State the relationship between percent light transmitted and the number of bacteria in a broth sample.

Procedure

1. Perform a serial dilution of a bacterial sample according to instructions in the lab manual and plate out samples of each dilution using the spin-plate technique.

Results

1. Using a colony counter, count the number of colonies on a plate showing between 30 and 300 colonies and, by knowing the dilution of this plate, calculate the number of CFUs per ml in the original sample.

12. BIOCHEMICAL TEST FOR IDENTIFICATION OF BACTERIA

INTRODUCTION

Staining provides valuable information as to bacterial morphology, gram reaction, and presence of such structures as capsules and endospores. Beyond that, however, microscopic observation gives little additional information as to the genus and species of a particular bacterium.

To identify bacteria, we must rely heavily on biochemical testing. The types of biochemical reactions each organism undergoes act as a "thumbprint" for its identification. This is based on the following chain of logic:

- Each different species of bacterium has a different molecule of DNA (i.e., DNA with a unique series of nucleotide bases).
- Since DNA codes for protein synthesis, then different species of bacteria must, by way of their unique DNA, be able to synthesize different protein enzymes.

- Enzymes catalyze all the various chemical reactions of which the organism is capable. This in turn means that different species of bacteria must carry out different and unique sets of biochemical reactions.

When identifying a suspected organism, you inoculate a series of differential media. After incubation, you then observe each medium to see if specific end products of metabolism are present. This can be done by adding indicators to the medium that react specifically with the end product being tested, giving some form of visible reaction such as a colour change. The results of these tests on the suspected microorganism are then compared to known results for that organism to confirm its identification.

Different bacteria because of their unique enzymes, are capable of different biochemical reactions. It will also show the results of the activity of those enzymes. In later labs we will use a wide variety of special purpose differential media frequently used in the clinical laboratory to identify specific pathogenic and opportunistic bacteria.

In general, we can classify enzymes as being either exoenzymes or endoenzymes. Exoenzymes are secreted by bacteria into the surrounding environment in order to break down larger nutrient molecules so they may enter the bacterium. Once inside the organism, some of the nutrients are further broken down to yield energy for driving various cellular functions, while others are used to form building blocks for the synthesis of cellular components. These later reactions are catalyzed by endoenzymes located within the bacterium.

A. STARCH HYDROLYSIS

Introduction

Starch is a polysaccharide which appears as a branched polymer of the simple sugar glucose. This means that starch is really a series of glucose molecules hooked together to form a long chain. Additional glucose molecules then branch off of this chain as shown below:

GLU
|
(—GLU-GLU-GLU-GLU-GLU-GLU-GLU—)n

Some bacteria are capable of using starch as a source of carbohydrate but in order to do this, they must first hydrolyze or break down the starch so it may enter the cell. The bacterium secretes an exoenzyme which hydrolyzes the starch by breaking the bonds between the glucose molecules. This enzyme is called a diastase.

(—GLU / GLU / GLU / GLU / GLU / GLU / GLU—)n
action of diastase

The glucose can then enter the bacterium and be used for metabolism.

Medium and Organisms. Trypticase Soy broth cultures of *Bacillus subtilis* and *Escherichia coli.*

Procedure

1. Using a wax marker, draw a line on the bottom of a Starch agar plate so as to divide the plate in half. Label one half *B. subtilis* and the other half *E. coli.*
2. Make a single streak line with the appropriate organism on the corresponding half of the plate.

3. Incubate at 37°C until the next lab period.
4. Next period, iodine will be added to see if the starch remains in the agar or has been hydrolyzed by the exoenzyme diastase. Iodine reacts with starch to produce a dark brown or blue/black colour. If starch has been hydrolized there will be a clear zone around the bacterial growth. If starch has not been hydrolized, the agar will remain a dark brown or blue/black colour.

B. PROTEIN HYDROLYSIS

Introduction

Proteins are made up of various amino acids linked together in long chains by means of peptide bonds. Many bacteria can hydrolyze a variety of proteins into peptides (short chains of amino acids) and eventually into individual amino acids. They can then use these amino acids to synthesize their own proteins and other cellular molecules or to obtain energy. The hydrolysis of protein is termed proteolysis and the enzyme involved is called a protease. In this exercise we will test for bacterial hydrolysis of the protein casein, the protein that gives milk its white, opaque appearance.

Organisms. Trypticase Soy broth cultures of *Bacillus subtilis* and *Escherichia coli.*

Procedure

1. Divide the Skim Milk agar plate in half and inoculate one half with *Bacillus subtilis* and the other half with *Escherichia coli* as done above with the above starch agar plate.
2. Incubate at 37°C until the next lab period. If casein is hydrolized, there will be a clear zone around the bacterial growth. If casein is not hydrolized, the agar will remain white and opaque.

C. FERMENTATION OF CARBOHYDRATES

Introduction

Carbohydrates are complex chemical substrates which serve as energy sources when broken down by bacteria and other cells. They are composed of carbon, hydrogen, and oxygen (with hydrogen and oxygen being in the same ratio as water; $[CH_2O]$) and are usually classed as either sugars or starches.

Facultative anaerobic and anaerobic bacteria are capable of fermentation, an anaerobic process during which carbohydrates are broken down for energy production. A wide variety of carbohydrates may be fermented by various bacteria in order to obtain energy and the types of carbohydrates which are fermented by a specific organism can serve as a diagnostic tool for the identification of that organism.

We can detect whether a specific carbohydrate is fermented by looking for common end products of fermentation. When carbohydrates are fermented as a result of bacterial enzymes, the following fermentation end products may be produced:

1. acid end products, or
2. acid and gas end products.

In order to test for these fermentation products, you inoculate and incubate tubes of media containing a single carbohydrate (such as lactose or maltose), a pH indicator (such as phenol red) and a durham tube (a small inverted tube to detect gas production). If the particular carbohydrate is fermented by the bacterium, acid end products will be produced which lowers the pH, causing the pH indicator to change colour (phenol red turns yellow). If gas is produced along with the acid, it collects in the durham tube as a gas bubble. If the carbohydrate is not fermented, no acid or gas will be produced and the phenol red will remain red.

Media. 3 tubes of Phenol Red Lactose broth and 3 tubes of Phenol Red Maltose broth.

Organisms. Trypticase Soy agar cultures of *Bacillus subtilis*, *Escherichia coli*, and *Staphylococcus aureus*.

Procedure

1. Label each tube with the name of the sugar in the tube and the name of the bacterium you are growing.
2. Inoculate one Phenol Red Lactose broth tube and one Phenol Red Maltose broth tube with *Bacillus subtilis.*
3. Inoculate a second Phenol Red Lactose broth tube and a second Phenol Red Maltose broth tube with *Escherichia coli.*
4. Inoculate a third Phenol Red Lactose broth tube and a third Phenol Red Maltose broth tube with *Staphylococcus aureus.*
5. Incubate all tubes **at 37°C** until next lab period.

D. INDOLE AND HYDROGEN SULPHIDE PRODUCTION

Introduction

Sometimes we look for the production of products produced by only a few bacteria. As an example, some bacteria use the enzyme tryptophanase to convert the amino acid tryptophan into molecules of indole, pyruvic acid and ammonia. Since only a few bacteria contain tryptophanase, the formation of indole from a tryptophan substrate can be another useful diagnostic tool for the identification of an organism. Indole production is a key test for the identification of *Escherichia coli.*

By adding Kovac's reagent to the medium after incubation we can determine if indole was produced. Kovac's reagent will react with the indole and turn **red.**

Likewise, some bacteria are capable of breaking down sulphur containing amino acids (cystine, methionine) or reducing inorganic sulphur-containing compounds (such as sulphite, sulphate, or thiosulphate) to produce hydrogen sulphide (H_2S). This reduced sulphur may then be incorporated into other cellular amino acids, or perhaps into coenzymes. The ability of an organism to reduce sulphur-containing compounds to hydrogen sulphide can be another test for identifying unknown organisms such as certain *Proteus* and *Salmonella.* To test for hydrogen sulphide production, a medium with a sulphur-containing compound and iron salts is inoculated and incubated. If the sulphur is reduced and hydrogen sulphide is produced, it will combine with the iron salt to form a visible **black** ferric sulphide (FeS) in the tube.

Medium. Three tubes of SIM (Sulphide, Indole, Motility) medium. This medium contains a sulphur source, an iron salt, the amino acid tryptophan, and is semi-solid in agar content (0.3%). It can be used to detect hydrogen sulphide production, indole production, and motility.

Organisms. Trypticase Soy agar cultures of *Proteus mirabilis, Escherichia coli,* and *Enterobacter cloacae.*

Procedure

1. Stab one SIM medium tube with *Proteus mirabilis.*
2. Stab a second SIM medium tube with *Escherichia coli.*
3. Stab a third SIM medium tube with *Enterobacter cloacae.*
4. Incubate at 37°C until the next lab period.
5. Next lab period add Kovac's reagent to each tube to detect indole production.

E. CATALASE ACTIVITY

Introduction

Catalase is the name of an enzyme found in most bacteria which initiates the breakdown of hydrogen peroxide (H_2O_2) into water (H_2O) and free oxygen (O_2).

During the normal process of aerobic respiration, hydrogen ions (H^+) are given off and must be removed by the cell. The electron transport chain takes these hydrogen ions and combines them with half a molecule of oxygen (an oxygen atom) to form water (H_2O). During the process, energy is given off and is trapped and stored in ATP. Water is then a harmless end product. Some cytochromes in the electron transport system, however, form toxic hydrogen peroxide (H_2O_2) instead of water and this must be removed. This is done by the enzyme catalase breaking the hydrogen peroxide into water and oxygen as shown above. Most bacteria are catalase-positive; however, certain genera that don't carry out aerobic respiration, such as *Streptococcus, Lactobacillus,* and *Clostridium,* are catalase-negative.

Materials

Trypticase Soy agar cultures of *Staphylococcus aureus* and *Streptococcus lactis,* 3% hydrogen peroxide.

Procedure

Add a few drops of 3% hydrogen peroxide to each culture and look for the release of oxygen as a result of hydrogen peroxide breakdown. This appears as foaming.

RESULTS OF A BIOCHEMICAL TESTS

A. Starch Hydrolysis

When iodine is added to starch, the iodine-starch complex which forms gives a characteristic dark brown or deep purple colour reaction. If the starch has been hydrolyzed into glucose molecules by the diastase exoenzyme, it no longer gives this reaction. Therefore, flood the surface of the Starch agar plate with gram's iodine. If the exoenzymes of the organism broke down the starch in the agar, a clear zone will surround the bacterial growth. If the organism lacks the exoenzyme to break down the starch, the agar around the growth should turn dark brown or blue/black colour due to the iodine-starch complex.

Record your results and indicate which organism was capable of hydrolyzing the starch (+ = hydrolysis; – = no hydrolysis).

B. Protein Hydrolysis

The protein casein exists as a colloidal suspension in milk and gives milk its characteristic white, opaque appearance. If the casein in the milk is hydrolyzed into peptides and amino acids, it will lose its opaqueness. Therefore, if the bacteria has the exoenzyme capable of casein hydrolysis, there will be a clear zone around the bacterial growth. If the organism lacks the exoenzyme to break down casein, the Skim Milk agar will remain white and opaque.

Record your results and indicate which organism was capable of hydrolyzing casein (+ = hydrolysis; – = no hydrolysis).

C. Fermentation of Carbohydrates

Phenol red pH indicator is red at neutral pH and yellow at acid pH. A change in colour in the tube from red to yellow indicates that the organism has fermented that particular carbohydrate, producing acid end products. Gas bubbles in the durham tube indicate gas was also produced from fermentation. (The results of fermentation may be acid alone or acid plus gas, but never gas alone.) If the phenol red remains red, no acid was produced and the carbohydrate was not fermented.

- Carbohydrate fermentation producing acid but no gas.
- Carbohydrate fermentation producing acid and gas.
- No carbohydrate fermentation. No acid or gas.

Carbohydrate Fermentation

Record your results below (+ = positive; – = negative).

Organism	*Phenol Red Maltose*	*Phenol Red Lactose*
Bacillus subtilis acid gas fermentation		
Escherichia coli acid gas fermentation		
Staphylococcus aureus acid gas fermentation		

D. Production of Indole and Hydrogen Sulphide

1. Observe the SIM tubes. A black colour indicates the organism has produced hydrogen sulphide.
2. Carefully add a dropper full of Kovac's reagent to each tube. A red colour indicates the production of indole.

SIM Medium

Record your results below (+ = positive; – = negative).

Organism	Indole	Hydrogen Sulphide
Escherichia coli		
Enterobacter cloacae		
Proteus mirabilis		

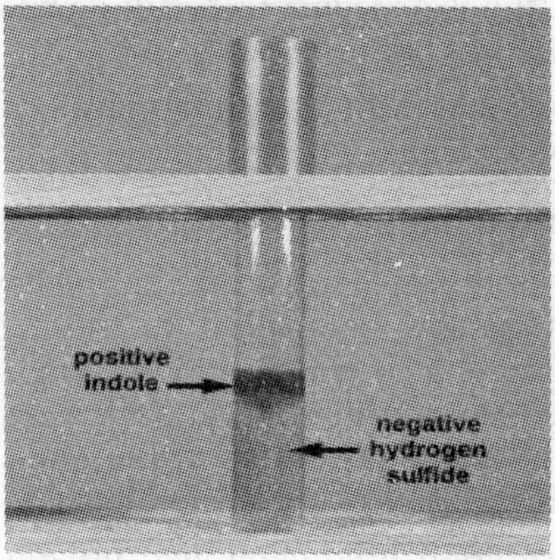

E. Catalase Activity

If the bacterium produces the enzyme catalase, then the hydrogen peroxide added to the culture will be broken down into water and free oxygen. The oxygen will bubble through the water causing a surface froth to form. A catalase-negative bacterium will not be able to break down the hydrogen peroxide, and no frothing will occur.

Catalase Test

Record your results below (foaming = positive; no foaming = negative).

PERFORMANCE OBJECTIVES

Introduction

1. State the chemical nature and function of enzymes.
2. Define endoenzyme and exoenzyme.

A. Starch Hydrolysis

Discussion

1. Describe a method of testing for starch hydrolysis and state how to interpret the results.

Results

1. Interpret the results of starch hydrolysis on a Starch agar plate that has been inoculated, incubated, and flooded with iodine.

B. Protein Hydrolysis

Discussion

1. Describe a method of testing for casein hydrolysis and state how to interpret the results.

Results

1. Interpret the results of casein hydrolysis on a Skim Milk agar plate after it has been inoculated and incubated.

C. Fermentation of Carbohydrates

Discussion

1. Name the general end products which may be formed as a result of the bacterial fermentation of sugars and describe how these end products change the appearance of a broth tube containing a sugar, the pH indicator phenol red, and a durham tube.

Results

1. Interpret the carbohydrate fermentation results in tubes of Phenol Red Carbohydrate broth containing a durham tube after it has been inoculated and incubated.

D. Indole and Hydrogen Sulphide Production

Discussion

1. State the pathway for the breakdown of tryptophan to indole.
2. State the pathway for the detection of sulphur reduction in SIM medium.
3. State three reactions that may be tested for in SIM medium and describe how to interpret the results.

Results

1. Interpret the hydrogen sulphide and indole results in a SIM medium tube after inoculation, incubation, and addition of Kovac's reagent.

E. Catalase Activity

Discussion

1. State the function of the enzyme catalase and describe a method of testing for catalase activity.

Results

1. Interpret the results of a catalase test after adding hydrogen peroxide to a plate culture of bacteria.

TRIPLE SUGAR IRON TEST

AIM

To detect whether the given test organism can ferment sugars and produce H_2S by triple sugar iron test.

Introduction

Triple sugar iron agar is used for differentiation of members of Enterobacteriaceae, according to their ability to ferment lactose, sucrose dextrose and to produce H_2S. The medium contains phenyl red as pH indicator. The indicator at different pH shows different colours in the medium. When oxidative decarboxylation of proteins take place, H_2S produced imparts a black precipitate resulting from the reaction between H_2S and ferrous sulphate present in the medium has produced (usually hydrogen, carbon dioxide) is visible as bubbles in the medium or cracking of medium.

Principle

The medium contains little amount of dextrose as compared to lactose and sucrose which results in the formation of very little amount of acid by dextrose fermentation, which is quickly oxidized by aerobic growth on the slant, that reverts back the colour of the slant to pink in absence of lactose and sucrose fermentation. Thus pink slant and yellow butt is dextrose fermentation.

Requirements

Test cultures test tubes, conical flasks, glass rod, inoculation loop and triple sugar iron agar.

Procedure

1. Prepare the TSI media and distribute among all the tubes.
2. Sterilise the media in the test tubes in an autoclave for 15 min at 151b/inch sq
3. Cool the tubes after sterilization and prepare the slants in such a way that a small but region remains in the bottom of the tube and an upper slant region.
4. Allow this to solidify.
5. Under aseptic conditions, using an inoculation loop, take the culture of the test organisms and pierce into the but region and strek in the slant region.
6. Incubate the tube for 24–48 hrs at 37°C and then record the observations.

STARCH HYDROLYSIS TEST (II METHOD)

AIM

To study the hydrolysis of starch by microorganisms by the production of the enzyme amylase.

Introduction

Starch is a polysaccharide found abundantly in plants and are usually deposited in the form of large granules in the cytoplasm of the cell. Starch granules can be isolated from the cell extracts by differential centrifugation. Starch consists of two components-amylase and amylopectin which are present in various amounts. The amylase consists of D-glucose units linked in a linear fashion by α-1,4 linkages. It has two non reducing ends and a reducing end. Amylopectin is a branched polysaccharide. In these molecules shorter chains of glucose units linked by α-1,4 are also joined to each other by α-1,6 linkages. The major component of starch can be hydrolysed by α-amylase which is present in saliva and pancreatic juice and participate in digestion of starch in gastro intestinal tract.

Principle

Starch is a polysaccharide made of two components, namely amylase and amylopectin. Amylose is not truly soluble in water but forms hydrated micelle which gives blue colour with iodine. Amylose gives a characteristic blue colour with iodine but the halide occupy a position in the interior of a helical coil of glucose units. This happens when amylase is suspended in water. Amylopectin yields a micellar which gives a violet colour with iodine.

Requirements

Petri plates, conical flasks, starch agar media, bacterial specimen and iodine.

Procedure

A. Preparation of starch agar

Beef extract – 3 g

Agar agar – 15 g

Starch – 3 g

Tryptone – 5 g

Distilled water – 1000 ml

pH-7

B. Steps

1. Soluble starch is dissolved in little water and is heated slowly with constant stirring. Than all the ingredients are added to it and transferred into a conical flask and sterilized by autoclaving at 121.5°C for 15 min.
2. The sterilized agar medium is poured into the sterilized Petri plates and allowed to solidify.
3. Each plate is inoculated at the center with the bacterial inoculum.
4. Plates are incubated at 37°C for 24-48 hrs.
5. To test the hydrolysis of starch each plate is flooded with iodine.

Result

Observe your experimental result.

GELATIN HYDROLYSIS TEST

AIM

To study the ability of microorganisms to hydrolyse gelatin by proteolytic enzyme gelatinase.

Principle

The ability of microorganisms to hydrolyse gelatin is commonly taken as the evidence of the organism to hydrolyse protein in general. But there are exceptions. Microorganisms vary from species to species with regard to their ability to hydrolyse protein. This feature is taken to characterize some species.

Gelatin is a protein obtained by the hydrolysis of collagen compound of connective tissues of animals. It is convenient as a substrate for proteolytic enzymes in microorganisms.

Gelatin is used in the media from the experiment which is liquid at room temperature and solidifies at –4 centigrade. If the gelatin has been hydrolyzed by the action of organism the media will remain liquid.

Requirements

Nutrient gelatin media, test organism, test tubes, inoculation loop.

Procedure

Preparation of Nutrient Gelatin Media
Composition
Peptone – 5 g
Gelatin – 20 g
Beef Extract – 3 g
Sodium Chloride – 5 g
Distilled Water – 1000 ml
pH–7.2.

Different Steps

1. Media is prepared according to the above given composition.
2. It is sterilized at 121°C for 15 min at 15 lb/inch square and poured into presterilized tubes.
3. Tubes were allowed to cool and then inoculated with test organisms, one inoculated tube used as control.
4. Tubes were incubated for 24 hrs and observed for liquefaction of gelatin after keeping in ice for half an hour.

Result

Observe your experimental result.

Discussion

Gelatin is an incomplete protein lacking many aminoacids such as tryptophan. When collagen is heated and hydrolysed denatured protein gelatin is obtained. Collagen accounts for 90-95% of organic matter in the cell. It is the most important protein rich in aminoacids. Microorganism like bacteria can use gelatin only if they are supplemented with other proteins. Bacteria produce gelatin hydrolyzing enzyme gelatinase. Since gelatine is a good solidifying agent at low temperature, its property of solidification can be used to distinguish between gelatin hydrolyzing and non hydrolyzing agent. Most of the enterobacteriaceae member are gelatin hydrolysis test negative. Bacteria like Vibrio, Bacillus, Pseudomonas are gelatin positive.

CATALASE TEST (II METHOD)

To study the organism that are capable of producing enzyme catalase.

Introduction

Most aerobic and facultative bacteria that utilize oxygen to produce hydrogen peroxide. This hydrogen peroxide that they produce is toxic to their own enzyme system. Thus hydrogen peroxide act as an antimetabolite.

Their survival in the presence of toxic antimetabolite is possible because these organism produce an enzyme called catalase. This enzyme converts peroxides into water and oxygen.

Principle

The enzyme catalase present in most microorganism is responsible for the breakdown of toxic hydrogen peroxide that could accumulate in the cell as a result of various metabolic activities into non-toxic substances, water and oxygen.

Reaction

The hydrogen peroxide formed by certain bacteria is converted to water and oxygen by the enzyme reaction. This best demonstrate that whether that organism produces catalase or not. To do this test all that is necessary is to place a few drop of 3% hydrogen peroxide on the organism present as a slant culture. If the hydrogen peroxide effervescence is their the organism is catalase positive.

Alternatively a small amount of culture to be tested is picked from NA and organism is placed on the top of hydrogen peroxide. The production of gas bubbles indicates +ve reaction.

Requirements

Glass wares, test tubes with slant bacterial culture.
CHEMICALS: 3% Hydrogen peroxide.

Procedure

(A) DIRECT TUBE TEST: The tube is held at an angle and a few drops of 3% hydrogen peroxide is allowed to flow slowly over the culture. The emergence of bubble from the organism is noted. The presence of bubble indicates a positive reaction indicating the presence of enzyme catalase. If no gas is produced this is a –ve reaction.

(B) SLIDE TECHNIQUE: With the help of a sterile platinum loop transfer a small amount of culture on to a clean slide. About 0.5 ml of 3% hydrogen peroxide is added on the culture.

Note: If bubbles are formed, it indicates a positive reaction *i.e.* the presence of the enzyme catalase.

Result

Observe your experimental result.

OXIDASE TEST

To test the oxidase producing microorganisms.

Principle

The oxidase determines whether micro be can oxidize certain aromatic amines, *e.g.* para amino dimethyl alanine to form coloured end product. This oxidation correlates with the cytochrome oxidase activity of some bacteria including the genera Pseudomonas and Nisseria. A positive test is important in identifying these genera and also useful in characterizing the Enterobacteria which are oxidase –ve.

Requirements

Glassware, sample, culture of Pseudomonas, Bacillus, *E. coli.* Staphylococcus aureus and Klebsiella.

Chemical. Tetramethyl phenyl diamine, dihydrochloride.

Procedure

A PLATE METHOD: Separate agar plates streaked with Pseudomonas, Klebsiella, Bacillus are taken and 1% reagent tetra methyl phenyl diamine hydrochloride was directly added to the plates and the reaction were observed.

Result

Fig. 1. EMB

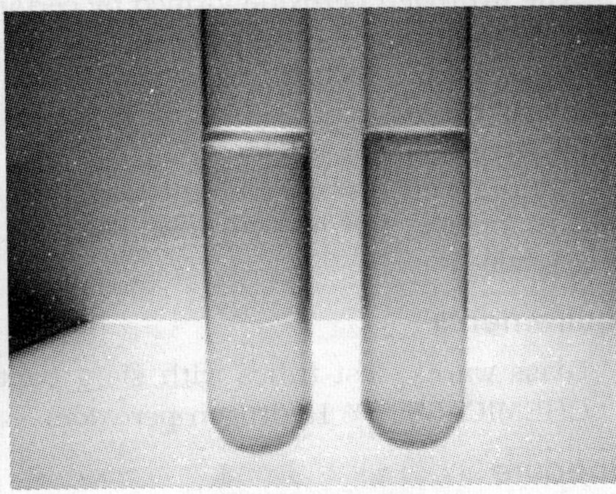

Fig. 2. Indole test

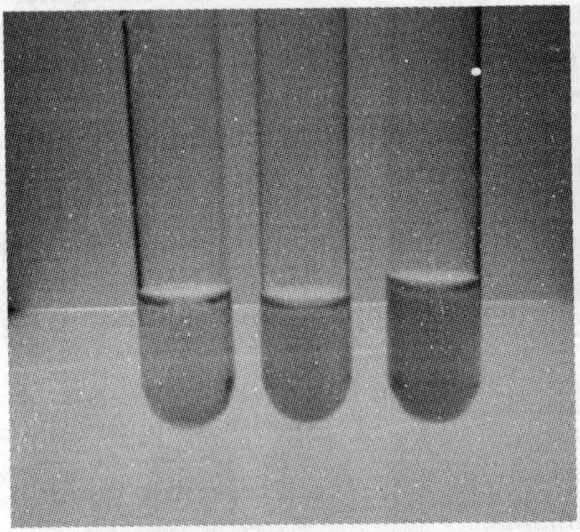

Fig. 3. Methyl red

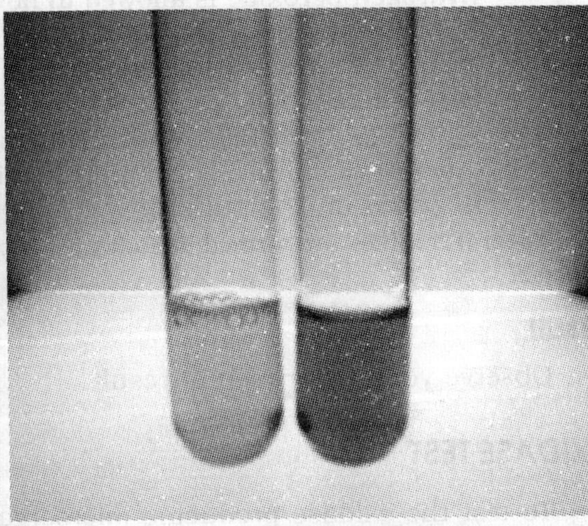

Fig. 4. Voges-proskauer

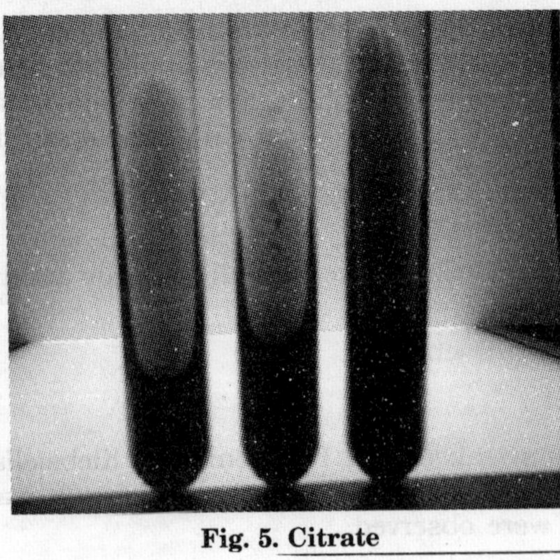

Fig. 5. Citrate

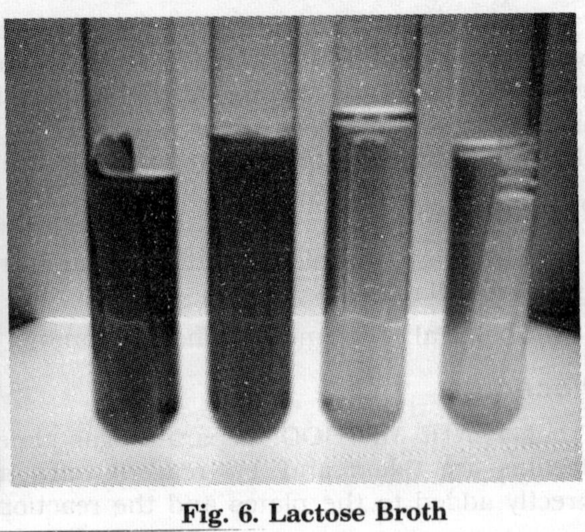

Fig. 6. Lactose Broth

Fig. 7. Mac Conkey

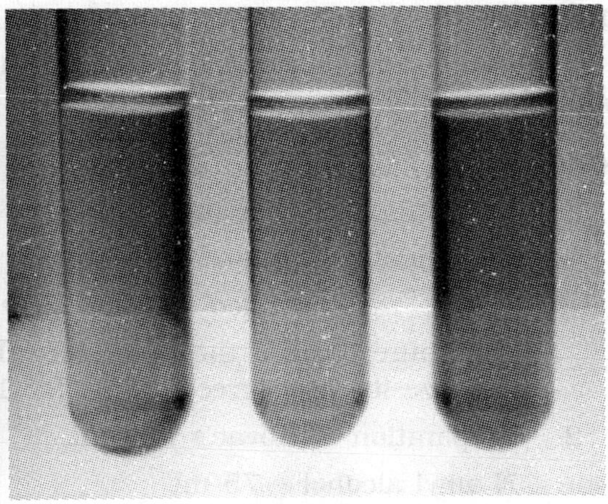

Fig. 8. Urease

13. IMVIC TEST

Differentiation of the principal groups of enterobacteriaceae can be accomplished on the basis of their biochemical properties and enzymeatic reactions in the presence of specific substrates. The IMVIC series of test indole, methyl red, Voges-Proskauer, and Citrate utilization can be used.

1. INDOLE TEST

To detect the production of Indole from the degradation of amino acid tryptophan.

Introduction

Amino acids are the basic constituents of many proteins that compose living organisms. Some microorganisms degrade amino acids to yield energy in a variety of end products of ammonia, indole acid, and water. Many reactions which involve degradation of amino acids are used for classifying the enterobacteriaceae.

Certain amino acids degrading bacteria like *E. coli* have the ability to degrade the amino acids tryptophan into indole and pyruvic acid. This hydrolysis is brought about by the enzyme tryptophanse.

Principle

Indole is a nitrogen containing compound from the degradation of the amino acid tryptophan. The indole test is important because only certain bacteria form indole. Indole can be easily detected with Kovac's reagent. After the addition of the reagent and mixing the contents the tube is allowed to stand. The alcohol layer is separated from this aqueous layer and upon standing the Redding of the alcohol layer is present in the culture. Thus, the formation of the red layer at the top of the culture indicates the positive test.

Pure tryptophan is not usually used in the test instead tryptone is used as a substrate because it contains typtophan.

Requirements

Sterilised test tubes, conical flasks, pipettes, glass rod, test culture, Kovac's reagent, tryptone broth.

Procedure

1. Preparation of tryptone broth:

   Ingredients: Tryptone – 10 gms.

   Distilled water – 1000 ml

   Distribute 5 ml of the broth into the test tubes and plug with cotton plugs. Sterilize it 121 degree Celsius for 15 minutes.
2. Preparation of Kovac's reagent:

   N-amyl alcohol – 75 ml

   Concentrated HCL – 25 ml

   P-dimethylaminebenzaldehyde – 5 gms.
3. Inoculate the tubes with the test bacterial culture.
4. Incubate all the tubes for 48 hours at 37°C.
5. Test for Indole – Add 0.3 ml of Kovac's reagent to each test tube. Mix well by rotating the tubes between your hands. The formation of red layer at the top of the culture indicates a positive test.

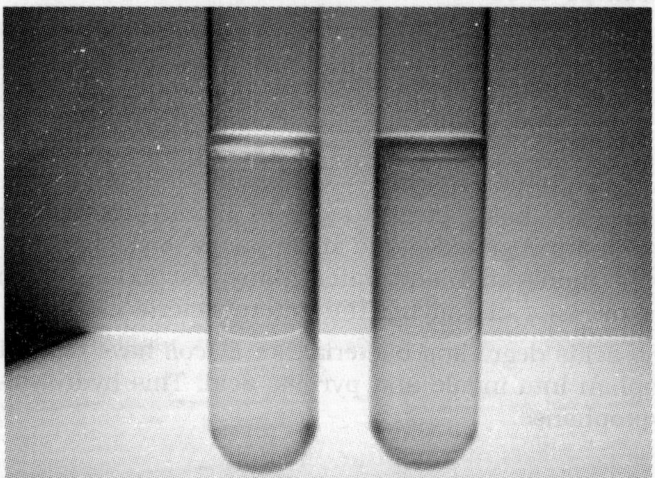

Fig. 26. Indole

Result

Observe your experimental result.

Discussion

One large family of gram negative bacteria that exhibits a considerable degree of relatedness is the enterobacteriaceae. This group is probably the most common one isolated in clinical specimens, sometimes as normal flora, sometimes as agents of diseases. The most important ones are *E.coli*, Klebsiella, Proteus, Enterobacter etc.

The IMVIC test is used to identify the level of genus *i.e.* Indole, Methyl red, Voges Proskaver test and citrate test. The "I" is for pronunciation. This is a traditional panel that can be used to differentiate among several genera.

The Indole test indicates the capacity of an isolate to cleave a compound indole of the amino acid tryptophan. Here a bright ring is formed on the surface of the tube on addition of Kovac's reagent if it is a positive test or else the tube remains yellow.

2. METHYL RED TEST

To detect the production of acids during the formation sugars.

Introduction

Bacteria such as *E.coli,* Proteus, ferment glucose to produce a large amount of lactic acid, acetic acids, succinic acid, and formic acid. Carbon dioxide, hydrogen, and ethyl alcohol is also formed. When the organisms do not convert the acidic product to neutral product the pH of the medium is 5 or less such acids are called Mixed acid fermenters.

Methyl Red Voges Proskaver medium is essentially a glucose broth with some buffered peptone and dipotassium phosphate is inoculated with test organisms. Methyl red indicates (methyl red indicator is yellow in colour at pH 5.2 and red at pH 4.5) is added to the culture. If methyl red indicator turns red then it indicates that the organism is a mixed acid fermenter-a positive reaction. The formation of yellow colour after the addition of methyl red indicator indicates a negative reaction.

Requirements

Methyl red indicator, sterilized test tubes, conical flasks, glass rods, cotton plugs and test culture and MRVP media.

Procedure

1. Preparation of methyl red indicator-Dissolve about 0.2 gm of methyl red in 500 ml of 95% ethyl alcohol and add 500 ml of distilled water and filter.
2. Preparation of MRVP media – (glucose phosphate broth)

   Dipotassium hydrogen phosphate (K_2HPO_4) – 5 gms

   Peptone – 5 gms

   Glucose – 5 gms

   Distilled water – 1000 ml

   Suspend all the ingredients in distilled water and gently warm. Do not alter pH.5 ml of media is distributed in test tubes which are plugged. Sterilize at 121 degree C for 15 mins.
3. Innoculate the tubes with the test bacterial culture except the control tube.
4. Incubate all the tubes at 37 degree for 48 hrs.
5. Add 3-4 drops of MR indicator into each tube. A distinct red colour formation indicates the positive test; yellow colour formation indicates a negative test.

Result

Observe the your result and interpret it.

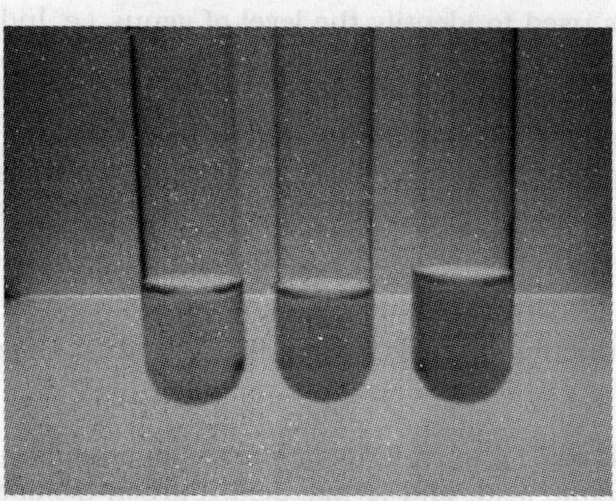

Fig. 27. Methyl Red test.

Discussion

Studies of recent years have emphasized the complexity of coliform group. The general practice is to classify the members. Classification is based on the results of the 4 tests– Indole, MR test, VP test and sodium citrate test.

Enterobacter cultures ferment lactose with the formation of acids and pH becomes less than 5. Klebsiella do not produce 2, 3-butylene glycol but others do.

E.coli, Bacillus, Proteus, and Staphylococcus gave a positive result as they oxidized the glucose to organic acid and stabilize the organic acid concentration. Klebsiella shows a negative result.

3. VOGES PROSKAVER TEST

To detect the production of acetyl methyl carbinol (acetoin) from glucose.

Introduction

All species of Enterobacter and Serratia as well as some species of Bacillus, Aeromonas produce a lot of 2, 3-butanediol and ethyl alcohol instead of acids. These organisms are called Butanediol Fermenters. The production of these non acid and products results in the less lowering of the pH in Methyl Red Voges Proskaver medium and thus the test is negative.

Principles

Butanediol fermenters produces 2, 3-butanediol for which there is no satisfactory test. 2, 3-butanediol is formed from acetyl methyl carbinol or acetoin ($CH_3CHOHCOCH_3$) which can be easily detected with the Barrit's reagent. This test is valid as acetoin and 2, 3-butanediol are present simultaneously. In VP test the formation of acetyl methyl carbinol is from glucose. In the presence of alkali atmospheric oxygen and the reagent small amounts of acetyl methyl carbinol present in the medium is oxidized to diacetyl which produces a red colour with guanidine residue in the media, thus the formation of the red or pink colour indicates the positive VP reaction.

Requirements

Methyl Red Voges Proskaver medium (or glucose phosphate broth), test tubes, pipettes, test cultures, cotton plugs, Barrit's reagent.

Procedure

1. (*a*) Preparation of MRVP medium or glucose phosphate broth:

   Dipotassium hydrogen phosphate (K_2HPO_4) - 5 gms

   Peptone - 5 gms

   Glucose - 5 gms

   Distilled water - 1000 ml

   Suspend all the ingredients in distilled water and gently warm. Do not alter the pH.3 ml of the media is distributed into test tubes which are plugged. Sterilized at 121 Degree Celsius.

   (*b*) Preparation of Barrit's reagent:

   Barrit's reagent consists of two solutions *i.e.* solution A and B.

   Solution A is prepared by dissolving 6 gms of alpha naphthol in 100 ml of 95% ethyl alcohol.

   Solution B is prepared by dissolving 16 gms of Potassium hydroxide in 100 ml of water.
2. Inoculate the tubes with the bacterial culture.
3. Incubate for 48 hrs at 37 degree Celsius.
4. Pipette 1 ml from each culture tube into clean separate tubes. Use separate pipettes for each tubes.
5. Add 18 drops (0.5 ml) of Barrit's solution A (α- naphthol) to each tube that contain the media. Add an equal amount of solution B into the same tube.
6. Shake the tubes vigorously every 30 sec. A positive reaction is indicated by the development of a pink colour which turns red in 1-2 hours after vigorous shaking. It is a very important step to achieve complete aeration.

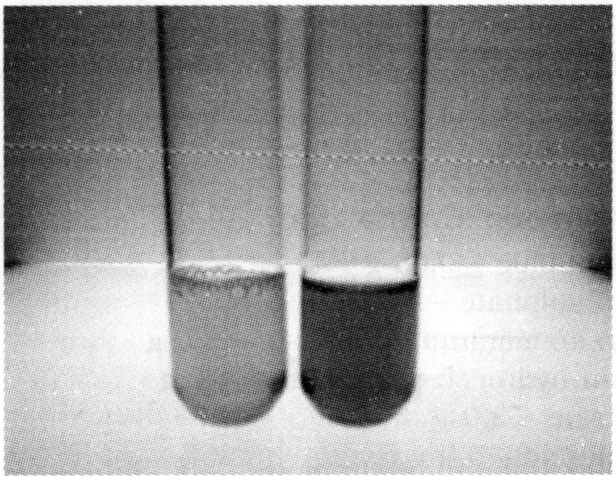

Fig. 28. Voges–Proskaver test

Result

Observe your experimental result.

Discussion

Voges and Proskaver found that the addition of KOH to the cultures of organisms of the haemorrhagic septicemia by Pasteurella group resulted in the development of a pinkish red colour, if allowed to stand for 24 hours or longer.

Harden and Walpole found that distinct differences exist in carbohydrate metabolism of typical enterobacters. The fermentation of glucose by the two organisms yielded varying amounts of products like acids, alcohol, carbon dioxide etc. It was due to formation of 2, 3-butylene glycol and acetyl methyl carbinol by Klebsiella but not by *E.coli.* Acetyl Methyl carbinol in the presence of KOH and air is further oxidized to diacetyl which in presence of peptone gives an eosin like colour. This colour is due to guanidine nucleus of amino acids present in it.

Thus, this test causes considerable significance in testing various samples because it disintegrates to a high degree between related enterobacters.

4. CITRATE UTILISATION TEST

To detect the ability to utilize Citrate by microorganisms.

Introduction

Certain bacteria such as Salmonella typhi and *Escherichia aerogen* are able to use citrate as a sole source of carbon. This ability to utilize Citrate as the sole source of carbon and energy is also used as a distinguishing test for certain G-rods.

Principle

Simmons citrate agar containing sodium citrate as the sole source of carbon is used to detect if the organism can utilize citrate or not. This agar contains the indicator Bromothymol blue which changes from green to blue when the growth of organisms causes alkalinity growth on the media and the consequent changing of the colour from green to blue indicates a positive test.

Requirements

Sterilised test tubes, conical flask, glass rod, cotton plug, Simmons Citrate agar media.

Procedure

A. Preparation of Simmon Citrate agar

Bromothymol blue – 0.08 g
Magnesium sulphate – 0.2 g
Dihydrogen ammonium phosphate – 1.0 g
Dipotassium hydrogen phosphate – 1.0 g
Sodium citrate – 2.0 g
Sodium chloride – 5.0 g
Agar agar – 15 g
Distilled water – 1000 ml
Bromothymol blue – 40 ml
pH – 6.8-7.0

B. Steps

1. All the ingredients are dissolved in distilled water and heated to melt the agar and distributed into test tubes.
2. Test tubes are sterilized at 121.5°C for 15 min. Then the hot test tubes are placed in the slanted position.
3. Test tubes are inoculated with the given bacterial culture.
4. The inoculated test tubes are incubated at 37°C for 24-48 hrs.
5. Growth during incubation results in the change of colour of the media from green to blue. This change in colour is regarded as a positive test.

Result

Observe your experimental result.

Fig. 29. Citrate utilization test

14. EXTRACTION OF BACTERIAL DNA

OBJECTIVES

- describe the DNA within bacterial cells.
- perform a DNA extraction isolating a DNA molecule.

Introduction

In this activity, you will extract a mass of DNA visible from bacterial cells visible to the naked eye.

1. The preparation of DNA from any cell type, bacterial or human, involves the same general steps:
   (a) disrupting the cell (and nuclear membrane, if applicable),

(*b*) removing proteins that entwine the DNA and other cell debris, and

(*c*) doing a final purification.

(*i*) These steps can be accomplished in several different ways, but are much simpler than you probably thought. The method chosen generally depends upon how pure the final DNA sample must be and how accessible the DNA is within the cell.

(*ii*) Bacterial DNA is protected only by the cell wall and cell membrane; there is no nuclear membrane as in eukaryotic cells. Therefore, the membrane can be disrupted by using dishwashing detergent which dissolves the phospholipid membrane, just as detergent dissolves fats from a frying pan. (The process of breaking open a cell is called cell lysis.) As the cell membranes dissolve, the cell contents flow out, forming a soup of nucleic acid, dissolved membranes, cell proteins, and other cell contents that is referred to as a cell lysate. Additional treatment is required for cells with walls, such as plant cells and bacterial cells that have thicker more protective cell walls (such as Gram positive or acid fast organisms). Additional treatments may include enzymatic digestion of the cell wall or physical disruption by means such as blending, sonication, or grinding.

2. After cell lysis, the next step involves purifying the DNA by removing proteins (histones) from the nucleic acid. Treatment with protein-digesting enzymes (proteinases) and/or extractions with the organic solvent phenol are two common methods of protein removal. Because proteins dissolve in the solvent but DNA does not, and because the solvent and water do not mix, the DNA can be physically separated from the solvent and proteins.

3. In this activity you will not attempt any DNA purification: your goal is simply to see the DNA. You will lyse *E. coli* with detergent and layer a small amount of alcohol on top of the cell lysate. Because DNA is insoluble in alcohol, it will form a white, web-like mass (precipitate) where the alcohol and water layers meet. Moving a glass rod up and down through the layers, allows you to collect the precipitated DNA. But this DNA is very impure mixed with cell debris and protein fibers.

4. Before you begin the DNA isolation, make sure you know the procedure to follow. Draw out a flow chart below including the amount of each reagent and the time for that part of the procedure.

Materials

disposable test tubes deionized water dishwashing detergent (50% mixture) glass rod.	Stock Cultures *E. coli*	water bath set at 60 – 70°C ice bath.

Methods

1. Apply your PPE including eye protection for this lab. Locate the water baths and the ice cold ethanol. Determine a method for timing the various steps.
2. Label a 5 ml disposable tube and fill it with exactly 3 ml of distilled water. Using a swab inoculate *E. coli* from the stock culture and agitate it in the 3 ml of distilled water.
3. Add 3 ml of the detergent to the suspension of *E.coli.* Mix each tube by gently shaking.
4. Place each tube into the water bath for 15 min. *Note:* Maintain the water bath temperature above 60°C but below 70°C. A temperature higher than 60°C is needed to destroy the enzymes that degrade DNA.
5. Cool the tube in an ice bath until it reaches room temperature.
6. The next step involves precipitating the DNA by using solvent. Carefully pipetting 3 ml of ICE COLD ethanol (it may be in the freezer) on top of the detergent and *E. coli* suspension mixture. The alcohol should float on top and not mix. (It *will* mix if you stir it or squirt it in too fast, so be careful.) Water-soluble DNA is insoluble in alcohol and precipitates when it comes in contact with it.
7. By carefully placing a clean glass rod through the alcohol into the suspension a web like mass will become evident; this mass is precipitated DNA. The rod carries a little alcohol into the suspension, precipitating and attaching to the DNA. Do not totally mix the two layers.

15. MEDICALLY SIGNIFICANT GRAM POSITIVE COCCI

OBJECTIVES

- Perform, interpret and define the relevance of the catalase test.
- Perform, interpret and define the relevance of the coagulase test.
- Perform, interpret and define the relevance of the test for hemolysis on BAP.
- Describe the Gram stain and arrangement of major GPC families.
- Determine the identification of an unknown GPC organism using the above tests.
- Relate the medical significance of each of the GPC covered in the lab.

Medically significant Gram Positive cocci are represented by two main families: *Micrococcaceae* (including the genera Staphylococcus and Micrococcus) and *Streptococcaceae* (including the genera Streptococcus and Enterococcus).

I. Micrococcaceae

Catalase +, Gram positive cocci in clusters.

(a) Genus Micrococcus. These bacteria are rarely associated with disease and are common environmental contaminants. They Gram stain as GPC in tetrads and produce yellow (*Micrococcus luteus)* or rose (*M. roseus)* coloured pigments on enriched media.

(b) Genus Staphylococci—Salt tolerant GPC

1. *Staphylococcus aureus*—These are pathogenic bacteria causing wound infections, abscesses, carbuncles, bacteremia, septicemia, osteomyelitis. Associated with purulent discharges and capable of producing a wide range of exotoxins (including hemolysins and DNAse), characterized as highly invasive. Causes food poisoning by the production of a heat resistant toxin. Important in nosocomial infections especially MRSA (Methicillin or Multiple Resistant *Staphylococcus aureus*). Nasal carriers are important.
2. *Staphylococcus epidermidis*—Normal skin flora, nonhemolytic, coagulase negative.

   Opportunistic pathogen isolated from catheters and I.V. lines and associated with transplant and immunosuppressed patients.
3. *Staphylococcus hemolyticus*—Normal skin inhabitant, beta hemolytic but coagulase negative. Opportunistic pathogen associated with UTI's.

II. Streptococcaceae

GPC in chain and pairs, easily over decolourized. This family produces a large number of exotoxins including hemolysins, erythrogenic toxins, nephrotoxins, and cardiohepatic toxins. Pathogenesis depends on species, strain, portal of entry and immune response. The Streptococcaceae are fastidious requiring blood agar for growth and producing typical characteristics as the blood cells are destroyed (hemolysis) for nutrients.

(*a*) *Streptococci.* Chains and pairs, some encapsulated, bile esculin and salt negative

1. *Strep. pyogenes*—Group A strep, beta hemolytic, causing strep throat, scarlet fever, peurperal fever (post natal sepsis), skin infections such as impetigo, and pneumonia. Post infection sequela such as glomerulonephritis and rheumatic fever represent serious syndromes if infections are not treated immediately.
2. *Strep. agalactiae*—Group B beta hemolytic strep causing neonatal meningitis thought to be associated with asympotomatic vaginal carriers. Recently reported in AIDS patients.
3. *Strep. pneumoniae*—Alpha hemolytic, mucoid, and lancet-shaped. Virulent strains are encapsulated. Causes classis pneumonia, ear and eye infections.
4. *Alpha strep*—Variety of nonpathogenic normal flora found on the skin and in the mouth. Occasionally associated with bacterial endocarditis. Many of these alpha strep are found on the respiratory tract as normal flora. *Strep mutans* is one of these strep associated with dental caries. *Strep. sanguinis* and *Strep. parasanguinis* are normal oral flora.

(*b*) *Enterococci.* Salt tolerant bile esculin positive strep. Normal flora of the G.I. tract. Opportunistic pathogens infecting decubiti (bedsores), causing UTI'S and associated with IV contamination. The enterococci are medically significant due to growing antibiotic resistance, they are referred to as VRE (Vancomycin Resistant Enterococci). Many species make up this group of strep including *Enterococcus faecalis* and *Enterococcus faecium.*

The table below represents a simple differentiation between these genera of bacteria. Refer to the Atlas for sample test results.

Family	*Micrococcaceae*		*Streptococcaceae*	
Test results	*Staphylococcus*	*Micrococcus*	*Streptococcus*	*Enterococcus*
Arrangement in broth	grape-like clusters	tetrads sarcina	chains (some pairs)	paired (occasional chains)
hemolysis on Blood agar	Staph. epidermidis - Staph. aureus +	negative	alpha, beta, gamma	alpha or gamma
growth on nutrient agar	good	good	poor or none	good
Catalase	positive	positive	negative	negative
MSA-Mannitol Salt Agar-NaCl tolerance	growth Staph. epidermidis - Staph. aureus +	negative	negative	growth
Coagulase	Staph. epidermidis - Staph. aureus +	negative	negative	negative

Materials

Hydrogen peroxide	Stock cultures:
dropper	Labeled 1-4
slide	*Staphylococcus aureus*
Coagulase tube	*Staphylococcus epidermidis*
MSA plate	*Micrococcus luteus*
Blood Agar plate	*Streptococcus sanguinis* or *Strep. parasanguinis*

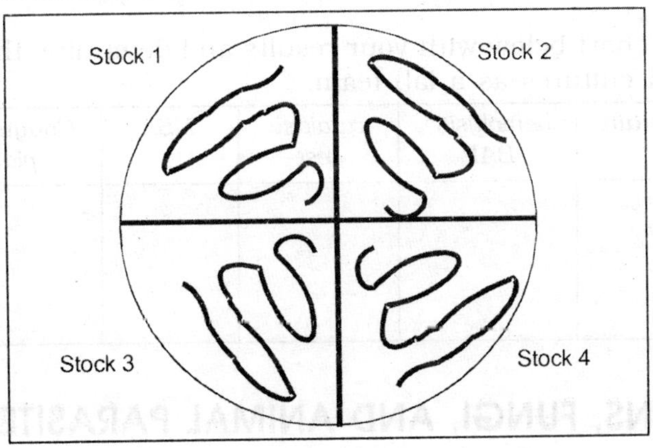

Fig. 30

Procedure

1. Gram stain each of the stock cultures and record the results.
2. Divide a Blood Agar plate into 4 sections and label them. Inoculate each section with a streak from each stock culture.

3. Perform a catalase test as described in the Atlas, by transferring a loopful of bacteria from each stock culture to a glass slide. Using a dropper apply a drop of hydrogen peroxide to each bacterial smear and record the results. Bubbling = positive
4. Divide an MSA plate into quadrants and inoculate it in the same fashion that the BAP was inoculated.
5. Perform a coagulase test, using the pre-made tubes and inoculating a tube with each of the stock cultures. Cover the tube with parafilm and incubate for 24-48 hours.
6. Incubate the BAP plate, MSA plate and coagulase at 37C. During the next lab period read and record the results.
7. Determine from the identification chart on page 1 of this lab, the identity of each stock culture.

Result

Observe your experimental result and interpret it.

Exercise

1. Define the following terms and describe a positive test.

   beta hemolysis -

   alpha hemolysis -

   gamma hemolysis -
2. Describe a positive coagulase test and identify the species of organism that would produce a positive test.
3. What advantage might coagulase give to a pathogen as it invades the human body?
4. Fill in the chart below with your results and determine the identity of each of your stock cultures as a lab team.

Stock #	*Gram stain*	*hemolysis -BAP*	*catalase test*	*MSA*	*Coagulase plate*	*Identification*
1						
2						
3						
4						

16. PROTOZOANS, FUNGI, AND ANIMAL PARASITES

OBJECTIVES

- Identify microscopic characteristics of organism belonging to the Kingdom Protozoa.
- Identify macroscopic and microscopic characteristics of an organism belonging to the Kingdom Fungi.

- Identify macroscopic and microscopic characteristics of an organism belonging to the Kingdom Animalia particularly helminths and arthropods.
- Compare and contrast methods for controlling health risks and treating each of these types of infections.
- Describe the important role of biomagnification and evaluate its impact on disease.

In an attempt to understand the diversity of life biologists have placed organisms into groups. This is similar to the way that chemists have created the periodic table where each column represents elements that behave in a similar manner. However, the number of organisms are unimaginable. We are still discovering new species every year, in fact, approximately 3 to 5 new bird species and many new mammal species are found EVERY year. We have also found that some organisms that look very similar externally do not have similar ancestors, and are physiologically very different. Scientists feel that the closest relations between organisms is represented by the ability of the organism to mate with another. We refer to organisms related this closely as the same **species.**

The purpose of this lab is to compare and contrast characteristics of organisms in Domain Eucarya, which will include Protozoans, Fungi and some Arthropod and Helminths in the Animal Kingdom. An important thing to observe is the size of the individual cell and their relationship to one another. Much of the classification of these organisms is based upon their nutritional style and easily observable characteristics.

The largest and most encompassing grouping of organisms is called a Kingdom. These groups have broad similarities, although there are debates as to how many kingdoms there should be we will use the widest accepted classification which is a five kingdom classification including:

Monera—Bacteria, surviving either as photosynthetic, chemosynthetic or decomposing organisms. Relatively simple, single-celled prokaryotic organisms.

Protista—Very diverse eukaryotic organisms, usually single-celled. Includes photosynthetic organisms, heterotrophs, and parasites. Classified by pigment or movement.

Fungi—Multicellular eukaryotes. Heterotrophic representing mainly decomposers-some pathogens and parasites. Classified by reproductive methods.

Plantae—Diverse multicellular eukaryotes. Photoautotrophs. Classified by tissue structure and reproductive methods.

Animalia—Animals. Diverse eukaryotic organisms. Heterotrophic including predators and parasites.

PART 1 PROTOZOA

Use the Atlas to identify the following Protozoa. Observe a prepared slide of the organisms with the intent of being able to identify them on a lab exam.

Organism (draw a picture and give scientific name)	*Drawing*	*Information on Human Infection*
Malaria *Plasmodium sp.*		
Sleeping Sickness and Chagas' Disease *Trypanosoma sp.*		
Amebic Dysentery		

PART 2 FUNGI

Use the Atlas to identify the following Fungi. Observe a prepared slide of the organisms with the intent of being able to identify them on a lab exam.

Organism (draw a picture and give scientific name)	*Drawing*	*Information on Human Infection*
Rhizopus		
Aspergillus or Penicillium		
Saccharomyces make a wet mount of the broth culture Stain with methylene blue		

PART 3 ARTHROPODS AND HELMINTHS

Organism (draw a picture and give scientific name)	*Drawing*	*Information on Human Infection*
Tick or Lice		
Tapeworm (draw a scolex and proglottid)		
Pinworm		

17. THE FUNGI, PART 1–THE YEASTS

INTRODUCTION

Fungi are eukaryotic organisms and include the yeasts, molds, and fleshy fungi. Yeasts are microscopic, unicellular fungi; molds are multinucleated, filamentous fungi (such as mildews, rusts, and common household molds); the fleshy fungi include mushrooms and puffballs.

All fungi are **chemoheterotrophs,** requiring organic compounds for both an energy and carbon source, which obtain nutrients by absorbing them from their environment. Most live off of decaying organic material and are termed **saprophytes.** Some are **parasitic,** getting their nutrients from living plants or animals.

The study of fungi is termed **mycology** and the diseases caused by fungi are called mycotic infections or **mycoses.**

In general, fungi are **beneficial** to humans. They are involved in the decay of dead plants and animals (resulting in the recycling of nutrients in nature), the manufacturing of various industrial and food products, the production of many common antibiotics, and may be eaten themselves for food. Some fungi, however, damage wood and fabrics, spoil foods, and cause a variety of plant and animal diseases, including human infections.

Discussion

Yeasts are unicellular, oval or spherical fungi which increase in number asexually by a process termed budding. A bud forms on the outer surface of a parent cell, the nucleus divides with one nucleus entering the forming bud, and cell wall material is laid down between the parent cell and the bud. Usually the bud breaks away to become a new daughter cell but sometimes, as in the case of the yeast *Candida*, the buds remain attached forming fragile branching filaments called pseudohyphae. Because of their unicellular and microscopic nature, yeast colonies appear similar to bacterial colonies on solid media. It should be noted that certain dimorphic fungi can grow as a yeast or as a mold, depending on growth conditions.

Yeasts are facultative anaerobes and can therefore obtain energy by both aerobic respiration and anaerobic fermentation. Most yeasts are nonpathogenic and some are of great value in industrial fermentations. For example, *Saccharomyces* species are used for both baking and brewing.

The yeast *Candida* is normal flora of the gastrointestinal tract and is also frequently found on the skin and on the mucous membranes of the mouth and vagina. *Candida* is normally held in check in the body by normal immune defenses and normal flora bacteria. Therefore, they may become opportunistic pathogens and overgrow an area if the host becomes immunosuppressed or is given broad spectrum antibiotics that destroy the normal bacterial flora.

Any infection caused by the yeast *Candida* is termed candidiasis. The most common forms of candidiases are oral mucocutaneous candidiasis (thrush), vaginitis, onychomycosis (infection of the nails), and dermatitis (diaper rash and other infections of moist skin). However, antibiotic therapy, cytotoxic and immunosuppressive drugs, and immunosuppressive diseases such as diabetes, leukemias, and AIDS can enable *Candida* to cause severe opportunistic systemic infections involving the skin, lungs, heart, and other organs. In fact, *Candida* now accounts for 10% of the cases of septicemia. Candidiasis of the esophagus, trachea, bronchi, or lungs, in conjunction with a positive HIV antibody test, is one of the indicator diseases for AIDS.

The most common *Candida* species causing human infections is *C. albicans*. This organism is usually oval and nonencapsulated, but under certain culture conditions may produce pseudohyphae, elongated yeast cells 4-6 μm in diameter that remain attached after budding to produce filament-like structures similar to the hyphae of molds. The pseudohyphae help the yeast to invade deeper tissues after it colonizes the epithelium. Asexual spores called blastospores develop in clusters along the pseudohyphae, often at the points of branching. Under certain growth conditions, thick-walled survival spores called chlamydospores may also form at the tips or as a part of the pseudohyphae.

A lesser known but often more serious pathogenic yeast is *Cryptococcus neoformans*. Like many fungi, this yeast can also reproduce sexually and the name given to the sexual form of the yeast is *Filobasidiella neoformans*. It appears as an oval yeast 5-6 μm in diameter, forms buds with a thin neck, and is surrounded by a thick capsule. It does not produce pseudohyphae and chlamydospores. The capsule enables the yeast to resist phagocytic engulfment.

Cryptococcus infections are usually mild or subclinical but, when symptomatic, usually begin in the lungs after inhalation of the yeast in dried bird feces. It is typically associated with pigeon and chicken droppings and soil contaminated with these droppings. *Cryptococcus*, found in soil, actively grows in the bird feces but does not grow in the bird itself. Usually the infection does not proceed beyond this pulmonary stage. In the immunosuppressed host however, it may spread through the blood to the meninges and other body areas, often causing cryptococcal meningoencephalitis. Any disease by this yeast is usually called cryptococcosis.

Dissemination of the pulmonary infection can result in a very severe and often fatal *cryptococcal meningoencephalitis.* Cutaneous and visceral infections are also found. Although exposure to the organism is probably common, large outbreaks are rare, indicating that an immunosuppressed host is usually required for the development of severe disease. Extrapulmonary cryptococcosis, in conjunction with a positive HIV antibody test, is another indicator disease for AIDS.

Cryptococcus can be identified by preparing an India ink or nigrosin negative stain of suspected sputum or cerebral spinal fluid in which the encapsulated, budding, oval yeast cells may be seen. It can be isolated on Saboraud Dextrose agar and identified by biochemical testing. Direct and indirect serological tests may also be used in diagnosis.

Pneumocystis carinii, once thought to be a protozoan but now considered a yeast-like fungus belonging to the fungal class Ascomycetes, causes an often lethal disease called Pneumocystis Carinii Pneumonia (PCP). It is seen almost exclusively in highly immunosuppressed individuals such as those with AIDS, late stage malignancies, or leukemias. PCP and a positive HIV-antibody test is one of the more common indicators of AIDS.

In biopsies from lung tissue or in tracheobronchial aspirates, both a unicellular organism about 1-3 µm in diameter with a distinct nucleus and a *cyst form* between 4-7 µm in diameter with 6-8 intracystic bodies, often in rosette formation, can be seen.

- *P. carinii* cysts from bronchoalveolar larvage
- *P. carinii* cysts from the lungs

Today we will use three agars to grow our yeast: Saboraud Dextrose agar (SDA), Mycosel agar, and Rice Extract agar. *Saboraud Dextrose Agar (SDA),* is an agar similar to trypticase soy agar but with a higher sugar concentration and a lower pH, both of which inhibit bacterial growth but promote fungal growth. SDA, therefore, is said to be selective for fungi. Another medium, Mycosel agar, contains chloramphenicol to inhibit bacteria and cycloheximide to inhibit most saprophytic fungi. Mycosel agar, therefore, is said to be selective for pathogenic fungi. Rice Extract agar with polysorbate 80 stimulates the formation of pseudohyphae, blastospores,, and chlamydospores, structures unique to C. albicans, and may be used in its identification. The speciation of *Candida* is based on sugar fermentation patterns.

Materials

Coverslips, alcohol, forceps, and one plate each of Saboraud Dextrose agar, Mycosel agar, and Rice Extract agar.

Organisms

Trypticase Soy broth cultures of *Candida albicans* and *Saccharomyces cerevisiae.*

Procedure

1. With a wax marker, divide a Saboraud Dextrose agar and a Mycosel agar plate in half. Using a sterile swab, inoculate one half of each plate with *C. albicans* and the other half with *S. cerevisiae.* Incubate the two plates at 37°C until the next lab period.
2. Using a sterile swab, streak two straight lines of *C. albicans* into a plate of Rice Extract agar plate. Pick up a glass coverslip with forceps, dip the coverslip in

alcohol, and ignite with the flame of your gas burner. Let the coverslip cool for a few seconds and place it over a portion of the streak line so that the plate can be observed directly under the microscope after incubation. Repeat for the second steak line and incubate the plate at room temperature until the next lab period.

Results

1. In the table below, describe the appearance of Candida albicans and Saccharomyces cerevisiae on Saboraud Dextrose agar.
2. Remove the lid of the Rice Extract agar plate and put the plate on the stage of the microscope. Using your yellow-striped 10X objective, observe an area under the coverslip that appears "fuzzy" to the naked eye. Reduce the light by moving the iris diaphragm lever almost all the way to the right. Raise the stage all the way up using the coarse focus (large knob) and then lower the stage using the coarse focus until the yeast comes into focus. Draw the pseudohyphae, blastospores, and chlamydospores.
3. Observe and make drawings of the demonstration yeast slides.

PERFORMANCE OBJECTIVES

After completing this lab, the student will be able to perform the following objectives:

Introduction

1. Define mycology and mycosis.
2. State three ways fungi may be beneficial to humans and three ways they may be harmful.

Discussion

1. Describe the typical appearance of a yeast and its usual mode of reproduction.
2. Describe yeasts in terms of their oxygen requirements.
3. State two ways the yeast *Saccharomyces* is beneficial to humans.
4. Name two yeasts that commonly infect humans.
5. Name four common forms of candidiasis.
6. Describe two conditions that may enable *Candida* to cause severe opportunistic systemic infections.
7. Describe pseudohyphae, blastospores, and chlamydospores.
8. State the usefulness of Saboraud Dextrose agar, Mycosel agar, and Rice Extract agar.
9. State how *Cryptococcus neoformans* is transmitted to humans, where in the body it normally infects, and possible complications.
10. State the primary method of identifying *Cryptococcus neoformans.*
11. State what disease is caused by *Pneumocystis carinii* and indicate several predisposing conditions a person is normally seen to have before they contract the disease.

Results

1. Describe the appearance of *Saccharomyces cerevisiae* and *Candida albicans* on Saboraud Dextrose agar and on Mycosel agar.
2. When given a plate of Mycosel agar showing yeast-like growth and a plate of Rice Extract agar showing pseudohyphae, blastospores, and chlamydospores, identify the organism as *Candida albicans*.
3. Recognize the following observed microscopically:
   (*a*) *Saccharomyces cerevisiae* and *Candida albicans* as yeasts in a direct stain preparation
   (*b*) A positive specimen for thrush by the presence of budding *Candida albicans*
   (*c*) *Cryptococcus neoformans* in an India ink preparation
   (*d*) A cyst of *Pneumocystis carinii* in lung tissue.

EXERCISE 18. THE FUNGI, PART 2–THE MOLDS

INTRODUCTION

Molds are multinucleated, filamentous fungi composed of hyphae. A hypha is a branching, tubular structure from 2-10 μm in diameter and is usually divided into cell-like units by crosswalls called septa. The total mass of hyphae is termed a mycelium. The portion of the mycelium that anchors the mold and absorbs nutrients is called the vegetative mycelium; the portion that produces asexual reproductive spores is termed the aerial mycelium.

Molds possess a rigid polysaccharide cell wall composed mostly of chitin and, like all fungi, are eukaryotic. Molds reproduce primarily by means of asexual reproductive spores such as conidiospores, sporangiospores, and arthrospores. These spores are disseminated by air, water, animals or objects and upon landing on a suitable environment, germinate and produce new hyphae. Molds may also reproduce by means of sexual spores such as ascospores and zygospores, but this is not common. The form and manner in which the spores are produced, along with the appearance of the hyphae and mycelium, provide the main criteria for identifying and classifying molds.

A. Nonpathogenic Molds

To illustrate how morphological characteristics such as the type and form of asexual reproductive spores and the appearance of the mycelium may be used in identification, we will look at three common nonpathogenic molds.

The two most common types of asexual reproductive spores produced by molds are conidiospores and sporangiospores. Conidiospores are borne externally in chains on an aerial hypha called a conidiophore; sporangiospores are produced within a sac or sporangium on an aerial hypha called a sporangiophore.

Penicillium and *Aspergillus* are examples of molds that produce conidiospores. *Penicillium* is one of the most common household molds and is a frequent food contaminant. The conidiospores of *Penicillium* usually appear grey, green, or blue and are produced in chains on finger-like projections called sterigmata. *Aspergillus* is another common contaminant. Although usually nonpathogenic, it may become opportunistic in the respiratory tract of a compromised host and, in certain foods, can produce mycotoxins.

The conidiospores of *Aspergillus* appear brown to black and are produced in chains on the surface of a ball-like structure called a vesicle.

- Scanning electron micrograph of the conidiospores of Penicillium.
- Scanning electron micrograph of the conidiospores of Aspergillus.
- *Rhizopus* is an example of a mold that produces sporangiospores. Although usually nonpathogenic, it sometimes causes opportunistic wound and respiratory infections in the compromised host. The sporangiospores of *Rhizopus* appear brown or black and are found within sacs called sporangia. Anchoring structures called rhizoids are also produced on the vegetative hyphae.

Rhizopus can also reproduce sexually. During sexual reproduction, hyphal tips of a (+) and (−) mating type join together and their nuclei fuse to form a sexual spore called a zygospore. This gives rise to a new sporangium producing sporangiospores having DNA that is a recombination of the two parent strain's DNA.

Nonpathogenic molds are commonly cultured on fungal-selective or enriched media such as Saboraud Dextrose agar (SDA), Corn Meal agar, and Potato Dextrose agar.

B. Dermatophytes

The *dermatophytes* are a group of molds that cause superficial mycoses of the hair, skin, and nails and utilize the protein keratin. Infections are commonly referred to as *ringworm* or *tinea* infections and include tinea capitis (head), tinea barbae (face and neck), tinea corporis (body), tinea cruris (groin), tinea unguium (nails), and tinea pedis (athlete's foot).

The three common dermatophytes are *Microsporum*, *Trichophyton*, and *Epidermophyton*. These organisms grow well at 25°C. They may produce large leaf or club-shaped asexual spores called macroconidia as well as small spherical asexual spores called microconidia, both from vegetative hyphae.

Microsporum commonly infects the skin and hair, *Epidermophyton*, the skin and nails, and *Trichophyton*, the hair, skin, and nails. Dermatophytic infections are acquired by contact with fungal spores from infected humans, animals, or objects. On the skin, the dermatophytes cause reddening, itching, edema, and necrosis of tissue as a result of fungal growth and a hypersensitivity of the host to the fungus and its products. Frequently there is secondary bacterial or *Candida* invasion of the traumatized tissue.

To diagnose dermatophytic infections, tissue scrapings can be digested with 10% potassium hydroxide (which causes lysis of the human cells but not the fungus) and examined microscopically for the presence of fungal hyphae and spores. To establish the specific cause of the infection, fungi from the affected tissue can be cultured on Dermatophyte Test Medium (DTM) and Saboraud Dextrose agar (SDA).

Dermatophyte Test Medium (DTM) has phenol red as a pH indicator with the medium yellow (acid) prior to inoculation. As the dermatophytes utilize the keratin in the medium, they produce alkaline end products which raise the pH, thus turning the phenol red in the medium from yellow or acid to red or alkaline. On SDA, the types of macroconidia and microconidia can be observed. Many dermatophyte species produce yellow to red pigmented colonies on SDA and the most common species of *Microsporum* fluoresce under ultraviolet light.

C. Dimorphic Fungi

Dimorphic fungi may exhibit two different growth forms. Outside the body they grow as a mold, producing hyphae and asexual reproductive spores, but inside the body they grow in a nonmycelial form. Dimorphic fungi may cause systemic mycoses which usually begin by inhaling spores from the mold form. After germination in the lungs, the fungus grows in a nonmycelial form. The infection usually remains localized in the lungs and characteristic lesions called granuloma may be formed in order to wall-off and localize the organism. In rare cases, usually in an immunosuppressed host, the organism may disseminate to other areas of the body and be life threatening. Examples of dimorphic fungi include *Coccidioides immitis*, *Histoplasma capsulatum*, and *Blastomyces dermatitidis*.

Coccidioides immitis is a dimorphic fungus that causes coccidioidomycosis, a disease endemic to the southwestern United States. The mold form of the fungus grows in arid soil and produces thick-walled, barrel-shaped asexual spores called **arthrospores** by a fragmentation of its vegetative hyphae. After inhalation, the arthrospores germinate and develop into endosporulating spherules in the lungs. Coccidioidomycosis can be diagnosed by culture, by a coccidioidin skin test, and by indirect serologic tests.

Histoplasma capsulatum is a dimorphic fungus that causes histoplasmosis, a disease commonly found in the Great Lakes region and the Mississippi and Ohio River valleys. The mold form of the fungus often grows in bird or bat droppings or soil contaminated with these droppings and produces large tuberculate macroconidia and small microconidia. After inhalation of these spores and their germination in the lungs, the fungus grows as a budding, encapsulated yeast. Histoplasmosis can be diagnosed by culture, by a histoplasmin skin test, and by indirect serologic tests.

Symptomatic and disseminated histoplasmosis and coccidioidomycosis are seen primarily in individuals who are immunosuppressed. Along with a positive HIV antibody test, both are indicator diseases for the diagnosis of AIDS.

Blastomycosis, caused by Blastomyces dermatitidis produces a mycelium with small **conidiospores** and grows actively in bird droppings and contaminated soil. When spores are inhaled or enter breaks in the skin, they germinate and the fungus grows as a yeast having a characteristic thick cell wall. Blastomycosis is common around the Great Lakes region and the Mississippi and Ohio River valleys. It is diagnosed by culture and by biopsy examination.

PROCEDURE

A. Nonpathogenic Molds

1. Using a dissecting microscope, observe the SDA plate cultures of *Penicillium*, Aspergillus, and Rhizopus. Note the colony appearance and colour and the type and form of the asexual spores produced.
2. Observe the prepared slides of Penicillium, Aspergillus, and Rhizopus under high magnification. Note the type and form of the asexual spores produced.
3. Observe the prepared slide showing the zygospore of *Rhizopus* produced during sexual reproduction.

B. Dermatophytes

1. Observe the dermatophyte Microsporum growing on DTM. Note the red colour (from alkaline end products) characteristic of a dermatophyte.
2. Microscopically observe the SDA culture of *Microsporum*. Note the macroconidia and microconidia.
3. Observc the photographs of dermatophytic infections.

C. Dimorphic Fungi

1. Observe the prepared slide of *Coccidioides immitis* arthrospores.
2. Observe the pictures showing the mold form and endosporulating spherule form of *Coccidioides immitis.*
3. Observe the pictures showing the mold form and yeast form of *Histoplasma capsulatum.*
4. Observe the photographs of systemic fungal infections.

RESULTS

A. Nonpathogenic Molds

Make drawings of the molds as they appear microscopically under high magnification and indicate the type of asexual spore they produce. Also note their colour and appearance on SDA.

Type of asexual spore = Colour on SDA =	Type of asexual spore = Colour on SDA =
Type of asexual spore = Colour on SDA =	

B. Dermatophytes

1. Describe the results of *Microsporum* growing on **Dermatophyte Test Medium (DTM):**
   - Original colour of DTM =
   - Colour following growth of Microsporum =
   - Reason for Colour Change =
2. Draw the **macroconidia and microconidia** seen on the SDA culture of *Microsporum.*

C. Dimorphic Fungi

1. Draw the arthrospores of *Coccidioides immitis.*
2. Draw the mold form and endosporulating spherule form of *Coccidioides immitis.*
3. Draw the mold form and yeast form of *Histoplasma capsulatum.*

PERFORMANCE OBJECTIVES

Discussion

1. Define the following: hypha, mycelium, vegetative mycelium, and aerial mycelium.
2. Describe the principle way molds reproduce asexually.
3. State the main criteria used in identifying molds.

A. Nonpathogenic Molds

1. Describe conidiospores and sporangiospores and name a mold that produces each of these.

2. Recognize the following genera of molds when given an SDA plate culture and a dissecting microscope and state the type of asexual spore seen:
   (a) *Penicillium*
   (b) *Aspergillus*
   (c) *Rhizopus*
3. Recognize the following genera of molds when observing a prepared slide under high magnification and state the type of asexual spore seen:
   (a) *Penicillium*
   (b) *Aspergillus*
   (c) *Rhizopus*
4. Recognize *Rhizopus* zygospores.

B. Dermatophytes

1. Define dermatophyte and list three common genera of dermatophytes.
2. Name four dermatophytic infections and state how they are contracted by humans.
3. Describe macroconidia and microconidia.
4. Describe how the following may be used to identify dermatophytes: potassium hydroxide preparations of tissue scrapings, DTM, and SDA.
5. Recognize a mold as a dermatophyte and state how you can tell when given the following:
   (a) a flask of DTM showing alkaline products
   (b) an SDA culture (under a microscope) or picture showing macroconidia.
6. Recognize macroconidia and microconidia.

C. Dimorphic Fungi

1. Define dimorphic fungi and state how they are usually contracted by humans.
2. Name three common dimorphic fungal infections found in the United States, state how they are transmitted to humans, and indicate where they are found geographically.
3. Describe the mold form and the nonmycelial form of the following:
   (a) *Coccidioides immitis*
   (b) *Histoplasma capsulatum*
   (c) *Blastomyces dermatitidis*
4. Recognize *Coccidioides immitis* and its arthrospores when given a prepared slide and a microscope.

19. VIRUSES: THE BACTERIOPHAGES

INTRODUCTION

Viruses are infectious agents with both living and nonliving characteristics.

1. *Living characteristics of viruses*
   (a) They reproduce at a fantastic rate, but only in living host cells.
   (b) They can mutate.

2. *Nonliving characteristics of viruses*
   (*a*) They are acellular, that is, they contain no cytoplasm or cellular organelles.
   (*b*) They carry out no metabolism on their own and must replicate using the host cell's metabolic machinery. In other words, viruses don't grow and divide. Instead, new viral components are synthesized and assembled within the infected host cell.
   (*c*) They possess DNA or RNA but never both.

Viruses are usually much smaller than bacteria. Most are submicroscopic, ranging in size from 10-250 nanometers.

Structurally, viruses are much more simple than bacteria. Every virus contains a **genome** of single-stranded or double-stranded DNA or RNA that functions as its genetic material. This is surrounded by a protein shell called a capsid or core which is composed of protein subunits called capsomeres. Many viruses consist of no more than nucleic acid and a capsid, in which case they are referred to as nucleocapsid or naked viruses.

Most animal viruses have an envelope surrounding the nucleocapsid and are called enveloped viruses. The envelope usually comes from the host cell's membranes by a process called budding, although the virus does incorporate glycoprotein of its own into the envelope.

Bacteriophages are viruses that infect only bacteria. In addition to the nucleocapsid or head, some have a rather complex tail structure used in adsorption to the cell wall of the host bacterium.

Since viruses lack organelles and are totally dependent on the host cell's metabolic machinery for replication, they cannot be grown in synthetic media. In the laboratory, animal viruses are grown in animals, in embryonated eggs, or in cell culture. (In cell culture, the host animal cells are grown in synthetic medium and then infected with viruses.) Plant viruses are grown in plants or in plant cell culture. Bacteriophages are grown in susceptible bacteria.

Today we will be working with bacteriophages since they are the easiest viruses to study in the lab. Most bacteriophages, such as Coliphage T4 that we are using today, replicate by the lytic life cycle and are called lytic bacteriophages.

The lytic life cycle of Coliphage T4 consists of the following steps:

1. *Adsorption.* Attachment sites on the bacteriophage tail adsorb to receptor sites on the cell wall of a susceptible host bacterium.
2. *Penetration.* A bacteriophage enzyme "drills" a hole in the bacterial cell wall and the bacteriophage injects its genome into the bacterium. This begins the eclipse period, the period in which no intact bacteriophages are seen within the bacterium.
3. *Replication.* Enzymes coded by the bacteriophage genome shut down the bacterium's macromolecular (protein, RNA, DNA) synthesis. The bacteriophage genome replicates and the bacterium's metabolic machinery is used to synthesize bacteriophage enzymes and bacteriophage structural components.
4. *Maturation.* The bacteriophage parts assemble around the genome.
5. *Release.* A bacteriophage-coded lysozyme breaks down the bacterial peptidoglycan causing osmotic lysis of the bacterium and release of the intact bacteriophages.

6. *Reinfection.* From 50-200 bacteriophages may be produced per infected bacterium and they now infect surrounding bacteria.

Some bacteriophages replicate by the lysogenic life cycle and are called temperate bacteriophages. When a temperate bacteriophage infects a bacterium, it can either 1) replicate by the lytic life cycle and cause lysis of the host bacterium, or it can 2) incorporate its DNA into the bacterium's DNA and assume a noninfectious state. In the latter case, the cycle begins by the bacteriophage adsorbing to the host bacterium and injecting its genome, as in the lytic cycle. However, the bacteriophage does not shut down the host bacterium. Instead, the bacteriophage DNA inserts or integrates into the host bacterium's DNA. At this stage, the virus is called a *prophage.* Expression of the bacteriophage genes controlling bacteriophage replication is repressed by a repressor protein and the bacteriophage DNA replicates as a part of the bacterial nucleoid. However, in approximately one in every million to one in every billion bacteria containing a prophage, spontaneous induction occurs. The bacteriophage genes are activated and bacteriophages are produced as in the lytic life cycle.

Today you will infect the bacterium Escherichia coli B with its specific bacteriophage, Coliphage T4.

In the first part of the lab you will perform a plaque count. A plaque is a small, clear area on an agar plate where the host bacteria have been lysed as a result of the lytic life cycle of the infecting bacteriophages. As the bacteria replicate on the plate they form a "lawn" of confluent growth. Meanwhile, each bacteriophage that adsorbs to a bacterium will reproduce and cause lysis of that bacterium. The released bacteriophages then infect neighboring bacteria, causing their lysis. Eventually a visible self-limiting area of lysis, a plaque, is observed on the plate.

The second part of the lab will demonstrate viral specificity. *Viral specificity* means that a specific strain of bacteriophage will only adsorb to a specific strain of susceptible host bacterium. In fact, viral specificity is just as specific as an enzyme-substrate reaction or an antigen-antibody reaction. Therefore, viral specificity can be used sometimes as a tool for identifying unknown bacteria. Known bacteriophages are used to identify unknown bacteria by observing whether or not the bacteria are lysed. This is called *phage typing.*

Phage typing is useful in identifying strains of such bacteria as *Staphylococcus aureus, Pseudomonas aeruginosa* and *Salmonella* species. For example, by using a series of known staphylococcal bacteriophages against the *Staphylococcus aureus* isolated from a given environment, one can determine if it is identical to or different from the strain of *Staphylococcus aureus* isolated from a lesion or from a food. This can be useful in tracing the route of transmission.

A. PLAQUE COUNT

Materials

Bottle of sterile saline, sterile 10.0 ml and 1.0 ml pipettes and pipette fillers, sterile empty dilution tubes (7), Trypticase Soy agar plates (3), bottle of melted Motility Test medium from a water bath held at 47°C.

Cultures

Trypticase Soy broth culture of *Escherichia coli* B, suspension of Coliphage T4.

Procedure

1. Take 2 tubes containing 9.9 ml of sterile saline, 2 tubes containing 9.0 ml of sterile saline, and 3 sterile empty dilution tubes and label the tubes as shown in.
2. Dilute the Coliphage T4 stock as described below.
   (*a*) Remove a sterile 1.0 ml pipette from the bag. Do not touch the portion of the pipette that will go into the tubes and do not lay the pipette down. From the tip of the pipette to the "0" line is 1 ml; each numbered division (0.1, 0.2, etc.) represents 0.1 ml; each division between two numbers represents 0.01 ml.
   (*b*) Insert the cotton-tipped end of the pipette into a blue 2 ml pipette filler.
   (*c*) Flame the sample of Coliphage T4, insert the pipette to the bottom of the tube, and withdraw 0.1 ml of the sample by turning the filler knob towards you. Reflame and cap the tube.
   (*d*) Flame the first (10^{-2}) dilution tube and dispense the 0.1 ml of sample into the tube by turning the filler knob away from you. Draw the liquid up and down in the pipette several times to rinse the pipette and help mix. Reflame and cap the tube.
   (*e*) Mix the tube thoroughly by holding the tube in one hand and vigorously tapping the bottom with the other hand. This is to assure an even distribution of the bacteriophage throughout the liquid.
   (*f*) Using the same pipette and procedure, aseptically withdraw 0.1 ml from the first (10^{-2}) dilution tube and dispense into the second (10^{-4}) dilution tube and mix.
   (*g*) Using the same pipette and procedure, aseptically withdraw 1.0 ml (up to the "0" line); from the second (10^{-4}) dilution tube and dispense into the third (10^{-5}) dilution tube. Mix as described above.
   (*h*) Using the same pipette and procedure, aseptically withdraw 1.0 ml from the third (10^{-5}) dilution tube and dispense into the fourth (10^{-6}) dilution tube. Mix as described above. Discard the pipette.
3. Take the 3 remaining empty, sterile tubes and treat as described below.
   (*a*) Using a new 1.0 ml pipette and the procedure described above, aseptically remove 0.1 ml of the 10^{-6} bacteriophage dilution and dispense into the third (10^{-7}) empty tube.
   (*b*) Using the same pipette and procedure, aseptically remove 0.1 ml of the 10^{-5} bacteriophage dilution and dispense into the second (10^{-6}) empty tube.
   (*c*) Using the same pipette and procedure, remove 0.1 ml of the 10^{-4} bacteriophage dilution and dispense into the first (10^{-5}) empty tube. Discard the pipette.
4. Using a new 1.0 ml pipette, add 0.5 ml of E. coli B to the 0.1 ml of bacteriophage in each of the 3 tubes from step 4 and mix.
5. Using a new 10.0 ml pipette, add 2.5 ml of sterile, melted Motility Test medium to the bacteria-bacteriophage mixture in each of the 3 tubes from step 3 and mix.

6. Quickly pour the motility medium-bacteria-bacteriophage mixtures onto separate plates of Trypticase Soy agar and swirl to distribute the contents over the entire agar surface.
7. Incubate the 3 plates right-side-up at 37°C until the next lab period.

B. VIRAL SPECIFICITY

Materials

Trypticase Soy agar plates (2)

Cultures

Trypticase Soy broth cultures of 4 unknown bacteria labelled #1, #2, #3, and #4; suspension of Coliphage T4.

Procedure

1. Using a wax marker, draw a line on the bottom of both Trypticase Soy agar plates dividing them in half. Number the 4 sectors 1, 2, 3, and 4, to correspond to the 4 unknown bacteria.
2. Draw a circle about the size of a dime in the center of each of the 4 sectors.
3. Using a sterile inoculating loop, streak unknown bacterium #1 on sector 1 of the first Trypticase Soy agar plate by streaking the loop through the circle you drew. Be careful not to streak into the other half of the plate.
4. Using the same procedure, streak the three remaining sectors with their corresponding unknown bacteria.
5. Using a sterile Pasteur pipette and rubber bulb, add 1 drop of Coliphage T4 to each sector in the area outlined by the circle.
6. Incubate the 2 plates right-side-up at 37°C until the next lab period.

RESULTS

A. Plaque Count

Observe the 3 plates for plaque formation and make a drawing.

- 1/100,000 (10^{-5}) dilution
- 1/1,000,000 (10^{-6}) dilution
- 1/10,000,000 (10^{-7}) dilution

B. Viral Specificity

Make a drawing of your results and state which of the unknowns (#1, #2, #3, or #4) was *E. coli*.

PERFORMANCE OBJECTIVES

Discussion

1. Define the following: bacteriophage, plaque, and phage typing.
2. Describe the structure of the bacteriophage coliphage T4.
3. Describe the lytic life cycle of bacteriophages.
4. Define viral specificity.

Results

1. Recognize plaques and state their cause.
2. Interpret the results of a viral specificity test using Coliphage T4.

20. SEROLOGY, PART 1–DIRECT SEROLOGIC TESTING

A. INTRODUCTION TO SEROLOGIC TESTING

The immune responses refer to the ability of the body (self) to recognize specific foreign factors (nonself) that threaten its biological integrity. There are two major branches of the immune responses:

1. *Humoral Immunity:* humoral immunity involves the production of antibody molecules in response to an antigen (ANTIGEN: 'A substance that reacts with antibody molecules and antigen receptors on lymphocytes. An immunogen is an antigen that is recognized by the body as nonself and stimulates an adaptive immune response) and is mediated by B-lymphocytes.
2. *Cell-Mediated Immunity:* Cell-mediated immunity involves the production of cytotoxic T-lymphocytes, activated macrophages, activated NK cells, and cytokines in response to an antigen and is mediated by T-lymphocytes.

To understand the immune responses we must first understand what is meant by the term antigen. Technically, antigen is defined as a substance that reacts with antibody molecules and antigen receptors on lymphocytes. An immunogen is an antigen that is recognized by the body as nonself and stimulates an adaptive immune response. For simplicity, both antigens and immunogens are usually referred to as antigens.

Chemically, antigens are large molecular weight proteins (including conjugated proteins such as glycoproteins, lipoproteins, and nucleoproteins) and polysaccharides (including lipopolysaccharides). These protein and polysaccharide antigens are found on the surfaces of viruses and cells, including microbial cells (bacteria, fungi, protozoans) and human cells.

As mentioned above, the B-lymphocytes and T-lymphocytes are the cells that carry out the immune responses. The body recognizes an antigen as foreign when that antigen binds to the surfaces of B-lymphocytes and T-lymphocytes by way of antigen-specific receptors having a shape which corresponds to that of the antigen, similar to interlocking pieces of a puzzle. The antigen receptors on the surfaces of B-lymphocytes are antibody molecules called B-cell receptors or sIg; the receptors on the surfaces of T-lymphocytes are called T-cell receptors (TCRs).

The actual portions or fragments of an antigen that react with receptors on B-lymphocytes and T-lymphocytes, as well as with free antibody molecules, are called epitopes or antigenic determinants. The size of an epitope is generally thought to be equivalent to 5-15 amino acids or 3-4 sugar residues. Some antigens, such as polysaccharides, usually have many epitopes, but all of the same specificity. This is because polysaccharides may be composed of hundreds of sugars with branching sugar side chains, but usually contain only one or two different sugars. As a result, most "shapes" along the polysaccharide are the same. Other antigens such as proteins usually have many epitopes of different specificities. This is because proteins are usually hundreds of amino acids long and are composed of 20 different amino acids. Certain amino acids are able to interact with other amino acids in the protein chain and this causes the protein to fold over upon itself and assume a complex three-dimensional shape. As a result, there are many different "shapes" on the protein. That is why proteins are more immunogenic than polysaccharides; they are chemically more complex.

A microbe, such as a single bacterium, has many different proteins on its surface that collectively form its various structures, and each different protein may have many different epitopes. Therefore, immune responses are directed against many different parts or epitopes of the same microbe.

In terms of infectious diseases, the following may act as antigens:

1. microbial structures (cell walls, capsules, flagella, pili, viral capsids, envelope-associated glycoproteins, etc.); and
2. microbial exotoxins

Certain noninfectious materials may also act as antigens if they are recognized as "nonself" by the body. These include:

1. allergens (dust, pollen, hair, foods, dander, bee venom, drugs, and other agents causing allergic reactions);
2. foreign tissues and cells (from transplants and transfusions); and
3. the body's own cells that the body fails to recognize as "normal self" (cancer cells, infected cells, cells involved in autoimmune diseases).

Antibodies or immunoglobulins are specific protein configurations produced by B-lymphocytes and plasma cells in response to a specific antigen and capable of reacting with that antigen. Antibodies are produced in the lymphoid tissue and once produced, are found mainly in the plasma portion of the blood (the liquid fraction of the blood before clotting). Serum is the liquid fraction of the blood after clotting.

There are 5 classes of human antibodies: IgG, IgM, IgA, IgD, and IgE. The simplest antibodies, such as IgG, IgD, and IgE, are "Y"-shaped macromolecules called monomers composed of four glycoprotein chains. There are two identical heavy chains having a high molecular weight that varies with the class of antibody. In addition, there are two identical light chains of one of two varieties: kappa or gamma. The light chains have a lower molecular weight. The four glycoprotein chains are connected to one another by disulfide (S-S) bonds and noncovalent bonds. Additional S-S bonds fold the individual glycoprotein chains into a number of distinct globular domains. The area where the top of the "Y" joins the bottom is called the hinge. This area is flexible to enable the antibody to bind to pairs of epitopes various distances apart on an antigen.

The two tips of the "Y" monomer are referred to as the Fab portions ('Fab:', 'Antigen-binding fragment. The actual portions of an antibody molecule that have a shape corresponding to that of an epitope of an antigen. The amino acid sequences of the Fab portion of antibody molecules is hypervar of the antibody. The first 110 amino acids or first domain of both the heavy and light chain of the Fab region of the antibody provide specificity for binding an epitope (The actual portions or fragments of an antigen that react with receptors on B-lymphocytes and T-lymphocytes, as well as with free antibody molecules. Usually equivalent to 5-15 amino acids or 3-4 sugar residues)', on an antigen. The Fab portions provide specificity for binding an epitope on an antigen. The bottom part of the "Y" is called the Fc portion and this part is responsible for the biological activity of the antibody (see diagram of IgG). Depending on the class of antibody, biological activities of the Fc portion of antibodies include the ability to activate the complement pathway (IgG & IgM), bind to phagocytes (IgG, IgA), or bind to mast cells and basophils (IgE).

Two classes of antibodies are more complex. IgM is a pentamer, consisting of 5 "Y"-like molecules connected at their Fc portions, and secretory IgA is a dimer consisting of 2 "Y"-like molecules.

Serology refers to using antigen-antibody reactions in the laboratory for diagnostic purposes. Its name comes from the fact that serum, the liquid portion of the blood where antibodies are found is used in testing. Serologic testing may be used in the clinical laboratory in two distinct ways: to identify unknown antigens (such as microorganisms) and to detect antibodies being made against a specific antigen in the patient's serum. There are two types of serologic testing: direct and indirect.

(a) Direct serologic testing is the use of a preparation known antibodies, called antiserum, to identify an unknown antigen such as a microorganism.

(b) Indirect serologic testing is the procedure whereby antibodies in a person's serum being made by that individual against an antigen associated with a particular disease are detected using a known antigen.

B. USING ANTIGEN-ANTIBODY REACTIONS IN THE LABORATORY TO IDENTIFY UNKNOWN ANTIGENS SUCH AS MICROORGANISMS

This type of serologic testing employs known antiserum (serum containing specific known antibodies). The preparation of known antibodies is prepared in one of two ways: in animals or by hybridoma cells.

1. Preparation of known antisera in animals

Preparation of known antiserum in animals involves inoculating animals with specific known antigens such as a specific strain of a bacterium. After the animals immune responses have had time to produce antibodies against that antigen, the animal is bled and the blood is allowed to clot. The resulting liquid portion of the blood is the serum and it will contain antibodies specific for the injected antigen.

However, one of the problems of using antibodies prepared in animals (by injecting the animal with a specific antigen and collecting the serum after antibodies are produced) is that up to 90% of the antibodies in the animal's serum may be antibodies the animal has made "on its own" against environmental antigens, rather than those made against the injected antigen. The development of monoclonal antibody technique has largely solved that problem.

2. Preparation of known antibodies by monoclonal antibody technique

One of the major breakthroughs in immunology occurred when monoclonal antibody technique was developed. Monoclonal antibodies are antibodies of a single specific type. In this technique, an animal is injected with the specific antigen for the antibody desired. After appropriate time for antibody production, the animal's spleen is removed. The spleen is rich in plasma cells and each plasma cell produces only one specific type of antibody. However, plasma cells will not grow artificially in cell culture. Therefore, a plasma cell producing the desired antibody is fused with a myeloma cell, a cancer cell from bone marrow which will grow rapidly in cell culture, to produce a hybridoma cell. The hybridoma cell has the characteristics of both parent cells. It will produce the specific antibodies like the plasma cell and will also grow readily in cell culture like the myeloma cell. The hybridoma cells are grown artificially in huge vats where they produce large quantities of the specific antibody.

Monoclonal antibodies are now used routinely in medical research and diagnostic serology and are being used experimentally in treating certain cancers and a few other diseases.

3. The concept and general procedure for direct serologic testing

The concept and general procedure for using antigen-antibody reactions to identify unknown antigens are as follows:

- *Concept.* This testing is based on the fact that antigen-antibody reactions are very specific. Antibodies usually react only with the antigen that stimulated their production in the first place, and are just as specific as an enzyme-substrate reaction. Because of this, one can use known antiserum (prepared by animal inoculation or monoclonal antibody technique as discussed above) to identify unknown antigens such as a microorganisms.
- *General Procedure.* A suspension of the unknown antigen to be identified is mixed with known antiserum for that antigen. One then looks for an antigen-antibody reaction.

Examples of serologic tests used to identify unknown microorganisms include the serological typing of *Shigella* and *Salmonella*, the Lancefield typing of beta streptococci, and the serological identification of meningococci. Serological tests used to identify antigens which are not microorganisms include blood typing, tissue typing, and pregnancy testing.

4. Detection of antigen-antibody reactions in the laboratory

Antigen-antibody reactions may be detected in the laboratory by a variety of techniques. Some of the commonly used techniques for observing in vitro antigen-antibody reactions are briefly described below.

(a) *Agglutination.* Known antiserum causes bacteria or other particulate antigens to clump together or agglutinate. Molecular-sized antigens can be detected by attaching the known antibodies to larger, insoluble particles such as latex particles or red blood cells in order to make the agglutination visible to the naked eye.

(b) *Precipitation.* Known antiserum is mixed with soluble test antigen and a cloudy precipitate forms at the zone of optimum antigen-antibody proportion.

(c) *Complement-fixation.* Known antiserum is mixed with the test antigen and complement is added. Sheep red blood cells and hemolysins (antibodies that lyse the sheep red blood cells in the presence of free complement) are then added. If the complement is tied up in the first antigen-antibody reaction, it will not be available for the sheep red blood cell-hemolysin reaction and there will be no hemolysis. A negative test would result in hemolysis.

(d) *Enzyme-linked immunosorbant assay or ELISA (also known as Enzyme immunoassay or EIA).* Test antigens from specimens are passed through a tube (or a membrane) coated with the corresponding specific known antibodies and become trapped on the walls of the tube (or on the membrane). Known antibodies to which an enzyme has been chemically attached are then passed through the tube (or membrane) where they combine with the trapped antigens. Substrate for the attached enzyme is then added and the amount of antigen-antibody complex formed is proportional to the amount of enzyme-substrate reaction as indicated by a colour change.

(e) *Radioactive binding techniques.* Test antigens from specimens are passed through a tube coated with the corresponding specific known antibodies and become trapped on the walls of the tube. Known antibodies to which a radioactive isotope has been chemically attached are then passed through the tube where they combine with the trapped antigens. The amount of antigen-antibody complex formed is proportional to the degree of radioactivity.

(f) *Fluorescent antibody technique.* A fluorescent dye is chemically attached to the known antibodies. When the fluorescent antibody reacts with the antigen, the antigen will fluoresce when viewed with a fluorescent microscope.

C. EXAMPLES OF DIRECT SEROLOGIC TEST TO IDENTIFY UNKNOWN ANTIGENS

As stated above, this type of serologic testing uses known antiserum (antibodies) to identify unknown antigens.

1. Serological Typing of *Shigella*

There are four different serological subgroups of Shigella, each corresponding to a different species:

- subgroup A = *Shigella dysenteriae*
- subgroup B = *Shigella flexneri*
- subgroup C = *Shigella boydii*
- subgroup D = *Shigella sonnei*

Known antiserum is available for each of the 4 subgroups of *Shigella* listed above and contains antibodies against the cell wall ("O" antigens) of *Shigella*. The suspected Shigella (the unknown antigen) is placed in each of 4 circles on a slide and a different known antiserum (A, B, C or D) is then added to each circle. A positive antigen-antibody reaction appears as a clumping or agglutination of the *Shigella*.

2. Serological Typing of Streptococci

Many of the streptococci can be placed into serological groups called Lancefield groups based on carbohydrate antigens in their cell wall. Although there are 20 different Lancefield groups of streptococci, the groups A, B, C, D, F, and G are the ones usually associated with human infections. The Slidex Strepto-Kit® system is a commercial kit for typing the 6 Lancefield groups of streptococci that commonly infect humans. To make the reaction more visible, since the antigens for which one is testing are only fragments of the bacterial cell wall, the known monoclonal antibodies have been adsorbed to latex particles. In this way, when the known monoclonal antibodies react with the streptococcal cell wall antigens, agglutination of the latex particles will occur and can be easily seen with the naked eye.

3. Serological Testing to Diagnose Pregnancy

The hormone human chorionic gonadotropin (HCG), produced by the placenta, appears in the serum and urine of pregnant females. The HCG is composed of two subunits - alpha and beta. The CARDS O.S.® HCG-Serum/Urine is a one step pregnancy test that detects measurable levels of HCG as early as early as 7-10 days after conception.

HCG, the unknown antigen for which one is testing, is identified in the urine by using known monoclonal antibody against human HCG.

This test uses a colour immunochromatographic assay to detect the antigen-antibody reaction. Inside the plastic card is a membrane strip along which the urine flows and on which the reaction occurs. The urine is placed in the "add urine" well on the right side of the card and flows along the card from right to left. The membrane just to the left of the sample well is coated with red latex beads to which known antibodies against the beta chain of human HCG have been attached. If there is HCG in the urine, the beta subunit of the HCG will react with the known anti-beta HCG antibody/red latex conjugate and this complex of HCG-antibody/red latex will become mobilized and flow with the urine towards the left side of the card. In the "read results" window of the card is a vertical line to which is immobilized known antibodies against the alpha subunit of human HCG. As the urine containing the antibody/red latex conjugate bound to the beta subunit of HCG flows past the vertical line, the alpha subunit of the HCG binds to the immobilized antibodies located on the line, trapping the complex and causing a vertical red line to appear. The vertical red line crosses the horizontal blue line preprinted in the "read results" window to form a (+) sign.

If the woman is not pregnant and there is no HCG in the urine, then there will be no antigen to react with the anti-beta HCG antibody/red latex conjugate to the left of the sample well and likewise, no reaction with the anti-alpha HCG antibodies immobilized along the vertical line in the "read results" window. The antibody/red latex conjugate will continue to flow to the left of the slide until it reaches the "test complete" window. Since no vertical red line forms, a (–) sign appears in the "read results" window.

4. Identification of Microorganisms Using the Direct Fluorescent Antibody Technique

Certain fluorescent dyes can be chemically attached to the known antibody molecules in antiserum. The known fluorescent antibody is then mixed with the unknown antigen, such as a microorganism, fixed to a slide. After washing, to remove any fluorescent antibody not bound to the antigen, the slide is viewed with a fluorescent microscope.

If the fluorescent antibody reacted with the unknown antigen, the antigen will glow or fluoresce under the fluorescent microscope. If the fluorescent antibody did not react with the antigen, the antibodies will be washed off the slide and the antigen will not fluoresce.

Many bacteria, viruses, and fungi can be identified using this technique.

PROCEDURE

A. Serologic Typing of *Shigella*

1. Using a wax marker, draw two circles (about the size of a nickel) on each of two clean glass slides. Label the circles A, B, C, and D.
2. Add one drop of the suspected Shigella (unknown antigen) to each circle. (The *Shigella* has been treated with formalin to make it noninfectious but still antigenic.)
3. Now add one drop of known Shigella subgroup A antiserum to the "A" circle, one drop of known *Shigella* subgroup B antiserum to the "B" circle, one drop of known *Shigella* subgroup C antiserum to the "C" circle, and one drop of known *Shigella* subgroup D antiserum to the "D" circle.

4. Rotate the slide carefully for 30-60 seconds.
   Agglutination of the bacteria, indicates a positive reaction.
   No agglutination is negative.
5. Dispose of all pipettes and slides in the disinfectant container.

B. Serologic Typing of Streptococci

1. The cell wall antigens of the unknown *Streptococcus* used in this test are extracted by mixing the organism with extraction enzyme. This step has been done for you.
2. Place one drop of the appropriate known streptococcal monoclonal antibody/latex conjugate (groups A, B, C, D, F, and G) on the corresponding six circles of the slide.
3. Add one drop of the extracted antigen from the unknown *Streptococcus* prepared in step 1 to each circle.
4. Spread the antigen-antibody mixtures over the entire circles using separate applicator sticks for each circle.
5. Rock the slide back and forth for no longer than one minute and look for agglutination.

C. Serologic Testing to Detect Pregnancy

1. Fill the disposable pipette to the line with urine and dispense the urine into the "add urine" well.
2. Shortly after the urine is added, a blue colour will be seen moving across the "read results" window.
3. The test results can be read in the "read results" window when a distinct blue line appears in the "test complete" window (approximately 5 minutes). A (+) sign indicates a positive test; a (–) sign is a negative test.

D. The Direct Fluorescent Antibody Technique

Observe the demonstration of a positive direct fluorescent antibody test.

RESULTS

A. Serologic Typing of Shigella

Make a drawing of your results.

- Agglutination of bacteria is positive.
- No agglutination of bacteria is negative.

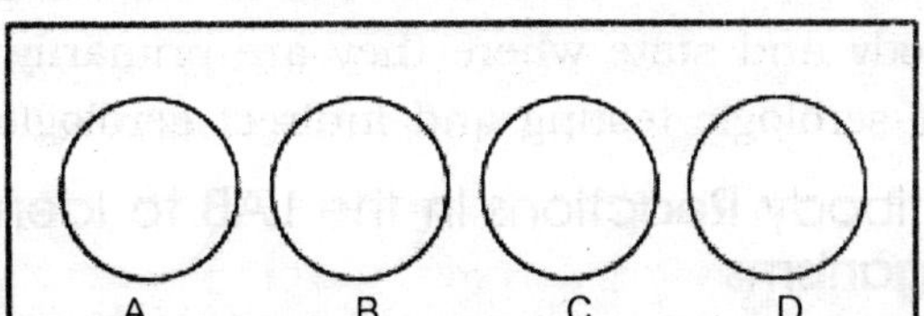

Fig. 31. Shigella typing slide B.

B. Serologic Typing of Streptococci

Make a drawing of your result.

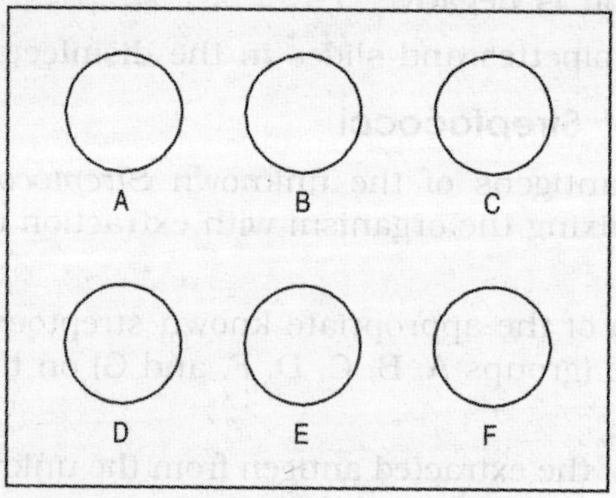

Fig. 32. Streptococcus typing slide.

C. Serologic Testing to Diagnose Pregnancy

Make a drawing of a positive pregnancy test.

Fig. 33. Positive pregnancy test.

D. The Direct Fluorescent Antibody Technique

Make a drawing and describe a positive direct fluorescent antibody test.

PERFORMANCE OBJECTIVES

After completing this lab, the student will be able, to perform the following objectives:

A. Introduction to Serological Testing

1. Define serology.
2. Define antigen and state what may act as an antigen.
3. Define antibody and state where they are primarily found in the body.
4. Define direct serologic testing and indirect serologic testing.

B. Using Antigen-Antibody Reactions in the LAB to Identify unknown Antigens such as Microorganisms

1. Define antiserum.
2. Describe two ways of producing known antiserum.

3. Describe the concept and general procedure for using serologic testing to identify unknown antigens (direct serologic testing).

C. Examples of Serologic Tests to Identify Unknown Antigens

1. Describe how to determine serologically whether an organism is a subgroup A, B, C, or D *Shigella*.
2. Describe how to determine serologically whether an organism is a Lancefield group A, B, C, D, F, or G *Streptococcus*.
3. Describe how to diagnose pregnancy serologically.
4. Briefly describe the direct fluorescent antibody technique.

RESULTS

1. Correctly interpret the results of the following serological tests:
   (*a*) serological typing of *Shigella*
   (*b*) serological typing of streptococci
   (*c*) serological testing for pregnancy
   (*d*) a direct fluorescent antibody test

21. SEROLOGY PART 2–INDIRECT SEROLOGIC TESTING

A. USING ANTIGEN-ANTIBODY REACTIONS IN THE LABORATORY TO INDIRECTLY DIAGNOSE DISEASE BY DETECTING ANTIBODIES IN A PERSON'S SERUM PRODUCED AGAINST A DISEASE ANTIGEN

Indirect serologic testing is the procedure whereby antibodies in a person's serum being made by that individual against an antigen associated with a particular disease are detected using a known antigen.

1. The Concept and General Procedure for Indirect Serologic Testing

The concept and general procedure for this type of serological testing are as follows:

- *Concept.* This type of testing is based on the fact that antibodies are only produced in response to a specific antigen. In other words, a person will not be producing antibodies against a disease antigen unless that antigen is in the body stimulating antibody production.
- *General Procedure.* A sample of the patient's serum (the liquid portion of the blood after clotting and containing antibodies against the disease antigen if the person has or has had the disease) is mixed with the known antigen for that suspected disease. One then looks for an antigen-antibody reaction.

  Examples of serologic tests to diagnose disease by the detection of antibodies in the patient's serum include the various serological tests for syphilis or STS (such as the RPR, the VDRL, and the FTA-ABS tests), the tests for infectious mononucleosis, the tests for the Human Immunodeficiency Virus (HIV), the tests for systemic lupus erythematosus, and tests for variety of other viral infections.

2. Qualitative and Quantitative Serologic Tests

Indirect serologic tests may be qualitative or quantitative. A qualitative test only detects the presence or absence of specific antibodies in the patient's serum and is often used for screening purposes. A quantitative test gives the titer or amount of that antibody in the serum. Titer indicates how far you can dilute the patient's serum and still have it contain enough antibodies to give a detectable antigen-antibody reaction. In other words, the more antibodies being produced by the body, the more you can dilute the person's serum and still see a reaction. Quantitative serological tests are often used to follow the progress of a disease by looking for a rise and subsequent drop in antibody titer.

3. Detection of Antigen-antibody Reactions in the Laboratory

Antigen-antibody reactions may be detected in the laboratory by a variety of techniques. Some of the commonly used techniques are briefly described below:

(*a*) *Agglutination.* Antibodies in the patient's serum cause the known particulate antigens or cells to clump or agglutinate. Molecular-sized known antigens can be attached to larger, insoluble particles such as latex particles, red blood cells, or charcoal particals in order to observe agglutination with the naked eye.

(*b*) *Precipitation.* The patient's serum is mixed with soluble known antigen and a cloudy precipitate forms at the zone of optimum antigen-antibody proportion.

(*c*) *Complement-fixation.* The patient's serum is mixed with the known antigen and complement is added. Sheep red blood cells and hemolysins (antibodies that lyse the sheep red blood cells in the presence of free complement) are then added. If the complement is tied up in the first antigen-antibody reaction, it will not be available for the sheep red blood cell-hemolysin reaction and there will be no hemolysis. A negative test would result in hemolysis.

(*d*) *Enzyme immunoassay (EIA).* The patient's serum is placed in a tube or well coated with the corresponding known antigen and becomes trapped on the walls of the tube. Enzyme-labeled anti-human gamma globulin or anti-HGG (antibodies made in another animal against the Fc portion of human antibody and to which an enzyme has been chemically attached), is then passed through the tube where it combines with the trapped antibodies from the patient's serum. Substrate for the enzyme is then added and the amount of antibody-antigen complex formed is proportional to the amount of enzyme-substrate reaction as indicated by a colour change.

(*e*) *Radioactive binding techniques.* The patient's serum is passed through a tube coated with the corresponding known antigen and becomes trapped on the walls of the tube. Radioisotope-labeled anti-human gamma globulin or anti-HGG (antibodies made in another animal against the Fc portion of human antibody and to which a radioactive isotope has been chemically attached), is then passed through the tube where it combines with the trapped antibodies from the patient's serum. The amount of antibody-antigen complex formed is proportional to the degree of radioactivity measured.

(*f*) *Fluorescent antibody technique.* The patient's serum is mixed with known antigen fixed to a slide. Fluorescent anti-human gamma globulin or anti-HGG (antibodies made against the Fc portion of human antibody and to which a fluorescent dye has been chemically attached) is then added. It combines with the antibodies from the patient's serum bound to the antigen on the slide causing the antigen to fluoresce when viewed with a fluorescent microscope.

B. EXAMPLES OF INDIRECT SEROLOGIC TESTS TO DETECT ANTIBODIES IN THE PATIENT'S SERUM

1. The RPR Test for Syphilis

Syphilis is a sexually transmitted disease caused by the spirochete Treponema pallidum. The RPR (Rapid Plasma Reagin) Card® test is a presumptive serologic screening test for syphilis. The serum of a person with syphilis contains a nonspecific antilipid antibody (traditionally termed regian), which is not found in normal serum. The exact nature of the antilipid (reagin) antibody is not known but it is thought that a syphilis infection instigates the breakdown of the patient's own tissue cells. Fatty substances which are released then combine with protein from *Treponema pallidum* to form an antigen which stimulates the body to produce antibodies against both the body's tissue lipids (nonspecific or nontreponemal) as well as the T. pallidum protein (specific or treponemal). The RPR Card® test detects the nonspecific antilipid antibody and is referred to as a nontreponemal test for syphilis.

It must be remembered that tests for the presence of these nonspecific antilipid antibodies are meant as a presumptive screening test for syphilis. Similar reagin-like antibodies may also be present as a result of other diseases such as malaria, leprosy, infectious mononucleosis, systemic lupus erythematosus, viral pneumonia, measles, and collagen diseases and may give biologic false-positive results (BFP). Confirming tests should be made for the presence of specific antibodies against the *T. pallidum* itself. The confirming test for syphilis is the FTA-ABS test discussed below. Any serologic test for syphilis is referred to commonly as an STS (Serological Test for Syphilis).

The known RPR antigen consists of cardiolipin, lecithin, and cholesterol bound to charcoal particles in order to make the reaction visible to the naked eye. If the patient has syphilis, the antilipid antibodies in his or her serum will cross-react with the known RPR lipid antigens giving a visible clumping of the charcoal particles.

We will do a quantitative RPR Card® test today in lab. Keep in mind that a quantitative test allows one to determine the **titer** or amount of a certain antibody in the serum. In this test, a constant amount of RPR antigen is added to dilutions of the patient's serum. The most dilute sample of the patient's serum still containing enough antibodies to give a visible antigen-antibody reaction is reported as the titer.

2. Serologic Tests for Infectious Mononucleosis

During the course of infectious mononucleosis, caused by the Epstein-Barr virus (EBV), the body produces nonspecific heterophile antibodies which are not found in normal serum. As it turns out, these heterophile antibodies will also cause horse or sheep erythrocytes (red blood cells) to agglutinate.

The infectious mononucleosis serologic test demonstrated today is a rapid qualitative test for infectious mononucleosis that uses specially treated horse erythrocytes (acting as the "known antigen") that are highly specific for mononucleosis heterophile antibodies. Agglutination of erythrocytes after adding the patient's serum indicates a positive test.

Quantitative tests may be done to determine the titer of heterophile antibodies and follow the progress of the disease.

3. Serologic Tests for Systemic Lupus Erythematosus (SLE)

Systemic lupus erythematosus or SLE is a systemic autoimmune disease. Immune complexes become deposited between the dermis and the epidermis, and in joints, blood vessels, glomeruli of the kidneys, and the central nervous system. It is four times more common in women than in men. In SLE, autoantibodies are made against components of DNA. This test is specific for the serum anti-deoxyribonucleoprotein antibodies associated with SLE. The known antigen is deoxyribonucleoprotein adsorbed to latex particles to make the reaction more visible to the eye. This is a qualitative test used to screen for the presence of the disease and to monitor its course.

4. Detecting Antibody Using the Indirect Fluorescent Antibody Technique: The FTA-ABS Test for Syphilis

The **indirect** fluorescent antibody technique involves three different reagents:

(*a*) the patient's serum (containing antibodies against the disease antigen if the disease is present)

(*b*) known antigen for the suspected disease

(*c*) fluorescent anti-human gamma globulin antibodies (antibodies made in another animal against the Fc portion of human antibodies by injecting an animal with human serum. A fluorescent dye is then chemically attached to the anti-human gamma globulin (anti-HGG) antibodies.

The FTA-ABS test (Fluorescent Treponemal Antibody Absorption Test) for syphilis is an example of an indirect fluorescent antibody procedure. This is a confirming test for syphilis since it tests specifically for antibodies in the patient's serum made in response to the syphilis spirochete, Treponema pallidum.

In this test, killed *T. pallidum,* (the known antigen), is fixed on a slide. The patient's serum is added to the slide. If the patient has syphilis, antibodies against the *T. pallidum* will react with the antigen on the slide. The slide is then washed to remove any antibodies not bound to the spirochete.

To make this reaction visible, a second animal-derived antibody made against human antibodies and labelled with a fluorescent dye (fluorescent anti-human gamma globulin) is added. These fluorescent anti-HGG antibodies react with the patient's antibodies which have reacted with the *T. pallidum* on the slide. The slide is washed to remove any unbound fluorescent anti-HGG antibodies and observed with a fluorescent microscope. If the spirochetes glow or fluoresce, the patient has made antibodies against *T. pallidum* and has syphilis.

5. The EIA and Western Blot Serologic Tests for Antibodies Against the Human Immunodeficiency Virus (HIV)

In the case of the current HIV antibody tests, the patient's serum is mixed with various HIV antigens produced by recombinant DNA technology. If the person is seropositive (has repeated positive antigen-antibody tests), then HIV must be in that person's body stimulating antibody production. In other words, the person must be infected with HIV. The two most common tests currently used to detect antibodies against HIV are the enzyme immunoassay or EIA (also known as the enzyme-linked immunosorbant assay or ELISA) and the Western blot or WB. A person is considered to be seropositive for HIV infection only after an EIA screening test is repeatedly reactive and another test such as the WB has been performed to confirm the results.

The **EIA** is less expensive, faster, and technically less complicated than the WB and is the procedure initially done as a screening test for HIV infection. The various EIA tests give a spectrophotometric reading of the amount of antibody binding to known HIV antigens.

The EIA test kit contains plastic wells to which various HIV antigens have been adsorbed. The patient's serum is added to the wells and any antibodies present in the serum against HIV antigens will bind to the corresponding antigens in the wells. The wells are then washed to remove all antibodies in the serum other than those bound to HIV antigens. Enzyme-linked anti-human gamma globulin (anti HGG) antibodies are then added to the wells. These antibodies, made in another animal against the Fc portion of human antibodies by injecting the animal with human serum, have an enzyme chemically attached. They react with the human antibodies bound to the known HIV antigens. The wells are then washed to remove any anti-HGG that has not bound to serum antibodies. A substrate specific for the enzyme is then added and the resulting enzyme-substrate reaction causes a colour change in the wells. If there are no antibodies in the patient's serum against HIV, there will be nothing for the enzyme-linked anti-HGG to bind to and it will be washed from the wells. When the substrate is added, there will be no enzyme present in the wells to give a colour change.

If the initial EIA is reactive it is automatically repeated to reduce the possibility that technical laboratory error caused the reactive result. If the EIA is still reactive, it is then confirmed by the Western blot test.

The Western blot WB is the test most commonly used as a confirming test if the EIA is repeatedly positive. The WB is technically more complex to perform and interpret, is more time consuming, and is more expensive than the EIAs.

With the WB, the various protein and glycoprotein antigens from HIV are separated according to their molecular weight by gel electrophoresis (a procedure that separates charged proteins in a gel by applying an electric field). Once separated, the various HIV antigens are transferred to a nitrocellulose strip. The patient's serum is then incubated with the strip and any HIV antibodies that are present will bind to the corresponding known HIV antigens on the strip. Enzyme-linked anti-human gamma globulin (anti HGG) antibodies are then added to the strip. These antibodies, made in another animal against the Fc portion of human antibodies by injecting the animal with human serum, have an enzyme chemically attached. They react with the human antibodies bound to the known HIV antigens. The strip is then washed to remove any anti-HGG that has not bound to serum antibodies. A substrate specific for the enzyme is then added and the resulting enzyme-substrate reaction causes a colour change on the strip. If there are no antibodies in the patient's serum against HIV, there will be nothing for the enzyme-linked anti-HGG to bind to and it will be washed from the strip. When the substrate is added, there will be no enzyme present on the strip to give a colour change.

It should be mentioned that all serologic tests are capable of giving occasional false-positive and false-negative results. The most common cause of a false-negative HIV antibody test is when a person has been only recently infected with HIV and his or her body has not yet made sufficient quantities of antibodies to give a visible positive serologic test. It generally takes between 2 weeks and 3 months after a person is initially infected with HIV to convert to a positive HIV antibody test.

PROCEDURE

A. The RPR® Card Test for Syphilis

1. Label 6 test tubes as follows: 1:1, 1:2, 1:4, 1:8, 1:16, and 1:32.
2. Using a 1.0 ml pipette, add 0.5 ml of 0.9% saline solution into tubes 1:2, 1:4, 1:8, 1:16, and 1:32.
3. Add 0.5 ml of the patient's serum to the 1:1 tube (undiluted serum).
4. Add another 0.5 ml of serum to the saline in the 1:2 tube and mix. Remove 0.5 ml from the 1:2 tube and add it to the 1:4 tube and mix. Remove 0.5 ml from the 1:4 tube, add to the 1:8 tube and mix. Remove 0.5 ml from the 1:8 tube, add to the 1:16 tube and mix. Remove 0.5 ml from the 1:16 tube, add to the 1:32 tube and mix. Remove 0.5 ml from the 1:32 tube and discard.
5. Using the capillary pipettes provided with the kit, add a drop of each serum dilution to separate circles of the RPR card. Spread the serum over the entire inner surface of the circle with the tip of the pipette, using a new pipette for each serum dilution.
6. Using the RPR antigen dispenser, add a drop of known RPR antigen to each circle. Do not let the needle of the dispenser touch the serum. Using disposable stirrers, mix the known RPR antigen with the serum in each circle.
7. Place the slide on a shaker and rotate for a maximum of 4 minutes.
8. Read the results as follows:
   - A definite clumping of the charcoal particles is reported as reactive (R).
   - No clumping is reported as nonreactive (N).

The greatest serum dilution that produces a reactive result is the titer. For example, if the dilutions turned out as follows, the titer would be reported as 1:4 or 4 dils.

1:1	1:2	1:4	1:8	1:16	1:32
R	R	R	N	N	N

B. The Serologic Tests for Infectious Mononucleosis

1. Place one drop of each of the patient's serum in circles on the test slide.
2. Add one drop of treated horse erythrocytes (the known antigen) to each circle and mix with disposable applicator sticks.
3. Rock the card gently for 1 minute, then leave undisturbed for 1 minute, and observe for agglutination of the red blood cells. Agglutination indicates the presence of heterophile antibodies.

C. The Serologic Tests for Systemic Lupus Erythematosus (SLE)

1. Add one drop of each of the patient's serum to separate circles on the test slide.
2. Add one drop of the Latex-Deoxyribonucleoprotein reagent (the known antigen, deoxyribonucleoprotein adsorbed to latex particles) to each serum sample and mix with disposable applicator sticks.
3. Rock the slide gently for 1 minute and observe for agglutination. Agglutination indicates the presence of antinuclear antibodies associated with SLE.

D. The FTA-ABS Test for Syphilis (Indirect Fluorescent Antibody Technique

Observe the 35 mm slide of a positive FTA-ABS test.

E. The EIA and WB Tests for HIV Antibodies

Observe the illustrations of the EIA and the WB tests for antibodies against HIV.

RESULTS

A. RPR Card® Test for Syphilis (Quantitative)

Detects nontreponemal antilipid antibodies (reagin)

Record your results in the table.

Dilution	*Result*
1:1	
1:2	
1:4	
1:8	
1:16	
1:32	
titer	

R = reactive (distinct clumps)

N = nonreactive (no clumps)

B. MONO-TEST for Infectious Mononucleosis (Qualitative)

Detects heterophile antibodies.

Draw the results of a positive and a negative test.

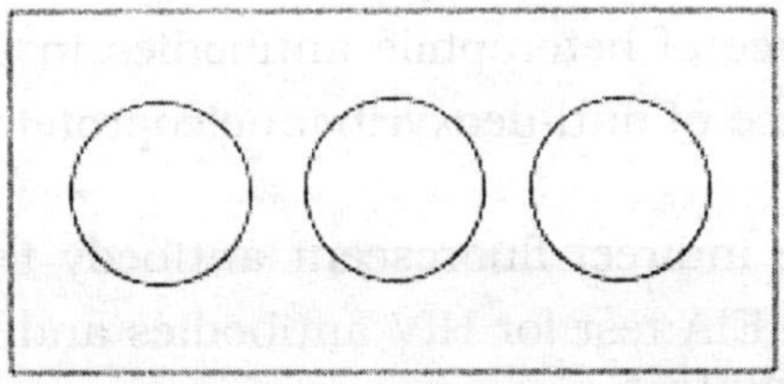

Fig. 34. Infectious mononucleosis test slide

+ = agglutination of RBCs

– = no agglutination of RBCs

C. Serologic test for SLE (Qualitative)

Detects anti-deoxyribonucleoprotein antibodies.

Draw the results of a positive and a negative test.

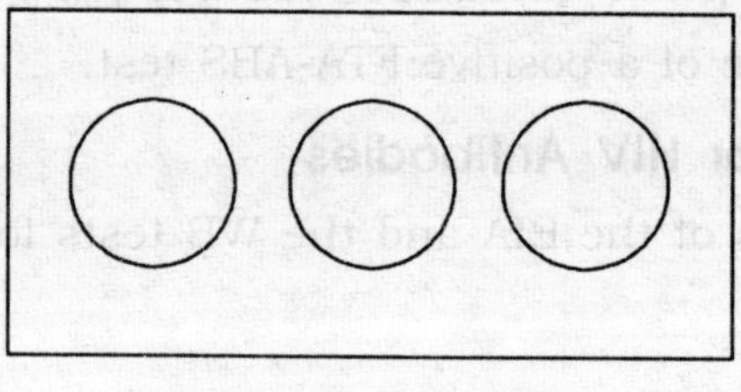

Fig. 35. SLE test slide

+ = agglutination

– = no agglutination

D. FTA-ABS Test for Syphilis (Confirming)

Detects antibodies against *Treponema pallidum*

Draw the results of a positive FTA-ABS test.

PERFORMANCE OBJECTIVES

Discussion

1. State the principle and the general procedure behind indirect serologic testing.
2. State the difference between a qualitative serological test and a quantitative serological test.
3. Define titer.
4. State what disease the RPR and the FTA-ABS procedures test for. Indicate which of these is a presumptive test, which is a confirming test, and why.
5. State the significance of nontreponemal antilipid (reagin) antibodies in serological testing.
6. State the significance of heterophile antibodies in serological testing.
7. State the significance of anti-deoxyribonucleoprotein antibodies in serological testing.
8. Briefly describe the indirect fluorescent antibody technique.
9. Briefly describe the EIA test for HIV antibodies and state the significance of a positive HIV antibody test.
10. State the most common reason for a false-negative HIV antibody test.

RESULTS

1. Interpret the results of the following serological tests:
   (*a*) serologic test for infectious mononucleosis
   (*b*) serologic test for SLE
   (*c*) FTA-ABS test
2. Determine the titer of a quantitative RPR Card® test.

7 CELL BIOLOGY AND GENETICS

BASIC STAINING TECHNIQUES OF CHROMOSOMES

CELL CYCLES

Introduction

The onion root tip and the whitefish blastula remain as the standard introduction to the study of mitosis. The onion has easily observable chromosomes, and the whitefish has one of the clearest views of the spindle apparatus. The testis of the grasshopper and the developing zygote of the roundworm Ascaris are the traditional materials used for viewing the various stages of meiosis.

In a single longitudinal section of a grasshopper testis one can usually find all of the stages of meiotic development. The stages are also aligned from one pole of the testis. Few other meiotic samples are as convenient. For most material, meiosis occurs in a more randomly distributed pattern throughout the testis.

Ascaris is utilized to observe the final stages of development in eggs (oogenesis). The Ascaris egg lies dormant until fertilized. It then completes meiosis forming two polar bodies while the sperm nucleus awaits fushion with the female nucleus. When this phenomenon is coupled to the large abundance of eggs in the Ascaris body, it makes an ideal specimen for observing the events of fertilization, polar body formation, fusion of pronuclei and the subsequent division of the cell (cytokinesis).

Interphase G1-S-G2

The stages of mitosis were originally detailed after careful analysis of fixed cells. More recently, time lapse photography coupled with phase contrast microscopy has allowed us to visualize the process in its entirety, revealing a dynamic state of flux.

In early work, so much emphasis was placed on the movement of the chromosomes that the cell was considered to be "at rest" when not in mitosis. As significant as mitotic division is, it represents only a small fraction of the life span of a cell. None the less, you may still come across the term "resting phase" in some older texts. This term is rarely used today, and the term interphase is sufficient for all activities between two mitotic divisions. The cell is highly active during interphase and most of the metabolic and genetic functions of the cell are reduced during the physical division of the nuclear and cellular materials (mitosis).

Presents our current view of a cell cycles. Note that interphase is divided into three sub-phases, G1, S and G2. The basis for this division is the synthesis of DNA. Note also that while the entire cycle may be as long as 24 hours, mitosis is normally less than one hour in length.

Because of the synthesis of DNA in interphase, the amount of DNA per nucleus is different depending on which sub phase of interphase the cell is in. DNA can be measured using the fuelgen reaction and a microspectrophotometer. The basic amount of DNA in a haploid nucleus is given the value C. A diploid nucleus would be 2C. A triploid and tetraploid cell would be 3C and 4C respectively.

However, when nuclei are actually measured, diploid cells in interphase can be divided into three groups; some are 2C, some are 4C while a few are at intermediate values between 2C and 4C. The conclusion is that the genetic material (DNA) and presumably the chromosomes must duplicate. The period within interphase and during which DNA is synthesized is termed the S phase (for Synthesis). The period of interphase preceding the S phase is the G1 phase (for 1st Growth Phase), while the period subsequent to the S phase is the G2 phase (for the 2nd Growth Phase). During the G1 period, the cell is generally increasing in size and protein content. During S, the cell replicates the chromosomes and synthesizes DNA. During G2, it continues to increase in size, but also begins to build a significant pool of ATP and other high energy phosphates, which are believed to be a significant part of the triggering mechanism for the subsequent karyokinetic and cytokinetic events of mitosis.

Mitosis returns the cells to the 2C state. Meiosis reduces the amount of DNA even further, to 1C. Meanwhile the number of chromosomes (designated with the letter N) is also changing. For a diploid cell, the number of chromosomes is twice that of a haploid, or 2N. During mitosis, a diploid cell would go from one 2N cell to two 2N cells. Since the daughter cells have the same chromosome number as the parent, mitosis is also referred to as equational division. If a diploid (2N) cell undergoes meiosis, it will result in four haploid cells, each 1N. Thus, meiosis is also referred to as reductional division. Refer figure for comparison of C and N values during division.

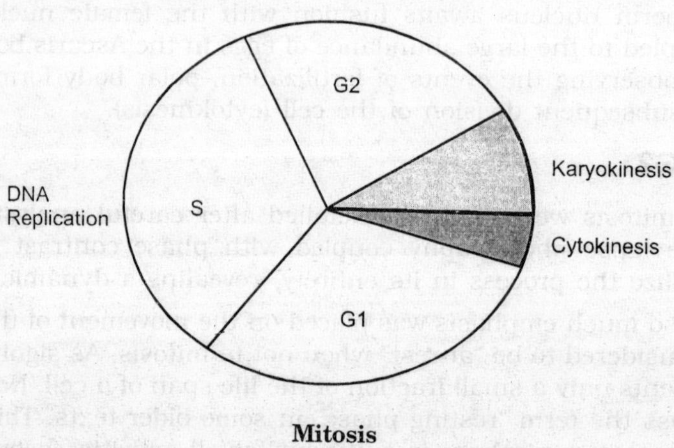

Mitosis

It is possible to visualize the process of DNA synthesis within either nuclei or chromosomes by the incorporation of a radioactive precursor to DNA into cells and subsequent detection by autoradiography. Incorporation of thymidine, a DNA precursor, will only occur during the S phase, and not during G1 nor G2. If a pulse (short period of exposure) of $_3$H-thymidine is presented to cells, those that are in the S period will incorporate this radioactive substance, while all others will not. Careful application of the pulse will allow the timing of the S phase. By knowing the timing for the entire cycle (from mitosis to mitosis), one can deduce the G1 and G2 periods.

Meiotic division differs from mitosis in that there are two division cycles instead on one. In the first cycle, interphase is the same as for mitosis. That is there is an S phase with corresponding G1 and G2. During the second interphase, however, no DNA synthesis occurs. Consequently there is no G1 or G2 in the second interphase. The result is that chromosomes are replicated prior to Meiosis, and do not replicate again during meiosis.

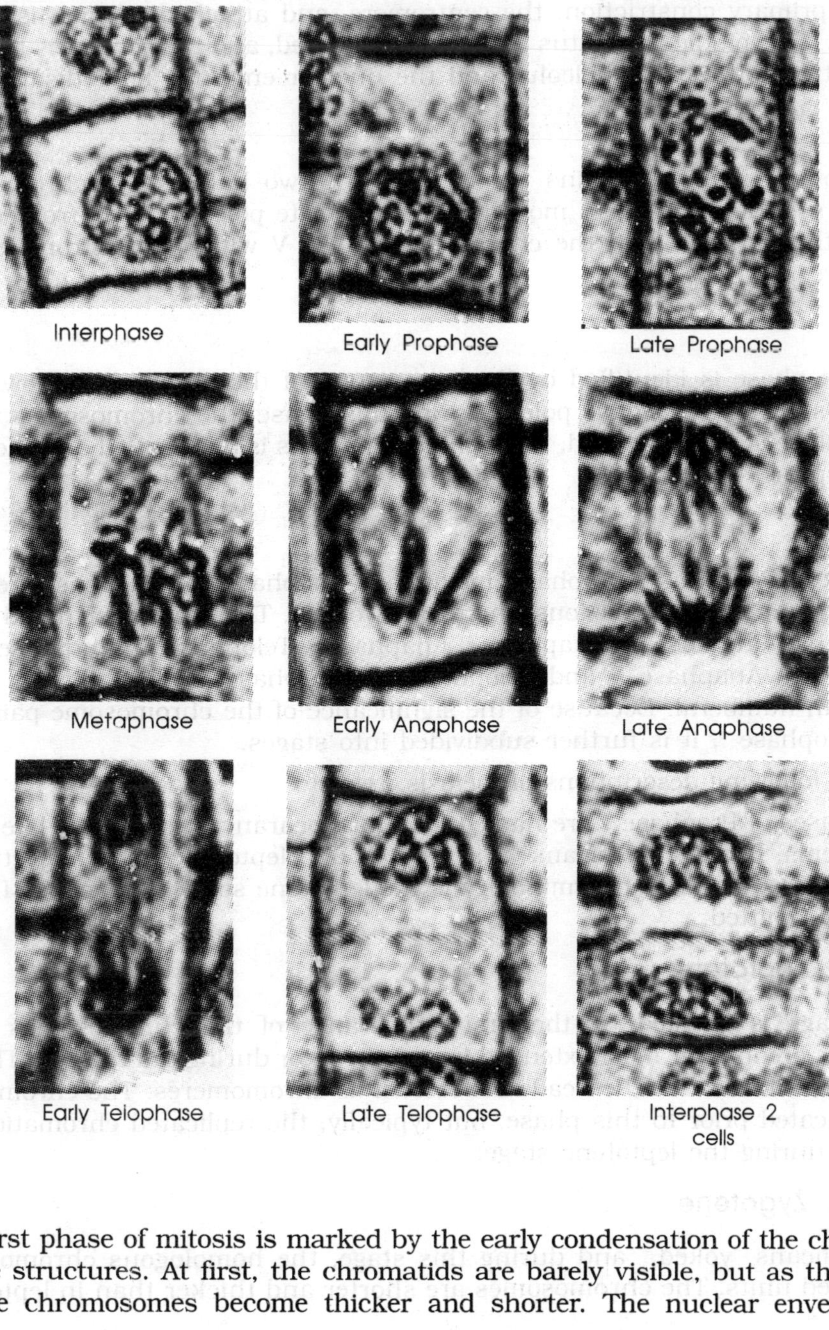

Interphase — Early Prophase — Late Prophase

Metaphase — Early Anaphase — Late Anaphase

Early Telophase — Late Telophase — Interphase 2 cells

Prophase

The first phase of mitosis is marked by the early condensation of the chromosomes into visible structures. At first, the chromatids are barely visible, but as they continue to coil, the chromosomes become thicker and shorter. The nuclear envelope is still present during this stage, as are any nucleolar structures. The centrioles are moving to the poles of the cell and spindle fibers are just beginning to form.

Metaphase

During the middle phase of karyokinesis, the chromosomes line up in the center of the cell, and form a metaphase plate. Viewed on edge, the chromosomes appear to be aligned across the entire cell, but viewed from 90 Degrees they appear to be spread throughout the entire cell (visualize a plate from its edge or from above). Each chromosome has a clear primary constriction, the centromere, and attached to each is a definitive spindle fiber. The spindle apparatus is completely formed, and the centrioles have reached their respective poles. The nucleolus and the nuclear envelope have disappeared.

Anaphase

The movement phase begins precisely as the two halves of a chromosome, the chromatids, separate and begin moving to the opposite poles. The centromere will lead the way in this process, and the chromatids form a V with the centromeres pointing toward the respective poles.

Telophase

The last phase is identified by the aggregation of the chromatids (now known as chromosomes) at the respective poles. During this phase, the chromosomes uncoil, the nuclear envelope is resynthesized, the spindle apparatus is dismantled and the nucleolus begins to appear.

Meiosis

For meiosis, the phases prophase, metaphase, anaphase and telophase are identified, but because there are two divisions, there are two sets. These are designated by Roman numberals; thus Prophase I, Metaphase I, Anaphase I, Telophase I, Interphase, Prophase II, Metaphase II, Anaphase II and Telophase II. Interphase is normally not designated with a Roman numberal. Because of the significance of the chromosome pairing which occurs in Prophase !, it is further subdivided into stages.

For the following descriptions of meiosis, refer to

The phases of Prophase I are named for the appearance of a thread-like structure, known as nema. Leptonema means "thin thread" and leptotene is the adjective applied to the term stage, i.e. proper terminology is the leptotene stage of prophase I. The word stage is often omitted.

Prophase I: Leptotene

This stage is marked by the first appearance of the chromosomes when the chromosomes are in their most extended form (except for during interphase). They appear to be a string with beads. The beads are known as chromomeres. The chromatids have already replicated prior to this phase, but typically, the replicated chromatids can not be observed during the leptotene stage.

Prophase I: Zygotene

Zygos means "yoked," and during this stage, the homologous chromosomes are seen as paired units. The chromosomes are shorter and thicker than in leptotene, and in some cells they remain attached to the nuclear envelope at the points near the aster. This gives rise to an image termed the "bouquet." This attachment is rare in invertebrates and absent in plants, where the chromosomes appear to be a tangled mass.

Prophase I: Pachytene

When the pairing of zygotene is complete, the chromosomes appear as "thick" strings, or pachynema. The chromosomes are about 1/4 the length they were in leptotene, and there are obviously two chromosomes, each with two chromatids in each bundle. The two chromosomes are referred to as a "bivalent," while the same structure viewed as four chromatids is known as a "tetrad."

Prophase I: Diplotene

This stage results as the gap between the two homologous chromosomes widens. The homologs have already paired during zygotene, recombined during pachytene and are now beginning to repel each other. During this stage, the chromosomes of some species uncoil somewhat, reversing the normal direction typical of prophase. As the chromosomes separate, they are observed to remain attached at points known as "chiasmata." These are believed to be the locations where genetic recombination of the genes has taken place.

Prophase I: Diakinesis

Prophase I ends as the homologs completely repel each other. The chromosomes will continue to coil tightly (reversing the slight uncoiling of the diplotene) and will reach their greatest state of contraction. As diakinesis progresses, chiasmata appear to move toward the ends of the chromosomes, a process known as "terminalization." Since this stage is the end of prophase, the nucleolus usually disappears, along with the nuclear envelope.

Metaphase I

The tetrads move toward the centre and line up on a metaphase plate. The nuclear envelope completely disappears. As the tetrads align themselves in the middle of the cell, they attach to spindle fibers in a unique manner. The centromeres of a given homolog will attach to the spindles from only one pole.

Anaphase I

The unique event occurring at this phase is the separation of the homologs. In contrast to mitotic anaphase, the centromeres of a given homolog do not divide, and consequently each homolog moves toward opposite poles. This results in a halving of the number of chromosomes, and is the basis of the reduction division that characterizes meiosis.

Telophase I: Interphase

Telophase in meiosis is similar to that of mitosis, except that in many species, the chromosomes do not completely uncoil. If the chromosomes do uncoil and enter a brief interphase, there is no replication of the chromatids. Remember that the chromatids have already been replicated prior to Prophase I.

Prophase II - Telophase II

These phases are essentially identical in meiotic and mitotic division: the only distinction is that the chromosome number is half of the number the cell had prior to meiosis. Each chromosome (homolog) is composed of two chromatids, and during Anaphase II, the two chromatids of each chromosome move apart and become separate chromosomes.

While the two chromatids remain attached at the centromere, they are known as chromatids. Immediately upon separating, each chromatid becomes known as a chromosome and is no longer referred to as a chromatid. This is the reason that a cell can divide one chromosome (with two chromatids) into two cells, each with a chromosome-the term applied to the chromatid is changed.

Damage Induced During Division

In 1949, Levan developed what was to become known as the Allium test for chromosome damage. Growing roots from onion bulbs were soaked in various agents and analyzed for their effect on mitosis. It was discovered that caffeine, for example, caused complete inhibition of mitosis, primarily through the inhibition of cell plate formation.

This test was later used extensively by B.A. Kihlman and extended to other higher plants. Kihlman found that 1 to 24 hour treatments of cells with caffeine and related oxypurines not only inhibited mitosis, but also induced significant chromosome alterations (aberrations). Specifically, this treatment induced "stickiness" and "pseudochiasmata." Stickiness is the clumping of chromosomes at metaphase and the formation of chromatin bridges at anaphase. Pseudochiasmata is the formation of side-arm bridges during anaphase. Caffeine also causes the formation of other chromosome and chromatid breaks and exchanges.

Colchicine, a drug which inhibits spindle fiber formation during mitosis, can be added to the growing cells to halt cell division at metaphase. This often will result in a doubling of the chromosome number, since colchicine typically inhibits cytokinesis, but not karyokinesis. The doubled chromosomes will fuse within a single nucleus, thus increasing the ploidy value of the nucleus.

Moreover, methylated oxypurines (caffeine, theophylline, 8-ethoxycaffeine) are inhibitors of cell plate formation. Treatment with these agents for 0.5-1 hour, with concentrations as low as 0.02-0.04%, results in the cell's failure to undergo cytokinesis; in addition, the nuclei do not fuse into a single unit. Thus, treatment with any of these agents should result in binucleate or multinucleate cells.

In addition, alkylating agents such as (Di-chloroethyl)methylamine or nitrogen mustard, Di-epoxypropyl ether (DEPE) and β-Propiolactone (BPL), Nitrosocompounds (N-Nitroso-N-methylurethan (NMU)), N-Methyl-phenyl-nitrosamine (MPNA), N-hydroxylphenylnitrosamine-ammonium (cupferron) and 1-Methyl-3-nitrosoguanidine (MNNG) have all been indicated as potent chromosome-breaking agents. Other compounds have included such things as maleic hydrazide, potassium cyanide, hydroxylamine, and dyes such as acridine orange in visible light.

The damages involve abnormal metaphases, isochromatid breaks, chromatid exchanges and anaphase bridges to name a few.

1. MEIOSIS IN FLOWER BUDS OF ALLIUM CEPA-ACETOCARMINE STAIN

Requirements

Onion flower, acetocarmine stain, glass slides, cover slips and blotting paper.

Procedure

1. Select appropriate flower buds of different size from the inflorescence.
2. Fix them in Carnoy's fluid, which is used as fixative.
3. Take a preserved flower bud and place it on a glass slide.
4. Separate the anthers and discard the other parts of the bud.
5. Put one or two drops of acetocarmine stain and squash the anthers.
6. Leave the material in the stain for 5 minutes.
7. Place a cover slip over them and tap it gently with a needle or pencil.
8. Warm it slightly over the flame of a spirit lamp.
9. Put a piece of blotting paper on the cover slip and apply a uniform pressure by your thumb.
10. Observe the slide under microscope for different meiotic stages.

Interphase

Before under going in meiosis-I, each cell will remain in an interphase during which the genetic materials are duplicated due to active DNA replication.

Prophase-I

In the first meiotic division, production in the chromosome number occurs without separation of chromatids. Prophase is the longest phase and has got five stages.

Leptotene

Chromosomes appear as long thread like structure interwoven together. Chromosomes give beaded appearance and are called as chromomeres. Ends of chromosomes are drawn towards nuclear membrane near centriole. In some plants, chromosomes may form synthetic knot.

Zygotene

The homologous chromosomes pair with one another, gene by gene, over the entire length of the chromosomes. The pairing of the homologous chromosomes is called Synapsis. Each pair of homologous chromosomes is known as Bivalent.

Pachytene

Each paired chromosomes become shorter and thicker than in earlier sub-stages and splits into two sister chromatids except at the region of the centromere. As a result of the longitudinal division of each homologous chromosome into two chromatids, there are four group of chromatids in the nucleus parallel to each other called Tetrads.

Diplotene

During diplotene stage, Chiasmata appear to move towards the ends of the Synapsed chromosomes in the process of Terminalization. Repulsion of homologous Chiasmata are very clear than in Pachytene because of the increased condensation of the chromosomes.

Diakinesis

The chromosomes begin to coil and so become shorter and thicker. Terminalization is completed. The nucleolus detaches from the nucleolar organizer and disappears completely. The nuclei envelop starts to degenerate and spindle formation is well under the way.

Metaphase-I

The bivalents orient themselves at random on the equatorial plate. The centromere of each chromosome of a terminalized tetrad is directed towards the opposite poles. The chromosomal microtubular spindle fibres remain attached with the centromeres and homologous chromosomes become ready to separate.

Anaphase-I

It is characterized by the separation of whole chromosomes of each homologous pair (Tetrad), so that each pole of the dividing cell receives either a paternal or maternal longitudinally double chromosome of each Tetrad. This ensures a change in chromosome number from diploid to monoploid or haploid in the resultant reorganized daughter nuclei.

Telophase-I

The chromosomes may persist for a time in the condensed state, the nucleolus and nuclear membrane may be reconstituted and cytokinesis may also occur to produce two haploid cells.

Metaphase-II

It is of very short duration. The chromosomes rearrange in the equatorial plate. Centromere lies in the equator while the arms are directed towards the poles. Centromeres divide and separate into two daughter chromosomes.

Anaphase-II

Daughter chromosomes start migrating towards the opposite poles and the movement is brought about by the action of spindle fibres.

Telophase-II

The chromosomes after reaching the opposite poles uncoil and become less distinct. Nuclear membrane and nucleolus reappear resulting in the formation of four daughter nuclei which are haploid.

Cytokinesis

This separates each nucleus from the others. Cell wall is formed and four haploid cells are produced.

2. MEIOSIS IN GRASSHOPPER TESTIS (POECILOCERUS PICTUS)

The testes of grasshopper are removed and are fixed in Carnoy's fluid. After 2-14 hours, the testes are transferred to 10% alcohol and are stored.

Squash Preparation

1. One or two lobes of the testes are removed.
2. The testes are placed on a glass slid.
3. Put one to two drops of acetocarmine stain.
4. With a sharp blade, the testes lobes are cut into minute pieces and are then kept for 10 minutes.
5. The slide is then gently covered with a cover slip taking care such that air bubbles are not formed.
6. Warm the slide gently and place the slide between two folds of filter paper.
7. Press the material with the tip of the finger and remove the excess stain, which comes out on the sides of cover slip.
8. The slide is observed under the microscope for observing the various stages of microscopes.

Interphase

This phase is present usually in animal cells. The cells in this stage are physiologically active. No DNA replication takes place.

Prophase-I

(*a*) *Leptotene.* The chromosomes are long, standard and uncoiled. They are densely formed on one side of the cell. Only one sex chromosome occurs in the males, which normally replicates later and hence appears as dark skin body.

(*b*) *Zygotene.* Homologous chromosomes pair by a process called synapsis. Pairing starts from many points on the chromosomes. The chromosomes are called bivalents. Bivalents become shortened and thickened by coiling and condensation. Synapsis of a chromosome is cemented by a complex called synaptonemal complex, which facilitates crossing over.

(*c*) *Pachytene.* Crossing over takes place between nonsister chromatids. Crossing over is accompanied by the chiasmata formation.

(*d*) *Diplotene.* Condensation of chromatid material is more. Each chromosome can be distinguished separately.

(*e*) *Diakinesis.* Homologous chromosomes begin to coil and so become shorter and thicker. Chromosomes are fully contracted and deeply stained. The 'X' chromosome is rod shaped, univalent, easily distinguishable from the rest of the chromosomes.

Metaphase-II

The chromosomes get oriented in the equatorial region of the spindle and their centromeres are attached to the chromosomal fibres. Each chromosome is easily seen. Maximum concentration occurs in this stage.

Anaphase-II

The spindle fibres contract and the homologous chromosomes separate and move towards the opposite poles. Each chromosome consists of two chromatids attached to one centromere.

Telophase-I

The separation of homologous chromosomes is completed. They reach the opposite poles. Two distinct daughter nuclei are formed. The daughter nuclei formed contain only half the number of chromosomes present in the parent cell. Cytokinesis may occur after the completion of telophase.

Prophase-II

The chromosomes with two chromatids become short and thick.

Metaphase-II

This is the stage of the second meiotic division. The nuclear membrane and the nucleolus are absent. Spindle is formed and the chromosomes are arranged on the equator.

Anaphase-II

Spindle is formed. The centromeres of the daughter chromosomes are attached to the spindle fibres. The two groups of the daughter chromosomes in each cell have started moving apart towards the opposite poles of the spindle.

Telophase-II

The two groups of daughter chromosomes in each haploid cell have reached the two poles of the spindle. The two haploid daughter cells formed as a result of first meiotic division divide again by the second meiotic division. Four haploid cells are formed from a single diploid cell.

3. MITOSIS IN ONION ROOT TIP (ALLIUM CEPA)

Requirements

Onion root tips, 1N HCL, Acetocarmine stain, glycerine, watch glass, slides and cover slip.

Procedure

1. Take 2-3 root tips on a watch glass.
2. Add 2 drops of 1N HCL and gently warm it.
3. Blot the HCL with blotting paper and then add 2-3 drops of acetocarmine stain and warm it.
4. Allow the root tips in stain for 5-10 minutes.
5. Take the stained root tips on a slide with 2-3 drops of glycerine.
6. Place the cover slip over the slide with a blotting paper squash it with thumb.

Mitosis

It is also called somatic cell division or equatorial division. The process of cell division whereby chromosomes are duplicated and distributed equally to the daughter cells is called Mitosis. It helps to maintain the constant chromosome number in all cells of the body.

Interphase

It is also called resting stage. This is a transitional phase between the successive mitotic divisions.

1. Replication of DNA takes place.
2. Volume of nucleus increases.
3. The chromosomes are thin coiled.

Prophase

1. Each chromosome consists of chromatids united by centromere.
2. Spindle formation is initiated.
3. The chromosomes shorten, thicken and become stainable.
4. The nuclear membrane and nucleolus start disappearing.

Metaphase

1. Disappearance of nucleolus and nuclear membrane.
2. Chromosomes are at their maximum condensed state.
3. Spindle formation is complete.
4. The chromosomes align in the equatorial position of the spindle and form the equatorial plate that is at right angle to the spindle axis.
5. The centromeres are arranged exactly at the equatorial plate and the arms are directed towards the poles.

Anaphase

1. The centromere of the chromosomes divides and the two chromatids of each pair separate.
2. Sister chromatids start moving towards the opposite poles due to the contraction of chromosomal fibres.
3. The daughter chromosomes assume V or J shape.

Telophase

1. It is reverse of prophase.
2. The chromosomes aggregate at the poles.
3. The spindle starts disappearing.
4. The new nuclear membrane starts to reappear around each set of chromosomes.
5. Nucleolus gets reorganized.

Cytokinesis

A cell plate is formed by the formation of phragnoplast from the Golgi complex. Later the primary, secondary wall layers are deposited and finally the cells are divided into two daughter cells. Cytoplasm gets divided into two.

4. DIFFERENTIAL STAINING OF BLOOD

To identify different stages of white blood cells in human blood.

Introduction

White blood corpuscles or leucocytes are colourless, actively motile, nucleated living cells. They are variable in size and shape and exhibit characteristic amoeboid movement. Under certain physiological and pathological conditions, WBCs can come out of the blood vessels by a process called diapedesis. The lifespan of WBCs is 12-15 days. Unlike red blood corpuscles, leucocytes or WBCs have nuclei and do not contain haemoglobin. They are broadly classified into two groups:

1. Granulocytes
2. Agranulocytes

Granulocytes develop from red bone marrow and agranulocytes from lymphoid and myeloid tissues. Granulocytes are sub-divided into three groups:

1. Eosinophils
2. Basophils
3. Neutrophils

Agranulocytes are sub-divided into two groups:

1. Lymphocytes
2. Monocytes

The main features of WBC's are:

1. Phagocytosis
2. Antibody formation
3. Formation and secretion of lysozyme, heparin etc.

Principle

WBCs are the type of blood cells which have a nucleus but no pigment. They are important in defending the body against diseases because they produce antibodies against any foreign particle or antigens. WBC can be divided into 2 types - granulocytes and agranulocytes. Granulocytes consists of neutrophils, eosinophils and basophils.

Neutrophils. Do not stain with either acidic or basic dyes. They have many lobed nucleus and are called polymorphonuclear leucocytes or polymorphs. They constitute about 76.9% of the total leucocytes found.

Eosinophils. A lobed nucleus is present with cytoplasmic granules that stain with acidic dyes. They constitute 1-4% of the total leucocyte count.

Basophils. It has a lobed nucleus and the cytoplasm contains granules which stain with basic dyes. It comprises 0-4% of total leucocyte count.

Agranulocytes. Consists of lymphocytes and monocytes.

Lymphocytes. This is a type of WBC with a very large nucleus. It is rich in DNA and a small amount of clear cytoplasm is present.

Procedure

1. The finger tip was cleaned with cotton dipped in alcohol and was pricked with a sterilized needle.

2. 1-2 drops of blood was placed on the right side of the slide and with the help of another clean slide, the blood was drawn such that a thin smear was formed and dried.

3. The dried smear was fixed using zinc acetone, 3 methyl alcohol or absolute alcohol for 5 min.
4. The smear was dried and stained with Giemsa stain or Leishman stain.
5. Subsequently, the distilled water (double the amount of water as that of the stain) was added onto the same smear slide.
6. The smear is mixed using a pipette for 10-15 min.
7. The slides are kept in running water to remove excess stain and then dried.
8. The slide was observed under oil immersion lens.

5. BUCCAL EPITHELIAL SMEAR AND BARR BODY

The study of Barr body from the (female) smear of Buccal epithelial cells of female.

Material Required

Buccal epithelial cells, Giemsa stain, Carnoy's fixative, slides, cover slip, microscope, etc.

Procedure

- Gently rub the inside of the check with a flat rounded piece of wood and transfer the scraping over a clean glass slide.
- Then, thin film of cells was made on the slide and kept for air-drying.
- Air-dried smear was kept in Carnoy's fixative for 30-35 minutes.
- Then the Giemsa stain was poured and allowed to stand for 20-25 minutes.
- After staining, the slide was washed with distilled water to remove the excess stain.
- Finally kept for air-drying and then observed under microscope.

Observations

We found that very lightly stained cells are scattered here and there in the smear. In the cells violet coloured Barr bodies are observed inside pink coloured nucleus. Barr body is nothing but an inactivated (heterochromatinized) X chromosome. It was first observed by Murray Barr in 1949. It is found only in female cells, because here one X chromosome is enough for metabolic activity. It is absent in male somatic cells, because here only one X chromosome is present and which is in active state.

Precautions

The smear or film should be uniform and thin over the glass slide so the cells will not overlap each other.

6. VITAL STAINING OF DNA AND RNA IN *PARAMECIUM*

Description

DNA and RNA are two types of nucleic acids which have different staining properties. DNA is acidic and stains to acidic dyes, while RNA along with proteins stain basically.

The two stains used are methyl green and pyronine. DNA stains blue or bluish green with methyl green and RNA stains pink with pyronine dyes.

DNA is present in the nucleus and stains blue or bluish green, while RNA is present in the cytoplasm and stains pink.

Materials Required

1. Cultured paramecia
2. Cavity slides
3. Plain slides
4. Cover slips
5. Methyl green and pyronine stains 0.5%

Procedure

1. Pipette out few paramecia onto a cavity slide.
2. Blot out excess water using filter paper.
3. Put 2-3 drops of methyl green pyronine stain and keep it for 5-10 minutes.
4. Transfer the paramecia onto a clean slide. put them under cover slip in aqueous medium and observe under the microscope.

Observation

DNA (nucleus) stained blue and RNA (cytoplasm) stained pink.

7. INDUCTION OF POLYPLOIDY

Cytological Techniques

This involves various steps like choice of treatment, pretreatment, fixation, staining and squashing, choice of the material. Healthy root tips are taken which are excised at about 11:30 am. After thorough washing these root tips are handled directly or can be subjected to pretreatment.

Pretreatment

It involves treating the materials for cytological studies with certain physical and chemical agents like colchicines, 8-hydroxy quinoline, para dichloro benzene, etc. pretreatment can be done before or after fixation of the materials.

Objectives of Pretreatment

- Removal of extra deposits on the cell walls specially waxy or oily deposits otherwise these extra deposits come in the way of fixation. Chloroform is recommended for the dissolution of the waxy substance.
- To clear the cytoplasm and to make it transparent. This is done by using 1N HCl. Certain enzymes like cellulose and pectinases are also used for dissolving certain substances.
- To soften the tissues, it involves the dissolution of middle lamella that connects the adjacent cells. In plant 1N HCl is used.

- To increase the frequency of nuclear division. This is done through colchicine which has the property of inducing nuclear division. It also destroys spindle.
- To bring about the differential condensation of chromosome at metaphase, this refers to the coiling of the chromosomes.
- Demonstration of the heterochromatin in the chromosome for this special treatment is needed i.e. low temperature treatment.

Mitotic Poisons

These are the chemicals that bring about the arrest of the mitotic apparatus in dividing cells and results in scattering of chromosomes. They do not affect the cell in any other way. A very large variety of chemicals are used. The best of them are colchicine, gammaxene and their derivatives, 8-hydroxy quinoline, para dichloro benzene.

Colchicine

It is a poisonous alkaloid that occurs in *liliacae* plant, colchicual autumnase (Autumn lilly). It is a small plant with small corn native of Europe and U.K. alkaloid is positive in underground corn and seed. Seeds are said to be the chief source of colchicine.

Action of Colchicine

It is believed that the organization of the mitotic apparatus depends upon balance between elements of cytoplasm and mitotic apparatus. Any chemical that disturbs this delicate balance will prevent the formation of the mitotic apparatus and colchicine is said to be having this property.

Preparation of Colchicine Solution

In case of onion root tips the strength of the solution prepared is 0.05% (*i.e.* 100 ml of H_2O, 0.5 mg of colchicine). This treatment to the root tips at room temperature varies from 1–1½ hours.

8. MOUNTING OF GENITALIA IN *DROSOPHILA MELANOGASTER*

Description

Genital plate is located in the abdominal region of the male and female flies but in the males the genital plate is more prominent and used as a copulatory pad. It is also called as Epandrium. Genital plate is horse shoe shaped and it is bent with a posterior region. It is divided into two parts namely the heel and the toe. Inside the arch a pair if anal plates and a pair of primary claspersis present. On the primary claspers 6-19 dark spines or bristles of interest are present which are very thick and their number is species specific. The main function of the genital plate is to hold the female to transfer the sperms into her genital organ during copulation.

Materials Required

Male flies, 1N HCl, cresosote solution, cavity slides, glycerine, cover slips.

Procedure

1. Remove the last abdominal segment of the male *Drosophila melanogaster.*

2. Transfer into 1N HCl taken in a cavity slide and allow it for 15 minutes.
3. Blot out the HCl and put 2-3 drops of cresote solution and allow it for about 20 minutes.
4. Remove the pellicle organ and clean the genital plat.
5. Transfer the genital plate onto a clean plain slide and mount the genital plate with glycerine.

9. MOUNTING OF GENITALIA IN THE SILK MOTH *BOMBYX MORI*

Description

Genital structures in silk moths are very important for copulation and transferring the gametes. Male genitalia is called Herpes and the female genitalia is called the labius. Herpes is present at the terminal part of the abdomen of the male and has two strong hooks to hold the female labius during copulation. Labius are terminally present which are fleshy with sensory hairs which help in copulation and transferring the gametes.

Materials Required

1. Live male and female moths
2. Dissection set
3. Needles
4. Blades
5. Slides.

Procedure

1. Remove the last segment of the abdomen with the help of blade or needle under the dissection microscope.
2. Keep the dissected segment in 10% KOH for 5 minutes.
3. Clean the tissue of the segment.
4. Observe the labius or herpes under the microscope.

Observation

Both herpes and labius were observed.

10. MOUNTING OF SEX COMB IN *DROSOPHILA MELANOGASTER*

Sex comb is a specialized structure present exclusively in the forelegs of the male flies. Location, size and structure varies from species to species. An adult *Drosophila* has three pairs of legs.

1. Forelegs
2. Midlegs
3. Hindlegs

In adult males the forelegs have a comb like structure with chitinized black teeth called sex comb. It is present in the first tarsal segment of the forelegs of the males. It is absent in the females. The sex comb helps the male to hold the female during copulation or mating.

Materials Required

Glass slides, cover slip, needles, male flies, glycerine, dissecting microscope etc.

Procedure

1. Etherize the flies and separate the males.
2. Separate the forelegs using needles under dissecting microscope.
3. Put a drop of glycerine on the forelegs and place the cover slip.
4. Search and observe the sex comb on the first tarsal segment.

11. MOUNTING OF MOUTH PARTS OF MOSQUITO

Mouthparts are long piercing and sucking tubular proboscis.

Labium

It forms the proboscis sheath. It is like a long gutter or a half tube, ending in a pair of white, pointed lobes, the labellae, bearing tactile hair. The labium bears a dorsal groove which lodges all thc other mouthparts modified into six needle shaped piercing stylets, all finer than hair, meant for puncturing the skin of the host.

Labrum-Epipharynx

The epipharynx, which is an outgrowth from the roof of the mouth, becomes completely fused with the labrum to form the labrum-epipharynx. This compound structure makes a long, pointed and stiff rod, which closes above the dorsal groove of the labium.

Hypopharynx

The ventral surface of labrum-epipharynx also bears a groove, which is closed below by a long, pointed and flattened plate, like a double edged sword called the hypopharynx. It is traversed by a minute median channel, the salivary duct.

Mandibles and Maxillae

The paired mandibles and the first maxillae form long and needle shaped stylets, the former ending in tiny blades and the latter into saw like blades. A pair of long tactile maxillary pulps projects from the sides at the base of the proboscis.

Only the females can suck the blood, as they possess well developed, piercing mouthparts. In the males the piercing organs are reduced, the mandibles are absent, but the sucking mouthparts are well developed so that they can suck up only plant juices.

12. NORMAL HUMAN KARYOTYPING

To study the chromosomal sets [Karyotype] of a Normal female Human being.

Introduction

Karyotype is based on the size and position of chromosomes and centromeres respectively. It was first developed by Levan in 1960. Based the centromeric position that is on the length of arms of chromosomes he divided chromosomes as:

- 1, 2, 3, 16, 19 and 20 - Metacentric
- 4–12, 17, 18 and X - Submetacentric
- 13–15, 21, 22 and Y - Telocentric

Later, Pataii Classified the Chromosomes into Different Families(Groups)

- Group A: 1-3 Chromosomes-Metacentric; longer than the all other chromosomes.
- Group B: 4 and 5 Chromosomes-Submetacentric.
- Group C: 6-12 and X Chromosomes-Submetacentric.
- Group D: 13-15 Chromosomes-Acrocentric.
- Group E: 16-18 - Chromosomes-16: Metacentric, 17 and 18: Submetacentric.
- Group F: 19 and 20 Chromosomes-Metacentric, smaller comparatively.
- Group G: 21, 22 and Y Chromosomes-Acrocentric and the smallest in size.

The chromosomes of group D and G have secondary constrictions.

13. KARYOTYPING

Principle

Karyotyping is a valuable research tool used to determine the chromosome complement within cultured cells. It is important to keep in mind that karyotypes evolve with continued culture. Because of this evolution, it is important for the interpretation of biochemical or other data, that the karyotype of a specific sub-line be determined. Numerous different technical procedures have been reported that produce banding patterns on metaphase chromosomes. A band is defined as that part of a chromosome which is clearly distinguishable from its adjacent segments by appearing darker or lighter. The chromosomes are visualized as consisting of a continuous series of light and dark bands. A G-staining method resulting in G-bands uses a Giemsa dye mixture or Leishman dye mixture as the staining agent. What follows is a brief description of the steps involved in assembling a karyotype.

Time Required

15-30 minutes to cut, arrange, glue and interpret one metaphase spread on a karyotype sheet.

Procedure

1. Count the number of chromosomes. Solid stained chromosomes or chromosomes treated with a trypsin and giemsa stain can be counted at the microscope with a 100X magnification. However, for analysis such as identification of marker chromosomes or determination of the number of copies of individual chromosomes, it is usually necessary to photograph and print the chromosome spreads. Refer to procedures in this manual for black and white photography and film development. Two prints should be made of each spread. One will be cut for karyotyping; the uncut print serves as a reference if questions arise about the interpretation of a certain chromosome.
2. Cut out each individual chromosome and arrange on a karyotype sheet. Chromosomes are ordered by their length, the position of the centromere, the position of the chromosome bands, and the relative band sizes and distributions.

3. In the construction of the karyotype the autosomes are numbered 1 to 22, in descending order of length. The sex chromosomes are referred to as X and Y. The symbols p and q are used to designate, respectively, the short and long arms of each chromosome. There are 7 groups identified in the karyotype and data pertaining to each group.
4. Secure chromosomes in place with glue. Pair the chromosomes closely together and align the centromeres (for easier band comparison and checking for structural chromosome aberrations). If possible, have a second technologist check the interpretation of the karyotype before chromosomes are secured in place.
5. A description of the karyotype should be recorded on the karyotype sheet. First record the number of chromosomes, including the sex chromosomes, followed by a comma (,). The sex chromosome constitution is given next. Any structural rearrangements and additional or missing chromosomes are listed next. Other information such as the cell line number, the date karyotype was prepared, the specimen type, and the technologist also should be recorded on the karyotype sheet.

14. BLACK AND WHITE FILM DEVELOPMENT AND PRINTING FOR KARYOTYPE ANALYSIS

Purpose

35 mm black and white film (Kodak Technical Pan film) is developed and print enlargements made from the negatives. The chromosomes from the prints are identified, cut out and arranged on the karyotype form.

Time Required

1. 5-20 minutes to develop 1-2 rolls of film
2. 45-60 minutes to print 20 quality prints.

Special Supplies

1. Kodak Technical Pan Film 2415
2. Kodak Polycontrast III Paper 5" × 7", RC plastic coated
3. Kodak Ektamatic S30 Stabilizer
4. Kodak S Activator
5. Kodak D19 Developer
6. Photo-Flo 200.

Special Equipment

1. Kodak Ektamatic Print Processor
2. Beseler Enlarger
3. Graylab 500 Timer.

Procedure

Developing the film

Note. It is important to develop the film at a constant temperature to prevent excess grain on the film. A pan filled with water is used to hold all solutions and water rinses at 23-25°C.

1. In total darkness (no safe light on), remove the film from the cassette and wind into the developing reel. It is important to wind the film correctly to prevent undeveloped areas on the film. Correctly wound film will have no edges protruding. Place the reel in the developing tank and twist the lid of the tank to close. It is now safe to turn on the room lights.
2. Pour 400 ml of Kodak developer D-19 into the top of the tank to develop the film. Agitate the tank periodically over 4.0 minutes to ensure the developer contacts all parts of the film (prevents uneven film development). After the 4.0 minute developing time, pour the developer into a small storage bottle. Developer may be used a total of 3 times.
3. Briefly rinse the tank with water, agitate and drain water completely.
4. Pour 400 ml fixer into the developing tank and fix for 4 minutes, agitating several times during this time period. Film should not be left in the fixer longer than 4 minutes, because the negative will bleach, or become saturated with fixer. Note: the fixer time should be for 2-4 minutes for high contrast films, while most continuous tone films should be fixed for 5-10 minutes.
5. Remove the lid and empty the developing tank. Rinse the film in the tank for two minutes with running water at 23-30°C. Because fixer is heavier than water, be certain to fill and empty the tank several times to prevent the fixer from remaining at the bottom of the tank. The amount of time the film stays wet should be kept to a minimum to prevent film deterioration, e.g., swelling and clumping of the grain may occur, which decreases the sharpness of the image.
6. Remove the film from the reel and add photo-flo solution to tank. Empty the tank after 30 seconds. This helps film dry without streaking. Drain film briefly, squeegee, and hang in a dust free area to dry.

Producing 5 × 7 prints

Note. Polycontrast paper by Kodak is a variable contrast paper and works well for negatives of low contrast or underexposed film. Polycontrast filters can also be used to improve the contrast of the prints. In the steps below.

7. Load the negative onto the negative carrier. This is done by moving the negative stage lever (16) downward, removing the negative carrier from the stage, opening the carrier, inserting the negative, closing the carrier and replacing the carrier onto the stage. Pull the stage lever back up to close.
8. Turn on the enlarger with the graylab timer. The negative stage guide (13) should be positioned on 35 mm and smaller formats. Using the negative stage adjustment knob, lower the stage to the 35 mm mark. Use the negative lock (18) to secure the stage at this position. The elevator motor/control box (3) moves the enlarger head up or down for changes in elevation. Use the elevation switch (4) to move the enlarger head until the image becomes clearly visible. Manual elevation control (5) can be used for precise elevation. By using the motor switch as a scale indicator, a record of the height can be recorded for repeat magnification at a later date. For enlarging a 35 mm negative onto a 5" × 7" print, position switch to approximately 4 inches on the reference scale. A grain focusing scope (microsight) should be used to ensure the film is at the best magnification to obtain a clear, fine detailed exposures. Place the microsight directly on a piece of white paper for focusing and move the manual elevation control (5) until the grains in the negative are clearly visible on the paper.

9. The lens on the enlarger is sharpest at an estimated f stop of f8 or f11. Set the aperture and test the exposure using a 'test strip', in which a piece of paper is used to mask portions of the print paper during a series of exposures. For example:

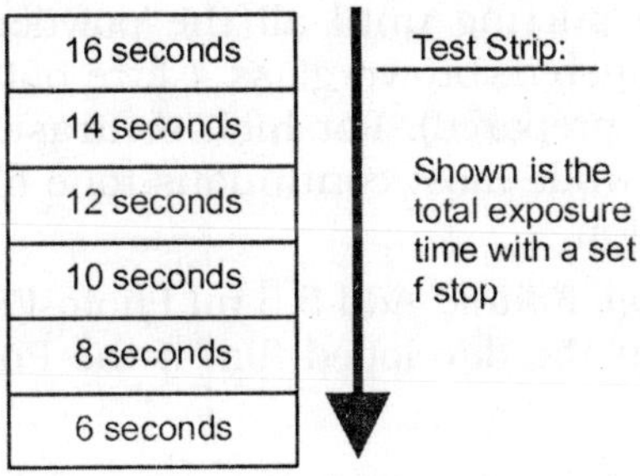

To expose the print paper, place the shiny side of the paper facing up, and center the paper on the enlarging easel. Set the timer for an exposure time (1-2 seconds). Expose the unmasked portion of the print paper by pressing the red button on the timer (Expose/Hold). The enlarger light will automatically stop when the timer goes off. Move the masking paper to reveal more of the print paper and re-expose. (The portion of the paper that was previously exposed now has two exposures.) For each time point, move the masking paper down the paper. Develop the test strip (follow step 4 below) and decide which exposure time is best to use for the negative.

The film has a much greater capacity for producing detail than the print paper, so it may be necessary to do some dodging (holding back light from the over-exposed dark areas) or burning-in (using more light on a particular area to increase the exposure and darken the print) to bring out more details of the chromosomes. For underexposed areas such as F and G group chromosomes, burning-in will make the arms and satellites more visible. A black piece of paper with a hole cut out can be used as a tool, or even your hand with your fingers spread apart to expose those areas that need darkening.

10. Place the print paper on the feed shelf of the Ektamatic film processor with the exposed surface facing downward. The rollers will pull the paper through the activator and the stabilizer sections and out to the other side of the processor. The print can now be viewed under room lights to check the exposure, focusing, etc.
11. Place the developed print in fixer solution (which is poured in a print tub to a depth that will cover prints) for 5 minutes.
12. Transfer the print to a washer tub and rinse for 1-2 minutes. Water will travel across the prints with whirlpool action, eliminating the fixer more effectively than by standing water alone.
13. Hang the prints to dry or dry them flat on a counter top at least 30 minutes.

Solutions

- *Developer, Kodak D-19.* Avoid breathing dust when preparing, may be harmful if swallowed.

Slowly add a package (595 g) of developer to 3.8 liters of water at 52°C, with stirring until chemicals are dissolved and the solution is completely mixed. Store in a 4 litre brown glass bottle at room temperature for up to 1 year (label bottle with the date prepared).

- *Kodak Fixer.* Slowly pour powdered fixer (680 g) into 3.8 litres of water (not above 26.5°C), with stirring until all the powder is dissolved. Fixer can be stored in a well stoppered, brown glass 4 litre bottle for up to 2 months (label bottle with the date prepared). For high contrast films, the fixer time should be for 2-4 minutes; while most continuous tone films should be fixed for 5-10 minutes with agitation.
- *Photo-Flo 200 solution, Kodak.* Add 5.5 ml Photo-Flo to 1.1 litres of water. Note: Scum will appear on the developed film if the Photo-Flo concentration is too high.

15. STUDY OF DRUM STICKS IN NEUTROPHILS OF FEMALES

Study of Drumsticks in neutrophils of females.

Introduction

The sex chromatin of poly morphic nuclear neutrophil of human blood is contained a specific 'drumstick' like nuclear appendages that have its head about 1.5 μm in diameter attached to the nucleus by a thread like stalk. Drumstick differs from sex chromatin of other cells in being extruded from the nucleus and also it is visible in only relatively small portion of cells [in about 1/40 neutrophils of a normal female].

Materials Required

Female blood sample, slides, needles, cover-slip, alcohol, cotton, Leishmann stain, microscope.

Procedure

1. The fingertip was cleaned with cotton soaked in alcohol and was pricked with sterilized needle.
2. 1 or 2 drops of blood was placed on the right side of the slide.
3. With the help of another clean slide, the blood is along the slide, such that a tongue shaped thin-layered smear was formed and was air-dried.
4. The dried smear was fixed with acetone free methanol or absolute alcohol for about five minutes.
5. The smear was then dried and stained with Leishmann stain.
6. Subsequently distilled water at the double amount of stain was added on the smear.
7. The smear was mixed using a pipette for 10-20 minutes.
8. The slide was kept in running water to remove excess stain and then air-dried.
9. The slide thus prepared was observed under microscope using oil immersion lens.

16. STUDY OF MALARIA PARASITE

Phylum: Protozoa

Class: Sporozoa

Genus: Plasmodium

It is an intra cellular parasite found in the blood of the man. They cause malaria. They are malarial parasites and have two hosts.

(*a*) Man-Primary host

(*b*) Female Anopheles-Secondary host.

Life cycle in man is sexual and in mosquito is asexual. There is an alteration of generation of asexual and sexual cycles. Malarial parasites are transmitted from person to person by the sites of adult female anopheles mosquito. Male mosquito does not play any role in the transmission of malaria because they do not feed on blood.

Life Cycle of Plasmodium

Includes three stages:

- Pre-erythrocytic or exoerythrocytic cycle in human.
- Erythrocytic or schizygotic cycle in RBC.
- Sexual or gametogenic in female mosquitoes.

In Man

When female Anopheles mosquito bites man and also infects the plasmodium into the blood of man. This infection stage of plasmodium is called *sporozoire.*

Exoerythrocytic Cycle

It is the life cycle of parasites inside the RBC of the host. This is an asexual phase which results in the production of gametocytes. This cycle starts with the entry of merozoire into the RBC. In the RBC the parasite enters resulting in resting period. It attains round shape called Trophozoire. In the Trophozoire a large vacuole develops and pushes the nucleus to one side. This stage of parasite is known as signet ring. The vacuole disappears; the parasite fills the RBC to become schizont ring.

The schizont mature and undergo fusion to form mesozoites. They are released into the blood stream by bursting of RBC. These mesozoite attaches fresh RBC and the cycle thus repeats. After few generations some of the mesozoites develop into gametocytes in the RBC. There are two types of gametocytes, among them one is macrogametocyte - which has small nucleus, large cytoplasm, and circular in shape.

Both the gametocytes do not undergo further development until they reach stomach of the Anopheles, if they cannot reach they disintegrate.

When an Anopheles mosquito bites malarial parasite patient for blood, in the stomach of the mosquito, only mature gametocytes survives to develop into gametes, others disintegrate the process of gamete formation from gametocytes called gametogamy.

Gametogamy

During gametogamy, the microgamete becomes active. It produces 6-8 slender nucleated bodies called male gametes by a process of exflagellation. The microgamete settles freshly in stomach of mosquito. The macro gametocyte undergoes maturation phase, develops into female.

Sporogony

The male gametes fuses with the female gamete to form a zygote in spherical and remains inactive for some time. Later it transforms into an elongated worm like mouth structure called ookinite. Ookinite pierces through wall of the stomach and bind to the outer layer of the wall. There it becomes round secretes a cyst wall and grows in size. This stage is called oocyst. Nucleus divides into bits each of which develops into slender sickle shape cell bodies called sporozoire. The mature cyst raptures to liberate the sporozoites into body cavity of the mosquito. Formation of sporozoites from zygote is called sporogony. These sporozoites are ready to reach the salivary glands and when a mosquito bites a healthy person sporozoites are released to his blood stream, cycle repeats.

17. VITAL STAINING OF DNA AND RNA IN PARAMECIUM

Description

DNA and RNA are two types of nucleic acids which have different staining properties. DNA is acidic and stains to acidic dyes, while RNA along with proteins stain basically. The two stains used are methyl green and pyronine. DNA stains blue or bluish green with methyl green and RNA stains pink with pyronine dyes.

DNA is present in the nucleus and stains blue or bluish green, while RNA is present in the cytoplasm and stains pink.

Materials Required

1. Cultured paramecia.
2. Cavity slides
3. Plain slides
4. Cover slips
5. Methyl green and pyronine stains 0.5%.

Procedure

1. Pipette out few paramecia onto a cavity slide.
2. Blot out excess water using filter paper.
3. Put 2-3 drops of methyl green pyronine stain and keep it for 5-10 minutes.
4. Transfer the paramecia onto a clean slide put them under cover slip in aqueous medium.
5. Observe under the microscope.

Observation

DNA (nucleus) stained blue and RNA (cytoplasm) stained pink.

18. SEX LINKED INHERITANCE IN DROSOPHILA MELANOGASTER

Description

In majority of cases the sex of an individual is determined by a pair of genes on the sex chromosome. Females are homozygous and the males are heterozygous.

Sex linked inheritance is defined as the inheritance of somatic characters which are linked with sex chromosome (X and Y).

The characters are described as sex linked characters. The phenomenon of sex linkage was first observed by T.H. Morgan in 1910 while experimenting on Drosophila. Morgan observed the appearance of white eye colour in males in the cultures of normal wild eye flies. He thus proposed the phenomenon of sex linkage.

In the present experiment we have taken yellow body mutant and crossed with O K strain to understand the pattern of sex linked inheritance.

Requirements

1. O K strains of *Drosophila melanogaster.*
2. Yellow body mutants
3. Media bottles
4. Anaesthetic ether
5. Etheriser

Procedure

Drosophila melanogaster O K strain and the yellow body flies were cultured in standard media separately. When the pupa were ready to close the bottles the bottles were cleaned by taking out all the flies present there. From the pupa the virgfin females were isolated and they were aged for 2-3 days and then the crosses were conducted as follows:

1. Yellow body females × O K males
2. Yellow body males × O K females

The progeny produced in the F1 were observed for phenotype and the data was recorded. The few of F1 progenies were inbred to get the F2 generation which were also observed and the data was recorded.

Direct Cross

Parents : yellow females × O K males

F1 : all F1 females were grey bodied and the males were yellow

Inbreed : F1 females × F1 males

F2 : grey female : grey male : yellow female : yellow male

Ratio : 46 : 61 : 51 : 50

Phenotype	Observed	Expected	Deviation	d^2	$d^2/E = X$
Grey body male	61	52	9	81	1.5
Grey body female	46	52	–6	36	0.6
Yellow body female	51	52	1	1	0.01
Yellow body male	50	52	–2	4	0.07

$\Sigma X^2 = 2.18$

Degree of freedom = 4 – 1 = 3

Reciprocal Cross

Parents : yellow males × O K females

F1 : all flies are grey bodied

Inbreed : F1 males × F1 females

F2 : grey body male : grey body female: yellow body male: yellow body female

Ratio : 63 : 52 : 48 : 48

Phenotype	Observed	Expected	Deviation	d^2	$d^2/E = X$
Grey bodied male And female	115	122.25	7.25	52.56	0.4
Yellow bodied male And female	48	40.75	–7.25	52.56	1.2

$\Sigma X^2 = 1.71$

Degree of freedom = 3 – 1 = 2

Analysis of Result

At 5% level of significance at 3 degree of freedom table value = 7.815 and at 2 degree of freedom it is 5.991

In direct cross the X^2 value = 2.18 which is less than table value and in reciprocal cross the calculated X^2 value = 1.71

Which is less than table value. In both the cases the deviations are not significant so null hypothesis is accepted.

19. PREPARATION OF SOMATIC CHROMOSOMES FROM RAT BONE MARROW

To study the structure of somatic chromosomes in the Bone marrow of rat.

Materials Required

0.05% colchicine, 0.56% KCl, centrifuge, centrifuge tubes, syringe, glass slides, fixatives, stain and microscope.

Procedure

0.05% colchinine is to be injected interperitonially to the rat 1½ hours before the experiment started. The volume to be injected varies with the weight of weight of the rat. If the rat is around 40 gms or more 1 ml of 0.05% colchinine is injected. Leave the rat for 1½ hours and then kill the rat by cervical dislocation. Then remove the femur bone. Flush the bone marrow in a clear petridish using 0.56% KCl, which serves as a hypotonic solution. Remove all large particles by strain the solution in a clear cheese cloth or muslin cloth. The filtrate is taken in centrifuge tube and spun it for 10 minutes at 1000 rpm. A cell button was obtained after pouring the supernatant. The fixative (1:3 Acetic alcohol) was added, which is made by mixing one part of acetic acid with three parts of acetone free methanol or absolute alcohol, to the cell button and gently mixed. Centrifuge again to get a fresh cell button and fresh fixative is added to get a cell suspension. This method is repeated twice. Finally drop the cell suspension to a clear slide kept preferably in clod alcohol. Allow the drops on the surface of the slide to dry. This technique is known as Air Drop Technique or Air Dry Preparation. Then the air-dried slides were taken for the staining. Freshly prepared air-dried slides give better results for Gimsa stain. For non-banded chromosomes the stocks standard Gimsa stain is diluted with 6.8 pH Phosphate buffer solution. 5 ml Phosphate buffer solution was taken and added to matured 1 ml of stock Gimsa stain in a coupling jar, mixed well. This is the active working Gimsa stain. Dip the slides in the stain for 15-20 minutes. Wash it under slow running water till the excess stain was removed and then rinse it with distilled water. Allow it to dry. Then the slides were observed under microscope.

Observation

Somatic chromosomes of Rat Bone Marrow were observed deep violet under microscope.

20. CHROMOSOMAL ABERRATIONS

The chromosome of each species has a characteristic morphology and number. But sometimes due to certain irregularities at the time of cell division, crossing over or fertilization some alterations in the morphology and number ,takes place. The slightest variation in the organization of chromosome is manifold phenotypically and is of great genetic interest.

Duplication

The presence of a part of the chromosome in excess of the normal compliment is known as duplication. During pairing the chromosome bearing the duplicated segment forms a loop.

Examples: 1. Bar eye in Drosophila.

This is characterized by narrow, oblong bar shaped eye with few facets. It is associated with duplication of a segment of the X-Chromosome called 16A. Each added sections intensified the bar phenotype.

2. A reverse repeat in the chromosome IV causes eyeless (Ey) individual.
3. A tandem duplication in chromosome III causes confluens(Co) resulting in thickened veins.
4. Another duplication causes hairy wing (Hw)

Deletion

Loss of a broken part of a chromosome is called deletion. Deletion may be terminal or intercalary. During pairing between normal chromosome and deleted chromosome, a loop is formed in the normal chromosome and this is known as the deletion loop.

Example: Notched wing mutation in Drosophila

In the presence of deletion a recessive allele of the normal homologous chromosome will behave like a dominant allele (Pseudo dominance).

Inversion

It involves a rotation of a part of a chromosome or a set of genes by 180° on its own axis. It essentially involves the occurrence of breakage and reunion. The net result of inversion is neither gain nor loss in the genetic material but simply a re-arrangement of the sequence.

During pairing an inversion loop is formed in which one chromosome is in the inverted order and its homologue is in the normal order.

Significance

Chromosome with inversions have practical applications in maintaining Drosophila stock. Crossing over is suppressed in such chromosomes and is possible to maintain a gene in the heterozygous state that would cause death when present in the homozygous condition.

Translocation

The shifting of a part of a chromosome or a set of genes to a non-homologous one is called translocation. During synapsis a cross-shaped configuration is formed.

H.J. Muller found one strain of Drosophila in which a group of genes including scarlet which normally is on the 3rd chromosome translocated to the 2nd chromosome. Cytological examination showed that the 3rd chromosome was much shorter than usual while the 2nd chromosome was longer than usual.

21. STUDY OF PHENOCOPY

The strength of environmental changes is sufficient to modify the effects of many genes. In some instances, specific environmental changes may modify the development of an organism so that its phenotype stimulates the effects of particular gene although this effect is not inherited. Such individuals are known as *phenocopies.*

Diabetics dependent on insulin are an example of phenocopy of normal individuals in the sense that drug environment prevents the effects of the disease. Should their offspring also inherit diabetes? The phenocopy treatment with insulin may have to be administered again to achieve the normal phenotype. In no sense, therefore is diabetic changed by the insulin treatment. There is only a phenotypic effect.

By subjecting *normal* Drosophila egg, larva, pupae to various stress condition like temperature shock, or others, we obtain a phenotype effect similar to that of mutant gene. These abnormal effects incurred through these agents are almost identical to specific gene mutations although they are not inherited. Such individuals are known as phenocopies. The phenocopy partly imitates the mutant gene.

An *experiment* on this problem was supported by Sang and McDonald [1954] and it allows the exposure of two different stocks of Drosophila flies to phenocopytreatment. One homozygous wild type and other heterozygous for wild type and mutant recessive gene.

The mutant recessive, if it were homozygous would give a morphological effect duplicated by the phenocopy. Therefore, after particular treatment, more phenocopies are noted in the heterozygous stock than in the homozygous. This may be considered as an evidence, that phenocopy is controlling the developmental action.

The mutant effect will be partly duplicated by the phenocopy agent, *sodium meteorite* is eyeless. Other effects that may also be noted in phenocopy are changes in antennae, forelegs.

Prepare sufficient media for Drosophila growth and add around 0.1 ml of *silver nitrate* to the 8 bottles and keep 2 bottles as control.

Transfer wild, coloured Drosophilas in to all 10 bottles. Permit the parents to lay eggs for about 3 gays and discard them. Later examine the progenies. The *silver nitrate* has an effect of phenocopy agent and changes the developing larvae to become colourless flies, which is not inherited.

22. STUDY OF MENDELIAN TRAITS

Survey of Human Heredity

Some variations in human beings are inheritable. These variations can be studied using genetic principle. Some examples are given below:

Colour of Eyes

The colour of iris is an example of Mendelian inheritance in man. Brown eyes are dominant over blue eyes. There are many variations in eye colour. The blue colour of eye is due to pigment present in retinal layer, but in those with darker eyes it is marked by the development of melanin, partially marks the reflection from the near of the iris.

The is also variations in distribution of melanin; it may be smoky distributed on in spots. Darker colours are dominant over blue and gray. In some albinos iris is pink because the blood in the retinal layer is visible. Black circle in the centre of iris is the result of sex linked recessive gene.

Ears

Variations in the size and form of the ears, their position on the head indicate multiple gene inheritance. A few of them correspond to variations in a single gene.

Free Ear Lobes

Free ear lobes are dominants over attached ear lobes but there is a variation of freedom of these, which are not attached. On the outer margin of the pinna is a rolled rim; there is a variation in its size in some it is almost lacking.

Nearly all persons have a enlarged portion of the cartilage which projects inwards from the rolled rim at a distinct point. It is called as *Darwin's point,* which is inherited as dominant, it is expressed only in some persons it is expressed only one ear. The inheritance of natural ear lobes is the expression of autosomal dominant gene. Many persons show this character expressed on only one side.

Widow's Peak

In case of human beings at the centre of the forehead, the hair like may diploid to form a Q point which is called as Widow's peak. This character in some families.

This one of the most important Mendelian trait found in human beings.

Tongue

The shape and size of the tongue is the result of many different genes. Some people have the ability to roll their tongue in to U shape. This is the expression of dominant gene. The gene gives a few individuals an ability to fold the tongue.

Hitchhikers Thumb

The ability to hyper extend the thumb (extended backwards at the last joint) is due to a recessive allele (n) and a straight thumb (N) is dominant to hitchhikers thumb.

PTC Tasting

The ability to taste phenylthio carbamide [PTC] is dominant (T) to the inability to taste it (t). Your counselor will provide you with small pieces of paper that have been previously soaked in this harmless chemical.

Hypertrichesis

This character is controlled by 'helandric genes'. It is characterized by hair on the pinna. This character follows the linear pattern of inheritance that is transferred from father to son never to daughter.

23. ESTIMATION OF NUMBER OF ERYTHROCYTES (RBC) IN HUMAN BLOOD

Principle

To count the erythrocyte cells, RBC diluting fluid was used diluting fluid for RBC is an isotonic solution. This isotonic keeps the RBC cells unheamolysed. The RBC were counted using improved neubauer chamber, which has an area of square millimeter and depth of 0.1 mm. It is made up of 400 boxes. The cells in the four corner boxes and one box in the middle were counted.

Reagents Required

- EDTA: Ethylene diamine tetra acetic acid was used as anticoagulant.
- RBC diluting solution: This is prepared by dissolving 0.425 gms of NaCl in 50 ml of distilled water.

Material Required

Syringe, needle, alcohol, vials, RBC pipette, cotton, neubauer counting chamber, cover glass, microscope.

Procedure

Using a syringe venous blood was drawn and poured into vials containing anticoagulant [EDTA] and mixed well with the help of RBC pipette blood was drawn up to 0.5 mark. The tip of the pipette was cleaned with cotton and RBC diluting fluid was drawn up to 101 mark. The pipette with blood and fluid was shaken well avoiding bubbling and kept aside for 5 minutes. The counting chamber was charged using the posture pipette. After charging the chamber the cells were allowed to settle down to the bottom of the chamber. The chamber was placed on the stage of the microscope and using 45X objective the RBC cells in the smallest square were counted.

Comment

The normal range of RBC in adult male is 50-60 lakhs and in female is 40-50.5 laks/mm^3. In case of polycythemia, the number of RBC increases, this disease is associated with heart disease. Low count of RBC results in anemia polycythemia may be pathologic due to [red blood cells] tumour in bone marrow and the number of red cells may reach 11 million per cubic mm of blood.

The present determination of total RBC count shows about ____/mm^3 of blood, which is nearly equal to normal range.

Calculation

No of cells in 1 mm^3 of blood

$$= \frac{\text{No. of cells counted} \times \text{Dilution Factor} \times \text{Depth Factor}}{\text{Area of chamber counted}}$$

$$= \frac{\text{No. of cells counted} \times 200 \times 10}{1/5}$$

$$= \text{No. of cells counted} \times 10{,}000$$

24. ESTIMATION OF NUMBER OF LEUCOCYTES (WBC) IN HUMAN BLOOD

Principle

In order to count the WBC the lysing of RBC or erythrocytes should be done. For this 1.5-2% acetic acid solution with little quantity of crystal violet was used as diluting fluid. Acetic acid destroys the erythrocytes leaving behind the leucocytes. Crystal violet was used as stain. These stains, stain the nucleolus of leucocytes. This enables the leucocytes to be easily identifiable and to be counted. The number of leucocytes is in a dilute fluid can be count by using a 'neubauer' counting chamber. This chamber has specified ruling as shown in the figure 3. (Exercise 3 in microbiology section).

Reagents Required

- EDTA: Ethylene diamine tetra acetic acid was used as anticoagulant.
- WBC diluting Solution: This was prepared by using 1.5 ml of 1.5-2% acetic acid solution of crystal violet and 98.5 ml of distilled water.

Materials Required

Syringe, needle, alcohol, cotton, vials, WBC pipette, neubauer counting chamber, cover glass and microscope.

Procedure

Using a syringe venous blood was drawn and poured into a vial containing anticoagulant [EDTA] and mixed well. EDTA acts by chelating and preserves cellular elements with help of WBC pipette this well mixed venous blood was drawn.

Both the fluid and blood was mixed gently avoiding bubbling. The tube was kept aside for 5 minutes without disturbing. In the mean time cover slip was placed on the counting chamber at the right place on the counting chamber at the right place. The fluid blood mixture was shaked and transferred using a fine bore posture pipette on to the counting chamber. This is called as charging the chamber. While charging the chamber the care was taken not to overflow. Whenever there is a overflow it was washed and dried and it was recharged again. After charging the chamber the cells were allowed to settle down to the bottom of the chamber for 2 minutes. It was seen that the fluid did not get dried up. For counting the under part of the chamber was cleaned and placed on the stage of microscope using 10X or low power objective. The leucocytes were counted uniformly in the four larger corner squares and cells present on the outermost lines are counted on one side and those present on the lines opposite were not counted.

Comment

Present determination of total WBC count shows that there were about 7,100 WBC/cm^3 of blood. This is well with in the normal range (6000-8000 WBC/cm^3). In some cases of parasitic infection, the total WBC count increases. However in case of leukemia, the number of WBC increases to more then 15,000 cells/cm^3 of blood. Normally it would be about 4000-7000 WBC/mm^3 of blood. If the WBC count decreases then that condition is called Leukemia.

Calculation

Number of cells in 1 mm^3 of blood

$$= \frac{\text{Cells Counted} \times \text{Dilution Factor} \times \text{Chamber Depth}}{\text{Area of chamber counted}}$$

$$= \frac{\text{Cells counted} \times 20 \times 10}{4}$$

$$= \text{Cells counted} \times 50.$$

25. CULTURING TECHNIQUES AND HANDLING OF FLIES

Drosophila like other animals requires an optimum temperature for its survival, growth and breeding (20-25°C). The temperature around and above 31°C makes the flies sterile and reduces the oviposition and may also result in death. At lower temperature, the life cycle is prolonged and the viability may be impaired.

The routinely used food media for the maintenance of Drosophila is "Cream of wheat agar" medium.

The ingredients of this media for preparing culture bottles are as follows:

1. Distilled water – 1000 ml
2. Wheat flour (rava) – 100 gm
3. Jaggery – 100 gm
4. Agar agar – 10 gm
5. Propionic acid – 7.5 ml
6. Yeast granules.

Wheat flour and jaggery are boiled in distilled water till a paste is formed. To that agar agar and yeast granules are added after cooling. Propionic acid is added to avoid fungal infection of the medium.

Heat sterilized bottles should be used for preparing cultures. Similarly, sterilized cotton has to be used to plug the bottles. As the condition of the medium deteriorates with time, the flies have to be transferred from the old to new culture medium at least once in three weeks.

Whenever the flies have to be analyzed, either for routine observations or for experiments, they have to be anaesthetized to make them inactive. The procedure is to transfer the flies from the media bottle to another empty wide mouthed bottle, referred to as etherizer. The mouth of this bottle is to be covered with a stopper sprayed with ether. It takes about a minute to anaesthetize the flies. After this, the flies are transferred to a glass plate for observation under a stereo zoom microscope. If the etherized flies revive before the completion of the observation, they have to be re-etherized using a re-etherizer (ether soaked filter paper fitted in a petriplate, which has to be placed over the flies on the glass plate). The over etherized flies will have their wings and legs extended at right angles to the body and such flies are considered to be dead.

The flies should be handled with a fine painting brush. In the process of handling the flies, care should be taken not to damage the flies and the flies should be discarded after observation.

26. LIFE CYCLE OF MOSQUITO (*Culex pipiens*)

Classification

Phylum	:	Arthropoda
Class	:	Insecta
Order	:	Diptera
Family	:	Culicidae
Genus	:	*Culex*
Species	:	*pipiens*

Mosquitoes are holometabolous insects. The complete life cycle of mosquitoes takes about 13-15 days to complete.

Egg

Anopheles lays eggs horizontally and singly on the water surface. Eggs are boat shaped having two lateral air floats, which help in floatation. In Culex, the eggs are laid in clusters on the water surface forming rafts. The egg in Culex is cigar shaped without lateral air floats.

Larva

The eggs hatch after two to three days, and a small transparent larva measuring about 1 mm emerges. The larva of mosquitoes is popularly known as a wriggler. The larva of mosquito is popularly known as wriggler. Body is divided into head, thorax and abdomen. The head bears a pair of compound eyes, a pair of simple eyes, a pair of two jointed antennae and the chewing mouthparts. The thorax is slightly broader than the head and bears three pairs of lateral tufts of hair. Abdomen is 9 segmented. The larva contains long respiratory siphons. It undergoes four moults and five instars.

Pupa

Pupa or tumblers are motile. The head and the thorax are found to form the cephalothorax, which has a pair of trumpet shaped breathing tube. The body is comma shaped. Abdomen consists of 9 segments, having palmate hair and caudal fins on the

8th segment for swimming. The pupa remains at the surface for about a day before the adult emerges. After 48 hours of pupal stage, the pupal skin splits and the mosquito emerges as an adult.

Adult

The adult uses air pressure to break open the pupal case, and crawls to a protected area and rests while its external skeletal hardens, spreading its wings to dry. Males are smaller; mouthparts are not adapted for sucking type. The abdomen is smaller. The female mosquitoes are bigger in size than the males. Their mouthparts are adapted for blood sucking which is essential for the development of ovary/eggs. The males feed on nectars from flowers. Thorax bears three pairs of legs one on each segment and a pair of wings on mesothorax. Wings on metathorax are modified to halters. Adult females live for a month and adult males for a week.

27. LIFE CYCLE OF SILKWORM (*Bombyx mori*)

Classification

Phylum	:	Arthropoda
Class	:	Insecta
Sub-Class	:	Pterygota
Division	:	Endopterygota
Order	:	Lepodoptera
Genus	:	Bombyx
Species	:	*mori*

Various life cycle stages of silkworm are egg, larva, pupa and adult male and female.

Egg

Females lays 300-500 eggs in clusters upon mulberry leaves.

Larva

Egg hatches into a larva called as caterpillar larva. It is 6 mm long, rough and wrinkled, whitish or grayish body made of 12 segments. The head bears mandibulate mouthparts. The thorax has 3 pairs of jointed true legs. Abdomen has 5 pairs of unjointed, stumpy pseudolegs. Larva undergoes 4 moults and the 5th instar larva spins the silken cocoon around it.

Pupa

With in a fortnight caterpillar larva inside the cocoon becomes pupa or chrysalis. Body becomes shortened. Silk threads wrap around the body of the larva to give rise to cocoon.

Adult

In 3 or 4 months, pupa metamorphoses into adult. A single egg either develops into male or female. Males are smaller than female. Males die after copulation and females after laying eggs.

STAGES OF SILKWORM IN DETAIL

Egg

Soon after fertilization each female lays about 300-500 eggs in cluster upon the leaves of the mulberry tree. The female covers the eggs by a gelatinous secretion, which sticks them to the leaves. The small, smooth and spherical eggs are first yellowish white and become darker later on. In tropical countries, the silk moth lays non-diapause eggs which enables to raise two to seven generations with in a year. In temperate countries diapause eggs are laid, so that a single generation is produced per year.

Larva

The silkworm, which hatches from egg, is known as the caterpillar. It is a tiny creature about 6 mm long and moves about in characteristic looping manner.

Head

It bears mandibulate mouthparts.

Thorax

It has 3 distinct segments with 3 pairs of jointed true legs.

Abdomen

It is 10 segmented with 5 pairs of unjointed, stumpy pseudolegs, a short dorsal and horn on the 8th segment and a series of respiratory spiracles or ostia on either lateral side.

Pupa

It is also called as Chrysalis. The caterpillar stops feeding and returns to a corner among the leaves. It begins to secrete the sticky fluid of its salivary glands through a narrow pore, called spinneret. The sticky substance turns into a fine, long and solid thread of silk. The thread becomes wrapped around the body of the caterpillar forming a pupal case or covering known as the cocoon. Within a fortnight the silkworm transforms into tubular brownish organism, the pupa or chrysalis.

Adult

The adult moth is about 25 mm long with a wing size of 40-50 mm. It is robust and creamy white in colour. The body is divided into 3 portions *i.e.*, head, thorax and abdomen.

Head

It bears a pair of compound eyes, a pair of branched or plumed antennae and the mandibulate mouth part.

Throax

The thorax bears 3 pairs of legs and 2 pairs of wings. The cream coloured wings are about 25 mm long and are marked by several faint or brown lines.

Abdomen

It consists of 10 segments.

IDENTIFICATION OF MALE AND FEMALE SILKMOTH

Feature	Male	Female
Body	small	large
Antennae	large and dark colour	small and light colour
Abdomen	slender and small	stout
Activeness	very active	less active
Genitalia	herpes	labium

28. VITAL STAINING OF EARTHWORM OVARY

To mount and stain earthworm ovary using Janus green stain.

Requirements

Well grown earthworm, dissection set, Janus green stain, clean glass slide.

Procedure

Earthworm is a hermaphrodite. A pair of ovary is situated in the 13th segment. They are pyriform, semi-transparent, hanging freely into coelom and attached by their broad ends to septum of the 12th and 13th segments. Each ovary is white compact mass made up of finger like processes. Each ovary contains ovary in a linear series. Each ovum is large, having a distinct nucleus and nucleolus.

Take a full size earthworm and open it dorsally. Cut the intestine, invert it and ventrally below the heart the ovaries are seen as white dots. Pick them up carefully with the help of forceps and place it on the glass slides. Put 2 drops of Janus green stain. Leave it for 5 minutes and observe under the microscope.

29. CULTURING AND OBSERVATION OF PARAMECIUM

Materials Required

Micropipette, hay stems, wheat, paramecium.

Procedure

It is abundantly found in the ponds and ditches and in decaying vegetation.

1. Boil 20 gms of wheat and 20-25 hay stems in 500cc of distilled water for about 10 mins.
2. Keep the culture in dark and cool place for about 4 days and then inoculate it with few paramecia by a micropipette.
3. Within few days the culture contains numerous paramecia.
4. Observe paramecium under microscope.

PARAMECIUM

Classification

Phylum : Protozoa

Class : Ciliata
Order : Holotricha
Family : Paramecidae
Genus : Paramecium
Species : Caudatum, aurelia

Paramecium is best-known ciliate, found in fresh water ponds, rivers, lakes, ditches, streams and pools etc. It has cosmopolitan distribution. It is commonly called as slipper animalcule. Anterior end is bluntly rounded, while posterior end is pointed. Paramecium caudatum measures 80-350 microns. Paramecium aurelia measures 120-290 microns. Cilia cover the entire body. The ventral surface is marked by the presence of obliquely longitudinal groove, the oral groove or peristome.

Reproduction is by binary fission. Conjugation constitutes the sexual part of the reproduction.

30. CULTURING AND STAINING OF *E.COLI* (GRAM'S STAINING)

E.coli colonies were isolated from water and inoculated in nutrient agar.

Composition of Nutrient Agar

Peptones – 5 gm

Beef extract – 3 gm

NaCl – 5 gm

Agar Agar – 20 gm

Distilled water – 300 ml

And pH (7.0 ± 0.2)

The above medium is prepared and is autoclaved at 15 lbs. The solution is poured into clean sterilized petridishes. After cooling the petridishes are kept in the refrigerator for 10-20 hrs. Then the petridish is kept for incubation for 2-4 hrs for 37°C. The isolated *E.coli* inoculum is treated with nutrient Agar plate. The plates are inverted and incubated at 37°C for 24 hrs. After 24 hrs Whitish streaks indicate growth of *E.coli.* A portion of the colony is stained and mounted.

STAINING OF *E.COLI*

Objective

To stain bacteria (*E.coli*) using gram staining method.

Requirements

E.coli culture, crystal violet, gram iodine solution, ethyl alcohol (98%), saffranine, staining tray, dropper, glass slide, microscope, etc.

Procedure

1. Prepare a smear of the culture on a glass slide and heat fix it.
2. Flood the smear with crystal violet for a minute.

3. The rinses are washed under slow running water and flood the smear with iodine solution and keep for a minute.
4. Rinse the slide under slow running water and wash with 98% ethyl alcohol and keep for a minute.
5. Rinse the slide under slow running water and add a few drops of saffranine to the smear and keep it for 30 seconds.
6. Wash the smear and dry the slide.
7. Put a cover slip, put a drop of cedar wood oil over the cover slip and observe under oil immersion lens.

Result

If the material is gram negative, it is decolorized by alcohol and takes saffranine. If it is positive, it takes gram iodine stain.

31. BREEDING EXPERIMENTS IN *DROSOPHILA MELANOGASTER*

Life cycle of *Drosophila melanogaster*

Drosophila melanogaster is a common fruit fly as a test system has contributed to the establishment of the basic principles of heredity. It is also called as "Cinderella of Genetics".

Drosophila melanogaster is a dipterous, holometabolous insect. *Drosophila melanogaster* has a characteristic larval stage preceded by the egg and succeeded by the pupal stage.

Egg

Egg is about 0.5 mm in length, ovoid in shape and white in colour. Extending from the anterior dorsal surface, there is a pair of egg filaments. The terminal portion of this filaments are flattened into spoon like floats. This floats keep the egg from sinking into the semi liquid medium.

Larva

The larva hatches out from the egg. It is white segmented and worm like. The head is narrow and has black mouth parts (jaw hooks). The larva undergoes two moults. So that the larval phase consists of three instars. After this stage the larvae crawls out of the medium and finally attaches to the inner drier surface of the bottle. This culminates in pupation.

Pupa

Soon after the formation of "Pupal horn" from anterior spiracle, the larval body is shortened and the skin becomes hardened and pigmented. The pupa is considered as a reorganization stage. During this process most of the adult structures are developed from the imaginal disc. A fully transformed adult fly emerge out through the anterior end of the pupal case.

At the times of eclosion, the fly is greatly elongated and light in colour with wings yet to be unfolded. Immediately after this the wings unfold and the body gradually turns dark and brown. After six hours of emergence the adult fly attains the potency to participate in reproduction.

Adult

Body is divided into head thorax and abdomen. Head bears a pair of compound eyes and a pair of antennae. Thorax is divided into 3 segments- prothorax, meso and meta thorax, each bearing a pair of legs. The mesothorax has a pair of wings and metathorax a pair of halters. Abdomen is 4-5 segmented in males and 6-7 in females. Abdominal tip of males are darkly pigmented.

MORPHOLOGY OF *DROSOPHILA MELANOGASTER*

The body of an adult *Drosophila melanogaster* is divided into three parts namely - Head, thorax and abdomen.

(*i*) *Head.* The head is composed of six fused segments, a pair of antennae with plumose aristae and a proboscis of licking type without mandibles. On the dorsal side of the head between the compound eyes are three simple eyes called ocelli. Bristles are found on the head.

(*ii*) *Thorax.* It is composed of 3 fused segments namely prothorax, mesothorax and metathorax. All the three segments have a pair of wings. The metathorax contains halters (reduced wings).

(*iii*) *Abdomen.* The abdomen consists of 7 or 8 visible segments in the female and 5 or 6 segments in male.

DIFFERENTIATION BETWEEN MALE AND FEMALE DROSOPHILA

S. No.	*Characters*	*Male*	*Female*
1.	Body size	small	large
2.	Dorsal side of abdomen	3 separate dark bands	5 separate dark bands
3.	Abdominal tip pigmentation	present	absent
4.	Abdominal tip shape	round	pointed
5.	Sex comb in foreleg	present	absent

Media Preparation

Drosophila melanogaster like other animals requires an optimum temperature for its survival, growth and breeding. The optimum temperature for the maintenance of *Drosophila melanogaster* is between 20°C to 25°C. The temperature around and above 31°C makes the flies sterile and reduces the oviposition and may result in death. At lower temperature, the life cycle is prolonged and the viability may be impaired.

The routinely used food media for the maintenance of *Drosophila melanogaster* is cream of white agar medium. The ingredients of this media are

1000 ml of distilled water

100 gm of wheat flour (sooji)

100 gm of Jaggery

10 gm of Agar agar

7.5 ml of Propionic acid, and

Yeast granule.

Heat sterilized bottles should be used for preparing culture. Similarly sterilized cotton has to be used to plug the bottles. As the condition of the medium deteriorates with time, the flies have to be transferred from old to new bottles with fresh culture medium at least once in three weeks.

Procedure

Take a clean vessel and 1000 ml of water and boil it. Then add 10 gm of jaggery and stir well. To this add 10 gm of Agar agar which acts as solidifying agent. Once it boils add 100 gm of sooji. Then add 7.5 ml of Propionic acid which acts as an antimicrobial agent. By constant stirring the medium becomes a viscous fluid. The hot mixture is transferred into the culture bottle. The bottles are left for cooling, yeast is added and plugged with cotton. Bottles are ready to use only after the date of adding the yeast solution.

Etherisation

When flies have to be analysed either for routine observation or for experiments, they are anaesthetized to make inactive. The procedure is to transfer flies from the media bottle to another empty wide mouthed bottle, referred to as etheriser. The mouth of this bottle is to be covered with a stopper sprayed with ether. It takes a minute or so to anaesthetize flies. After this if etherized flies revive before completion of observation, they have to be re-etherised by using re-etheriser. The re-etheriser is an ether soaked filter paper fitted in a petri plates which has to be placed over the flies on the glass plate.

Sexing

Adult flies are 2-3 mm long while females are slightly larger than males. The males carry sex-comb on the first tarsal segment of the first leg. Males can also be identified by the presence of black pigmentation at the tip of the rounded abdomen. The tip of the female's abdomen is pointed and non-pigmented. After the seperation of the two sexes the unwanted flies should be discarded immediately into the morgue (a bowl of mineral oil or detergent in water).

Isolation of Virgins

In many experiments with Drosophila it is essential that the sperm of a particular genotype (male) is used for fertilization of a particular female. To ensure this it is often essential to isolate virgin females. The females can store and utilize the sperms from one insemination for large part of the reproductive life. Females which have any chance of being non-virgins should not be used for crosses. For ensuring virginity females are removed before 6 hrs of their emergence and males are removed from the culture bottle before 12 hrs of emergence.

Making and Conducting Crosses

To make crosses between different strains 1-10 virgins from first strain are mated in a culture bottle with corresponding no of males from other strain. The reciprocal mating of the two strain are also done. As soon as the cross is made the bottle is marked with the nature of cross and date of crossing. If the larva does not appear even after 5-7 days then the culture is discarded. If culture is successful then parents are discarded and the already laid eggs are allowed to develop into adults.

Analysis of the Progeny

The aim is to understand the pattern of inheritance of a character from parents to offspring and to subsequent generation. Therefore each progeny (F1, F2 and test cross) has to be carefully analysed and classified according to the phenotype and sex of each individual. Utmost care must be taken to record the number of flies in each category. From each experiment bottles the counting must be restricted to first seven days from third day of eclosion. A minimum of 200 flies must be analysed from each of F1, F2 and test cross progenies.

Statistical Test and Confirmation of Results

The data obtained from the analysis of the progeny has to be tested with and an established hypothesis. This has to be done to ascertain whether the observed data fixed into hypothesis or not. The routinely used statistical test for such experiment is Chi_square test

$$\chi^2 = \in \frac{(0 - E)^2}{E}$$

Chi_square test obtained from this test and the degree of freedom (*df*) $df = n - 1$ was to be checked for the level of significance in the chi-square probability table.

If the calculated value of chi-square value is less than the table value at particular level of significance the difference between the observed and expected frequency is not significance. Then we have to accept the hypothesis.

On the other hand when the calculated value is more than the table value the difference between the observed and expected value is significant, then we have to reject the hypothesis.

32. PREPARATION OF SALIVARY GLAND CHROMOSOMES

To prepare salivary gland chromosome in *Drosophila melanogaster*

Requirements

Stereo zoom/Dissecting microscope
Third instar larva
1N HCl
Physiological saline (0.7% NaCl)
Lacto aceto orcein
45% Acetic acid
Nail polish or wax for sealing
Slides and cover slips.

Procedure

(*i*) Dissect out the salivary gland of the third instar in physiological saline
(*ii*) Place it in 1N HCl for 2-3 min

(*iii*) Transfer it to 2% Lacto Aceto orcein stain for 30 min

(*iv*) Squash it with freshly prepared 45% Acetic acid

(*v*) Seal the edges of cover slips with nail polish or wax

(*vi*) Observe under the microscope for polytene chromosome.

Description

1. Balbiani in 1881 observed salivary gland chromosome in *Chironomous* larva
2. Painter discovered the same in *Drosophila melanogaster*
3. The polytene chromosomes are the largest chromosomes available for cytological studies
4. These chromosomes are clearly seen in third instar larva of *Drosophila melanogaster*
5. The salivary gland cells all such chromosomes undergo somatic pairing and endoduplication without separation
6. This multistranded chromosome contains 1024 chromosomal fibrils
7. When stained, chromonema shows bands and interbands
8. Along the length there are bulged regions called Balbiani rings or puffs which are the sites of gene action
9. Thus this chromosome has common chromocentre with 5 long ares radiating outwards.

33. OBSERVATION OF MUTANTS IN *DROSOPHILA MELANOGASTER*

Introduction

Mutations is a heritable change in the genetic material. A mutant phenotype is a heritable deviant from the standard phenotype and caused due to mutation. A mutation is said to be dominant if it expresses in the heterozygous condition. A mutation is said to be recessive if it requires a homozygous state for its expression. A recessive mutation on X chromosome in male is expressed since the Y chromosome does not carry the corresponding allele and this is referred to as hemizygous condition. However the same recessive mutation on X chromosome needs homozygous condition in the female for expression.

Yellow Body

Symbol : y

Location : 1-0.0

Phenotype

The body colour is yellow and with body hairs and bristles brownish in colour but with yellow tips. The wings, hairs and veins are yellow. The larval mouth parts are yellow to brown.

Ebony Body

Symbol : e

Location : 3-30.7

Phenotype

The body colour is black in adult. Larvae show arkened spiracles sheath compared to wild type.

Vestigial Wing

Symbol : vg

Location : 2-67.0

Phenotype

Wings and balancers are greatly reduced.

Curly Wing

Symbol : Cy

Location : 2-6.1

Phenotype

Associated with curly inversion on the left arm(2L) of second chromosome. Wings are strongly curved upward and forward. The homozygote is lethal.

White Eye

Symbol : w

Location : 1-1.5

Phenotype

White eye colour, colourless ocelli, Malphigian tubules and testis.

Sepia Eye

Symbol : Se

Location : 3-2.6

Phenotype

Brownish red eye colour which darkens to sepia and finally to dark. The ocelli are wild type.

Brown Eye

Symbol : bw

Location : 2-104.5

Phenotype

Brownish wine eye colour which becomes purplish with age.

34. ABO BLOOD GROUPING AND Rh FACTOR IN HUMANS

Introduction

The ABO system of blood group was introduced by Landsteiner in 1900. He found two types of antigens present on the RBCs. They are antigen A and antigen B. Similarly there are two types of antibodies present in the plasma called antibody A & B

Based on the presence or absence of antigen and antibodies human blood is classified into 4 groups A, B, AB & O.

- A group person contains antigen A and antibody B.
- B group person contains antigen B and antibody A.
- AB group person contains both antigens A & B and no antibodies.
- group person contains no antigen and both antibodies A & B.
- ABO blood group is inherited by a set of multiple alleles.

Presence of a particular factor is given by Rh factor discovered by Weiner from the rabbits immunized with the blood of the monkey called *Macaca rhesus.*

35. DETERMINATION OF BLOOD GROUP AND Rh FACTOR

Materials Required

Blood sample

Applicator sticks

Antisera A,B & Rh

Glass slide

Cotton

Spirit and Disposable needles

Procedure

1. Clean a glass slide thoroughly
2. Place a drop of antiserum A on the left side, antiserum B on the right side and antiserum D at the centre of the slide
3. Clean the tip of index finger with cotton soaked in sprit
4. Prick the finger tip with the help of sharp sterilized disposable needle
5. Place 3 drops of blood near the antisera
6. Mix the blood and antiserum using application sticks.

Result : Observe your experimental result.

Conclusion : Give conclusion base on your result.

36. DEMONSTRATION OF LAW OF INDEPENDENT ASSORTMENT

Description

Mendels second law is known as the law of independent assortment which states that 2 different sets of genes assort independently of each other during formation of gametes through meiosis.

The cross conducted taking 2 contrasting pair of characters is known as the dihybrid cross and it gives 9:3:3:1 in the F2 generation. In drosophila this was stated by Morgan. In this experiment pairs of contrasting characters such as sepia eye and vestigial eye mutants were taken to study whether inheritance patterns follow the Mendelian laws or not. Vestigial wing is characterized by nearly reduced wings and balancers and sepia eye mutant is characterized by brown eye colour.

Requirements

1. *Drosophila melanogaster*
2. Sepia vestigial mutant
3. Media bottles
4. Anaesthetic ether
5. Etheriser.

Procedure

Normal *D. melanogaster* strain and sepia vestigial wing strains were taken in standard media bottles separately. When the flies were ready to emerge from pupa the original stock were discarded. The newly emerged male and female virgins were isolated. Flies which were collected were aged for 3-5 days and were crossed with each other by conducting a reciprocal cross also. The 2 crosses were also in the two following manners.

(*a*) normal females × sepia vestigial males.

(*b*) normal males × sepia vestigial females.

The progenies produced in the F1 were observed for phenotypic expression and the data was collected and recorded. By taking a few of the F1 flies inbreeding was carried out to obtain F2 generation. The phenotype of F2 flies were observed and the data was recorded.

Direct Cross

Parents: sepia vestigial females × normal males

F1: all were normal eyed and normal winged

Inbreed: F1 females × F1 males

F2 : normal winged vestigial winged normal wing vestigial wing

And red eyed : and red eyed : and sepia eye: and sepia eye

Ratio: 177 : 60 : 65 : 20

PHENOTYPE OBSERVED EXPECTED DEVIATION d2 d2/E

Normal winged 177 181.12 – 4.12 16.97 0.07

And red eye

Vestigial winged 60 60.37 – 0.37 0.137 0.002

And red eye

Normal wing 65 60.37 + 4.63 21.43 0.35

And sepia eye

Vestigial wing 20 20.12 – 0.12 0.0144 0.0007

And sepia eye

Degree of freedom = 4 – 1 = 3

Sx_2 = 0.4427

Reciprocal Cross

Parents : normal females × sepia vestigial males

F1 : all were normal eye (red) and normal wing.

Inbreed : F1 female × F1 male

F2 : red eyed and normal wing : red eyed vestigial wing : sepia eyed and normal wing : sepia eyed and vestigial Wing

Ratio : 212 : 70 : 66 : 22

PHENOTYPE OBSERVED EXPECTED DEVIATION d2 d2/E = X2

Red eyed and 212 208.12 3.88 15.05 0.0723 normal wing

Red eyed and 70 69.71 0.625 0.390 0.0055

Vestigial wing

Sepia eyed and 66 69.375 3.375 11.396 0.164

Normal wing

Sepia eyed and 22 23.12 1.125 1.265 0.0547

Vestigial wing

Sx_2 = 0.029689

Degree of freedom = 4 – 1 = 3 Analysis of Result.

At 5% level of significance at 3 degree of freedom, table value = 7.815 and at 2 degree of freedom it is 5.991

At 4 degree of freedom it is 9.48

In direct cross the X2 value = 0.4427 and is less than deviation. In reciprocal cross the X2 value = 0.29689 and is

Less than table value.

In both the cases, the deviation is not significant so null hypothesis is accepted.

37. DEMONSTRATION OF LAW OF SEGREGATION

Description

Heredity or the inheritance of parental character is offspring's has long been the subject of great deal of experimental work in biology. Gregor Mendel an Austin monk carried out an extensive series of experiments on the common edible pea (Pisum sativum) to find out the inheritance pattern.

The law of segregation is one among the laws proposed by Mendel, which states that the genes or alleles present in the F1 will not blend or contaminate or influence one another, rather they segregate in the same pure form which they arrived from the parents. The members of an allele pair separate from each other without influencing each other, when an individual forms haploid germ cells.

To understand the pattern of inheritance of vestigial wing mutant of *Drosophila melanogaster* to understand the law of segregation.

Materials Required

1. *Drosophila melanogaster*
2. Vestigial wing mutant of *Drosophila melanogaster.*
3. Bottles with standard medium
4. Anaesthetic ether.
5. Etherizer
6. Re-etheriser
7. Needles
8. Brushes
9. Yeast granules
10. Glass plate

Procedure

Drosophila melanogaster normal and vestigial wing mutant flies were cultured in standard media bottles separately. When pupae in the cultured bottles were ready, the bottles were cleaned by taking out all the flies present there. The enclosed males and females virgin were isolated, aged for 2-3 days and then, by taking these virgins, crosses were made. They were crossed in the following way to get F1 generation:

1. Normal female X vestigial wing virgin male.
2. Normal female X vestigial male

The progeny produced in the F1 generation were observed for the phenotypic expression and the data was proposed.

Then F1 males and females were inbred and the resulting F2 phenotypes were observed and the data was recorded.

Observation

Parents: phenotype normal female X ebony males

F1 progeny: phenotype all normal body coloured flies.

Inbreed: F1 females X F1 males.

F2 progeny: phenotype normal flies and ebony flies.

Phenotypic ratio: 163: 52

PHENOTYPE OBSERVED EXPECTED DEVIATION d2 d2/E = X2

Normal females

And males 163 161.25 1.75 3.06 0.189

Ebony females

And males 52 53.75 – 1.75 3.06 0.056

Sx_2 = 0.075

Degree of freedom = 2 – 1 = 1

Analysis of result

At 5% level of significance at 10 of freedom table value is 385.

In direct crosses the calculated value of x_2 = 0.075 which is less than the table value.

In reciprocal crosses the calculated x_2 value is less than table value.

In both the crosses, the deviation was not significant, so null hypothesis is accepted.

Reciprocal cross

PARENTS: phenotype ebony female X normal male.

F1 : all normal body coloured flies

Inbreed : F1 female × F1 male

F2 : normal flies and ebony flies

Phenotypic ratio: 180:48

PHENOTYPE OBSERVED EXPECTED DEVIATION d2 d2/E

Normal male
And female 180 171 9 81 0.47

Ebony male and
Female 46 57 9 81 1.42

Sx_2 = 189

Degree of freedom = 2 – 1 = 1.

8 MOLECULAR BIOLOGY

1. THE CENTRAL DOGMA

INTRODUCTION

The Central Dogma of modern biology is the conversion of the genetic message in DNA to a functional mRNA (transcription) and subsequent conversion of the copied genotype to a phenotype in the form of proteins.

The process of conversion of a mRNA to a functional protein is known as translation. It involves the attachment of a messenger RNA to the smaller subunit of a ribosome, the addition of the larger subunit, plus initiation by a host of other factors. The entire process can be accomplished in the absence of a cell, if all of the necessary factors are present. Unfortunately, studies on translation and post-translational changes in protein structure are rather complex. They require a heavy investment in time and equipment. To some extent, the electrophoretic identification of proteins is part of this process, and the appearance of specific proteins can be monitored during any of the developmental processes identified in Chapter Thirteen. Chapter Fourteen lists several means of studying both the DNA and RNA contents of cells during changes.

To study the process of translation in any meaningful way requires that reasonably purified sources of mRNA, ribosomes and all of the amino acids be available. In addition, there is a requirement for various factors responsible for peptide chain initiation on the ribosome.

Consequently this chapter will deal with only one exercise. It is definitely an advanced technique, and requires mastery of many of the techniques presented in previous chapters. It must be performed on an independent basis, since the extensive time commitment does not lend itself to typical laboratory periods.

M1—PROTEIN SYNTHESIS IN CELL FREE SYSTEMS

Materials

- Suspension culture of fibroblast cells (1 litre)
- 35 mM Tris-HCl, pH 7.4, 140 mM NaCl (TBS buffer)
- 10 mM Tris-HCl, pH 7.5, 10 mM KCl, and 1.5 mM magnesium acetate (TBS-M)
- 10X TBS-M: 200 mM Tris-HCl, pH 7.5, 1200 mM KCl, 50 mM magnesium acetate and 70 mM β-mercaptoethanol

- 10X solution of 20 amino acids
- Teflon homogenizer
- Refrigerated preparative centrifuge
- Sat. $(NH_4)_2SO_4$
- TBS-M plus 20% (v/v) glycerol
- 1X TBS-M buffer containing 0.25 M sucrose
- 1X TBS-M buffer containing 1.0 M sucrose
- Sephadex G-25 column equilibrated with 1X TBS-M buffer
- Liquid nitrogen storage
- Reaction mixture for protein synthesis, containing the following in a total volume of 50 µl

Tris-HCl, pH 7.5	1.5 µm
Mg acetate	0.15-0.20 µm
KCL	4.0-5.0 µm
β-Mercaptoethanol	0.25 µm
ATP	0.05 µm
GTP	0.005 µm
Creatine phosphate	0.50 µm
Creatine kinase	8.0 µg
Each of 19 amino acids(-leucine)	2.0 nmol
$_{14}$C-leucine (150 mCi/mmol)	0.125 µCi
Ribosome fraction	1 to 2 A_{260} units
Viral mRNA or Globin 9S mRNA	2.0 to 5.0 µg
or	
Poly U	10.0 µg

Procedure

1. Chill the suspension culture (app. 10_9 cells) rapidly in an ice bath. Collect the cells as a pellet, by centrifugation at 600 xg for 10 minutes at 4° C. Resuspend the cells in TBS buffer and wash them three times with cold TBS buffer.
2. Suspend the final pellet in two volumes of TBS-M for 5 minutes at 0° C and homogenize the cells with 10 to 20 strokes in a tight fitting Teflon homogenizer.
3. For each 0.9 ml of homogenate, add 0.1 ml of concentrated 10X TBS-M buffer. Centrifuge the mixture at 10,000 xg for 10 minutes at 4° C.
4. Decant and collect the supernatant extract and adjust the extract such that the following are added to yield final concentration:

ATP to 1.0 mM ATP

GTP to 0.1 mM GTP

Creatine phosphate to 10 mM

Creatine kinase to 160 µg/ml

Amino acids to 40 µm each.

5. Incubate the mixture for 45 minutes at 37° C.
6. Centrifuge the mixture at 10,000 xg for 10 minutes at room temperature. Cool the supernatant and pass the supernatant through a Sephadex G-25 column at 4° C.
7. Turn on a UV spectrophotometer and adjust the wavelength to 260 nm. Blank the instrument with TBS buffer.
8. Centrifuge the filtrate excluded from the Sephadex column at 165,000 xg for 90 minutes at 4° C.
9. Precipitate the proteins within the supernatant by the addition of saturated $(NH_4)_2SO_4$ to yield a final 60% $(NH_4)_2SO_4$. Collect the precipitate by centrifugation.
10. Dissolve the precipitate in TBS-M buffer and dialyze it against the same buffer containing glycerol.
11. Suspend the resulting ribosome pellet in 1X TBS-M buffer containing 0.25 M sucrose. Place 5 ml of TBS buffer with 1.0 M sucrose into the bottom of a centrifuge tube and layer the suspended ribosomes on top. Centrifuge at 216,000 xg for 2.5 hours at 4° C.
12. Wash the resulting pellet with TBS-M buffer, and resuspend it in the same buffer with 0.25 M sucrose.
13. Determine the ribosome concentration using a UV spectrophotometer to measure the A_{260}. The extinction coefficient for ribosomes is 12 A units per mg per ml at 260 nm.
14. The ribosomes may be frozen and stored in liquid nitrogen, or used for *in vitro* protein synthesis. If frozen, they should be thawed only once prior to use.

    To test for protein synthesis, prepare the reaction mixture for protein synthesis.
15. Incubate the reaction mixture at 37° C for 60 minutes. Terminate the reaction by pipetting 40 μl of the mixture onto a 2.5 cm disk of Whatman 3 MM filter paper. Dip the disk into cold 10% TCA for 15 minutes and then in 5% TCA at 90° C for 15 minutes.
16. Rinse the disk twice in 5% TCA for 5 minutes, once in alcohol:ether (1:1), and then dry it.
17. Place the disk into a scintillation vial and add a toluene-based fluor. Refer to Appendix H for details on the use of a scintillation counter.
18. Measure the amount of radioactively labelled amino acid incorporated into protein.
19. Graph the protein synthesized versus time.

Optional

For advanced work, compare the activity of ribosomes isolated from the fibroblast cultures to those isolated from a prokaryote culture, a plant (yeast or pea seedlings) and from genetic μ tants known to alter the structure of either rRNA or any of the ribosome structural proteins.

CHROMOSOMES

Introduction

The interphase chromosomes of eucaryotic cells are complex molecular structures composed primarily of a DNA core and a protein matrix complexed into a long thread-like structure. This basic chromosome thread is then coiled through several layers of organization and ultimately gives rise to a structure that can be visualized with a light microscope.

Chemically, the interphase nucleus is composed of a substance known as "chromatin," which is further subdivided into euchromatin and heterochromatin. The distinction between these subdivisions is based on quantitative distribution of the basic chromosome fiber, with a higher concentration found in heterochromatin. Heterochromatin, therefore, will stain more intensely than euchromatin, since the fiber is packed tighter within a given volume.

Proteins extracted from chromatin have been classified as either basic or acidic in nature. The basic proteins are referred to as "histones" and the acidic as "non-histone proteins". Histones play an integral part in the structural integrity of a eucaryotic chromosome. They are organized into specific complexes, known as nucleosomes, and around which the DNA molecule is coiled. Acidic proteins within the nucleus compose many of the DNA replication and RNA transcription enzymes and regulatory molecules. They vary in size from small peptides of a few amino acids to large duplicase and replicase enzymes (respectively DNA and RNA polymerases).

Transcription of DNA on the chromosome fiber results in the presence of a host of RNA species found within the nucleus of the cell. When the RNA is transcribed from the "nucleolar organizer" region of a genome and complexed with ribosomal proteins, granules are formed which collectively produce a "nucleolus," visible at the light microscope level of resolution. When transcribed from other portions of the genome, the RNA is either in the form of pre-transfer RNA, or heterogeneous nuclear RNA (hnRNA). The precursor tRNA must be methylated and altered before becoming functional within the cytoplasm, and the hnRNA will also be significantly modified to form functional mRNA in the cytoplasm.

Thus, a chemical analysis of chromosomes will yield DNA, RNA and both acidic and basic proteins. It is possible to extract these compounds from an interphase nucleus (i.e. from chromatin) or to physically isolate metaphase chromosomes and then extract the components. For the former, the nuclear envelope will be a contaminating factor, as will the nucleolus. For isolated chromosomes, many of the regulatory molecules may be lost, since the chromosomes are essentially non-functional during this condensation period.

EXERCISE 1.1. POLYTENE CHROMOSOMES OF DIPTERANS

Materials

- Prepared slides of *Drosophila* (fruit fly) salivary gland chromosomes
- Genetic map of polytene chromosome bands.

Procedure

1. Examine the slides for the presence of bands. Select a single chromosome spread demonstrating all four chromosomes and draw the complete structure.

2. Label each of the four chromosomes of the fruit fly, as well as the chromocenter of the connected chromosomes.
3. Compare your drawings to the genetic map for *Drosophila*.

Notes

The glands of dipterans (flies) have a useful characteristic for analysis of gene location on chromosomes. During their mitotic division, the normal division of the chromosomes is aborted and the replicated chromosomes remain as an integral unit. The chromosome content thus increases geometrically and produces "giant" polytene chromosomes. The chromosomes remain attached at a point where the centromeres fuse, at the chromocenter. This is clearly observed in the chromosomes of the fruit fly salivary gland tissue. The fruit fly chromosomes are ideal specimens since they are in a near constant state of prophase and are incapable of further division. Because they have been extensively analyzed for their genetic composition, co-linear maps of genes within genetic linkage groups have been produced and correlated with the physical location of a band on the chromosomes.

EXERCISE 1.2. SALIVARY GLAND PREPARATION (SQUASH TECHNIQUE)

Materials

- Fruit fly larva (wild type and tandem duplication mutants)
- Ringers insect saline
- Fine forceps and probe
- Aceto-orcein
- Dissecting and regular microscopes
- Slides, coverslips
- Small dish of melted paraffin and paintbrush.

Procedure

1. Select a third instar larva, for which the cuticle has not yet hardened, from a wild-type culture of *Drosophila*. Place it into a drop of Ringer's saline solution on a slide.
2. Place the slide on the stage of a dissecting microscope and view the larva with low power. Grasp the anterior of the larva with a fine point forceps and pin down the posterior portion with a probe. Gently pull the head off and discard the tail of the larva.
3. Locate the salivary glands and their attached fat bodies. The glands are semitransparent and attached by ducts to the digestive system. The fat bodies are white and opaque. Tease away the fat bodies and discard.
4. Place a drop of aceto-orcein on the slide next to the Ringer's and move the salivary glands into the stain. Blot away any excess Ringer's.
5. Place a coverslip over the preparation and allow it to stand for 1-3 minutes (it will take a few trials to obtain properly stained chromosomes). Gently squash the gland preparation in the following manner:

- o Place the slide between several layers of paper toweling.
- o Place your thumb on the top of the towel immediately over the coverslip and gently roll your thumb while exerting a small amount of pressure (as though you were making a finger print). Do not move your thumb back and forth. One gentle roll is sufficient.
- o Remove the slide from the towels, and seal the edges of the coverslip by using a paint brush dipped in melted paraffin.

6. Examine the slide with the microscope and diagram the banding patterns that are observed.
7. Compare your squash preparation to that of the prepared slides examined in experiment 10.1.
8. Repeat the squash technique using larva from a genetic variant known to be the result of a deletion and/or tandem duplication. Determine the location of the deleted or duplicated bands on the chromosomes.

EXERCISE 1.3. EXTRACTION OF CHROMATIN

Materials

- Bovine or porcine brain
- 0.25 M sucrose containing 0.0033 M calcium acetate
- 2.0 M sucrose with 0.0033 M calcium acetate
- 0.075 M NaCl with 0.024 M EDTA, pH 8.0
- Tris-HCl buffer, pH 8.0 with the following molarities, 0.05 M, 0.002 M, 0.0004 M
- TCA (Trichloroacetic acid)
- Tissue homogenizer
- Cheesecloth
- Refrigerated preparative centrifuge
- Bradford or Lowry protein assay
- UV spectrophotometer (Optional).

Procedure

Homogenize approximately 30 gms of bovine or porcine cerebellar tissue in a teflon-glass homogenizer in 9 volumes of cold 0.25 M sucrose containing 0.0033 M calcium acetate.

1. Filter the resulting brei through several layers of cheesecloth and obtain crude nuclear pellets by centrifuging at 3,500 xg for 20 minutes.
2. Resuspend the nuclear pellet in 80 ml of cold 0.25 sucrose containing 0.0033 M calcium acetate.
3. Obtain three cellulose nitrate centrifuge tubes and place 25 ml, aliquots of the resuspended nuclear pellet in each. Carefully pipette 5.0 ml of 2.0 M sucrose-

0.0033 M calcium acetate into the bottom of each tube. Insert a pipette with the 2.0 M sucrose through the suspended nuclei and allow the viscous sucrose to layer on the bottom of the tube. Centrifuge the tubes at 40,000 xg for 60 minutes.

4. Using the resulting nuclear pellet is just above the dense sucrose layer. It is used to extract chromatin proteins. Carefully remove the supernatant above the pellet with a pipet. Then, insert the pipet through the nuclear layer and remove the bottom sucrose layer. The nuclear pellet will remain in the tube. Resuspend the pellet in 40 ml of 0.075 M NaCl-0.024 M EDTA, pH 8.0 and centrifuge at 7700 xg for 15 minutes.
5. Remove and discard the supernatant, resuspend the pellet once again in 40 ml of 0.075 M NaCl-0.024 M EDTA, pH 8.0 and centrifuge again at 7700 xg for 15 minutes. Repeat this process one more time.
6. Resuspend the nuclear pellet in 40 ml of 0.05 M Tris-HCl, pH 8.0 and centrifuge at 7700 xg for 10 minutes.
7. Repeat step 7 to thoroughly wash the nuclei and then wash 2x each in 0.01 M Tris pH 8.0, 0.002 M Tris pH 8.0 and 0.0004 M Tris pH 8.0.
8. Resuspend the final washed nuclear pellet in ice cold distilled water to a final volume of 100 ml and allow to swell overnight at 4° C. Gently stir the mixture on the following day. This solution is the pure chromatin to be used for subsequent analysis.
9. Determine the purity of the chromatin sample within the nuclear pellet using one of the following:
   - Determine the protein concentration by Lowry or Bradford procedure.
   - Measure the optical absorbance at 230 nm (UV). The absorbance of a 1 mg/ml concentration of pea bud histone at 230 nm equals 3.5 OD units. The absorbance follows the Beer-Lambert law, and is linear with histone concentration. Since it is merely an optical reading, the sample is not destroyed in the measurement.
   - Measure the turbidity of the solution. Add trichloroacetic acid to a final concentration of 1.1 M and wait exactly 15 minutes. Read the OD_{400}. A 10 ug/ml solution of pea bud histone has an OD = 0.083 at 400 nm. This technique is excellent for readings between 0 and 0.15 OD. The TCA precipitates some proteins and thus this procedure is more specific to histones than B. It can also be performed without a UV spectrophotometer.
   - Measure by non-destructive fluorometry. Histones can be detected by an excitation wavelength of 280 nm and a fluorescence measurement at 308 nm. Non-histones can be detected in the same sample by excitation at 290 nm and measurement at 345 nm. Of course, this procedure requires the use of a fluorescence spectrophotometer.

Notes

The extraction of chromatin proteins starts with the isolation of a good nuclear fraction. Nuclear pellets and chromatin should be extracted one day before the laboratory period. If DNA, RNA, and both Histone and Non-Histone proteins are to be separated.

EXERCISE 1.4. CHROMATIN ELECTROPHORESIS

Materials

- 14 M Urea
- 6 M NaCl
- 0.05 M and 0.9 M Acetic acid
- Dialysis tubing
- Electrophoresis apparatus
- Prepared gels
- 10 M urea-0.9 N acetic acid-0.5 M β-mercaptoethanol
- 0.25% Coomasie blue.

Procedure

1. To the chromatin suspension from experiment 10.3, add concentrated urea and concentrated NaCl separately to yield a final concentration of 7 M urea and 3 M NaCl.
2. Centrifuge the clear solution at 85,500 xg for 48 hours at 4° C to pellet extracted DNA.
3. Collect the supernatant and dialyze it against 0.05 M acetic acid (three changes, 6 litres each at 4° C). Remove the dialyzed protein solution and lyophilize it to dryness.
4. Meanwhile, set up a standard polyacrylamide gel, using 15% acrylamide (15%T:5%C) in 2.5 M urea and 0.9 M acetic acid. Set up the gel in the electrophoresis unit and run the gel at 2 mA/gel for 2 hours with no sample, using 0.9 M acetic acid for the running buffer.
5. Dissolve the lyophilized protein from step 3 in 10 M urea-0.9 N acetic acid-0.5 M β-mercaptoethanol (to a final concentration of 500 micrograms protein per 100 μl of buffer) and incubate at room temperature for 12-14 hours prior to the next step.
6. Apply 20 μl samples of the redissolved protein extract to 0.6 × 8.0 cm polyacrylamide prepared as in step 4.
7. The gels are run against 0.9 M acetic acid in both upper and lower baths for approximately 3 hours at 100 V.
8. Stain the gels for 1 hour in Coomasie blue, rinse with water, destain and store in 7% acetic acid.
9. If densitometry measurements are made, 5 μg of pea bud fraction II a protein has a density of 1.360 density units x mm with a 95% confidence limit of 10%. By comparison, the density value can be used to quantitate the concentration of protein fractions in μg of your sample.

EXERCISE 1.5. EXTRACTION AND ELECTROPHORESIS OF HISTONES

Chromosomes

Table 1 Properties of chromatin

Morphotype	Activated chromatin Euchromatin	Non-activated chromatin Facultative and obligate heterochromatin, euchromatin
Structural organization	Less condensed, unfolding of functional domains (2040 kbp) exhibit	Highly condensed
DNA methylation (CG sites)	DNase I sensitive sites mv-sites unmethylated	mv-sites methylated
Nucleosomes	DNase I sensitive	DNase I resistant
Histones	H1-deprivated; core histones highly acetylated	H1-enriched; association with special H1 isofores, e.g., H5; H2A/H2B under-acetylated; eventually H2A modified by ubiquitin
HMG 14/17	Present	Absent
HMG 1/2	Present	Absent
Transcription RNAP/RNP	Presence of RNAP and RNP depends on the actual transcription state	No

Materials

- Saline Citrate (1/10 SSC)
- 1.0 N H_2SO_4
- Refrigerated preparative centrifuge
- Absolute ethanol
- β-mercaptoethanol
- 0.01 M sodium phosphate buffer, pH 7.0 + 1% (w/v) SDS + 0.1% (v/v) β-mercaptoethanol
- 10% acrylamide gels (10%T:5%C) with 0.1% (w/v) SDS
- 7% (w/v) Acetic acid
- 0.25% Coomasie blue.

Procedure

Dissolve crude chromatin in cold dilute saline citrate (0.015 M NaCl + 0.001 M sodium citrate) to a final DNA concentration of 500 μg/ml.

1. Stir the solution on ice and slowly add 1/4 volume of cold 1.0 N H_2SO_4. Continue stirring for 30 minutes.
2. Centrifuge the suspension at 12,000 xg for 20 minutes at 4° C. Save the supernatant. For maximum yield, break up the pellet, resuspend in fresh, cold 0.4 N H_2SO_4, re-extract, centrifuge, and add the resulting supernatant to the first.

3. Add 4 volumes of cold absolute ethanol to the supernatant and store for 24 hours at –10° C to precipitate the histone-sulfates.
4. Collect the precipitate by centrifugation at 2,000 xg for 30 minutes.
5. Decant as much of the alcohol as possible, and resuspend the pellet in cold absolute ethanol.
6. Centrifuge at 10,000 xg for 15 minutes.
7. Collect the pellet and freeze dry for later analysis.

   To continue with the electrophoresis, carefully weigh the histone protein sample and dissolve in 0.01 M sodium phosphate buffer, with a pH 7.0 and containing 1% sodium dodecyl sulfate and 0.1% β-mercaptoethanol; final volume should contain approximately 300 µg of protein in 100 µl of buffer.
8. Prepare the electrophoresis chamber with a 10% acrylamide gel with 0.1% SDS.
9. Add separately 25 µl of the dissolved protein and 25 µl of protein standards to:

   50 µl of 0.1% SDS, 0.1% β-mercaptoethanol in Buffer

   5 µl of β-mercaptoethanol

   1 µl of 0.1% bromophenol blue in water
10. Mix thoroughly and apply the histone extract and protein standards to separate wells of the electrophoresis gel.
11. Separate the proteins in the anode direction (Anionic system).
    - The addition of SDS anions to the proteins results in negatively charged proteins which will separate according to molecular weight.
    - Electrophoresis is carried out in the standard manner following the basic steps given in Chapter 4. The buffer utilized is that of Laemmli.
    - 0.025 M Tris-0.192 Glycine and 0.1% SDS, pH 8.3.
    - Proteins are separated by a current of 3-4 mA per gel until the bromophenol marker reaches the bottom of the tube (about 7 hours at 3 mA, and 4 hours at 4 mA).
12. Stain the gels with 0.25 % coomasie blue for 2 hours.
13. Destain and store in 7 % acetic acid.
14. Scan the gels and determine the molecular weights of each component.

Notes

Preparation of a total histone fraction from nuclei is normally accomplished by extraction with a dilute acid or a high molarity salt solution. The acidic extraction removes histones from DNA and non-histones immediately, while the dissociation of chromatin in salt solutions will require further purification. In either event, the histones themselves are subdivided into five major types, designated as H1, H2, H3, H4 and H5. H2 dissociates into two peptides, which are thus designated as H2A and H2B. The classification of histones is based on their electrophoretic mobility.

Non-histone proteins can also be extracted and separated by electrophoresis. Whereas histones have only 5 major types, non-histones are extremely heterogeneous and up to 500 different proteins have been identified from one cell type, while the major proteins comprise less than 20 types.

The extraction of chromatin DNA was possible with the 7 M urea-3 M NaCl extraction performed in experiment 10.4. Further analysis of DNA will be undertaken as part of a later lab exercise (on Transcription), and the DNA sample from this lab may be kept lyophilized and frozen until that time.

For our current needs it is sufficient to note that the genes are composed of DNA, and that various specific regions of the DNA/Genetic information can be physically isolated to a specific locus on a chromosome. This in turn is readily observed and correlated with banding patterns, such as those in the fruit fly polytene chromosomes.

EXERCISE 1.6. KARYOTYPE ANALYSIS

Materials

- Fresh venous blood
- Heparinized syringes
- Eagle's spinner modified media with PHA
- Culture flasks
- Tissue culture grade incubator at 37° C
- 10 μ g/ml Colcemid
- Clinical centrifuge and tubes
- 0.075 M KCl
- Absolute Methanol and glacial acetic acid (3:1 Mixture, prepared fresh)
- Dry ice
- Slides, coverslips and permount
- Alkaline solution for G-banding
- Saline-Citrate for G-banding
- Ethanol (70% and 95% (v/v))
- Giemsa stain.

Procedure

Draw 5 ml of venous blood into a sterile syringe containing 0.5 ml of sodium heparin (1000 units/ml). The blood may be collected in a heparinized Vacutainer, and transferred to a syringe.

1. Bend a clean, covered 18 gauge needle to a 45° angle and place on the syringe. Invert the syringe (needle pointing up, plunger down), and stand it on end for 1-1/2 to 2 hours at room temperature.

   During this time the erythrocytes settle by gravity, leaving approximately 4 ml of leukocyte-rich plasma on the top, and a white buffy coat of leukocytes in the middle.

2. Carefully tip the syringe (do not invert) and slowly expel the leukocyte-rich plasma and the buffy coat into a sterile tissue culture flask containing 8 ml

of Eagle's spinner modified media supplemented with 0.1 ml of phytohemaglutin (PHA).

3. Incubate the culture for 66-72 hours at 37° C. Gently agitate the culture once or twice daily during the incubation period.
4. Add 0.1 ml of colcemid (10 micrograms/ml) to the culture flasks and incubate for an additional 2 hours.
5. Transfer the colcemid-treated cells to a 15 ml centrifuge tube and centrifuge at 225 xg for 10 minutes.
6. Aspirate and discard all but 0.5 ml of the supernatant. Gently tap the bottom of the centrifuge tube to resuspend the cells in the remaining 0.5 ml of culture media.
7. Add 10 ml of 0.075 M KCl, dropwise at first, and then with gentle agitation to the centrifuge tube. Gently mix with each drop. Start timing the next step immediately with the first drop of KCl.
8. Let the cells stand exactly 6 minutes in the hypotonic KCl. The hypotonic solution should not be in contact with the cells in excess of 15 minutes from the time it is added.
9. Centrifuge the cells at 225 xg for 6 minutes. Aspirate the KCl and discard all but 0.5 ml of the supernatant. Gently resuspend the cells in this small volume of fluid.
10. Add 10 ml freshly prepared fixative dropwise at first and then with gentle agitation. Gentle and continuous agitation is important at this step to prevent clumping of the cells. If the cells were not properly resuspended in step 10, the cells will clump beyond any further use.
11. Allow the cells to stand in fixative at room temperature for 30 minutes.
12. Centrifuge at 200 xg for 5 minutes and remove all but 0.5 ml of supernatant. Resuspend the cells in fresh fixative.
13. Wash the cells twice more in 10 ml volumes of fixative. Add the fixative slowly, recentrifuge, and aspirate the fixative as previously directed. the fixed, pelleted cells may be stored for several weeks at 4°C.
14. Resuspend the pellet of cells in just enough fixative to give a slightly turbid appearance.
15. Prop a piece of dry ice against the side of a styro foam container and lace a clean slide onto the dry ice to chill the slide.

    Use a *siliconized* pasteur pipette to draw up a few drops of the suspended cells and drop the cells onto the surface of the chilled slide. Spreading of the chromosomes may be enhanced by dropping the cell suspension from a height of at least 12 inches. As soon as the cells strike the slide, blow hard on the slide to rapidly spread the cells.
16. Remove the slides from the dry ice and allow them to air dry. Perform the desired banding and/or staining procedures.

    Procedure : Preparation of chromosomes for karyotype analysis can be performed in a number of ways and each will yield differing pieces of information. The chromosomes may be stained with aceto-orcein, feulgen or a basophilic dye such as toluidine blue or methylene blue if only the general morphology is desired.

Procedure : If more detail is desired, the chromosomes can be treated with various enzymes in combination with stains to yield banding patterns on each chromosome. These techniques have become common place and will yield far more diagnostic information than giemsa stain alone (the most commonly used process). A band is an area of a chromosome which is clearly distinct from its neighboring area, but may be lighter or darker than its neighboring region. The standard methods of banding are the Q, G, R, and C banding techniques. These are defined as follows:

- Q-banding
  1. Quinacrine stain
  2. Fluorescence microscopy
- G-banding
  1. Giemsa stain
  2. Additional Conditions
     - (*a*) Heat hydrolysis
     - (*b*) Trypsin treatment
     - (*c*) Giemsa at pH 9.0
- R-banding
  1. Giemsa or acridine orange
  2. Negative bands of Q and G reversed
  3. Heat hydrolysis in buffered salt
- C-banding
  1. Giemsa stain
  2. Pretreatment with BaOH or NaOH followed by heat and salt.

The following directions are for a G-banding:

- Treat fixed and flamed slides in alkaline solution, room temperature for 30 seconds.
- Rinse in saline-citrate solution, 3 changes for 5-10 minutes each.
- Incubate in saline-citrate solution, 65° C for 60-72 hours.
- Treat with 3 changes of 70% ethanol and 3 changes of 95% ethanol (3 minutes) each.
- Air dry.
- Stain in buffered Giemsa for 5 minutes.
- Rinse briefly in distilled water.
- Air dry and mount.

17. Photograph appropriate spreads and produce 8 × 10 high contrast photographs of your chromosome spreads.
18. Cut each chromosome from the photograph and arrange the chromosomes according to size and position of the centromere.

19. Use given table to identify the specific chromosomes.

20. Tape or glue each chromosome to the form supplied for this purpose.

EXERCISE 1.7. IN SITU HYBRIDIZATION

A modern approach to the specific location of genes on chromosomes is a technique for the hybridization of DNA and RNA "in situ." With this procedure, specific radioactive RNA or DNA (known as probes) can be isolated (or synthesized "in vitro") and then annealed to chromosomes which have been treated in such a manner that their basic double stranded DNA has been "melted" or dissociated.

In theory, and fortunately in practice, when the DNA is allowed to re-anneal, the probe competes for the binding, but only where it mirrors a complimentary sequence. Thus, RNA will attach to the location on the chromosome where the code for its production is to be found. DNA will anneal to either RNA which is still attached to a chromosome, or to the complimentary sequence DNA strand within the chromosome. Since the probe is radioactive, it can be localized via autoradiographic techniques.

Finally, it is possible to produce an RNA probe that is synthesized directly from repetitive sequences of DNA, such as that found within the nucleolar organizer region of the genome. This RNA is known as cRNA (for copied RNA) and is a convenient source of a probe for localizing the nucleolar organizer gene within the nucleus, or on a specific chromosome.

The use of in situ hybridization begins with good cytological preparations of the cells to be studied, and the preparation of pure radioactive probes for the analysis. The details depend upon whether the hybridization is between DNA (probe) and DNA (chromosome), DNA (probe) and RNA (chromosome), or between RNA (probe) and DNA (chromosome).

Preparation of the Probe

Produce radioactive RNA by incubating the cells to be measured in the presence of $_3$H-uracil, a specific precursor to RNA. Subsequent to this incubation, extract rRNA from the sample and purify through differential centrifugation, column chromatography or electrophoresis. Dissolve the radioactive RNA probe in 4X Saline-Citrate containing 50% formamide to yield a sample that has 50,000 to 100,000 counts per minute, per 30 microlitre sample, as determined with a scintillation counter. Add the formamide is added to prevent the aggregation of RNA.

Preparation of the Slides

Fix the materials to be studied in either 95% ethanol or in 3 : 1 methanol : water, attach to pre-subbed slides (as squashes for chromosomes) and air dry.

Hybridization

Place the air dried slides into a moist chamber, usually a disposable petri dish containing filter paper and carefully place 30 microlitres of RNA probe in 4X SSC-50% formamide onto the sample.

Carefully add a cover slip (as in the preparation of a wet mount), place the top on the container and place in an incubator at 37° C for 6-12 hours.

Washing

1. Pick up the slides and dip into 2X SSC so that the coverglass falls off.
2. Place the slides in a coplin jar containing 2X SSC for 15 minutes at room temperature.
3. Transfer the slides to a treatment with RNase (50 microgram/ml RNase A, 100 units/ml RNase T1 in 2X SSC) at 37° C for 1 hour.
4. Wash twice in 2X SSC, 15 minutes each.
5. Wash twice in 70% ethanol, twice in 95% ethanol and air dry.

Autoradiography

Add photographic emulsions to the slides and after a suitable exposure period, develop the slides, counterstain and add cover slips.

Analyze the slides by determining the location of the radioactive probe on the chromosomes or within the nuclei.

EXERCISE 1.8. CULTURING PERIPHERAL BLOOD LYMPHOCYTES

Principle

Peripheral blood lymphocytes are incubated in a defined culture medium supplemented with serum, phytohemaglutinin, and other additives for the purpose of preparing metaphase chromosomes.

Time Required

72 to 96 hours.

Special Reagents

1. McCoy 5A (TissueSupport Centre)
2. Fetal Bovine Serum (Tissue Support Centre)
3. L-Glutamine, 200 mM, 100X, (Tissue Support Centre)
4. Phytohemaglutinin
5. Gentamicin Reagent Solution (50mg/mL), liquid
6. Vacutainer, green top, sodium heparin
7. Heparin, sodium salt 300 USP.

Safety Considerations

Because we do not test the cell lines or blood samples, always work under the assumption that the cells carry infectious agents, e.g., HIV virus, Hepatitis B, etc. Keep the samples isolated, work only in the biological safety hoods, and always wear gloves. Autoclave all waste materials.

Procedure

Day 1

1. Collect blood by venipuncture in one or two 7 ml sterile venipuncture tubes coated with sodium heparin. Rotate the tubes to prevent clotting. Label tubes with individual's name, date and time of drawing the specimen. Store at room temperature (never place in refrigerator or on ice).
2. Label 8-10, 15 ml centrifuge tubes, then pipette 10 ml McCoy 5A growth medium to each labelled tube. Add 9-12 drops of whole blood to each, using a sterile 5 3/4" pasteur pipette. Cap the centrifuge tubes tightly and mix gently by inverting tubes. Loosen the caps on the 15 ml tubes and incubate at 37°C in a centrifuge rack positioned at a 45 degree angle for approximately 72 hours.

Days 2-3

1. Mix the blood by inverting the tubes twice daily during the 3 day incubation period.

Note. If clumps form at the bottom of a culture tube, 5-10 drops of heparin (300 USP) can be added with a 1 cc syringe.

Solutions

- McCoy 5A growth medium:

McCoy 5A	1000.0 ml
Phytohemaglutinin	10.0 ml
L-Glutamine	10.0 ml
Fetal Bovine Serum	100.0 ml
Gentamicin	1.2 ml
	1121.2 ml

- Filter sterilize with 0.22 μm cellulose acetate membrane; store medium at 4°C for up to 2 weeks; or freeze medium in 50 ml centrifuge tubes at –80°C for up to 1 year.
- *Phytohemaglutinin,* M Form, PHA, lyophilized

  Rehydrate with 10 ml sterile deionized distilled water (ddH$_2$O). Store at +2°C to +8°C. After reconstitution, the phytohemaglutinin solution is good for 14 days.
- *Heparin,* sodium salt 300 USP:

  Rehydrate with 5 ml of McCoy growth medium. Store at 4°C for up to 2 weeks.

EXERCISE 1.9. MICROSLIDE PREPARATION OF METAPHASES FOR IN-SITU HYBRIDIZATION

Purpose

Metaphases are prepared from lymphocytes for in-situ hybridization experiments. The lymphocytes, already suspended in fixative, are dropped from a given height onto cleaned slides. The best preparations of metaphases are obtained when the cells are dropped onto slides immediately after the last wash of the harvesting procedure. The slides are then air dried, labelled, rinsed in alcohol, and stored in slide boxes at –80°C.

Time Required

In 3-4 hours approximately 80-100 microslides can be prepared for in-situ hybridization.

Special Reagents and Supplies

1. Microslides, 25 × 75 mm
2. Reagent absolute alcohol
3. Solution A or Phosphate buffered saline
4. Diamond scriber
5. Microslide Box, for 100 slides
6. Coplin staining dishes, Quanity 4
7. 9" × 9" Technicloth wipes
8. Forceps.

Special Equipment

- Nikon TMS inverted microscope (Frank E. Fryer Company) equipped with 20X and 40X phase contrast objectives and a green filter. The microscope is also equipped with L20 and L40 phase annulus for the LWD condenser.

Procedure

1. Presoak 6-8 microslides (slides) in absolute alcohol a coplin staining dish (coplin jar) for several minutes.
2. Aspirate the old fixative above the cell pellet and resuspend the pellet in 1.5 ml of "fresh" fixative.
3. Using forceps, remove 1 slide from the coplin jar and wipe dry with a 9" × 9" wipe.
4. Scratch 2 parallel lines approximately 22 mm apart on the lower third of unfrosted "back" side of slide:

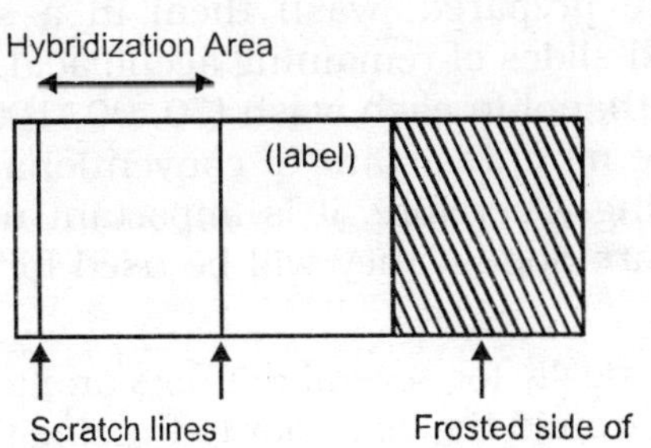

The scratches will define the area of hybridization of probe DNA and metaphase DNA.

Note. The room temperature, the barometric pressure, and the humidity will affect the drying time of cells and the quality of the metaphases. It is essential to try to control all the variables and strive to achieve the optimum conditions. The use of vaporizers on days when the air is extremely dry and the use of air conditioners when the room temperature is above 65°F, will help to produce high quality metaphases. *Optimum conditions are: room temperature between 60–65°F, > 50% humidity (a rainy day) and low barometric pressure.*

5. Drop 100 µl of the cell suspension from 3-4" above the slide surface onto the frosted side of the slide. Drop the suspension slowly, 1 drop at a time, moving the pipetman so that no two drops fall exactly on the same surface area (but all 100 µl should fall within the scratched area on slide). The fixative will quickly evaporate from under and around the cell and the cell will flatten completely, forcing the chromosomes to spread. Different methods of making slides may speed up or slow down this evaporative process, causing more or less spreading. Throughout the whole process, the cell membrane is present. If a cell breaks open during any stage of the harvest or slide making, some or all of the chromosomes will spill out, causing hypodiploidy with random loss of chromosomes. In severe cases, the slides will have "scattered" chromosomes or completely lack chromosome spreads. To shorten the evaporation time and prevent "scattering" of chromosomes, hold the slide at a 45° angle. Well spread chromosomes will also have less cytoplasm around each chromosome.
6. Air dry the slide either at a slant if the chromosomes are well spread, or horizontal if the chromosomes are not well spread. Quickly scan every slide, using the 20X objective, green filter and L20 phase annulus. Be careful when viewing the slides; scratching the surface will ruin chromosome material. Check to see if the chromosomes are spread sufficiently and that no cytoplasm is observed (halo or dark area around chromosomes). If cytoplasm is detected, disperse it by washing the cells several more times with "fresh" fixative before preparing more slides. Also, check to see if the suspension it too concentrated or is too dilute. Concentrated cell suspensions will produce underspread metaphases, while dilute suspensions are very time consuming to scan for metaphases.
7. In a coplin jar, wash 5-6 slides at a time in 50 ml of PBS for 5 minutes. Always be careful not to scratch the slides.
8. After the slides are prepared, wash them in a series of ethanol rinses, 5 minutes each, to rid slides of remaining acetic acid. This is done by increasing the percentage of ethanol in each wash (70, 90, 100%). The chromosomes will harden and become more resistant to conventional banding procedures, *e.g.*, G or Giemsa banding. Therefore, it is important not to wash the slides after slide preparations are done if they will be used for procedures other than in-situ hybridization.
9. Air dry the slides vertically for several minutes and transfer them to a microslide box. Wrap parafilm around the box prior to freezing the slides at –80°C (prevents moisture from entering). Slides can be stored for 1 year at –80°C.

Note: Do not store slides at –20°C or at +4°C, because the quality of the chromosomes deteriorates.

EXERCISE 1.10. STAINING CHROMOSOMES (G-BANDING)

Method

To stain metaphase chromosomes with Giemsa or Leishman's stain to elicit a banding pattern throughout the chromosome arms, designated G-Bands. This G-Banding technique requires a chromosomal pretreatment step of trypsin to induce chromosome bands.

Time Required

30-60 minutes

Special Reagents

- Leishman stain
- Gurr Buffer Tablets

Procedure

1. Prepare the staining solution the day prior to use. Also, slides should be aged at least 7-10 days or placed in a 55-65°C oven for 45 minutes before staining, to ensure excellent banding patterns. Aging the slides helps to eliminate fuzzy banding and increases contrast of the bands.
2. Exact timing is important; therefore, a maximum of 5 slides should be stained at one time. Optimum time in the stain appears to be between 2.5-4 minutes. It is necessary to determine the approximate staining times for each bottle of stain solution. The exact time will vary by several seconds depending on the source of cells, age of slides, the cell concentration on the slide, etc. (refer to the table below).

	Trypsin Time (seconds)	*Staining Time (minutes)*
Cell Source		
Lymphoblastoid	30	4.0
Blood Lymphocytes	15	3.0
Age of Oven Dried Slides		
0-3 days	15	3.0
3-20 days	30	3.5
20+ days	45	4.0
Previously Banded	45	4.0
Cell Concentration		
< 20 mitosis	15	3.0
20-50 mitosis	30	3.5
50+ mitosis	45	4.5

Note: These are approximate times, and test slides need to be done to determine trypsin and staining time for each cell line.

3. Mix 1 ml 0.25% stock trypsin with 49 ml 0.85% NaCl (Working Salt Solution). Wait 4 minutes before beginning to stain to allow the trypsin to dissolve.
4. Dip oven-dried slides that have cooled to room temperature in the trypsin solution for 5-30 seconds. The time in trypsin is dependent on slide preparation

conditions, harvesting conditions, material being banded, etc. Stain a test slide first to determine optimum conditions.

5. Rinse slides in 50 ml working salt solution.
6. Use a graduated cylinder to mix 15 ml Leishman stain and 45 ml Gurr buffer just prior to staining, and pour into a coplin staining jar. Stain the slides for 3-4 minutes.
7. Rinse slides in distilled water and air dry with compressed air or use a blow dryer.
8. Check the slides using a Zeiss Microscope, 100X plan-apochromatic oil objective, brightfield. See "brightfield photography".
   - Over-trypsinized chromosomes appear fuzzy; somewhat difficult to recognize exact bands.
   - Under-trypsinized chromosomes will have indistinct bands, decreased contrast; very difficult or impossible to determine bands.
   - Adequately trypsinized chromosomes will show telomeres not overly digested and G-Bands will appear sharp and in contrast.
9. If slides are under-treated with trypsin, destain the slides before rebanding. Quickly dip the slides in 3:1 methanol:acetic acid 2-3 times, or until all stain is removed. Rinse in distilled water, air dry and reband.
10. 10 (optional) coverslip with 24 × 50 mm #1 coverslip using permount.
11. Transfer slides to microslide box or suitable storage container.

Solutions

- *Trypsin (1:250) USP Grade; porcine parvovirus tested.* Prepare a 0.25% (1X) solution by dissolving 0.25g trypsin in 100 ml PBS. Aliquot into 1 ml quantities and store at –20°C for up to 1 year. Thaw just before using. Lowering the temperature of the trypsin in the working salt solution will lower its activity, thus requiring longer exposure times and also increasing the life of the solution. Test slides should be run to determine optimum exposure times.
- *Gurr Buffer Solution: (1 litre).* 1 tablet is dissolved in 1 litre distilled water. Produces a solution of pH approximately 6.8 at 20°C. Use solution at room temperature.
- *Working Salt Solution: (1 litre).* Dissolve 8.5g of NaCl in 1 litre distilled water. Use at room temperature.
- *Leishman Stain Solution.* Dissolve 1g Leishman stain into a 500 ml bottle of methanol. Stir for 4 hours and allow stain to age at least one day prior to use. If stain is not dissolved, filter stain through a 0.45 µm filter.

2. NUCLEIC ACIDS

Introduction

The previous exercise dealt with differentiation, and specifically with the events leading to differential gene activity. In molecular terms, this process involves regulation using DNA as a primer molecule for the selective synthesis of RNA. DNA primed RNA synthesis is termed "transcription."

Transcription is a complex series of reactions which involve the use of an RNA polymerase enzyme, known as transcriptase. If the reaction occurs in reverse, that is with an RNA primer synthesizing DNA, the process is known as reverse transcription, and the enzyme is reverse transcriptase. This latter process is important for RNA virus replication in general, and is most significant when examining oncornagenic virus. Oncornagenic virus are RNA-containing virus that are also causal agents for some forms of cancer (primarily in birds). Reverse transcription is also important in the development of amphibians, and in the process of gene amplification.

EXERCISE 2.1. EXTRACTION OF DNA FROM BOVINE SPLEEN

Materials

- Bovine spleen
- Saline Citrate Buffer (SSC)
- Chilled blender
- Refrigerated preparative centrifuge
- 2.6 M NaCl
- 95% (v/v) ethanol.

Procedure

1. Weigh out approximately 15 grams of frozen bovine spleen. Record the exact weight for future reference.

   Weight of the liver ________ gm

2. Drop the pieces of spleen one at a time into a chilled blender containing 150 ml of cold citrate-saline buffer (SSC). Continue to homogenize the tissue until it is thoroughly macerated. Do not over-homogenize and allow the contents to warm up. Proper procedure should take about 30-60 seconds of blending.
3. Pour the homogenate into nalgene centrifuge tubes and centrifuge at 4,000 xg for 15 minutes at 4°C.
4. Decant the supernatant and discard.

   The supernatant contains most of the materials that are soluble in physiological buffer. RNA, protein and many carbohydrates are found in this portion. The pellet contains most of the DNA, but it is complexed and in the form of DNP. The pellet also contains any cell debris and unbroken cells resulting from the homogenization.
5. Resuspend the pellet in about 20 ml of saline/citrate buffer by gently stirring with a glass rod. If the pellet is packed hard and will not disperse easily, you may use a vortex mixer to aid in the dispersion.

6. Recentrifuge as in step 3. Discard the supernatant.
7. Add 20 ml of cold 2.6 M NaCl. Break up the pellet with a glass rod, close the centrifuge tube with a tight fitting cap and shake vigorously. DNA is soluble in cold NaCl and will also dissociate from the protein. Pour off the liquid portion from this procedure and save for next step. Add another 20 ml of cold 2.6 M NaCl and shake vigorously. Continue to do this for two more extractions. It is important that the salt be kept cold (use an ice bath) and that the shaking be vigorous. Breaking the pellet with a glass rod may also help.
8. Combine all four extractions from above and centrifuge at 20,000 xg for 20 minutes. This will pellet the insoluble proteins.
9. Pour the supernatant carefully into a litre beaker and slowly add 2.3 volumes of cold 95% ethanol allowing it to pour down the side of the beaker and layer on top of the aqueous supernatant.
10. Collect the DNA by gently stirring the mixture together. DNA, if it is highly polymerized, will "spool" onto a clean glass rod as the salt solution is mixed with the alcohol. It can then be removed from the solution in the beaker, washed twice with cold 70% ethanol and placed into 70% ethanol or lyophilized for storage.

Notes

DNA spooled by this procedure is impure. Before it can be used for further analysis, care must be taken to remove contaminating protein, RNA, and carbohydrate. There are a number of means to accomplish this, most involving either enzyme digestions (pronase, amylase, and RNAse) or differential salt solubility or combinations of these techniques.

EXERCISE 2.2. PURIFICATION OF DNA

Materials

- Spooled DNA from exercise 2.1
- SSC buffer
- Pancreatic Ribonuclease A (100 µg/ml)
- Pronase
- Sodium lauryl sulfate (SLS)
- Sodium perchlorate
- Chloroform : isoamyl alcohol (24 : 1)
- 95% and 70% (v/v) ethanol
- Refrigerated centrifuge, rotor and tubes.

Procedure

1. Decant off the alcohol, and dissolve the extracted DNA in 30 ml of diluted SSC buffer (0.1 X SSC) in a 125 ml erlenmeyer flask. This will require some time, as polymerized DNA dissolves slowly. Gentle swirling of the material will help.
2. Add pancreatic ribonuclease A to a final concentration of 100 micrograms/ml and agitate slowly at 37°C for 1 hour.

3. Add pronase to a final concentration of 50 micrograms/ml and again agitate slowly at 37°C for another hour.
4. Add sodium lauryl sulfate (SLS) to make a 1% concentration (w/v) and sodium perchlorate to a final concentration of 1 M. Agitate for 30 minutes at room temperature.
5. Extract the solution with chloroform : isoamyl alcohol (24 : 1 v/v) by adding an equal volume and shaking vigorously for at least 15 minutes.
6. Place the solution into appropriate centrifuge tubes and centrifuge for 5 minutes at 800 xg and 4°C.
7. Remove the upper aqueous phase, add 2.3 volumes of 95% cold ethanol and respool the DNA from this solution onto a glass rod.
8. Wash the spooled DNA twice with cold 70% ethanol and store for future analysis.

EXERCISE 2.3. CHARACTERIZATION OF DNA

Materials

- DNA sample
- SSC buffer
- UV spectrophotometer and quartz cuvettes.

Procedure

1. Dissolve a small quantity of your extracted DNA in 3.0 ml of 0.1X SSC.
2. Turn on and blank a UV spectrophotometer at 220 nm (use 0.1X SSC as the blank). Determine the absorbance of your sample DNA at 230 nm.
3. Change the wavelength to 230 nm, reblank the spectrophotometer and measure the absorbance of the sample at 230 nm.
4. Increment the wavelength by 10 nm and repeat blanking and measuring the absorbance until readings are taken through 300 nm.
5. Compute the absorbance ratio 260 nm to 280 nm. Pure DNA (without protein or RNA) will have a 260 : 280 absorbance ratio of 1.85. RNA will have a 260 : 280 ratio of 2.0.
6. Plot the absorbance spectrum of your sample and indicate the 260 : 280 ratio, as well as the amount of protein contamination on the graph.

Wavelength	*Absorbance*
220	
230	
240	
250	
260	
270	
280	
290	
300	

EXERCISE 2.4. DNA-DISCHE DIPHENYLAMINE DETERMINATION

Material

- Lyophilized DNA standard
- Sample DNA
- SSC
- Dische diphenylamine reagent
- Spectrophotometer.

Procedure

1. Weigh out 15.0 mg of commercial lyophilized DNA and prepare a stock solution of 3.0 mg/ml by dissolving the DNA in 5.0 ml of SSC. This material will be used to prepare a standard curve for the diphenylamine reaction.

   Note that lyophilized, highly polymerized DNA is extremely slow to go into solution. It will require preparation at least one day in advance of lab with constant shaking.

2. Prepare a series of known standard solutions by serially diluting the stock solution of DNA. Set up a series of test tubes containing 2.0 ml of SSC each. Pipette 2.0 ml of stock solution into tube #1, mix and pipette 2.0 ml of the resulting mixture into tube #2 and so on. This will yield a series of tubes containing 1.5, 0.75 and 0.375 mg/ml of DNA. Your original stock solution is 3.0 mg/ml and SSC should be used for the blank.

3. Remove and discard 2.0 ml of the final dilution. To each of the five tubes containing in step 2 (each should contain only 2.0 ml), add exactly 4.0 ml of Dische diphenylamine reagent and mix well.

   This Reagent contains glacial acetic acid. It is caustic and should be handled with care.

4. Place a marble on the top of each test tube (it should not fall into the test tube, as it will act as a reflux to prevent evaporation, while allowing for pressure changes). Place the tubes in a boiling water bath for 10 minutes, remove from the bath and immediately immerse in an ice bath to cool.

5. Turn on a spectrophotometer and adjust the wavelength to 650 nm. Use the tube containing no DNA from step 2 to blank the instrument and measure the absorbance of each of your standards.

   Plot the absorbance against DNA concentration, perform a linear regression of the data and compute the extinction coefficient.

6. Dissolve your extracted or sample DNA in 10 ml of SSC. Make serial dilutions of 1/10, 1/100 and 1/1000 with SSC. Measure the absorbance of your extracted or sample DNA dilutions and calculate the concentration of DNA in the sample. Use the dilution which gives an absorbance in the 0.1 to 1.5 range.

EXERCISE 2.5. MELTING POINT DETERMINATION

Materials

- DNA
- SSC
- UV spectrophotometer (preferably with temperature control)

Procedure

Dissolve your DNA preparation in SSC to give a final concentration of approximately 20 μg DNA/ml.

1. Place the dissolved DNA in an appropriate quartz cuvette along with a second cuvette containing SSC as a blank.
2. Place both cuvettes into a dual beam temperature regulated UV spectrophotometer and measure the absorbance of the sample at 260 nm at temperatures ranging from 25° C to 80° C. Continue to increase the temperature slowly and continue reading the absorbance until a sharp rise in absorbance is noted.

   Alternatively:

   (*a*) Place the cuvettes into a waterbath at 25°C and allow to temperature equilibrate. Remove the blank, wipe the outside dry and rapidly blank the instrument at 260 nm. Transfer the sample to the spectrophotometer (be sure to dry and work rapidly) and read the absorbance.

   (*b*) Raise the temperature of the bath to 50° C and repeat step (a).

   (*c*) Raise the temperature sequentially to 60° C, 65° C, 70° C, 75° C and 80° C and repeat the absorbance measurements.

   (*d*) Slowly raise the temperature above 80° and make absorbance measurements every 2° until the absorbance begins to increase. At that point, increase the temperature, but continue to take readings at 1°C intervals.
3. Correct all of the absorbance readings for solvent expansion relative to 25°C.
4. List the corrected values as A_t
5. Plot the value of A_t / A_{25} vs temperature and calculate the midpoint of any increased absorbance. This midpoint is the melting point (Tm) for your DNA sample.
6. Calculate the GC content of your sample using the formula

$$\text{Percent of G + C} = k(Tm - 69.3) \times 2.44$$

Notes

Single strand DNA absorbs more UV light than double strands. Moreover, double strands can be separated by heat (melted) and the temperature at which the strands separate (Tm) is related to the number of guanine-cytosine residues (each having three hydrogen bonds as opposed to the two in adenine-thymine). This has led to the development of a rapid test for an approximation of the GC/AT ratio using melting points and the change in UV_{260} absorbance (known as "hyperchromicity" or "hyperchromatic shift"). Of course, the separation is also dependent upon environmental influences, particularly the salt concentration of the DNA solution. To standardize this, all Tm measurements are made in SSC buffer. DNA melts between 85° and 100° C in this buffer (as opposed to 25° C in distilled water).

EXERCISE 2.6. CsCl-DENSITY SEPARATION OF DNA

Materials

- DNA
- CsCl

- 0.3 N NaOH
- 0.2 M Tris-HCl buffer, pH 7.0
- Ultracentrifuge and rotor
- UV spectrophotometer and cuvettes.

Procedure

1. Determine the G+C content of the sample DNA exercise 2.5
2. Once the G+C content is determined, the bouyant density of the DNA can be determined from the formula:

$$p = 1.660 \text{ g/cm}^3 + 0.098 \times (\text{G+C fraction})$$

Using the bouyant density and the CsCl Table in Appendix F, determine the concentration of CsCl salts to use for dissolution of the DNA.

3. Dissolve approximately 100 micrograms of DNA in 4.2 ml of the appropriate CsCl solution in 0.3 N NaOH.
4. Load the dissolved DNA/CsCl solution onto a centrifuge tube suitable for a 4.2 ml sample and speeds of 30-40,000 RPM (Beckman SW39 rotor, or equivalent).
5. For the Beckman SW39 rotor, centrifuge the material at 35,000 RPM for 65 hours at 22°C.
6. Collect the fractions in 0.1 ml steps.
7. Add 0.2 M Tris-HCl, pH 7.0 to each fraction and measure the A_{260} for each fraction. If available, a continuous flow system using a fraction collecting device may be used.

EXERCISE 2.7. PHENOL EXTRACTION OF rRNA (RAT LIVER)

Materials

- Rat liver (fasted rat)
- Liquid nitrogen
- p-Amino-salicic acid
- Phenol mixture
- Homogenizer or blender
- Refrigerated preparative centrifuge
- NaCl
- 95% and 70% (v/v) ethanol.

Procedure

1. Obtain a rat which has been fasted for 24 hours (to remove glycogen from the liver), decaptitate, exsanguinate and remove the liver as rapidly as possible.
2. Weigh the liver, don't allow it to dehydrate.
3. Immediately drop the liver into a container of liquid nitrogen.

   Caution: Liquid nitrogen will cause severe frostbite !

4. Using the weight of the liver as an indication of the volume (1 gm of liver equivalent to 1 ml), add 15 volumes of freshly prepared 6% para-amino-salicylate (pAS) to a chilled blender or homogenizer.
5. Add an equal volume (equal to the pAS) of phenol mixture to the blender and turn on the blender for a short burst to mix the pAS and phenol.

   Caution: Phenol is extremely caustic.

   Phenol causes severe skin burns, yet it is a local anesthetic. You will be unaware of the burn at first, except for tell-tale discoloration of the skin and blisters. You will become aware of the burn as the anesthetic properties wear off. Phenol also readily dissolves most countertops and all rubber compounds.
6. Stop the blender and add the frozen liver (handle the liver with long forceps, or tongs). Blend the entire mixture (pAS, phenol and liver) for 30 seconds at full speed. Do not blend for longer periods or you will sheer the RNA.
7. Carefully transfer the homogenate to a beaker and continue to stir the mixture for 10 minutes at room temperature.
8. Transfer the homogenate to nalgene centrifuge tubes and centrifuge the mixture at 15,600 xg at 4° C for 20 minutes.
9. Remove the centrifuge tubes and carefully separate the upper aqueous layer from the lower phenol layer. Take care that none of the white interphase material is mixed into the aqueous layer. The upper layer can most efficiently be removed by using a large hypodermic equipped with a long, large bore, square tipped needle. Should some of the interphase material be stirred into the aqueous phase, it will be necessary to repeat step 8.
10. Measure the volume of the aqueous layer and discard the phenol layer and interphase material.
11. Add 3.0 grams of NaCl per 100 ml of aqueous phase and stir until dissolved.
12. Add 0.5 volumes of phenol mixture to the aqueous phase, place into a suitable flask and shake vigorously for about five minutes. Recentrifuge as in step 8 above, but for 10 minutes.
13. Separate the aqueous phase and add 2.3 volumes of cold 95% ethanol. Allow the mixture to stand in the freezer until a precipitate forms.
14. Collect the RNA precipitate by centrifugation, wash once in 70% ethanol and store in 70% ethanol at 0-5° C.

Notes

Knowledge of transcription is based on our ability to extract "native" or functional RNA molecules from cells, with subsequent use of those molecules "in vitro." One of the earliest methods for this type of analysis is a phenol-detergent extraction of RNA coupled with separation of the various sized molecules of RNA with centrifugation in a gradient.

This basic procedure remains useful today, although there have been myriad additions and alterations to the procedure using a host of extraction techniques and separation procedures (such as electrophoresis or column chromatography).

EXERCISE 2.8. SPECTROPHOTOMETRIC ANALYSIS OF rRNA

Materials

- RNA
- UV spectrophotometer and cuvettes
- Alkaline distilled water.

Procedure

1. Dissolve 10 mg of commercial RNA in 250 ml of slightly alkaline distilled water. Use a volumetric flask and proper analytical technique. This will give a standard solution of 40 micrograms RNA/ml.
2. Prepare a series of dilutions so that you have 40, 20, 10, 5 and 2.5 micrograms of RNA per ml.
3. Turn on a UV spectrophotometer and adjust the wavelength to 260 nm. Use the alkaline water to blank the spectrophotometer at 260 nm.
4. Read the A_{260} of each of the standards. Plot the A_{260} vs the concentration of RNA and calculate the extinction coefficient.
5. Dissolve your isolated, precipitated RNA in 10.0 ml of alkaline water. Prepare a serial dilution for 1/10, 1/100, 1/1000 and 1/10000. Measure the absorbance of each at 260 nm and, using the dilution which gives a reading between .1 and 1.5 absorbance units, compute the concentration of RNA in your sample.

EXERCISE 2.9. DETERMINATION OF AMOUNT OF RNA BY ORCINOL METHOD

Materials

- RNA
- Alkaline distilled water
- Acid-orcinol reagent
- Boiling water bath
- Spectrophotometer and cuvettes.

Procedure

1. Prepare a series of serially diluted RNA standards as in Exercise 2.8, but having a range from 1.0 mg/ml down to 0.125 mg/ml.
2. Prepare a serial dilution of your sample RNA as in Exercise 2.8.
3. Place 3.0 ml of each standard and 3.0 ml of each serial dilution of the sample RNA into separate test tubes. Place 3.0 ml of alkaline water in a separate tube.
4. Add 3.0 ml of acid-orcinol reagent to each tube and mix well.
5. Add 0.3 ml of alcohol-orcinol reagent to each tube and mix well.
6. Place the tubes in a boiling water bath for 20 minutes with marbles placed on top of each tube to prevent evaporation. Cool the tubes by immersion in an ice bath at the end of the 20 minute period.

7. Turn on a spectrophotometer and adjust the wavelength to 660 nm. Blank the spectrophotometer with the alkaline water/orcinol tube. Measure the A_{660} of each of the remaining standards and diluted samples.
8. Plot the absorbance of the standards against the known concentrations. Calculate the extinction coefficient, and calculate the concentration of RNA in your sample. Use the dilution yielding an absorbance between 0.1 and 1.5 absorbance units.

EXERCISE 2.10. SUCROSE DENSITY FRACTIONATION

Materials

- 10% and 40% (w/v) Sucrose
- 0.02 M sodium acetate, pH 5.1, containing 0.1 M NaCl and 1 mM
- EDTA
- RNA
- Ultracentrifuge, swinging bucket rotor and tubes
- Centrifuge tube fractionating device
- UV spectrophotometer.

Procedure

1. Refer to Chapter Three for details of fractionation, and form a 10-40% linear sucrose gradient in a nitrocellulose tube.
   (*a*) Close all valves on the gradient device.
   (*b*) Place exactly 15 ml of 40% sucrose in the right chamber and 15 ml of 10% sucrose in the left chamber of the gradient device.
   (*c*) Open the flow from the right chamber to the centrifuge tube. Be sure there is a tube in place, that the flow is directed down the inside of the tube and that the magnetic stirrer is functioning.
   (*d*) Immediately open the valve on the left chamber and insure that sucrose is flowing from left to right, thereby diluting the 40% sucrose with 10% sucrose.
   (*e*) Allow the flow to continue until all of the sucrose enters the centrifuge tube.
2. Dissolve the RNA in 0.02 M sodium acetate solution to yield a final concentration of 250 µg/ml. The low pH of the solution helps to inhibit RNAse, while the salts will keep the RNA from forming large polymeric aggregates. Carefully layer 2.0 ml of the dissolved RNA onto the top of a sucrose gradient. This should be done by slowly allowing the solution to run down the side of the tube and onto the gradient. Be careful not to disturb the gradient.
3. Load the ultracentrifuge with the prepared tubes and centrifuge for the equivalent of 18,700 RPM for a Beckman SW27 rotor, for 20 hours at 4°C.
4. At the completion of centrifugation, remove the tubes and fractionate the contents into 1.0 ml fractions.

(a) Insert the nitrocellulose centrifuge tube into the plexiglass holder, but be careful NOT to puncture the bottom.

(b) Have a series of 30-35 test tubes ready to accept the effluent from the tubes. Each tube should be labelled and kept in order.

(c) Push down on the centrifuge tube (Gently!) in order to puncture the bottom of the tube and immediately begin to collect the effluent.

(d) Count the drops that fall from the device, and place 40 drops into each tube.

5. Using a UV spectrophotometer and microcuvettes, read the A_{260} of each fraction. Calculate the amount of RNA in each fraction.
6. Plot the concentration of RNA in each fraction against the fraction number. Based on the density of sucrose in each fraction, compute the density of the RNA in that fraction. Based on the relative size (greater density) of the RNA, determine the nature of the fractions (i.e. rRNA, tRNA).

EXERCISE 2.11. NUCLEOTIDE COMPOSITION OF RNA

Materials

- RNA sample
- 1 N and 0.1 N HCl
- Boiling water bath
- Whatman #1 filter paper (for chromatography)
- Chromatography tank
- 20 µl micropipette
- Acetic acid : butanol : water (15 : 60 : 25) solvent
- UV light source
- UV spectrophotometer.

Procedure

1. Place a portion of your RNA sample (approximately 40 mg, hydrated) into a heavy walled pyrex test tube. Add 1.0 ml of 1 N HCl and seal the tube.
2. Heat the tube in a boiling water bath for 1 hour.
3. Cool the tube, open it and place the contents into a centrifuge tube. Centrifuge the contents at 2,000 RPM in a clinical centrifuge to remove any insoluble residue. The supernatant contains your hydrolyzed RNA.
4. Prepare Whatman filter paper No. 1 for standard one-dimensional chromatography.
5. Using a micropipette, spot 20 µl of your hydrolyzate onto the paper, being careful to keep the spots as small as possible (repeated small drops are better than one large drop). Allow the spots to completely dry before proceeding.
6. Place the paper chromatogram into your chromatography tank and add the solvent (acetic acid:butanol:water). Allow the system to function for an appropriate time (approximately 36 hours for a 20 cm descending strip of

Whatman #1). Remove the paper and dry it in a circulating air oven at 40° C for about 2 hours.

7. Locate the spots of nucleotides by their fluorescence under an ultra-violet light source. Expose the paper chromatogram to a UV light source and outline the spots using a light pencil. The order of migration from the point of origin is guanine (light blue fluorescence), adenine, cytilic acid and finally, uridylic acid.

   Do not look directly at the UV light source. Use a cabinet designed to shield from harmful UV radiation.

8. After carefully marking the spots, cut them out with scissors and place the paper cut outs into separately labelled 15 ml conical centrifuge tubes. Add 5.0 ml of 0.1 N HCl to each tube and allow the tubes to sit for several hours to elute the nucleotides from the paper.
9. Pack down the paper with a glass rod (centrifuge in a clinical centrifuge if necessary) and remove an aliquot of the liquid for spectrophotometric assay.
10. Measure the absorbance of each of the four nucleotides at the indicated UV wavelength (having first blanked the instrument with 0.1 N HCl).

Base	*Wavelength*	*Molar Extinction Coefficient*
Guanine	250 nm	10.6
Adenine	260 nm	13.0
Cytidylic acid	280 nm	19.95
Uridylic acid	260 nm	9.89

11. Use the molar extinction coefficients to determine the concentration of each base in the sample. Calculate the percent composition of each base, and the purine/pyrimidine ratio.

3. ISOLATION OF GENOMIC DNA

EXERCISE 3.1. DNA EXTRACTION PROCEDURE-GENERAL

1. Grow cells overnight in 500 ml broth medium.
2. Pellet cells by centrifugation, and resuspend in 5 ml 50 mM Tris (pH 8.0), 50 mM EDTA.
3. Freeze cell suspension at –20°C
4. Add 0.5 ml 250 mM Tris (pH 8.0), 10 mg/ml lysozyme to frozen suspension, and let thaw at room temperature. When thawed, place on ice for 45 min.
5. Add 1 ml 0.5% SDS, 50 mM Tris (pH 7.5), 0.4 M EDTA, 1 mg/ml proteinase K. Place in 50°C water bath for 60 min.
6. Extract with 6 ml Tris-equilibrated phenol and centrifuge at 10,000X g for 15 min. Transfer top layer to new tube (avoid interface). Re-do this step if necessary.
7. Add 0.1 vol 3M Na acetate (mix gently), then add 2 vol 95% ethanol (mix by inverting).
8. Spool out DNA and transfer to 5 ml 50 mM Tris (pH 7.5), 1 mM EDTA, 200 g/ml RNase. Dissolve overnight by rocking at 4C.

9. Extract with equal volume chloroform (mix by inverting) and centrifuge at 10,000X g for 5 min. Transfer top layer to a new tube.
10. Add 0.1 vol 3M Na acetate (mix gently), then add 2 vol 95% ethanol (mix by inverting).
11. Spool out DNA and dissolve in 2 ml 50 mM Tris (pH 7.5), 1 mM EDTA.
12. Check purity of DNA by electrophoresis and spectrophotometric analysis.

EXERCISE 3.2. ISOLATION OF GENOMIC DNA FROM BACTERIAL CELLS

To isolate genomic DNA from the bacterial cells and visualizing the same DNA by gel electrophoresis.

Introduction

Genomic DNA preparation differs from the plasmid DNA preparation. Genomic DNA is extracted from bacterial cells by immediate and complete lysis where as plasmid DNA is isolated by Slow cell lysis to form a sphacroplast.

Principal

The procedure of genomic DNA extraction can be divided into 4 stages:

1. A culture of bacterial cell is grown and harvested
2. The cells are broke open to release their contents
3. This cell extracted is treated to remove all components except the DNA
4. There sulting DNA is then controlled

   The cell is lysed by adding guanidium thiocyanate and a detergent comprising solution A. It is then centrifuged to separate the RNA and proteins. The resulting supernatant mainly consists of genomic DNA and sometimes RNA. The DNA is precipitated using alcohol.

Materials Required

(*a*) Materials to be stored at –20°C. Bacterial cell pellet (in 1.5 ml tube)

(*b*) Materials to be stored at 40°C. Solution A, Solution B, control DNA (run only one control along with 5 samples)

(*c*) Materials to be kept at room temperature

1.5 ml capillaries, dispenser with tubing ethanol. Microcentrifuge, Irans illuminator.

Storage and Handling

1. Storage material according to the labelled temperatures in refrigerator, freezer compartments or room temperature
2. Store bacterial pellets in freezer/freezer compartments of the fridge on arrinal
3. Handle solution A with care, as it is corrosive in nature
4. All the reagenets are stable for a period of 4 months if stored under recommended conditions.

Procedure

1. Thaw bacterial cells to room temperature
2. Resuspend the cells in 700 ml of solution 'A' at room temperature
3. Stand at RT for 5 minutes and spin for 10 minutes at 10,000 rpm
4. Collect 500 ml of supernatant, avoid decanting the pellet
5. Add 1 ml of distilled ethanol (1000 ml) to 500 ml of supernatant. The tubes were mixed by investing till white strands of DNA was seen to precipitate.
6. Spun for 4 minutes at maximum speed and the supernatant or spool precipitated DNA was discarded with the help of a pipette tip and transferred into a fresh tube
7. The DNA pellet was washed with 95% alcohol and again ethanol was added and decoanted. Washes were repeated and a final wash was done with 75% ethanol and air dried for 5 minutes.
8. 100 ml of solution B was added and incubated for 5 minutes at 55.66°C to increase the stability of genomic DNA.
9. Spun 10 minutes at maximum speed to remove insoluble protrins. The supernatant was pipetted out into a fresh tube.
10. 25 ml of freshly extracted DNA is taken and 10 ml of gel loading dye is added and loaded into the wells.
11. 10 ml of control DNA was loaded and electrophorised along with experimental samples in 1% agarose gel. (Two controls can be run along with 10 samples).

Preparation of 1% Agarose Gel and Electrophoresis

1. 1X TAE is prepared by diluting appropriate amount of 50X TAE buffer with distilled water.
2. 0.5 gm of agarose was added to 5 ml of 1 X TAE in a 250 ml conical flask. It was boiled to dissolve agarose and cool to a warm liquid.
3. The combs of the electrophoresis set was placed such that the comb is about 2 cm away.
4. When the agarose gel temperature was around 600C, the cooled agarose solution was poured in the gel tank. Make sure that the agarose gel is poured only in the centre part of the gel tank and is 0.5-0.9 cm thick, without air bubbles. Keep the set undisturbed till. The agarose solidifies.
5. Once the gel has got solidified, 1X TAE buffer is poured solmly into the gel till the buffer level stands at 0.5-0.8 cm above the gel surface.
6. Wells are formed by gentle lifting the comb.
7. Power cords should be connected in such a way that red cord to the red electrode and black cord to the black electrode. Power should not be switched on before loading.
8. The samples are loaded into the wells. Recording which samples are bring loaded into which wells as lane 1, 2, etc. has been done. Power concentration is started after loading and the voltage has been set to 50 V.
9. The gel should run till the second dye from the well has reached 3/4th of the gel. Stained dye has been used for staining the gel after electrophoresis.

10. The staining dye used has been diluted in the ratio 1 : 6 units distilled water, before use.

Visualising DNA

1. After the run was completed the power supply was switched off and the cords were disconnected.
2. The gel was slowly removed by running a spatula along the walls of the gel tank and the gel bank was inverted on to our palm ensuring that the palm totally covers the central square. The palm should be held close to the gel to avoid breaking of the gel.
3. Ethidium bromide was used, but the gel in a small tray and the staining dye was poured on it. It should be made sure that the gel is completely immersed and the tray should be shaken slowly.
4. Staining dye has removed in a container and the gel was destined by washing with tap water several times till the DNA was visible as a dark hand against a light blue background.

Result

The genomic DNA has been successfully extracted and when it was run on a gel the distinct bands were visible.

Interpretation

The molecular weight of control DNA provided was around 50 Kb in size. Genomic DNA has a high molecular weight and so runs along the control DNA. If the shrinking has occurred during extraction, DNA band run below the control DNA. If RNA is present along with extracted DNA, it will be seen between the blue and purple dye.

EXERCISE 3.3. PREPARATION OF GENOMIC DNA FROM BACTERIA

Materials

- TE buffer
- 10% (W/V) sodium dodecyl sulfate (SDS)
- 20 mg/ml proteinase K
- Phenol/chloroform (50 : 50)
- Isopropanol
- 70% ethanol
- 3M sodium acetate PH 5.2 phase Lock gel (5 prime, 3 prime, Inc)

Procedure

1. Grow *E. coli* culture overnight in rich broth. Transfer 1.5 ml to a microcentrifuge tube and spin 2 min. Decant the supernatant. Repeat with another 1.5 ml of cells. Drain well onto a Kimwipe.
2. Resuspend the pellet in 467 µl TE buffer by repeated pipetting. Add 30 µl of 10% SDS and 3 µl of 20 mg/ml proteinase K, mix, and incubate 1 hr at 37°C.

3. Add an equal volume of phenol / chloroform and mix well by inverting the tube until the phases are completely mixed.

   Caution: Phenol causes severe burns, wear gloves goggles, and lab coat and keep tubes capped tightly.

   Carefully transfer the DNA/phenol mixture into a Phase Lock Gel™ tube (green) and spin 2 min.
4. Transfer the upper aqueous layer phase to a new tube and add an equal volume of phenol/ chloroform. Again mix well and transfer to a New Phase Lock Gel™ tube and spin 5 min. Transfer the upper aqueous phase to a new tube.
5. Add 1/10 volume of sodium acetate mix.
6. Add 0.6 volume of isopropanol and mix gently until the DNA gets precipitates.
7. Spool DNA onto a glass rod (or Pasteur pipette with a heat-sealed end).
8. Wash DNA by dipping end of rod into 1ml of 70% ethanol for 30 sec.
9. Resuspend DNA in 100-200 µl TE buffer. Complete resuspension may take several days.
10. Store DNA at 4EC short term, –20 or –80°C long term.
11. After DNA has dissolved, determining the concentration by measuring the absorbance at 260 nm.

EXERCISE 3.4. EXTRACTION OF GENOMIC DNA FROM PLANT SOURCE

Principle

The major given protocol describes a rapid method for the isolation of plant DNA without the use of ultra centrifugation. The DNA produced is of moderately high molecular Wright, serves as a suitable for most restriction end nucleases and is also suitable for genomic blot analysis (Delta portal et. al 1983).. This method also helps to extract DNA from small amount of tissues and large number of plant samples besides being very rapid.

Material Required

Plant tissue, young leaves and callus of hibiscus, pestle and motor, ice box and ice, DNA extraction buffer, 5M potassium acetate, 3M sodium acetate buffer, phenol choloroform, iso amyl alcohol, 100% ethanol (ice cold) 70% ice cold ethanol, sterile eppendorf, microcentrifuge and glass powder.

Procedure

1. 500 mg of young plant tissue was (stem/leaf/leaf callus of controlled or transformed plant), in a pre-coated mortor and pestle.
2. Grinding was carried out inside the ice box with ice, using 0.8 ml of extraction buffer and glass powder.
3. Rest of the buffer added, the ground extraction was tehn poured into a test tube.
4. 0.2 ml of 20% SDS was added and shaken and inculated at 65C-75C for 10-15 minutes.

5. 1.5 ml of potassium acetate (5 m) was added.
6. The mixture was shaken vigorously, vortexed kept on ice for 20 minutes.

Result

The DNA from plant cells has become extranded and a clear hands are seen when the DNA was seen on the gel.

Isolation of DNA from coconut endosperm

- *Introduction:* The distinctive property of DNA is its behavior upon denaturation. The native form of cellular DNA is a helical double stranded structure. When the native DNA is disrupted, the molecule looses its highly ordered structure and random coils results. Significant molecular dimensions accompany the disruption.
- *Principle:* DNA, the generic material can be isolated from the cells by phenol extraction method. Cell wall is denatured and chelated by SDS (Sodium dodecyl sulphate) or SLS (Sodium lauryl sulphate). The tris acts as a buffer. The change in pH of the medium is very less (about 0.3) because of tris. Action of nucleases is minimized by using EDTA. Phenol-chloroform treatment denatures DNA. Isopropanol is necessary for the proper precipitation of DNA, which after being renatured by sodium acetate is precipitated by cold absolute ethanol.

Materials Required

1. Lysis Buffer: 10 mM Tris (pH 8)
   5 mM EDTA
   1% SDS
2. Phenol
3. Chloroform
4. Isopropanol
5. 3 M Sodium acetate
6. Cold absolute alcohol
7. Sample (Coconut endosperm)

Procedure

1. Homogenize 250 mg of coconut endosperm in a mortar and pestle with lysis buffer and make up the volume to 5 ml.
2. Incubate at 50°C for five minutes.
3. Add 5 ml of 1 : 1 phenol-chloroform mixture.
4. Centrifuge for 10 minutes at 3000 rpm.
5. To the supernatant and equal volume of 24 : 1 chloroform-isopropanol mixture.
6. To the organic phase, add 120th volume of 3M sodium acetate and two volume of cold absolute alcohol.
7. A white turbidity is developed.
8. By stirring it with glass rod, fibrous DNA strand is collected on the glass rod.

Result

White DNA is observed in form of fibers. This is then confirmed with help of electrophoresis. On electrophoresis it has given a single band.

EXERCISE 3.5. EXTRACTION OF DNA FROM GOAT LIVER

Principle

DNA which is the genetic material can be isolated from cells and tissues by high salt or phenol extraction. The material gives fairly under graded DNA. Here chromatin and proteins are dissociated by treatment with SDS (Sodium dodecyl sulphate). The proteins are denatured by phenyl chloroform treatment. It becomes insoluble and can be prepared by the following chemicals.

Materials Required

Ethanol, Saline, EDTA, PH-8.0 (0.9% Nace and 0.01 M EDTA), 10% SDS, 80% phenol, 95% ethanol.

Procedure

1. 2.5 gm of lives tissue was homogenized in 25 ml of saline EDTA and transferred to a 250 mol conical flask.
2. To this, 2.5 ml of 10% SDS and 2.5 ml of 80% phenol was added and shaken well for 10 minutes.
3. The entire mixture was centrifuged at 10,000 rpm for 10 minutes. Aqueous layer was separated out.
4. To the remaining material (interphase + phenol phase) equal volume of chloroform, isoamyl alcohol (24 : 1) was added and shaken for 25 minutes.
5. It was again centrifuged at 10,000 rpm for 10 minutes.
6. The aqueous phase was collected.
7. To the aqueous phase, equal volume of 80% phenol was added and again centrifuged at 10,000 rpm for 10 minutes.
8. To the aqueous phase obtained, 95% ice cold ethanol was added and then the flask was shaken gently.
9. The DNA are precipitated as a white mass.

Result

By the addition of ice cold ethands DNA strands were obtained as white mass or turbidity.

EXERCISE 3.6. ISOLATION OF COTTON GENOMIC DNA FROM LEAF TISSUE

Collection and Storage of Cotton Leaf Tissue

1. Harvest cotton tissue of interest. Separate the different leaf sample in a plastic bag.
2. Chill the samples on ice for transport back to the laboratory.

3. Remove the midribs, if desired, then put the each sample in a 50 ml polypropylene tube (PPT).
4. Place the samples into –70°C until freeze-drying them.
5. Freeze-dry (See Lyophilization) the sample for 2 to 4 days.
6. Store the freeze-dried leaf tissue in a 50 ml PPT tightly-capped at –20°C or in a desiccator at room temperature.

Lyophilization of Cotton Leaves

1. Freeze cotton leaves sample at –70°C in a deep freezer.
2. Transport the sample in an ice chest box.
3. Check to see that drain plug is closed (left side of condenser) and be sure that (break) switch is off and ballast open.
4. Turn on bottom cool switch of the condenser.
5. Turn on top (pump, cool, control) switches of chamber and set "shelf temperature control" to –40 or –50°C. Then, wait for chamber temperature to drop to set level (it usually takes 1 to 3 hr).
6. Turn on bottom (pump) switch.
7. Place the frozen samples on trays. Arrange in even layers. Placing second layer of tubes in bottom 2 trays is OK. It may be fun to place the probes into samples to watch the changing temperature.
8. Close the chamber and wait for 1 to 2 minutes till the samples and chamber temperature equalize. During this time, ice crystals will coat metal parts of the chamber.
9. Turn on gauge switch and close ballast, and wait for vacuum to reach <100 mT (about 5 to 10 min). In the mean time, condenser temperature should be at least –40°C.
10. Set "shelf temperature control" to –20°C and let run for 3 hr at least or overnight.
11. Check the condenser periodically to be sure that the coil has not collected enough ice to lug it up (rarely a problem on a fresh run).
12. After the run is complete, turn off (gauge) switch and open ballast.
13. Turn on (break) and turn off vacuum (pump) switches simultaneously.
14. Remove the samples and turn off top (pump, cool, control) switches.
15. Turn off (break) switch.
16. Close the cap of samples tightly immediately.
17. Place samples in dry, air tight environment as soon as possible and store out of direct sunlight.

Procedure

1. Use fresh BEST or Lyophilize the leaf tissue.
2. Grind about 5 g leaf tissue with mortar and pestle in liquid N_2, transfer the powder into an 50 ml polypropylene tube (50 ml) and store at –20°C.
   (a) Add 25 ml of ice-cold *Extraction Buffer*, mix well, and put on ice until centrifugation at 3750 rpm for 20 minutes (4°C); remove supernatant [Use swinging bucket rotor].

(*b*) Add 10 ml *Lysis Buffer* and vortex to resuspend pellet; incubate in at 65°C water bath for 30 minutes. Mix the tubes periodically by stirring or rocking the tubes every 10 minutes.

(*c*) Add 12 ml of chloroform : octanol (24 : 1) and mix gently, inverting tube, until an emulsion forms.

(*d*) Centrifuge at 3750 rpm for 20 minutes (15°C). Transfer upper phase to new 50 ml tube through Miracloth.

(*e*) Add equal volume of cold isopropanol; mix gently; and let set at room temperature for more than 1 hour.

(*f*) Centrifuge at 2500 rpm for 10 minutes (RT). Decant supernatant, and let tubes air dry. Then add 10 ml cold 76% etahnol/0.2 M Na Acetate. Put at 4°C for 1 hour or overnight. *(good stopping point).*

(*g*) Centrifuge tubes (3750 rpm, 10 min). Carefully discard supernatent, and rinse the pellet with cold 70% EtOH. Let tubes dry upside down on paper towels (Keep track of tube labels!).

(*h*) Let pellets dissolve into 3 ml TE buffer. Vortex at speed 5, and incubate at 65°C for about 1 hour.

(*i*) Centrifuge at 3750 rpm for 10 minutes (15°C). Transfer supernatant to new 15 ml tube. Add RNase to a concentration of 20 μg/ml, mix gently and incubate at 37°C for 15 minutes.

(*j*) Add 0.1 volume of 3M Na Acetate and two volumes of 95% ethanol for DNA precipitation. Place at –20°C for 1-12 hours (overnight).

(*k*) Spool/scoop out the DNA on a 9" glass-hook pipette (new and unused) and let it air-dry under the hood or vacuum. If DNA cannot be spooled, centrifuge at 2500 rpm for 10 min to form DNA into pellet. Once pellet is dry, place pellet into 0.5 ml TE buffer in 1.5 ml microfuge tube for 30 min at 65°C.

(*l*) Determine both concentration and quality of the DNA with spectrophotometer with a 40μl/800μl dilution (1/20), and by running digested and undigetsted DNA in a 1% agarose gel.

Extraction Buffer (pH 6.0)

- 0.35 M glucose
- 0.1 M Tris HCl (pH 8.0)
- 5.0 mM Na-EDTA (pH 8.0)
- 2% PVP 40,000 MW
- 0.1% DIECA, diethyldithiocarbamic acid (disodium salt)
- 0.2% beta-mercapto-ethanol or $Na_2S_2O_5$ (add when used)

Lysis Buffer

- 0.1 M Tris HCl (pH 8.0)
- 1.4 M NaCl
- 20 mM Na EDTA (pH 8.0)
- 2% CTAB

- 2% PVP 40,000 MW
- 0.1% DIECA, diethyldithiocarbamic acid (disodium salt)
- 0.2% beta-mercapto-ethanol or $Na_2S_2O_5$ (add when used).

3.7. ARABIDOPSIS THALIANA DNA ISOLATION

Procedure: (all steps at 0-4°C unless indicated)

1. Harvest 100 g tissue (plants can be any age up to just bolting), which has been destarched by placing in the dark for 48 hr).
2. After washing (ice cold H_2O), chop into small pieces with a single-edge razor blade.
3. Add ice cold diethylether until it covers the tissue and stir for 3 min, pour into Buchner funnel to remove ether, rinse with cold H_2O.
4. Add 300 ml of buffer A (1 M sucrose, 10 mM Tris-HCl (pH 7.2), 5 mM $MgCl_2$, 5 mM 2-mercaptoethanol and 400 ug/ml ethidium bromide). Grind tissue with a Polytron (Brinkmann) at medium speed for 1-3 min until tissue is homogenized.
5. Filter through 4 layers of cheesecloth, then through 2 layers of Miracloth (Calbiochem).
6. Centrifuge 9000 rpm for 15 min in a Beckman JA-10 rotor.
7. Resuspend pellet in 50 ml of buffer A plus 0.5% Triton X-100 (Sigma) with a homogenizer (55 ml glass pestle unit).
8. Centrifuge in two 30 ml Corex tubes at 8000 rpm for 10 min in a Beckman JS-13 rotor.
9. Repeat step 7, centrifuge at 6000 rpm for 10 min in a JS-13 rotor.
10. Resuspend pellet in 10 ml of buffer A plus 0.5% Triton X-100. Layer crude nuclei over two discontinuous Percoll gradients constructed in 30 ml Corex tubes as follows: 5 ml layers containing from the bottom upward 60% (v/v) and 35% (v/v) percoll A:buffer A. Percoll A is made as follows: to 34.23 g sucrose add 1.0 ml of 1 M Tris-HCl (pH 7.2), 0.5 ml of 1 M $MgCl_2$, 34 µl 2-mercaptoethanol, and Percoll to a final volume of 100 ml Centrifuge in a JS-13 rotor at 2000 rpm after 5 min increase speed to 8000 rpm and centrifuge an additional 15 min. Starch will pellet, the bulk of the nuclei will band at 35%-65% interface and intact cloroplasts will band at the 0-35% interface.
11. The zone containing the nuclei is collected, diluted with 5-10 volumes of buffer A and pelleted by centrifugation at 8000 rpm for 10 min in a JS-13 rotor. Nuclei can be visualized by light microscopy after staining with 1/5-1/10 volume 1% Azure C (Sigma) in buffer A minus ethidium bromide.
12. Resuspend nuclei in 5-10 ml of 250 mM sucrose, 10 mM Tris-HCl (pH 8.0), 5 mM $MgCl_2$ by homogenization.
13. Add EDTA to 20 mM and TE (10 mM Tris HCl (pH 8.0), 1 mM EDTA) to a final volume of 20 ml. Add 1 ml of 20% (w/v) Sarkosyl. Add proteinase K to 50-100 ug/ml and digest at 55C until the solution clarifies (2 hr).
14. Add 21 g CsCl, when dissolved add 1 ml of 10 mg/ml ethidium bromide and transfer to two quick-seal Ti 70.1 tubes and centrifuge at 45000 rpm at 20C in a Beckman Ti70.1 rotor for 36-48 hours.

15. Remove banded DNA, extract with 1 volume of 3 M CsCl saturated isopropanol, repeat extraction until all ethidium bromide is removed, and dialyze four times against 1 L of TE plus 10 mM NaCl.

    Note: The inclusion of ethidium bromide is essential if high molecular weight DNA is desired.

EXERCISE 3.8. PLANT DNA EXTRACTION

A. thaliana has a very small haploid genome and this makes obtaining DNA somewhat difficult. The most notable problem is that DNA is usually contaminated with polysaccharide which inhibit restriction enzymes as well as other DNA modifying enzymes. This problem is most easily solved by using young plants which have not accumulated as much polysaccharide as older plants. The best results are obtained with plants that are two to three weeks post germinated.

1. Harvest plants using forceps-carefully remove any adhering soil by hand.
2. Grind up the following in a mortar and pestle until no large pieces of tissue remain:

   0.5-1.5 g plants

   0.5 g of glass beads (75-150 um) per gram of plants

   3 ml proteinase K buffer (0.2 M Tris (pH 8.0), 0.1 M EDTA, 1% Sarkosyl, 100 g/ml proteinase K)
3. Pour into 10 ml test tube. Incubate at 45-50 C for 1 hr.
4. Spin 10 min at top speed in table top centrifuge (~3000 rpm)
5. Decant supernatant to a fresh tube. Adjust volume to 3 ml with proteinase K buffer (with or without proteinase K).
6. Add 6 ml 100% ethanol at room temperature. Invert to mix.
7. Spin 10K rpm for 15 min in SS34 rotor. Discard supernatant.
8. Resuspend pellet in 3 ml Tris-Cl (pH 8.0), 1 mM EDTA (TE). Vortex to resuspend.
9. Extract with phenol, phenol:chloroform, chloroform.
10. Add 6 ml 100% ethanol. Invert to mix.
11. Spin 10K rpm for 15 min in SS34 rotor. Discard supernatant. Air-dry pellet briefly.
12. Resuspend in 4 ml TE. Vortex to resuspend.
13. Add 4,5 g CsCl, 10 mg/ml ethidium bromide and Mix.
14. Spin 53K rpm 16-20 hrs VTi65 20°C.

This protocol has been optimized for yield at the expense of high molecular weight DNA. The nuclear DNA can be separated from plastid DNA by running the gradients with Hoest dye, rather than ethidium bromide.

Plant DNA Isolation

1. Prepare 5-20 g of clean, frozen, young leaves taken from plants grown under controlled conditions and exposed to darkness for two days prior to isolation. Remove mid-ribs.
2. Grind leaves in stainless steel blender containing 150-200 ml of ice cold H buffer at maximum speed for 1 min.

3. Pour the homogenate in a 250 ml centrifuge bottle (on ice) while filtering through one layer of miracloth (Calbiochem) under four layers of cheesecloth (all previously wetted with 10 ml clod H buffer).
4. Centrifuge at 2000 g, 4 C, 20 min.
5. Discard the green supernatant and resuspend the pellet in 40 ml ice cold HT buffer.
6. Transfer to a 50 ml teflon tube (Oakridge OK) and centrifuge at 2000 g, 4 C, 10 min. Repeat until pellet of nuclei becomes greyish-white (1-3X). If anthocyanins are present in the plant, the pellet will be reddish-brown.
7. Resuspend the pellet thoroughly in 12 ml of HT buffer then add 12 ml of lysis buffer.
8. Immediately, add 23.28 g of powdered CsCl and incubate the tubes at 55-60° C for one hour with occasional inversion.
9. After solubilization of the CsCl, centrifuge the tubes at 28000 g, 15 C, 30 min.
10. Filter the supernatant through two layers of cheesecloth into a 38 ml quick-seal tube containing 1.47 ml EtBr solution using a 50 ml syringe and a 16 G needle as a funnel. Complete volume with CsCl solution.
11. DNA is recovered after centrifugation using standard procedures.
    - *1X TE:* 10 mM Tris-HCl (pH 8), 1 mM EDTA

      1X H: 10X H, 400 ml; sucrose, 684 g; β-mercaptoethanol, 8 ml; water to 4000 ml [pH 9.5]

      Lysis buffer: Na-sarcosine, 4 g; Tris base, 2.42 g; Na2EDTA, 2.98 g; water to 200 ml [pH 9.5]

      10X H: spermidine, 20.35 g; spermine, 27.8 g; Na4EDTA 83.24 g; Tris base, 24.2 g; KCl, 119.2 g; water to 1900 ml [pH 9.5]; Add Phenyl methyl sulfonyl fluoride (PMSF) solution [7 g in 100 ml of 95% ethanol]

      1X HT: 1X H, 1000 ml; Triton X100, 5 ml

      CsCl solution: CsCl, 97 g; 1X TE, 100 ml

      EtBr solution: EtBr, 1 g; water, 100 ml.

EXERCISE 3.9. PHENOL/CHLOROFORM EXTRACTION OF DNA

Materials

- Phenol:Chloroform (1:1)
- Chloroform

Procedure

1. Add an equal volume of buffer-saturated phenol:chloroform (1:1) to the DNA solution.
2. Mix well. Most DNA solution can be vortexed for 10 sec except for high molecular weight DNA which should be gently rocked. Spin in a microfuge for 3 min.
3. Carefully remove the aqueous layer to a new tube, being careful to avoid the interface.

4. Steps 1-4 can be repeated until an interface is no longer visible.
5. To remove traces of phenol, add an equal volume of chloroform to the aqueous layer.
6. Spin in a microfuge for 3 min.
7. Remove aqueous layer to new tube.
8. Ethanol precipitate the DNA.

EXERCISE 3.10. ETHANOL PRECIPITATION OF DNA

Materials

- 3 M sodium acetate pH 5.2 or 5 M ammonium acetate.
- DNA
- 100% ethanol.

Procedure

1. Measure the volume of the DNA sample.
2. Adjust the salt concentration by adding 1/10 volume of sodium acetate, pH 5.2 (final concentration of 0.3 M) or an equal volume of 5 M ammonium acetate (final concentration of 2.0–2.5 M). These amounts assume that the DNA is in TE only, if DNA is in a solution containing salt, adjust salt accordingly to achieve the correct final concentration. Mix well.
3. Add 2 to 2.5 volumes of cold 100% ethanol (calculated after salt addition). Mix well.
4. Place on ice or at –20EC for > 20 minutes.
5. Spin a maximum speed in a microfuge 10-15 min.
6. Carefully decant supernatant.
7. Add 1 ml 70% ethanol and mix Spin briefly.
8. Carefully decant supernatant.
9. Air dry or briefly vacuum dry pellet.
10. Resuspend pellet in the appropriate volume of TE or water.

EXERCISE 3.11. ISOLATION OF MITOCHONDRIAL DNA

1. Grind in mortar and pestle or Waring blender with 5-7 volumes buffer a per g tissue. Use MCE at 350 1/L, and if necessary, with 5 ml 1 M DIECA/L.
2. Squeeze through cheesecloth, two layers of Miracloth 10 min at 1000 g.
3. Decant supernatant and centrifuge 10 min at 15,900 g.
4. Resuspend each pellet in a few drops of buffer G with paint brush; combine; bring to about 10 ml/50 g, 15 ml/75 g.
5. 10 min at 1000 g; pour off most; swirl pellet to remove fluffy layer; combine.
6. Bring supernatant to 10 mM $MgCl_2$ (100 1 1M/10 ml). Bring to 20 g DNase/ ml (100 1 2 mg/ml/10 ml) 60 min 4°C.

7. Underlay shelf buffer, 20 ml/10-15 ml; always use 20 ml or more 20 min at 12000 g.
8. Resuspend in small volume shelf buffer with brush; bring to about 10 ml/50-100 g, 10 min at 15900 g.
9. Resuspend pellets in NN (lysis) buffer (4-5 ml/50-75 g).
10. Add SDS to 0.5% (250 l of 10%/5 ml NN). Swirl thoroughly.
11. Add proteinase K to 100 g/ml (20 mg/ml/5 ml NN). Swirl gently 60 min 37 C.
12. Add equal volume of 3:1 water-saturated phenol, chloroform-isoamyl alcohol mixture. Emulsify ca. 5 min 10 min at 7000 g.
13. Collect supernatant; repeat 17 and 18: 3 total extractions.
14. Final supernatant; add 0.1 volume 8 M Ammonium acetate; then add 2 volumes of absolute ethanol.
15. 60 min, –80°C; 10 min at 8000-9000 g; drain; add equal volume 70% ethanol; let sit 10 min; 10 min at 8000-9000 g; drain dry. Vacuum dry pellet, 30 min. Two small corex tubes are better than one 30 ml Corex.
16. Add 100-500 *l* 0.1X NTE, 10 *l* RNase mixture. Typically use 500 *l* per 50 g tissue.
17. Hydrate 30 min, 37°C.

A Plant Nuclear DNA Preparation

Procedure

This method can be used to prepare Petunia cv. Rose du Ciel nuclear DNA suitable for restriction enzyme analysis and cloning. Chloroplast DNA contamination will vary depending on the genotype. Greater chloroplast DNA contamination results when this method is used with the "Mitchell" line of Petunia.

1. Young leaves (20 g) from 1-3 month old Petunia plants are ground at 4°C with a minimal volume of 0.3 M sucrose, 5 mm $MgCl_2$, 50 mM Tris HCl (pH 8) (HB) in a mortar and pestle. The slurry volume is brought to 100 ml with HB and filtered through 2 layers, then 4 layers of cheesecloth.
2. The filtrate is centrifuged at 1000 g for 5 min and the pellet resuspended in 100 ml of HB and repelleted at 1000 g for 5 min. This pellet is resuspended in 100 ml of HB containing 2% Triton X-100 incubated at 4°C for 10 min and centrifuged at 1000 g for 5 min. The pellet is washed with 100 ml of BH with Tirton X100, recentrifuged, and the pellet cleaned of excess liquid with a paper towel.
3. The pellet is resuspended in 16 ml of 30 mM Tris-HCl (pH 8), 10 mM EDTA (RB), and transferred to a 100 ml flask. 2 ml of 10% (w/v) Sarkosyl is added along with 2 ml of 5 mg/ml pronase. The mixture is heated at 60°C for 5 min and incubated with gentle shaking at 37°C for 5-10 hrs. The solution is highly viscous and contains among other things, starch grains and nuclear debris.
4. After the incubation, measure the total volume of the lysate and bring it to a volume of 20 ml, Add 20 g of CsCl and dissolve completely over a period of about two hours at 4°C.
5. Bring the mixture to room temperature, add 2 ml ethidium bromide (10 mg/ml). Mix in gently and thoroughly.

6. Dispense in centrifuge tubes and centrifuge at 35K rpm for 30 hrs at 15 C. Collect the DNA band under long wave uv light.
7. Recentrifuge the DNA and collect, extract the ethidium bromide with RB-saturated isoamyl alcohol and dialyze against 5 mM Tris-HCl (pH 8), 0.25 mM EDTA. Store at 4°C.

EXERCISE 3.12. ISOLATION OF CHLOROPLAST DNA

1. Grind in blender with 7 vol LL for maize, 10 for sorghum. Use MCE at 350 g/L, and if necessary, 5 ml 1 M DIECA/L. Cut leaves into ca. 1" strips prior to blending.
2. Squeeze through cheesecloth, Miracloth.
3. Centrifuge in GSA/JA14 rotors, if necessary. 10 min at 500 rpm GSA; 10 min at 570 rpm JA14.
4. Pellet chloroplasts in GSA/JA14; 15 min at 3000 rpm GSA; 15 min at 3100 rpm JA14.
5. Resuspend with pipette, in LL without MCE, etc. (ca. 300 ml/250 g).
6. Centrifuge 10 min in SS34/SA600/JA17 rotors: SS34 600 rpm; SA600 530 rpm; JA17 560 rpm.
7. Pellet chloroplasts 15 min in SS34/SA600/JA17 rotors: SS34 3000 rpm; SA600 2700 rpm; JA17 2800 rpm.

   Note: Use of small rotors preferable.
8. Resuspend with pipette in LL, ca. 250 ml/100 g centrifuge as in 7.
9. Resuspend with pipette in LL, ca. 10 ml/100 g if possible.
10. Add $MgCl_2$ to 10 mM (100 l 1M/10 ml); DNase to 100 g/ml (250 l 2 mg/ml/ 10 ml). Incubate 60 min at room temperature.
11. Bring to about 250 ml/100 g with MM; centrifuge as in 7; resuspend with pipette in MM to ca. 80 ml; centrifuge as in 7.
12. Resuspend with pipette in NN; minimum volume, ca. 5 ml/100 g.
13. Add SDS to 0.5% (250 l of 10%/5 ml NN). Swirl.
14. Add Proteinase K to 200 g/ml (500 l of 20 mg/ml/5ml). Swirl.
15. Lyse 60 min at 37°C. (Go to Dellaporta for Sorghum).
16. Maize: 3 standard phenol:chloroform extractions, ethanol precipitation.

EXERCISE 3.13. DNA EXTRACTION OF RHIZOBIUM (CsCl METHOD)

1. Grow starter culture overnight in 5 ml GYPC medium.
2. Inoculate 200 ml GYPC and grow cells to mid-late log phase (visual).
3. Pellet cells by centrifugation. Resuspend cells in 5 ml 10 mM Tris (pH 8.0), 4 mM EDTA. Then add 150 ml more buffer to wash cells.
4. Re-pellet cells and resuspend in 4 ml 50 mM Tris (pH 8), 20 mM EDTA in a small flask. Place on ice for 15 min.
5. Add 0.4 ml proteinase K (2 mg/ml, fresh); swirl for 5 min. Then add 0.1-0.2 ml sarcosyl (10%). Swirl gently at 30-37C until cell lysis (about 30 min – 1 hr).

6. Extract 3X with equal volume of Tris-equilibrated phenol and centrifuge at 7,000X rpm for 15 min. Transfer top layer to new tube (avoid interface). Re-do this step if necessary.
7. Extract 2X with diethyl ether. Remove and properly discard top layer after centrifugation at 5,000X rpm for 5 min. Evaporate ether overnight in fume hood, or heat in 68°C water bath for 30 min.
8. Adjust aqueous phase to 8.7 ml with buffer. To this add 8.3 g CsCl and 0.9 ml ethidium bromide (10 mg/ml).
9. Extract with equal volume chloroform (mix by inverting) and centrifuge at 10,000X g for 5 min. Transfer top layer to a new tube.
10. Place in Quick-seal tubes and centrifuge in Ti50 rotor at 36,000 rpm for 48 hr.

EXERCISE 3.14. ISOLATION OF PLASMIDS

Introduction

Plasmids are extra chromosomal double stranded closed circular DNA present in many micro-organisms. Plasmids are usually present in the cell conferring extra ordinary properties to the cell like the ability to conjugate, conferring antibiotic resistance, degradation of xenobiotic substances. Production of substances to neutralize toxins etc.

In nature, Plasmids are large molecules with sizes ranging from 2Kb to 100Kb. Plasmids have origin of replication and multiplied by utilizing the long leaved enzymes of the host DNA polymerizes. However, the plasmids used in molecular cloning are different from the natural plasmids as they are small in size and offer multiple restriction sites.

Plasmid DNA needs to be extracted (rom bacterial hosts, mostly *E.coli* almost routinely in cloning experiments. Many methods have been described for successful extraction of Plasmid DNA, however they can be grouped into two categories, viz.

1. Miniprep method and

2. Two methods have been described here representing both the categories. The alkaline miniprep method is most useful for quick extraction of Plasmids mostly for analytical use, whereas, the large scale preparation, is most useful for preparative extraction.

Alkaline Miniprep Method

The principle of alkaline miniprep method involves the lysis of bacterial cell followed by SDS, NaOH treatment. The high pH of NaOH denatures the bacterial DNA but not the covalently closed circular Plasmid DNA. Neutralization of the high pH by sodium or potassium acetate whereby the bacterial DNA precipitates. The Plasmid DNA is then purified by organic solvent.

Growth of Bacteria and Amplification of Plasmids

Plasmids DNA can be isolated from bacterial culture, which is grown in liquid medium containing appropriate antibiotic. The bacterial culture should be grown in LB medium inoculated with a single colony picked from an agar plate. For low copy Plasmids like pBR322 chloramphenicol is to be added after the culture attains late log phase (A600=0.6) and shaken vigorously for several hours to amplify the Plasmid. However for very high copy number Plasmids like pUC series Plasmids, such amplification is not required.

Requirements

1. L-Broth: Trypton, 1%; Yeast extract, 0.5% NaCl, 1%
2. Antibiotics: Ampicillin, 50 mg/ml
3. Solution I : 50 mM glucose, 25 mM Tris, 10 mM EOT A, pH 8.0
4. Solution 11 : 0.2 M NaOH, 1% SDS
5. Solution III : 5M Potassium acetate, 60 ml; acetic acid, 11.5 ml; distilled water, 28.5 ml.
6. Phenol: Chloroform
7. Chloroform: isoamyl alcohol (24:1)
8. TE buffer: 10 mM Tris, 1 mM EDTA, pH 8.0
9. 70% and 100% ethanol.

Harvesting and Lysis of Bacteria

Harvesting. Transfer a single bacterial colony into 2 ml of LB medium containing the appropriate antibiotic in a loosely capped 15 ml tube.

Incubate the culture overnight at 37 degrees with vigorous shaking. Pour 1.5 ml of the culture into a centrifuge tube. Centrifuge at 12,000 g of 5 seconds at 4 degrees in a microphage.

Store the remainder of the culture at 4 degrees. Remove the medium by aspiration, leaving the bacterial pellet as dryas possible.

Lysis by Alkali

Resuspend the bacterial pellet in 100 ml of ice-cold Solution I by vigorous vortexing.

Solution I

50 mM glucose

25 mM Tris-CI (pH 8.0)

10 mM EDT A (pH 8.0)

Solution I can be prepared it hatches of approximately 100 ml autoclaved for 15 minutes at 10 lb/sq. and stored at 4°C.

Add 200 ml of freshly prepared Solution II.

Solution II

0.2 N NaOH (freshly diluted from 10 N stock)

1% SDS

Close the tube tightly and mix the contents by inverting the tube rapidly five minutes. Make sure that the entire surface of the tube comes in contact with Solution II. Do not vortex. Store the tube on ice.

Add 150 ml of ice-cold Solution III.

Solution III

5 M potassium acetate	60 ml
Glacial Acetic Acid	11.5 ml
Water	28.5 ml

The resulting solution is 3 M with respect to potassium Clod 5 M with respect to acetate.

Close the tube and vortex it gently in an inverted position for 10 seconds to disperse Solution III through the viscous bacterial lysate. Store the tube on ice for 3-5 minutes.

Centrifuge at 12, (XX) g for 10 minutes at 4°C in a microfuge. Transfer the supernatant to a fresh tube.

Add an equal volume of phenol:chloroform. Mix by vortexing. After centrifuging at 12, (XX) g for 10 minutes at 4° in a microfuge, transfer the supernatant to a fresh tube.

- Precipitate the double-stranded DNA with 2 volumes of ethanol at room temperature. Mix by vortexing. Allow the mixture to stand for 2 minutes at room temperature.
- Centrifuge at 12, (XX) g for 10 minutes at 4° in a microfuge.
- Remove the supernatant by gentle aspiration. Stand the tube in an inverted position on paper towel to allow all of the fluid to drain away. Remove any drops of fluid adhering to the walls of the tube.
- Rinse the pellet of double-stranded DNA with 1 ml of 70% ethanol at 4°. Remove the supernatant as described in previous step, and allow the pellet of nucleic acid to dry in the air for 1.0 minutes.
- Redissolve the nucleic acids in 50 ml of TE (pH 8.0) containing DNAase free RNAase 20 mg/ml. Vortex briefly. Store the DNA at –20°.

Important Notes

- The original protocol requires addition of lysozyme, before addition of solution 11, this is not necessary.
- Do not vortex the tubes after addition of Solution 11.
- If the Plasmid preparation is strictly for analytical purpose and non enzymatic manipulations are contemplated, then the phenol: chloroform step can be avoided.
- It is important to remove all the supernatant fluid after harvesting the bacterial pellet and all traces of ethanol etc. after precipitation.
- The phenol has to be tris saturated to pH 8.0 and of very good quality.
- Do not disturb the whitish interface while removing the upper aqueous phase after phenol and chloroform treatments.
- While washing with 70% ethanol do not break the DNA pellet. This step is meant for washing the pellet only to remove traces of ethanol and salts. If the pellet is disturbed at this stage it will be difficult to recover the DNA.
- This preparation will contain lot of RNA contamination. DNAse free RNAse may be added before the phenol: chloroform step to digest the RNA. Otherwise RNAse may be added along with the restriction enzyme during subsequent manipulations.

- Sometimes it may become difficult to dissolve the Plasmid preparation in IE. Keep in the freezer overnight. Next day DNA will easily dissolve.

EXERCISE 3.15. RNA ISOLATION

1. DEP treatment solutions.
2. Add 0.3 ml lysing buffer (0.1 M Hepes, 0.2 M EDTA, 10% SDS, 0.02 M EGTA (pH 7.5) [DEP-treated) to 3 ml bacterial solution.
3. Extract with 1 vol 1000:140:0.4 phenol:m-cresol:hydroxyquinoline.
4. Add directly to 2 volumes ethanol (~6.6 ml).
5. Precipitate for 16 hrs at –20°C.

RNA Preparation and Electrophoresis

Buffers needed:

A. 0.2 M MOPS (pH 7.0), 50 mM sodium acetate, 1 mM EDTA.

B. Formaldehyde, 37%.

C. Formamide - deionized till pH 7.0 and recrystallized at 0 C, stored at –20°C.

D. 50% glycerol, 1 mM EDTA, 0.4% bromphenol blue, 0.4% xylene cyanol.

PREPARATION OF VANADYL-RIBONUCLEOSIDE COMPLEXES WHICH INHIBIT RIBONUCLEASE ACTIVITY

Reagents

Vanadyl sulfate

Adenosine

Cytidine

Guanosine

Uridine

10 N NaOH

Procedure

For 100 ml vanadyl complex,

1. Make 10 ml 2 M vanadyl sulfate (0.366 g/ml)
2. To make 80 ml H_2O in a 250 ml flask, add 1.34 g adenosine, 1.45 g guanosine, 1.25 g uridine, and 1.25 g cytidine = 5 mM XRs
3. Boil in H_2O water bath with N_2 sparge (XRs dissolve but ppt. upon cooling).
4. Add vanadyl sulfate and continue to boil and sparge 1 min (solution should remain blue).
5. Turn off H_2O bath, continue to sparge, and add 8 ml 10 N NaOH, then 4 ml 1 N NaOH (ugly grey ppt.; redissolve when all NaOH added) check pH = 7-9 range.
6. Resulting 0.2 M vanadyl nucleoside solution stored at –20 C.
7. Upon thawing, redissolve ppt. at 65°C.

EXERCISE 3.16. RNA EXTRACTION METHOD FOR COTTON

1. Collect young expanding leaves (or other tissue) and freeze in liquid N2 and store at –80°C.
2. Pulverize tissue to a fine powder in a pre-cooled mortar and transfer to a glass homogenizer.
3. Heat RNA Extraction buffer to 80°C and add to frozen ground tissue at a ratio of 5:1 (buffer : tissue) and homogenize for 2 minutes.
4. Transfer homogenate to a 50 ml Oak Ridge tube containing 0.5 mg proteinase K per ml of extraction buffer and incubate with mild agitation on a rotary shaker at 100 rpm for 1.5 hr at 42°C.
5. Add KCl to 160 mM and chill on ice for 1 hour.
6. Centrifuge for 20 min at 12,000 g.
7. Filter supernatant through Miracloth and precipitate the RNA overnight in 2M LiCl at 4°C.
8. Centrifuge for 20 min at 12,000 g to collect RNA pellet. Wash pellet 2-3 times with 5 ml cold 2 M LiCl until supernatant is relatively colourless.
9. Resuspend RNA pellet in 2 ml 10 mM Tris-HCl (pH 7.5) and clarify by centrifugation for 10 min.
10. Add Potassium Acetate (KAc, pH 5.5) to 200 mM in RNA suspension and incubate for 15 min on ice.
11. Remove salt-insoluble material by centrifugation.
12. Precipitate RNA overnight by adding 2.5 volumes cold 100% ethanol and incubating at –20°C.
13. Pellet RNA, wash with 70% ethanol, dry briefly under vacuum, and suspend pellet in DEPC-treated deionized water of TE buffer.
14. Determine yield by observing absorbance spectra between 220 and 320 nm. Yield should be about 800-1200 μg/g (RNA/fresh tissue).

RNA Extraction Buffer

- 200 mM sodium borate decahydrate (Borax) pH 9.0
- 30 mM ethylene glycol bis(β-aminoethyl ether)-N,N´-tetraacetic acid (EGTA)
- 10 mM dithiothreitol (DDT)
- 2% polyvinylpyrolidone, Mr 4000 (w/v) (PVP)
- 1% sodium dodecyl sulfate (w/v) (SDS)
- 1% sodium deoxycholate (w/v)
- 0.5% Nonidet NP-40 (v/v) (NP-40)

Methods for Plant RNA Isolation

AMES/Chloroform extraction in our lab to extract total RNA from potato leaves for viroid analysis. The protocol is as follows:

AMES Buffer (for 200 mL): 11.7g NaCl, 160 mL dH_2O, Adjust pH to 6.0, 0.40 g $MgCl_2$, 8.21 g Na-acetate, 40 mL EtOH, 6.0 g SDS

We homogenize the plant material (usually leaves) and dilute the sap 1:5 in AMES. We then incubate for 30 min at 37°C followed by the addition of an equal vol.of chloroform. Vortex and store on ice. We then centrifuge the samples at 3000 rpm for 10 min. Total RNA is contained in the upper aqueous layer.

EXERCISE 3.17. ISOLATION OF RNA FROM BACTEROIDS

1. 3 g nodules (fresh or frozen in liquid N_2) were ground to a powder in mortar and pestle with liquid N_2.
2. To the powder was added ice cold 0.5 M mannitol (or sucrose), 0.05 M Tris (pH 7.5), 0.02 M Na succinate, 5 mM Na dithionite.
3. Pass the resuspension through one layer cheesecloth and rinse the cloth with 5 ml of the same buffer solution.
4. Spin filtrate in a sterile centrifuge tube and spin 5 min at 1500 rpm.
5. Respin supernatant for 5 min at 9000 rpm.
6. Resuspend pellet in LiCl buffer containing 10 mM vanadyl ribonucleosides and treat as stated above.

Reagents Required

1. 0.1 M LiCl,
2. 1% SDS,
3. 7 M urea,
4. 0.1 M Tris (pH 8.4),
5. 5 mM EDTA.

Extract with phenol:chloroform 3 times. Precipitate with 2 vol ethanol. Spin. Redissolve ppt. in 2 ml H_2O treated with 0.1% DEPC and autoclaved. Add 6 ml 4 M Na acetate (6) treated with DEPC. Incubate on ice 2 hr, spin 30 min at 8000 rpm, and wash ppt 3 times with ice cold 3 M Na acetate (6) treated with DEPC. Redissolve RNA in DEPC-treated H_2O, ppt. with 2.5 vol ethanol, and store as ppt. at –80°C.

EXERCISE 3.18. ISOLATION OF RNA FROM FREE-LIVING RHIZOBIA

Cells from 100 ml culture in early log phase GYPC were spun and resuspended in 1 ml sterile 25% sucrose. This solution was transferred to a sterile centrifuge tube and 6.5 ml M-STET, 0.5 ml lysozyme (10 mg/ml), and 0.4 ml 200 mM vanadyl ribonucleoides were added.

M-STET: 4% sucrose, 6% triton X100, 0.06 M Tris (pH 8), 0.06 M EDTA

Incubate on ice 5 min. Boil 2 min [or add SDS/sarkosyl, extract with phenol:chloroform, and ppt.], chill, and spin 10 min at 8000 rpm. To the supernatant add 0.7 vol isopropanol, chill, spin again, and resuspend pellet in 5 ml.

5X Running buffer for RNA (formaldehyde) gels: (for 500 ml) 2 ml 0.5 M EDTA, 6.8 g Na acetate, 12.8 g MOPS (free acid), 9.0 g MOPS (base).

1. Induce cells normally –10 ml of 0.2 OD600.

   Wash cells (grown O/N in ORS minimal plus yeast at 30 or 37 C) 2 times with nif media and diluting to 0.2 OD (10 mls each). Check 1 bottle for induction by acetylene reduction.

2. After induction add 0.5 ml vanadyl ribonucleosides directly through septum into vial with a syringe (vanadyl protects cells against oxygen).
3. Put cells in plastic tubes, spin a for 2 min at 8000 rpm in Sorvall.
4. Resuspend in 10 M urea lysis buffer with plastic pasteur pipette.
5. Add 0.5-0.75 ml of phenol:chloroform (50:50) ["phenol" is phenol: m-cresol:hydroxyquinoline] to microfuge tube.
6. Vortex, place at 65 C for 2-3 min, repeat, vortex and spin 2-3 min in microfuge.
7. Re-extract liquid phase 2 more times at room temp with phenol:chloroform. Avoid proteinaceous interphase when removing upper liquid phase.
8. Extract with chloroform 1 time.
9. Precipitate with 100% ethanol, 0.3 M Na acetate (5.2) at -20 C O/N.
10. Collect ppt., wash 1 time with 70% ethanol, 1 time with 100% ethanol, and resuspend in 10 mM Na acetate (5.2) - treat for gels.

ESTIMATION OF DNA PURITY AND QUANTIFICATION

(Characterization) By Spectrophotometric Method.

Introduction

The DNA isolated from living cells are usually contaminated with protein, RNA and salts used during the isolation process. The purity of DNA may be estimated by utilizing the property of the heterocyclic rings of the nucleotides of absorbing light strongly in the UV range. DNA absorbs maximum light energy at about 260 nm. An optical density of 1.0 corresponds to approximately 50 j.lg/ml of double stranded DNA. The ratio of absorbance viz. A260/ A2so and A2SO/ A260 provides an estimation regarding the purity of DNA. A typically pure preparation of good quality DNA should exhibit the following spectral properties:

A260/ A2so ~ 1.80

A2so/ A260 ~ 0.55

Requirements

1. Sample DNA
2. TE buffer : Tris-HCl, 10 mM EDT A, 1 mM; pH 8.0
3. Spectrophotometer and quartz cuvette.

Procedure

To find out the purity of DNA, make appropriate dilution with TE buffer and measure the absorbance at 260 nm and 280 nm.

Notes

Do not use glass or plastic cuvette, as lights in the UV range do not pass through these.

Observations

1. Calculate A260/ A280 and A280/ A260, and check if the values are within acceptable limit.

2. Calculate the Dt~A concentration as follows:

   Concentration of DNA = (A260 × 50 × dilution factor) in Mg/ml.

EXERCISE 3.19. FUNGAL DNA ISOLATION

DNA successfully isolated from fungal species of Cochliobolus, Aternaria, and Fusarium. The key elements in this prepare (1) the use of young lyophilized mycelial mats....young mats (4 days growth for C. carbonum)...yield less contaminating carbohydrates and other misc. junk (2) lots of proteinase K in the extraction buffer to kill DNases (final =0.3 mg/ml).

Procedure

1. Place 0.2-0.5 g (dry wt) lyophilized pad in a 50 ml disposable centrifuge tube, break up the pad with a spatula or glass rod, add c. carbonum 5 ml 3 mm glass beads and powder the pad by brief shaking.
2. Add 10 ml (for a 0.5 g pad) of CTAB extraction buffer(see recipe, below), gently mix to wet all the powdered pad
3. Place in 65 C water bath for 30 min
4. Cool, add equal vol. $CHCl_2$: IAA (24 : 1)
5. Mix, centrifuge in table top fuge 10 min at full speed
6. Transfer aqueous supernatant to a new tube
7. Add an equal volume of isopropanol
8. High mw DNA should spool out upon mixing spool out the DNA with a glass rod or hook, pour out the remaining supernatant
9. Rinse the spooled DNA with 70% ethanol
10. Air dry, add 1-5 ml TE containing 20 ug/ml RNAse A. To resuspend the samples, place in 65°C bath, or allow pellets to resuspend overnight at 4°C

Notes

If the spooled DNA is discoloured or has contaminating mycelial debris, phenol/ chlorofrom extract and precipitate w/ethanol.

This protocol can be scaled down using 0.1 g pad in a 2 ml eppendorf tube.

For Southerns we routinely cut 50-75 ul (2-4 ug) of a standard DNA prepare in 200 ul volumes, ETOH ppt, and resuspend in 30 ul.

Even after digestion the resuspended DNA can be very viscous at room temp. To load a Southern we keep the samples in a 50-60 C heat block while loading to keep the samples at a lower viscosity.

CTAB extraction buffer: O.1M Tris, pH 7.5, 1% CTAB (mixed hexadecyl trimethyl ammonium bromide), 0.7 M NaCl, 10 mM EDTA, 1% 2-mercaptoethanol. Add proteinase K to a final concentration of 0.3 mg/ml prior to use. Less prot K may be acceptable for different fungi, and in fact we haven't determined if we can use less, this conc. was calibrated for a different C. carbonum DNA extraction buffer.

Mini Plasmid Preparation

1. Growth O/N in 1.5 ml LMM or Terrific broth (see Reagents) with 75 ug/ml Amp
2. Pour into eppendorf tube and spin down cells at 7-8 K for 2 min

3. Aspirate s/n and resuspend in 50 ul 25 mM Tris pH 8, 10 mM EDTA; leave lids open
4. Add 100 ul of freshly prepared 1% SDS, 0.2M NaOH (5 ml = 100 ul 10M NaOH added to 4.4 ml DDW then 500 ul 10% SDS). Add it forcefully and you don't need to vortex
5. Add 75 ul KoAC solution and vortex
6. Add 100 ul of phenol/CHI_3, close lids, vortex
7. Spin 13K for 2 mins
8. Remove supernatant, add to 500 ul ethanol. Vortex and spin at 13 K for 5 min
9. Aspirate s/n, removing all ethanol
10. Resuspend in 50 ul TE
11. Digest 2-5 ul, adding 1 ul preboiled 10 mg/ml RNAse A (see Reagents for preparation)

KOAc solution:

60 ml 5M potasium acetate

11.5 ml glacial acetic acid

28.5 ml DDW

Note: Large scale plasmid preps can be made quickly by scaling all reagents of the "mini prep" procedure 1000-fold. Once the phenol step has been completed treat the supernatant with 50 ul 10 mg/ml RNA'se for 1 hr, isopropanol ppte (0.6 volumes) and resuspend DNA in 400 ul of TE, reprecipitate. If the ppte is very cloudy (residual RNA) hook the DNA out (fluffy ball) to separate it from RNA.

Large Scale Plasmid Preps

1. Prepare an o/n culture of 100-200 ml in Terrific broth with 75 ug/ml ampicillin.
2. Spin down cells 5K, 5 min.
3. Resuspend in 5 ml of 25 mM Tris, 10 mM EDTA pH 8.0 per 200 ml.
4. Add 10 ml of freshly prepared 1% SDS, 0.2M NaOH (50 ml = 1ml 10M NaOH added to 44 ml DDW then 5 ml 10% SDS). Swirl to lyse cells completely and then add 7.5 ml KoAc solution per 200 ml, Spin 7 K 10 min.
5. Pour supernatant through cheese cloth into SS34 tube and add 11 mls of isopropanol (for a 200 ml prep).
6. Spin 7K 10 min. Drain off all supernatant, being careful to wipe around the inside neck of the bottle to remove all the isopropanol.
7. Resuspend pellet in 2.4 ml TE. When dissolved thoroughly, add 4.4 g CsCl dissolve 1st (handy hint) then add and 40 ul 10 mg/ml EtBr. Make sure all CsCl is in solution.
8. Layer plasmid solution under 8.5 mls of CsCl/TE pl. 47 in a Ti70 tube and spin O/N at 50K. CsCl/TE [1.47] = Dissolve 76.68 gms of CsCl to every 100 mls TE.
9. Pull band in 1.0 ml or less. Extract 3 times against butanol saturated in TE (first extraction into ethidium waste, remainder into butanol waste).
10. Add 1.0 ml DDW, estimate volume and add 2.5 vol. ethanol (no salt). Do not chill. Spin at 3K in benchtop for 10 min (provided there was a good band to

begin with, otherwise in ultracentrifuge for 20' 20K). Pour off s/n and remove residue with a gilson, resuspend in 350ul TE. Read OD 260/280 on 5ul and reprecipitate remainder in ependorf with 35 ul 5M NaCl and 1.0 ml ethanol. Resuspend at 1 ug/1.

EXERCISE 3.20. METHYLENE BLUE DNA STAINING

Procedure

1. Load 2-5X the amount of DNA that would give bands of moderate intensity on an ethidium bromide stained gel. Typically this is something on the order of 0.5-2.5 µg of a 1 kb fragment on a 30 ml 1% mini gel. These numbers are guess-timates so your milage may vary.
2. Run the gel normally and then place in a 0.002% methylene blue (w/v, Sigma M-4159) solution in 0.1X TAE (0.004M Tris 0.0001 M EDTA) for 1-4 h at room temp (22°C) or overnight at 4°C. Diffusion of the DNA does not seem to be a problem for fragments as small as 100 bp (3% Nusieve: 1% agarose gel). This avoids background issues associated with staining with 0.02% methylene blue for 30-60 min and then destaining for what seems to be forever.
3. If destaining is needed to increase the visibility of the bands place the gel in 0.1X TAE with gentle agitation changing the buffer every 30-60 min until you are satisfied with the degree of destaining.

Notes

This method primarily eliminates the damage of DNA by UV irradiation. DNA isolated from MB stained gels should transform frozen competant *E. coli* (XL1-Blue and DH5) cells on the order of 20-50 fold more efficiently than EB isolated DNA. Factors influencing improved efficiency are: time factor (degradation, etc.), transilluminator wavelength and intensity, and the %AT of your DNA to mention a few. One of the advantages of MB staining is the elimination of several of the variables.

Both FMC GTG agarose and Nusieve GTG perform very well. Synergel is incompatible with MB (very high background). MB should be compatible with polyacrylamide (even less of a background problem).

Nuieve : Agarose (3 : 1, 4% final) gels stain very nicely and dsDNA as small as 75 bp is easily visualized.

Recovery of DNA from Low-melting-temperature Agarous Gels

Caution. Ultraviolet radiation and EtBr in a gel are dangerous. Wear protective goggles and gloves to protect the eyes and EtBr contamination.

1. Digest plasmid DNA (up to 20 µg) containing insert.
2. Pour a gel containing the appropriate concentration of low-melting- temperature agarous.
3. Mix the samples of DNA with tracking dye, heat shock at 65°C for 5 min, transfer in ice, and load onto the gel.
4. Carry out electrophoresis at 12 volt overnight.

   * DNA of a given size runs slightly faster through gels cast with low-melting-temperature agarous than through conventional gels.

5. Take a picture.
6. Using a hand-held UV light with long wavelength (to minimize the damage to the DNA), cut out insert bands using a scalpel. Cut the gel as close to band of interest as possible and transfer it to a clean 1.5 ml MFT. Check or draw the removing band on the picture for a record of which band was eluted.
7. Add appr. 5 volumes of H_2O to the slice of agarous.
8. Melt the agarous at 65°C for 1 to 5 min. Voltex for 20 sec and store at –20 C.

For Further Separation of DNA from Agarous

9. Spin the tubes (4K, 10 min, 20C).
10. Save the supernatent with Micropipette (P-20) into new MFT. The white substance at the interphase is powdered agarous.
11. Re-extract the agarous phase once with phenol:chloroform and once with chloroform.
12. Transfer the aqueous phase to a MFT, bring up to 105 µl with H_2O, add 35µl NH_4Ac and 1050 µl EtOH, and invert to mix.
13. Centrifuge 5 K, 20 min at 4°C.
14. Take supernatant, rains with 70% EtOH, dry and resuspended in 5–10 µl TE. And store at –20°C

EXERCISE 3.21. TRANSFORMATION

Introduction

Introduction of recombinant plasmid into cells is achieved by the transformation of competent cell. Competent cell are prepared by treating the cell with a divalent cation like calcium chloride. Once the cells are made competent, the plasmid DNA are mixed with the cells. The competent cells are then subjected to heat shock which allow the DNA to enter the cells. The cells are then plated on to medium containing antibiotics to allow identification of recombinants.

Competent Cells Preparation

The competent cells were prepared by following methods.

- Calcium Chloride Method
- Poly Ethylene Glycol 80(x) Method
- Calcium Chloride Method
- Material Calcium chloride (50 mM), *E. coli* cells, ice, 10% sterile glycerol, bullets, tips, LB
- Agar plate, LB broth, etc.

Procedure: E. coli strain cells were made competent for transformation by treating with calcium chloride as described by Sambrook et al (1989). Bacteria in glycerol frozen stock were streaked on a LB agar plate using a sterile platinum loop, grown at 37° for 16-20 hour. A single isolated colony from the plate was transferred into 25 ml of LB broth and grown for 16 hour at 37° with shaking until the culture reached an A600 nm of 0.4.

The culture was chilled on ice and the cells were harvested by centrifuging at 4000 rpm for 10 minutes, at 4° in a refrigerated centrifuge. The cell pellet was resuspended in 25 ml ice-cold sterile 50 mM calcium chloride and kept on ice for 20 minutes. Cells were recovered by centrifugation at 4° as described above. Finally the pellet was gently resuspended in 2.5 ml of ice-cold 50 mM calcium chloride containing 10% sterile glycerol, aliquoted (200 ml each) in 1.5 ml centrifuge tubes and stored at –70° for further use.

Transformation of Competent Cells

1. Mix 10 ml of ligation mix (or any other plasmid that is to be transformed) with 100 ml of competent cells and incubate on ice for 1 hour.
2. Give a heat shock at 42 degrees for 2 minutes.
3. Chill the tube to 0 degrees on ice immediately.
4. Add 800 ml of sac medium and incubate at 37°C for 60 minutes with slow shaking.
5. Prepare LB-agar plates by adding 2 ml Ampicillin (50 mg/ml), 10 ml IPTG and 10 ml X.
6. Stock per ml of melted LB agar.
7. Spread 100 1.11 of transformed cells onto the plates and incubate at 37° overnight.

Screening of Recombinant Ones

1. Selection of recombinants is based on the colour of the colony.
2. For a pUC vector, since the site used for insertion of foreign DNA is located within the lacZ gene, insertion of the foreign DNA is monitored by the loss of 13-galactosidase activity upon transformation. Cells with the intact lacZ gene produce functional 13-galactosidase, which converts the colourless substrates X-gal to blue chromophor in presence of an inducer IPTG and therefore produce blue colonies. Transformed cell with recombinant plasmid do not demonstrate 13-galactosidase activity and therefore cannot act on X-gal resulting in the production of white colonies.
3. Select white colonies as clones.

Storage of Clones

4. Transfer the white colonies one by one onto a fresh LB agar plate containing Ampicillin, IPTG and X-Gal.
5. Incubate the plate at 37° overnight. This is the master plate of the clones.
6. Inoculate the clones from the master plate to 1 ml of LB containing 100 1.11 of Ampicillin individually.
7. Incubate the tubes at 37° overnight with constant shaking.
8. Add 150 1.11 of 100% glycerol to the wells of the microtitre plate.
9. Add 850 1.11 of the overnight grown culture to glycerol in a microtitre plate. Use one well per clone. Mix well.
10. Freeze the plate, cover and store at –70°.

Notes

1. The competent cells prepared may be stored frozen at –70° without loss of

activity for long time. Store in small aliquots and take out a fresh tube and use for transformation whenever required.

2. The procedure for plating is given for plasmids having lacZ as insertional inactivation marker and Ampicillin as selection marker. For plasmids with other markers prepare the plating media accordingly.

Observations

Count white colonies as recombinant transformants and test for insert. Calculate the transformation efficiency in terms of the number of colony forming units (CFU) per microgram of transforming DNA as follows:

CFU on Plate

CFU/yg = ______________× 101 × dilution factor ng plasmid

DNA used in transformation

4. BLOTTING TECHNIQUES

SOUTHERN, NORTHERN, WESTERN BLOTTING

These are techniques for analyzing cellular macromolecules: DNA, RNA, and protein. These sections will describe how they work and how they can be used as analytical tools.

Theory: Complementarity and Hybridization

Molecular searches use one of several forms of complementarity to identify the macromolecules of interest among a large number of other molecules. Complementarity is the sequence-specific or shape-specific molecular recognition that occurs when two molecules bind together. For example: the two strands of a DNA double-helix bind because they have complimentary sequences; also, an antibody binds to a region of a protein molecule because they have complimentary shapes.

Complementarity between a probe molecule and a target molecule can result in the formation of a probe-target complex. This complex can then be located if the probe molecules are tagged with radioactivity or an enzyme. The location of this complex can then be used to get information about the target molecule.

In solution, hybrid molecular complexes hybrids of the following types can exist:

1. DNA-DNA. A single-stranded DNA (ssDNA) probe molecule can form a double-stranded, base-paired hybrid with a ssDNA target if the probe sequence is the reverse complement of the target sequence.
2. DNA-RNA. A single-stranded DNA (ssDNA) probe molecule can form a double-stranded, base-paired hybrid with an RNA (RNA is usually a single-strand) target if the probe sequence is the reverse complement of the target sequence.
3. Protein-Protein. An antibody probe molecule (antibodies are proteins) can form a complex with a target protein molecule if the antibody's antigen-binding site can bind to an epitope (small antigenic region) on the target protein. In this case, the hybrid is called an 'antigen-antibody complex' or 'complex' for short.

There are two important features of hybridization:

1. *Hybridization reactions are specific*, the probes will only bind to targets with complimentary sequence.

2. Hybridization reactions will occur in the presence of large quantities of molecules similar but not identical to the target. That is, a probe can find one molecule of target in a mixture of zillions of related but non-complementary molecules.

These properties allow you to use hybridization to perform a molecular search for one DNA molecule, or one RNA molecule, or one protein molecule in a complex mixture containing many similar molecules.

These techniques are necessary because a cell contains tens of thousands of genes, thousands of different mRNA species, and thousands of different proteins. When the cell is broken open to extract DNA, RNA, or protein, the result is a complex mixture of all the cell's DNA, RNA, or protein. It is impossible to study a specific gene, RNA, or protein in such a mixture with techniques that cannot discriminate on the basis of sequence or shape. Hybridization techniques allow you to pick out the molecule of interest from the complex mixture of cellular components and study it on its own.

Basic Definitions

Blots are named for the target molecule.

Southern Blot. DNA cut with restriction enzymes-probed with radioactive DNA.

Northern Blot. RNA-probed with radioactive DNA or RNA.

Western Blot. Protein-probed with radioactive or enzymatically-tagged antibodies.

Overview

The formation of hybrids in solution is of little experimental value - if you mix a solution of DNA with a solution of radioactive probe, you end up with just a radioactive solution. You cannot tell the hybrids from the non-hybridized molecules. For this reason, you must first physically separate the mixture of molecules to be probed on the basis of some convenient parameter.

These molecules must then be immobilized on a solid support, so that they will remain in position during probing and washing. The probe is then added, the non-specifically bound probe is removed, and the probe is detected. The place where the probe is detected corresponds to the location of the immobilized target molecule. This process is diagrammed below:

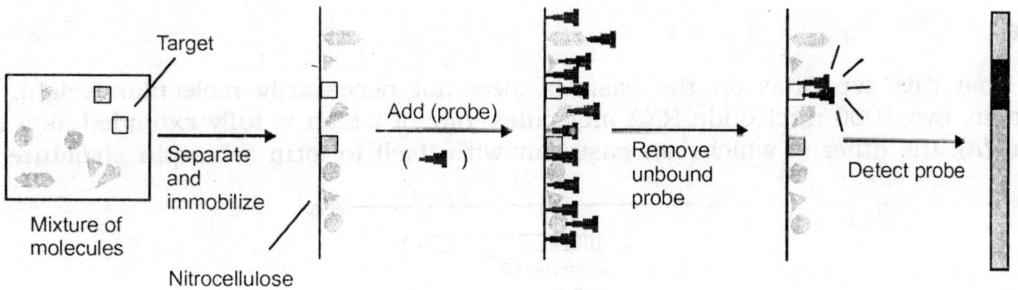

In the case of Southern, Northern, and Western blots, the initial separation of molecules is done on the basis of molecular weight.

In general, the process has the following steps, detailed below:

- Gel electrophoresis
- Transfer to Solid Support

- Blocking
- Preparing the Probe
- Hybridization
- Washing
- Detection of Probe-Target Hybrids.

Gel Electrophoresis

This is a technique that separates molecules on the basis of their size.

First, a slab of gel material is cast. Gels are usually cast from agarose or polyacrylamide. These gels are solid and consist of a matrix of long thin molecules forming sub-microscopic pores. The size of the pores can be controlled by varying the chemical composition of the gel. The gel is cast soaked with buffer.

The gel is then set up for electrophoresis in a tank holding buffer and having electrodes to apply an electric field:

The pH and other buffer conditions are arranged so that the molecules being separated carry a net (–) charge so that they will me moved by the electric field from left to right. As they move through the gel, the larger molecules will be held up as they try to pass through the pores of the gel, while the smaller molecules will be impeded less and move faster. This results in a separation by size, with the larger molecules nearer the well and the smaller molecules farther away.

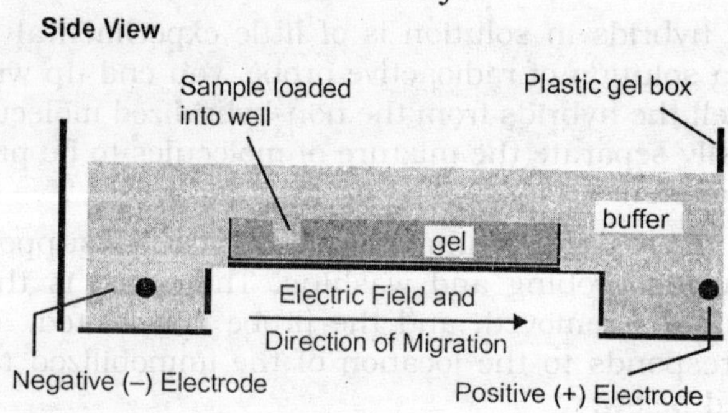

Notes

That this separates on the basis of size, not necessarily molecular weight. For example, two 1000 nucleotide RNA molecules, one of which is fully extended as a long chain **(A)**; the other of which can base-pair with itself to form a hairpin structure **(B)**:

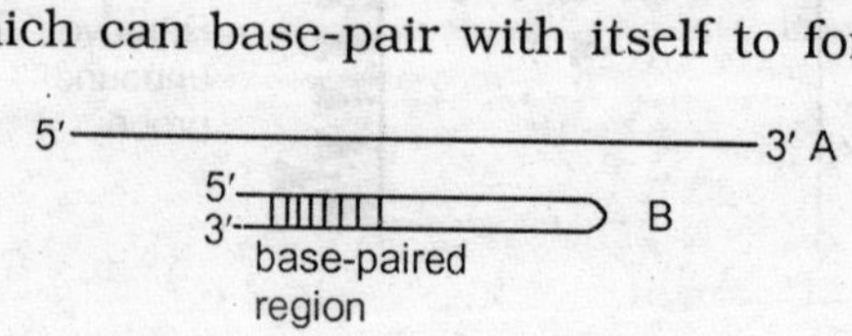

As they migrate through the gel, both molecules behave as though they were solid spheres whose diameter is the same as the length of the rod-like molecule. Both have the same molecular weight, but because **B** has secondary (2') structure that makes it smaller than **A, B** will migrate faster than **A** in a gel. To prevent differences in shape (2' structure) from confusing measurements of molecular weight, the molecules to be separated must be in a long extend rod conformation - no 2' structure. In order to remove any such secondary or tertiary structure, different techniques are employed for preparing DNA, RNA and protein samples for electrophoresis.

Preparing DNA for Southern Blots

DNA is first cut with restriction enzymes and the resulting double-stranded DNA fragments have an extended rod conformation without pre-treatment.

Preparing RNA for Northern Blots

Although RNA is single-stranded, RNA molecules often have small regions that can form base-paired secondary structures. To prevent this, the RNA is pre-treated with formaldehyde.

Preparing Proteins for Western Blots

Proteins have extensive 2' and 3' structures and are not always negatively charged. Proteins are treated with the detergent SDS (sodium dodecyl sulfate) which removes 2' and 3' structure and coats the protein with negative charges.

If these conditions are satisfied, the molecules will be separated by molecular weight, with the high molecular weight molecules near the wells and the low molecular weight molecules far from the wells. The distance migrated is roughly proportional to the log of the inverse of the molecular weight (the log of 1/MW). Gels are normally depicted as running vertically, with the wells at the top and the direction of migration downwards. This leaves the large molecules at the top and the smaller molecules at the bottom. Molecular weights are measured with different units for DNA, RNA, and protein:

- *DNA.* Molecular weight is measured in base-pairs, or *bp,* and commonly in kilobase-pairs (1000bp), or *kbp.*
- *RNA.* Molecular weight is measured in nucleotides, or *nt,* and commonly in kilonucleotides (1000nt), or *knt.* [Sometimes, bases, or *b* and *kb* are used.]
- *Protein.* Molecular weight is measured in Daltons (grams per mole), or *Da,* and commonly in kiloDaltons (1000Da), or *kDa.*

On most gels, one well is loaded with a mixture of DNA, RNA, or protein molecules of known molecular weight. These 'molecular weight standards' are used to calibrate the gel run and the molecular weight of any sample molecule can be determined by interpolating between the standards. Below is a gel stained with a dye: a coloured molecule which binds to a specific class of macromolecules in a sequence-independent manner (probes bind in a sequence-dependent manner).

Sample 1 contains only one size class of macromolecule - it could be a plasmid, a pure mRNA transcript, or a purified protein. In this case, you would not have to use a probe to detect the molecule of interest since there is only one type of molecule present. Blotting is usually necessary for samples that are not complex mixtures. By interpolation, its molecular weight is roughly 3.

Sample 2 is what a sample of total DNA cut with a restriction enzyme, total cellular RNA, or total cellular protein would look like in a gel stained with a sequence-independent stain. There are so many bands that it is impossible to find the one we are interested in. Without a probe (which acts like a sequence-dependent stain) we cannot get very much information from a sample like this.

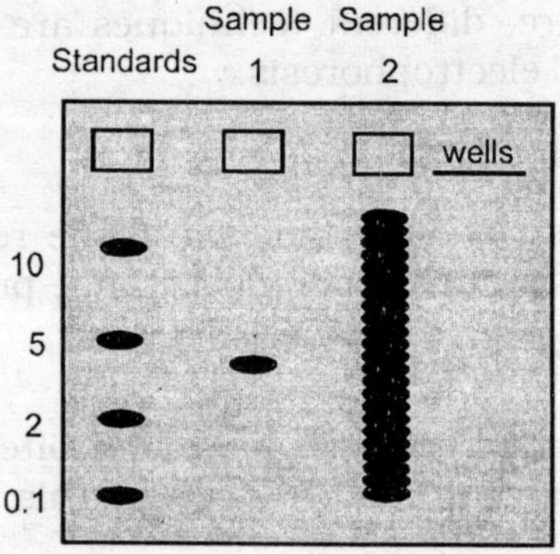

Different stains and staining procedures are used for different classes of macromolecules:

- *Staining DNA.* DNA is stained with ethidium bromide (EtBr), which binds to nucleic acids. The DNA-EtBr complex fluoresces under UV light.
- *Staining RNA.* RNA is stained with ethidium bromide (EtBr), which binds to nucleic acids. The RNA-EtBr complex fluoresces under UV light.
- *Staining Protein.* Protein is stained with Coomassie Blue (CB). The protein-CB complex is deep blue and can be seen with visible light.

Transfer to Solid Support

After the DNA, RNA, or protein has been separated by molecular weight, it must be transferred to a solid support before hybridization. (Hybridization does not work well in a gel.) This transfer process is called blotting and is why these hybridization techniques are called blots. Usually, the solid support is a sheet of nitrocellulose paper (sometimes called a filter because the sheets of nitrocellulose were originally used as filter paper), although other materials are sometimes used. DNA, RNA, and protein stick well to nitrocellulose in a sequence-independent manner.

The DNA, RNA, or protein can be transferred to nitrocellulose in one of two ways:

1. *Electrophoresis*, which takes advantage of the molecules' negative charge:

Side View

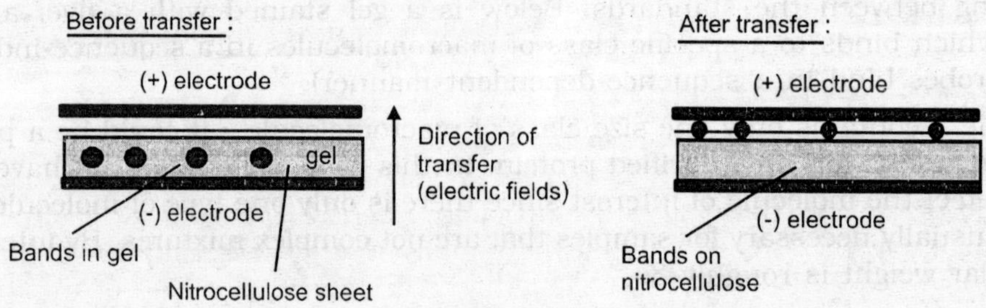

Note: All the layers are pressed tightly together.

2. *Capillary blotting*, where the molecules are transferred in a flow of buffer from wet filter paper to dry filter paper:

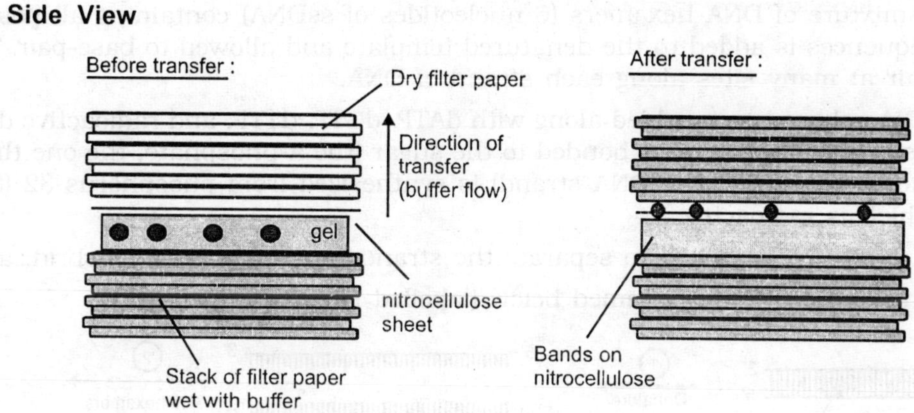

Notes

In a Southern Blot, the DNA molecules in the gel are double-stranded, so they must be made single stranded in order for the probe to hybridize to them. To do this, the DNA is transferred using a strongly alkaline buffer, which causes the DNA strands to separate this process is called denaturation - and bind to the filter as single-stranded molecules. RNA an protein are run in the gels in a state that allows the probe to bind without this pre-treatment.

Blocking

At this point, the surface of the filter has the separated molecules on it, as well as many spaces between the lanes, etc., where no molecules have yet bound. If we added the probe directly to the filter now, the probe would stick to these blank parts of the filter, like the molecules transferred from the gel did. This would result in a filter completely covered with probe which would make it impossible to locate the probe-target hybrids. For this reason, the filters are soaked in a blocking solution which contains a high concentration of DNA, RNA, or protein. This coats the filter and prevents the probe from sticking to the filter itself. During hybridization, we want the probe to bind only to the target molecule.

PREPARING THE PROBE

Radioactive DNA probes for Southerns and Northerns

The objective is to create a radioactive copy of a double-stranded DNA fragment. The process usually begins with a restriction fragment of a plasmid containing the gene of interest. The plasmid is digested with particular restriction enzymes and the digest is run on an agarose gel. Since a plasmid is usually less than 20 kbp long, this results in 2 to 10 DNA fragments of different lengths. If the restriction map of the plasmid is known, the desired band can be identified on the gel. The band is then cut out of the gel and the DNA is extracted from it. Because the bands are well separated by the gel, the isolated DNA is a pure population of identical double-stranded DNA fragments.

The DNA restriction fragment (template) is then labelled by Random Hexamer Labeling.:

1. The template DNA is denatured, the strands are separated by boiling.

2. A mixture of DNA hexamers (6 nucleotides of ssDNA) containing all possible sequences is added to the denatured template and allowed to base-pair. They pair at many sites along each strand of DNA.
3. DNA polymerase is added along with dATP, dGTP, dTTP, and radioactive dCTP. Usually, the phosphate bonded to the sugar (the a-phosphate, the one that is incorporated into the DNA strand) is synthesized from phosphorus-32 (32P), which is radioactive.
4. The mixture is boiled to separate the strands and is ready for hybridization.

This process is diagrammed below (labelled DNA shown in gray):

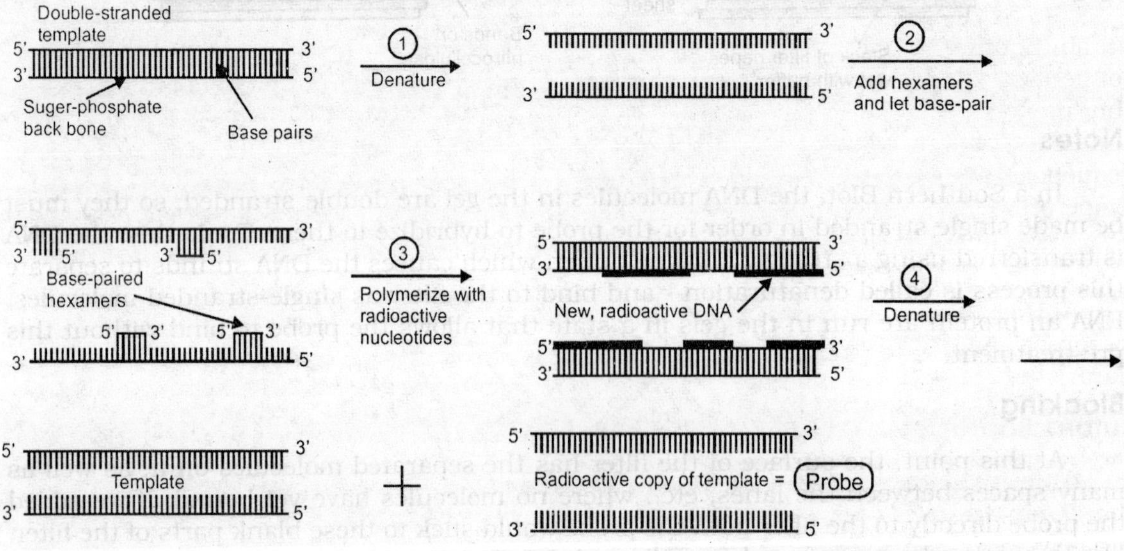

This produces a radioactive single-stranded DNA copy of both strands of the template for use as a probe.

Radioactive Antibodies for Westerns

Antibodies are raised by injecting a purified protein into an animal, usually a rabbit or a mouse. This produces an immune response to that protein. Antibodies isolated from the serum (blood) of that rabbit will bind to the protein used for immunization. These antibodies are protein molecules and are not themselves radioactive.

They are labelled by chemically modifying the side chains of tyrosines in the antibody with iodine-125 (125I), which is radioactive. A set of enzymes catalyzes the following reaction:

$$\text{antibody-tyrosine} + 125\text{I-} + H_2O_2 \longrightarrow H_2O + 125 \text{ iodo-tyrosine-antibody}$$

Enzyme-conjugated Antibodies for Westerns

Antibodies against a particular protein are raised as above and labelled by chemically cross-linking the antibody molecules to molecules of an enzyme. The resulting antibody-enzyme conjugate is still able to bind to the target protein.

Hybridization

In all three blots, the labelled probe is added to the blocked filter in buffer and incubated for several hours to allow the probe molecules to find their targets.

Washing

After hybrids have formed between the probe and target, it is necessary to remove any probe that is on the filter that is not stuck to the target molecules. Because the nitrocellulose is absorbent, some of the probe soaks into the filter and must be removed. If it is not removed, the whole filter will be radioactive and the specific hybrids will be undetectable.

To do this, the filter is rinsed repeatedly in several changes of buffer to wash off any un-hybridized probe.

Note

In Southerns and Northerns, hybrids can form between molecules with similar but not necessarily identical sequences (For example, the same gene from two different species). This property can be used to study genes from different organisms or genes that are mutated. The washing conditions can be varied so that hybrids with differing mismatch frequencies are maintained. This is called 'controlling the 'stringency'-the higher the wash temperature, the more stringent the wash, the fewer mismatches per hybrid are allowed.

Detecting the Probe-Target Hybrids

At this point, you have a sheet of nitrocellulose with spots of probe bound wherever the probe molecules could form hybrids with their targets. The filter now looks like a blank sheet of paper-you must now detect where the probe has bound.

Autoradiography

If the probe is radioactive, the radioactive particles that it emits can expose X-ray film. If you press the filter up against X-ray film and leave it in the dark for a few minutes to a few weeks, the film will be exposed wherever the probe bound to the filter. After development, there will be dark spots on the film wherever the probe bound.

Enzymatic Development

If an antibody-enzyme conjugate was used as a probe, this can be detected by soaking the filter in a solution of a substrate for the enzyme. Usually, the substrate produces an insoluble coloured product (a chromogenic substrate) when acted upon by the enzyme. This produces a deposit of coloured product wherever the probe bound.

Summary

The *procedure* for these three blots is summarized below:

Southern Blot	*Northern Blot*	*Western Blot*
1. Extract DNA from cells	1. Extract RNA from cells	1. Extract protein from cells
2. Cut with restriction enzyme	2. Denature with formalde-hyde	2. Denature with SDS
3. Run on gel (usually agarose)	3. Run on gel (usually agarose)	3. Run on gel (usually poly-acrylamide-called "SDSPAGE")
Denature DNA with alkali		
4. Transfer to nitrocellulose (usually by capillary action)	4. Transfer to nitrocellulose (uaually by capillary action)	4. Transfer to nitrocellulose (usually by electrophoresis)

5. Block with excess DNA	5. Block with excess RNA	5. Block with excess protein
6. Hybridize with labelled DNA probe	6. Hybridze with labelled DNA probe	6. Hybridize with labelled anitbody probe
7. Wash of unbound probe (use of controlled strin-gency)	7. Was off unbound probe (use controlled stringency)	7. Wash of unbound probe
8. Autoradiograph	8. Autoradiograph	8. Autoradiograph or develop with chromogenic substrate

The *important properties* of the three are shown below:

	Southern	*Northern*	*Western*
What is separated by molecular weight (target)	DNA cut with restri-tion enzymes	RNA denatured with formaldehyde	Protein denatured with SDS
Probe	Radioactive gene X DNA	Radioactive gene X DNA	Antibody against pro-tein X, labelled with radioactivity or enzyme
What do you learn from it ?	Restriction map of gene X in chromo-some	-how much gene X mRNA is present ? -how long is gene X mRNA ?	-how much protein X is present ? -how big is protein X ?

4.1. SOUTHERN BLOTTING (*Ist Method*)

Introduction

This method to know the presence or absence of a particular fragment in genomic DNA was first developed by E. D. Southern in 1975. The advent of Southern blotting technique was turning point in the field of molecular biology. It involves the capillary transfer of DNA fragments from an agarose gel to various types of membranes. RFLPs can be analyzed using the technique wherein DNA fragments are separated on agarose gels denatured in situ and transferred onto membranes for analysis.

Requirements

1. Denaturation solution: NaCl, 1.5 M and NaOH, 0.5 M
2. Neutralization solution: NaCl, 1.5 M; Tris-Cl (pH 7.5), 0.5 M and EDT A (pH 8.0), 1 mM
3. 20x SSC: NaCl, 1.5 M and trisodium citrate, 0.1 M
4. Depurinization solution: 0.25 N HCl
5. Nylon or nitrocellulose membrane

Procedure

After agarose gel electrophoresis, photograph the gel and soak the gel in 0.25 N HCl for 15 minutes at room temperature with gentle shaking.

Decant the acid solution and denature the DNA by soaking the gel in several volumes of Denaturation solution for 30 minutes at room temperature with constant shaking.

Neutralize the gel by shaking in several volumes of neutralization solution for 30 minutes at room temperature with shaking.

Wrap a piece of Whatman 3 mm paper around a glass plate. Place the wrapped support on a large plastic tray with the ends of 3 mm paper dipping into the 20x SSC solution in tray.

Invert the gel and place it on a damp 3 mm paper on the support. Make sure that there are no air bubbles between the 3 mm paper and the gel.

Cut a piece of nylon membrane slightly bigger than the gel. Use gloves and forceps to handle the membrane.

Float the membrane on 20X SSC until it wets completely.

Place the wet nylon membrane on top of the gel. Remove all the air bubbles that are trapped between the gel and tile membrane.

Wet two pieces of Whatman 3 mm paper, cut to exactly the Slime size as the gel in 10X SSC and place them on top of the membrane. Again remove 111 the air bubbles.

Cut a stack of coarse filter paper just smaller than the gel size. Keep on top of the Whatman filter papers.

Put a glass plate on the top and place (about 1 kg) on it to exert pressure.

Allow the transfer of DNA to proceed for about 12-24 hours.

Remove the stack of coarse filter papers and the 3 mm paper above the gel. Turn over the dehydrated gel and membrane and lay them gel side up on a dry sheet of 3 mm paper. Mark the position of the wells on the membrane with a soft pencil.

Peel off the gel. The transfer can be checked by restaining the gel. If the transfer is complete no DNA should be retained on the gel.

Soak the membrane in 6x SSC at room temperature for few minutes.

Allow excess fluid to drain off from the membrane and set it to dry at room temperature on a sheet of 3 mm paper.

Place the dried filter between two sheets of 3 mm paper.

Fix the DNA on the membrane by baking for 2 hours at 80 degrees under vacuum or cross linking on a UV transilluminator for a few minutes.

Wrap the membrane with saran wrap or keep in an envelope made up of Whatman No. 1 filter paper and store.

Notes

Nylon membranes are preferable over Nitrocellulose for transfer.

In case nitrocellulose is used, the DNA has to be fixed by baking at 80 degrees for 2 hours under vacuum. Nitrocellulose being combustible will become brittle if baked in presence of oxygen.

DNA can be fixed on nylon membrane by UV-cross linking using long wave UV rays or by baking at 120 degrees for 2 hours.

Care has to be taken so that the buffer passes to the filter paper through the gel only. A layer of para film may be put on the glass plate around the gel to avoid the filter papers touching the buffer directly.

While photographing the gel keep a fluorescent ruler alongside the gel for proper orientation later on.

Never touch the membrane with bare hand; any grease or powlier on the membrane will prevent transfer of DNA.

Nowadays charged and modified nylon membranes are available from Qlany suppliers, which gives better results. Follow the manufacturer's instruction for these membranes.

Depurinization of the DNA on the gel by using HCl is to facilitate transfer of large DNA fragments. For small DNA fragments this step may be avoided.

Observations

Restain the gel in ethidium bromide 5 mg/ml for 45 minutes and view on a UV transilluminator after proper washings. There should not be any DNA on the gel, as the entire DNA should have been transferred to the membrane.

Plasmid, there will be only one band in the lane. In case of genomic DNA a continuous smear should be visible, as digestion will result in many pieces of varying sizes.

4.2. SOUTHERN BLOTTING (*IInd Method*)

Procedure

1. Electrophoresis of DNA is carried out in a neutral agarose gel system. Prepare a 0.8-1% agarose gel containing 1X TAE buffer. Ethidium bromide can be added to a final concentration of 0.2 ug/ml.
2. Apply the samples to the gel.
3. Run the gel in 1X TAE buffer at 4V/cm until the bromophenol blue indicates that the sample has run for a sufficient distance.
4. Following electrophoresis, visualize the gel under UV transillumination and photograph with a ruler.
5. (*i*) Depurination, 10 minutes at room temperature with gentle agitation (optional). This step is necessary if target sequences are greater than 10 Kb in size.

   (*ii*) Denaturation, 25 minutes at room temperature with gentle agitation.

   (*iii*) Neutralization, 30 minutes at room temperature with gentle agitation.

   When using nitrocellulose membranes, the neutralization time should be extended to 45 minutes. Include a rinse in distilled water between each step.
6. Assemble the capillary blotting apparatus using 10X SSC as the transfer buffer. Allow the DNA to transfer overnight onto Hybond N+.
7. The following day, disassemble the apparatus, mark the membrane appropriately and fix the DNA to the membrane by UV crosslink or baking (2 hours at 80EC). For nitrocellulose membranes, bake for 2 hrs, at 80EC in a vacuum oven.

1. Hybridization buffer

   5X SSC

   1 in 20 dilution Liquid Block (Amersham) or other blocking reagent

   0.1% (w/v) SDS

   5% (w/v) Dextran sulphate

2. EDTA stock : 0.5M EDTA pH8.0
3. SDS stock : 10% or 20% (w/v) SDS
4. Depurination solution (for Southern blotting)
   250 mM HCl
5. Denaturation solution (for Southern blotting)
   1.5M NaCl
   0.5M NaOH
6. Neutralization solution (for Southern blotting)
   1.5M NaCl
   0.5M Tris-HCl
   pH adjusted to 7.5
7. 20X SSC : 0.3M Na (3) citrate, 3M NaCl

4.3. WESTERN BLOTTING

Aim

To separate the proteins through SDS-PAGE and detection followed by characterization of proteins through-Western Blotting.

Introduction

Western blotting (also known as protein or immunoblotting) is a sapid and sensitive assay for detection and characterization of proteins. It works by exploiting the specificity inherent in Ag-Ab recognition. It is used to identify specific antigens recognized by polyclonal or monoclonal antibodies. Western Blotting is carried out along with protein (antigen) separation in gel by electrophoresis and the blot development.

It is essentially a combination of three techniques :

Electrophoresis (PAGE)

Western Blotting

Immunochemical detection.

Principle

Identification of protein separated by gel electrophoresis is limited by the small pore size of the gel as macromolecule probe for protein analysis cannot permeate the gel. This limitation is overcome by blotting the protein analysis cannot permeate the gel. This limitation is overcome by blotting the protein into an adsorbent porous membrane. The apparatus consists of a tank containing buffer, in which is located a cassette. Clamping the gel and the membrane tightly together a current is applied from electrode repeated on either sides of the cassette to avoid heating effects. The proteins are separated according to their electrophoresis mobility and blotted onto the membrane identified using suitable immuno chemicals locate the protein of interest. The individual techniques are explained below.

SDS-PAGE

Stage 1. Prepare a PAGE gel slab and fix to a vertical electrophoretic apparatus as shown in the figure. Treat the sample with suitable buffer and load onto the gel slots.

Stage 2. Apply electric current, after a few minutes, protein in the sample migrate according to their electrophoretic mobility in the stacking gel the staking gel has a polyacrylamide concentration resulting in higher pore size and lower PH.7. This enables the protein to concentrate into sharp band due to isotachopharesis or band sharpending effect. At the end of the stacking gel it meets the separating gel which has a higher polyacrylamide concentration and higher pH. In the separating gel the protein travels according to their size.

Stage 3. When the dye front reach the bottom of the separating gel, the proteins in sample are resolved depending on their size. However the protein cannot be visualized directly. The gel needs to be stained with suitable stainer to visualize all the proteins. The identification of protein of interest can be done using a suitable probe and a developing system.

Western Blotting

Blotting is the transfer of resolved proteins from the gel to surface of suitable membrane. This is done commonly by electrophoresis (known as electroblotting). In this method, the transfer buffer has a low ionic strength which allows electrotransfer of proteins. Methane in the buffer increases binding of proteins to niotrocellulose and reduce smelling of gel during transfer. The use of membrane as a support for protein enables the case of manipulation efficient washing and faster reactions during the immunodetection as protein are more accessible for reaction.

(*a*) Membrane is in close contact with PAGE gel containing proteins. The proteins are electrotransferred to nitrocellulose membrane.

(*b*) At the end of electrotransfer, all proteins would have migrated to NC membrane.

The protein was transferred to the corresponding position on the membrane as on the gel. A mirror image of the gel was formed. However the protein location and detection can only be assessed after immunodetection.

Immunodetection

The transferred proteins are bound to surface of NC membrane and are accessible for reaction with immunochemical reagents. All the unoccupied sites on the membrane are blocked with inert proteins, detergents or other suitable blocking agents. The membrane is then probed with a primary antibody 20 antibody and a suitable substrate to the enzyme identifies the ag-ab complex form on the membrane.

Applications

To characterize proteins

To identify specific antigens for antibodies.

Preparation of Reagents

1. *Blotting Buffer.* Add 25 ml of blotting buffer component A and 25 ml of component B to 150 ml of distilled water.
2. *Other Buffers.* Dilute the required amount of buffer concentrate to 1X concentration with water.
3. *Antibody.* Diluted primary Ab and labelled secondary HRP conjugate in assay buffer.

4. *Substrate.* Dilute TMB/H_2O_2, 10X concentration ten times with heater just before use.

Procedure

1. SDS : Polyacrylamide gel electrophoresis was run.
2. Electroblotting
3. The blotting sandwich was assembled within the blotting cassette as shown in the figure. Care should be taken to avoid air bubbles between the gel and NC membrane.
4. The cassette was inserted into the apparatus filled with blotting buffer such that the gel faces the cathode.
5. The power supply was connected and voltage of 50 V for 5 hours was supplied for blotting.

Immunodetection

(*a*) The NC membrane was removed gently from the cassette and was placed in blocking buffer for 2 hours at room temperature or overnight in the cold.

(*b*) Primary antibody was suspended in 10 ml with the assay buffer using suitable tube.

(*c*) The blot was immersed in the 10 Ab solute and gently agitated for 30 minutes.

(*d*) The blot was washed by immersing in wash buffer for 3-5 minutes. This was repeated 2 more times.

(*e*) 1:1000 dilutions of labelled 20 Ab in assay buffer was prepared a sufficient (10 ml) volume of diluted Ab was prepared to cover the blot.

(*f*) The blot was immersed in 20 Ab solute and agitate gently for 30 minutes.

(*g*) The blot was washed in wash buffer for 3-5 minutes and this was repeated for 4 times.

(*h*) The washed blot was immersed in 10 ml of substrate solution with gentle shaking. Bands will develop sufficient colour within 5-10 min.

(*i*) The blot was removed and washed with distilled water and dried.

(*j*) Although the coloured hands fade with time, the rate of colour loss can be retarded if the blots are kept in the dark.

Interpretation and Result

The proteins separated through the SDS - PAGE has been successfully transferred onto the nitrocellulose membrane and the transferred proteins were detected by Immuno detection which was confirmed by the development of colour bands on the nitrocellulose membrane.

4.4. WESTERN BLOT ANALYSIS OF EPITOPED-TAGGED PROTEINS USING THE CHEMIFLUORESCENT DETECTION METHOD-FOR ALKALINE PHOSPHATASE CONJUGATED ANTIBODIES

1. Cut PVDF membrane to the appropriate size, activate with absolute methanol for 5 sec, and incubate in distilled water for 5 min. For electroblotting, equilibrate

in transfer buffer and follow the standard blotting procedure to transfer the proteins to the membrane. For dot blotting, keep membrane wet until ready to use.

2. After protein has been transferred to the membrane, wash again in absolute methanol for a few seconds and allow to dry at room temperature for 30 min. or more.
3. Block in 30 ml of 1X Western buffer (containing 0.1% Tween-20 and 0.2% I-Block), gently rocking, 1 hr, room temperature.
4. Add appropriate dilution of primary antibody (typically 1:5000 or 1:10,000) prepared in 1X Western buffer (containing 0.1% Tween-20 and 0.2% I-Block), incubate 30 min, room temperature, gently rocking.
5. Wash three times in 20 ml 1X Western buffer (containing 0.1% Tween-20 and 0.2% I-Block) for 5 min each. Add appropriate dilution of secondary antibody conjugated to alkaline phosphatase prepared in 1X Western buffer (containing 0.1% Tween-20 and 0.2% I-Block), gently rocking, 30 min, room temperature.
6. Wash as in step 5. Additionally, wash twice with 1X Western buffer *without I-block*.
7. At the end of the second final wash, leave some buffer in the container to keep the membrane moist. With the membrane facing protein-side up, add 0.5 ml of substrate solution directly into the remaining liquid, mix well, and pipette (with a p10000) the solution over the membrane to ensure the entire surface comes into contact with the substrate. Gently agitate for a few minutes, remove membrane to a paper towel and let dry completely. The substrate solution can be reused immediately for additional membranes.
8. Scan membrane using the molecular dynamics storm.

Western Blotting Solutions

- *10X Transfer buffer:* 240 mM Tris, pH = 8.0; 1.92 M Glycine. Use 0.5X for transfer in 20% methanol.
- *10X Western buffer:* 200 mM Tris pH = 7.5; 1.5 M NaCl. To prepare 1X Western Buffer, dilute 10X buffer to 1X, adding Tween-20 to 0.1%. Remove 50 ml and set aside for the last two washes. To the remainder, add I-Block to 0.2%, heating gently with constant stirring until dissolved. Bring to room temperature before using (containing 0.1% Tween-20 and 0.2% I-Block).
- *Primary antibody:* For his tagged proteins - Anti-His monoclonal antibody
- *Secondary antibody:* Goat anti-mouse alkaline phosphatase conjugated - Biorad #170-6520.
- *Substrate:* ECF chemifluorescent substrate - Mix substrate with accompanying buffer as per manufacturer's recommended instructions, prepare 1 ml aliquots and store at –20EC.

SOUTHERN BLOT

1. Run the gel as normal. Often for genomic southerns it is desirable to run long gels (18 cm) over 4-6 hrs.
2. Photograph the gel with a ruler adjacent to the molecular weight markers as a reference.

3. Alkali transfer buffer = 0.5M NaOH (20 g/lt), 1.5M NaCl (87.66 g/lt). Prepare 1 litre for one gel and another 750 ml for each additional gel. Beware this buffer is very dangerous, capable of causing severe eye damage. Use the large volumes involved in this procedure with care and wear protective glasses.
4. Add the gel to 250ml alkali transfer buffer, plus 125 ml for each additional gel.
5. Place gel on rocker with buffer solution for 20 mins. Note low percent agarose gels must be agitated slowly to prevent tearing.
6. Keep remaining 500 ml for the transfer tank.
7. Wear gloves for the following steps.
8. Cut 2 pieces of large 3MM paper (wicks) and 2 pieces about 2 mm smaller than the gel on each edge and one piece of nylon (Hybond N+, Amersham) the same size as the gel.
9. Large gel (HFI) Medium gel Mini gel
10. 2 (14 × 19 cm) 2 (15.2 × 10) 2 (9 × 5.2)
11. 2 (14 × 32 cm) 2 (15.2 × 32) 2 (18.5 × 5.2)
12. Cut a stack of paper toweling about 2 mm smaller than the gel. The stack needs to be about 6 cm thick.
13. Prewet nylon in DDW, then soak in alkali transfer buffer.
14. Add buffer to transfer tank to the level of the platform. Wet the long wicks in transfer buffer and place in the tank.
15. Place gel on platform and spoon on transfer buffer. Add the Nylon and smooth out any bubbles with the back of your finger.
16. Add the 2 slightly smaller 3MM filters, the first prewetted, the second dry.
17. Add the stack of paper towels. Note it is very important that the edges of the towel do not touch the wicks that the gel is sitting on or the transfer will be "shortcircuited".
18. Top the stack with a glass or plastic plate and a weight (a bottle with about 200-300 ml H_2O is ideal). Too much weight compresses the gel and terminates the transfer early.
19. Allow to transfer o/n and then remove the stack carefully. Mark the position of the wells on the filter with a biro (and note position of well #1) before removing the filter. Soak the filter in 0.5M Tris pH 7.5, 1.5M NaCl for 5min after removal from the gel.
20. Add filter to 200 ml 2 X SSC and allow thc soak without agitation for 5'.
21. Remove filter and blot dry and bake at 80 for 2 hr or place in autoclave for 10' when not operating but warm.
22. Prehybridige with 10 ml Aqua. hybridge (reagents), 1% SDS and 100 ug/ml boiled herring sperm DNA for 20-30 minutes (cloned DNA southern) to several hrs (genomic southern).
23. Boil probe for 3' and add to 3 ml of fresh hybridge solution/SDS/herring DNA. Squeeze Prehybridge from the bag and add probe. Seal, avoiding air bubbles and distribute the probe well. Incubate for 4-6 hr (cloned DNA) to 16-20 hr (genomic southern). Washes are performed in 2 to 0.2X SSC, 0.1% SDS depending on homology of probe to target.

4.5. SOUTHERN ANALYSIS OF MOUSE TOE/TAIL DNA

Material Required

- Tail buffer 50 ml

1% SDS	5 ml 10% SDS
10 mM Tris pH 7.5	0.5 ml 1M Tris pH 7.5
50 mM EDTA	5 ml 0.5M EDTA
150 mM NaCl	1.5 ml 5M NaCl
DDW	38 ml

- Phenol/chloroform (1:1 mixture)
- 0.5M EDTA
- 4M NH_4Ac
- Absolute EtOH
- 70% EtOH
- Tris/EDTA pH 7.5
- Proteinase K (20 mg/ml dissolved in dH_2O)

Procedure

1. This works best with mice that are at weaning age. Take one toe or 1mm of tail and place in a ependorf tube. Don't take more than this, it isn't necessary and too much material interfers further down the track. Also if amount taken are consistant between sample then the amounts used for Southern will be even between samples. If there are a number of samples to be collected, place tubes on ice.
2. Make up a mix of Tail Buffer and Proteinase K allowing 600 ul of Tail Buffer with 500ug/ml Proteinase K for each sample (plus an extra dose in case of in-accurate pipetting). Place in a waterbath at 55°C all day, or at least 2 hrs.
3. Samples are transferred to hot air shaker overnight at 62°C at a medium-shake speed.
4. To each eppendorf add 500ul of phenol/chloroform (lower phase of mix). Mix by inversion/shake (don't vortex as this shears the DNA). Spin at 13,000 rpm for 2 minutes. Collect supernatant, being careful to avoid the interphase.
5. Place supernatant in fresh eppendorf tube and add:
   - 2 ul 0.5M EDTA pH 8.0
   - 200 ul 4M NH_4Ac
   - 800 ul isopropanol
6. Mix by inversion/shake, Spin at 13000 rpm for 5 minutes. Remove and discard supernatant, being extremely careful not to disturb the pellet.
7. To the tube with pellet add 200 ul 70% EtOH and vortex, Spin at 13000 rpm for 2 minutes. Remove and discard supernatant (watch the pellet).
8. Add 200 ul TE and vortex, allowing DNA to resuspend at room temperature.

9. Once DNA has resuspendend, digests maybe set up. 30 ul of DNA/TE suspension per digest should be sufficient, RNAse is not necessary. Digest o/n, ppte with 12 ul 5 m NaCl, 600ul EtOH and resusp. pellet in 18 ul TE and 7 ul dye. Allow to dissolve for 20 min RT and heat to 37°C for 5-10 min before loading.

4.6. NORTHERN BLOTTING

A. Formaldehyde agarose gel electrophoresis

Note: Formaldehyde vapours are toxic and casting this gel should be performed in a fume hood. The gel tank must be covered when in use.

1. Preparation of equipment and reagents

Soak the gel tanks, combs etc. in 0.2 M NaOH for 15 minutes to destroy any contaminatig RNases before rinsing in distill water. It is not necessary to use DEPC treated water.

2. Make up 10 × Northern running buffer (containing MOPS)

Final conc:	*For 500 ml:*
0.2M MOPS	20.93 g
10 mM EDTA	10 ml of 0.5 M
50 mM Na Acetate	8.33 ml of 3M

The buffer needs to be brought to a pH of 7.0. For 500 ml of 10 X MOPS this means add approx 2 g of NaOH. Make up to 500 ml in DEPC-treated water to DEPC-treat solutions, add 0.1% DEPC to the solution in a bottle which can be autoclaved. Mix the solution well and allow it to stand, with the cap tightly closed, for at least 30 minutes. The loosen the cap and autoclave. This must be done in a fume hood as DEPC is very toxi. Note also DEPC is inactivated by water. Allow the DEPC stock solution bottle to warm to RT before opening to avoid condensation.

Autoclaving of the MOPS buffer is not necessary and turns the solution yellow. Store at 4°C.

3. Cast the gel

Cast a 14 cm, 0.7-1.0% agarose gel, which requires at least 100 ml of agarose solution.

Note: ethidium bromide is not added to the gel but rather to the sample buffer (see below). A thin, low percentage agarose (0.6-0.7%) gel is critical for good transfer of large MW RNA's (>4-5kb).

For 125 ml	
1.25 g	agarose (1% gel)
12.5 ml	10 × running buffer
102.5 ml	DDW
6 ml	35% formaldehyde (check that it is a fresh batch that doesn't contain precipitates)

Dissolve the agarose in the microwave, let the solution cool to less than 60¼C, then add the formaldehyde and cast the gel.

4. Prepare RNA samples for loading

RNA is stored at –80¼C as an ethanol precipitate. Determine the amount of RNA to be loaded in each well (*e.g.*, 2 ug of poly A(+) RNA). Based on your RNA yield estimations, precipitate the appropriate amount of RNA ethanol solution for at least 15 minutes at 13,000g at 4¼. Remove all supernatant and resuspend each sample in 12 ul of sample buffer.

RNA sample buffer (prepared fresh from frozen stocks): For 500 ul:

- 50 ul 10 × running buffer
- 250 ul deionised formamide
- 90 ul formaldehyde
- 108 ul water (DEPC-treated)
- 2 ul ethidium bromide (stock concentration, 10 mg/ml)

Heat the samples at 65¼C for 5 minutes then to each add 3 ul of RNA loading buffer. This consists of 0.25% bromophenol blue, 0.25% xylene cyanol in 20% Ficoll in DEPC-treated water.

5. Running the gel

Fill tank with 1 X MOPS (prepare 1X with DDW). Early protocols added formaldehyde to the running buffer but this is unnecessary. Run the gel at between 100 to 200 V for several hours, until the xylene cyanol dye front has migrated 3 to 4 cm into the gel and the bromophenol blue is about 2/3rds down the gel. Circulate the buffer from end to end every half an hour, especially if running the gel at 200V. When the RNA has run an appropriate distance, photograph the gel including a ruler aligned with the wells. Pre-soak the gel in 20 × SSC while setting up the transfer (about 15 minutes).

B. Northern transfer

1. Setting up transfer

The physical set-up for Northern transfer is identical to that for Southern transfer (see above) except that 20 × SSC is used as the transfer buffer (Note RNA is hydrolysed in strongly alkaline solutions within seconds!) and the membrane we use is Gene Screen Plus. Prewet membrane in DDW the 20 × SSC.

2. Post-transfer handling of the membrane

After overnight transfer, the position and orientation of the wells is marked on the membrane (#1 etc.) and it is rinsed gently in 2 × SSC for 5 minutes before being air dried in the fume hood. The membrane is baked at 80¼C for 2 hours and is then ready for pre-hybridisation and hybridisation.

3. Hybridize in Aqua hyb. with 0.1% SDS, 100 ug/ml herring testes DNA

See reagents.

HYBRIDIZING NORTHERN BLOTS WITH RIBOPROBES

This method is obtained from Bill Clouston's thesis and is essentially that described by Kalinyak and Perlman (JBC 262:460-4, 1987) with the addition of preservative-free sodium heparin to reduce non-specific hybridization.

Method

1. Run Northern in the usual way, transfer to Genescreen plus ( ) in 20X SSC and bake in oven at 80¼C for 2 hours.
2. Hybridization buffer is as follows: 50% Formamide, 3 × SSC, 10X Denhardts', 10 mM phosphate buffer, pH 8.0, 2 mM EDTA, 0.1% SDS, 200 ug/ml herring sperm DNA.
    1. 800 U/ml preservative-free sodium heparin.
    2. Pre-hyb membrane for 4-6 hours at 60¼C.
    3. Hybridize in fresh buffer for 18-24 hours at 65¼C-20 ng of riboprobe to the bag in a small volume of buffer (*e.g.*, 3-5 ml). Riboprobes should prepared exactly as described for hybridization histochemistry. This can include hydrolysis, but this may or may not make a difference to the degree of background.
    4. Wash at high stringency, *i.e.*, 0.1X SSC, 0.1% SDS at 65¼C. Be careful to wash all formamide-containing hybridization buffer off at low stringency first, *i.e.*, use a 2X SSC wash initially then increase stringency progressively. I obtained a good signal with little background using this washing regimen but lost the signal when I washed at 75¼C.
    5. Expose against X-ray film using an intensifying screen at –70¼C for 24 hours or longer. I obtained a good signal after 36 hours, equal to that obtained with Northerns using cDNA probes after 96 hours exposure.

5. RESTRICTION DIGESTION METHODS

RESTRICTION ENZYME DIGESTS

The unit definition of restriction enzyme activity is based on the amount of enzyme required to cut bacteriophage lambda DNA to completion in one hour's time. Assays developed by the manufacturer of the enzymes are most likely done using highly purified DNA (*i.e.*, plasmids or lambda phage DNA) as substrate, and assay conditions that produce the best results with their particular preparation. Often the laboratory conditions are not as ideal, and a slight excess of enzyme or a longer incubation period is used to help ensure complete digestion. There are many 'universal' enzyme buffers which will work with a variety of enzymes, but often they do not meet the most efficient requirements for any one enzyme. Because the universal buffers are not as efficient, we do not recommend their routine use in the lab (especially for complex genomic DNAs; digests of complex genomic DNAs are usually at high DNA concentrations which will inhibit the enzyme; also, the relatively crude DNA preps may contain inhibitors which requires more units enzyme and longer incubation times for complete digestion). It is always best to use the manufacturer's recommended assay conditions for restricition digestion. Some manufacturers have cloned enzymes which have very different requirements from other versions of the same enzyme, so check before using. Enzyme concentrations are always given on the side of the enzyme tube. Restriction enzymes used in the lab are always stored at –20°C (in a glycerol base), and should be kept as close to –20°C as possible to extend the life of the enzyme. When setting up digests, bring the reaction tube to the enzyme freezer in an ice bucket; remove the enzyme tube from the freezer and keep the tube on ice while working with the enzyme. Immediately return the tube to the freezer when finished. Use the pipetmen designated for restriction enzymes only and always use a fresh pipetman tip when removing enzyme from the stock tube.

The *lab* has prepared enzyme buffer stocks. Check the requirements of the enzyme and use the appropriate buffer. Nearly all the manufacturers of enzyme now supply buffers with the enzyme, and these can be found in the enzyme freezer. Never make up a digest having the enzyme as more than 1/10 the final volume; the glycerol will inhibit at higher concentrations. Always set up a control digest when using an enzyme. Use a substrate DNA which has known sites for the enzyme so that you can compare the control result to the expected pattern of fragments on an agarose gel. If preparing double enzyme digests, check the salt content of the buffers and use the enzyme with the lower salt requirements first, then adjust the reaction tube's salt concentration and digest with the second enzyme. Impurities present in some human DNA preparations may inhibit restriction digestion. If you cannot obtain complete digestion after adding additional enzyme, set up another digest and add spermidine to a final concentration in the digest of 2 mM.

Restriction Enzyme Buffers

Enzyme (source)	*concentration in digest*	*10X Buffer*	*recipe for 10X Buffer (50 ml)*	*Store at degrees C*
BamHI (NE Biolabs)	150 mM NaCl 6 mM Tris pH7.9 6 mM $MgCl_2$ 100 µg/ml BSA	1.5 M NaCl 60 mM Tris-HCl 60 mM $MgCl_2$ 1 mg/ml BSA H_2O	15 ml 5 M NaCl 3 ml 1 M Tris-HCl 3 ml 1 M $MgCl_2$ 1 ml 50 mg/ml BSA 28 ml H_2O	–20 to +4°C
BglII (NE Biolabs) = HinfI buffer = TaqI buffer = PstI buffer	100 mM NaCl 10 mM TrispH7.4 10 mM $MgCl_2$ 100 µg/ml BSA	1.0 M NaCl 100 mM Tris-HCl 100 mM $MgCl_2$ 1 mg/ml BSA H_2O	10 ml 5 M NaCl 5 ml 1 M Tris-HCl 5 ml 1 M $MgCl_2$ 1 ml 50 mg/ml BSA 29 ml dH_2O	–20 to +4°C
EcoRI (NE Biolabs)	50 mM NaCl 100 mM TrispH7.5 5 mM $MgCl_2$ 100 µg/ml BSA	0.5 M NaCl 0.82 M Tris-HCl 50 mM $MgCl_2$ 1 mg/ml BSA	5 ml 5 M NaCl 41.5 ml 1 M Tris-HCl 2.5 ml $MgCl_2$ 1 ml 50 mg/ml BSA	-20 to +4°C
HincII	100 mM NaCl	1 M NaCl	10 ml 5 M NaCl	+4°C

(NE Biolabs)	10 mM TrispH7.4 7 mM $MgCl_2$ 1 mM DTT 100 μg/ml BSA	100 mM Tris-HCl 70 mM $MgCl_2$ 10 mM DTT 1 mg/ml BSA +dH_2O	5 ml 1 M Tris-HCl 3.5 ml 1 M $MgCl_2$ 5 ml 100 mM DTT 1 ml 50 mg/ml BSA 25.5 ml dH_2O	
HindIII (MVO lab)	10 mM Tris-HCl 50 mM NaCl 10 mM $MgCl_2$ 10 mM 2-ME 100 μg/ml BSA	100 mM Tris-HCl 0.5 M NaCl 100 mM $MgCl_2$ 100 mM 2-ME 1 mg/ml BSA +dH_2O	5 ml 1 M Tris-HCl 5 ml 5 M NaCl 5 ml 1 M $MgCl_2$ 0.42 ml 12 M 2-ME 1 ml 50 mg/ml BSA 33.58 ml dH_2O	–20°C
MspI (NE Biolabs)	6 mM KCl 10 mMTrispH7.4 10 mM $MgCl_2$ 1 mM DTT 100 μg/ml BSA	60 mM KCl 100 mM Tris-HCl 100 mM $MgCl_2$ 10 mM DTT 1 mg/ml BSA +dH_2O	3 ml 1 M KCl 5 ml 1 M Tris-HCl 5 ml 1 M $MgCl_2$ 5 ml 100 mM DTT 1 ml 50 mg/ml BSA 31 ml dH_2O	+4°C
RsaI (NE Biolabs) = AluI = HaeIII	50 mM NaCl 6 mM TrispH8.0 12 mM $MgCl_2$ 6 mM 2-ME 100 μg/ml BSA	0.5 M NaCl 60 mM Tris-HCl 120 mM $MgCl_2$ 60 mM 2-ME 1 mg/ml BSA +dH_2O	5 ml 5 M NaCl 3 ml 1 M Tris-HCl 6 ml 1 M $MgCl_2$ 0.26 ml 12M 2-ME 1 ml 50 mg/ml BSA 34.74 ml dH_2O	–20°C

Stock solutions for restriction enzyme buffers

Always check the enzyme manufacturer's recommended buffer conditions before setting up digests. Prepare the 10X Buffers by combining all ingredients except the BSA. Filter sterilize, then add the sterile BSA. Prepare 1 ml aliquots, and store at the recommended temperatures. Buffers containing 2-ME can be stored in the refrigerator for a few weeks. If the mercaptoethanol is undetectible in refrigerated buffer, adding 6 μl of the 12M 2-ME stock per ml will restore the buffer

DTT (dithiothreitol), 100 mM stock: Prepare in dH_2O, filter sterilize and store at –20°C

NaCl, 5 M stock: Prepare in dH_2O, filter sterilize, store at room temperature.

KCl, 1 M stock: Prepare in dH_2O, filter sterilize, store at room temperature.

Tris-HCl, pH 7.5-8.3, 1 M stock: Filter sterilize lab stock and store at room temperature.

$MgCl_2$, 1 M stock: Filter sterilize lab stock and store at room temperature.

2-mercaptoethanol (2-ME), 12 M stock: Stored in the refrigerator. Pipette under the hood.

BSA (bovine serum albumin), 50 mg/ml stock: Purchase from BRL, store at –20°C

100 mM Spermidine stock :

1. Weigh out 0.25 g spermidine tetrahydrochloride.
2. Add ddH_2O to bring volume to 10 ml.
3. Filter sterilize and store at 4°C.

EXERCISE 5.1. RESTRICTION DIGESTION OF PLASMID, COSMID, AND PHAGE DNAs

Principle

The following is a generalized example of a restriction digest. Estimate the amount of DNA needed in your digest and scale up accordingly. To visualize a digest on an ethidium bromide-stained agarose gel, you will need to take the size of the fragments and the total size of the clone DNA into account (*e.g.*, 10-50 ng of intact lambda-sized genomes (~50 kb) are easily seen on gels but if cut into small (~1kb fragments), the relative proportion of the clone DNA in each fragment is ~ 1/50 and more DNA (500-1000 ng) should be loaded in order to see them).

If preparing a large number of digests at a time and the DNAs are at the same concentration, prepare a cocktail of the reaction mix then divide it among the tubes of DNA.

Time Required

1.5 hours

Procedure

A general rule of thumb is to use 1 µg clone DNA/ 10 µl in the final digest reaction mix (recommendation 1 µg/20 µl).

Reaction Mix:

1.0 µl	10 X enzyme buffer
0.5 µl	2-3 units restriction enzyme
1-8.5 µl	1 µg DNA (determine concentration on a gel before setting up digest) bring volume to 10 µl with sterile dH_2O

1. Put the water and buffer into the tube first, then add the enzyme (avoid putting enzyme into water first, as it may start to break down). Put the DNA in last and mix by tapping the tube with your finger.
2. Quickspin to remove bubbles (DNA will adhere to bubble surface and becomes inaccessible to the enzyme). Incubate at the recommended temperature for 1 hour.
3. Stop the reaction by adding 2.5 µl 5 X Ficoll dye mix if the sample is to be loaded directly onto a gel; otherwise stop the reaction by placing it at –20°C or add 0.5 µl of 0.5 M EDTA.

EXERCISE 5.2. MANUAL METHOD OF RESTRICTION DIGESTION OF HUMAN DNA

Principle

Human genomic restriction digests are used in the lab principally in generating Southern transfers, *i.e.*, screening, parent and family blots. Most of the digests are now prepared using the Biomek workstation. When only a few samples are needed, it may be faster to prepare them manually, instead of writing a new method for the Biomek.

Because of the complexity of the human genomic DNA (~ 3 billion bp), any restriction digest can be expected to produce what appears to be a continuum of fragment sizes, with no individual fragment bands visible. It is impossible to tell from the ethidium bromide stained pattern whether or not the DNA digest is complete or only partially complete. To help determine the extent of digestion, we set up a test digest. The test digest is prepared after the human genomic digest is set up: an aliquot of the human digest is combined with 1 μg lambda DNA (lambda being a relatively simple genome of ~48 kb in size). The complete digestion of the lambda genome produces a distinct banding pattern on a gel superimposed upon the faint smear of the human genomic DNA. We assume that if the lambda DNA had been digested to completion, the human DNA digest must also be complete. After the incubation period, the human-only digests are stored in the freezer until the result of the test digests is known.

The standard Southern blots in our lab uses 4 μg of genomic DNA per gel lane. We prepare digests for a minimum of two blots, and always include an additional μg of DNA for the test digest, (therefore the minimum amount of DNA to be digested for Southerns is 9 μg, and the following procedure is based on this amount). The human DNAs used in the lab are adjusted to a concentration of 200-250 μg/ml. The digests are prepared with the human DNA at a concentration of 167 μg/μl.

Time Required

2-3 days total time.

Procedure

Day 1. For each digest use (scale up accordingly):

36 μl human DNA (9 μg)

18 μl enzyme cocktail

54 μl total volume.

1. Enzyme cocktail: Prepare about 10% more of the ezyme cocktail than required, to allow for pipetting error, losses, etc.

   To prepare cocktail for one 9 μg digest:

   (scale up depending on the total number of digests), 5.4 μl 10X specific restriction enzyme buffer 18-45 units of enzyme (units = 2 – 5 × the # μgs), bring volume to 18 μl with sterile dH_2O.

2. Add 18 μl of the cocktail to 36 μl DNA in a labelled eppendorf tube. Mix by tapping the tube with your finger. Quick-spin to remove bubbles.

3. Prepare the test digest: remove 6 μl (= 1 μg human DNA) to another labelled tube containing 1 μg (1-2 μl) lambda DNA. Mix and quick-spin as before.

4. Prepare one control digest for each restriction enzyme used: combine 2 µl enzyme cocktail and 1µg lambda DNA in an eppendorf tube. Bring the volume to 8 µl using sterile dH_2O
5. Incubate both sets of digests in the same incubator at the appropriate temperature for 8 hours to overnight.

Day 2

1. Place the human-only digests in a –20°C freezer (store until the results of the test digest is known).
2. Add 1 µl 10X glycerol dye mix to the test and control digests and run the samples on an agarose gel. Also load a BRL 1kb ladder marker lane. Run the gel until the dyes are separated at least 1 inch. Stain and photograph.
3. Add another 3-5 units enzyme/µg DNA (= 24-40 µ) to the incomplete digests (be careful to keep the total amount of enzyme added less than 1/10 total digest volume). Incubate for another 6 hours to overnight. If the test digest appeared less than 80% complete, also prepare another test and control digest (otherwise, assume the extra enzyme and incubation will complete it).
4. Add 5 µl 10X glycerol stop mix to each of the complete digests, mix and quick-spin. Store digests in the –20°C freezer (good for a few months) or in the –80°C freezer for long-term storage.
5. For 4 µg/lane (southern gels), load 26 µl of the digest.

EXERCISE 5.3. PREPARATION OF HIGH MOLECULAR WEIGHT HUMAN DNA RESTRICTION FRAGMENTS IN AGAROSE PLUGS

Purpose

To prepare intact mammalian chromosomal DNA for use in pulsed-field electrophoresis mapping experiments using rare cutting restriction enzymes.

Time Required

4 days.

Special Reagents

- Phenylmethylsulfonyl fluoride (PMSF).

Procedure

Day 1-2

1. Place freshly grown mammalian cells in 15 ml Falcon tube and centrifuge in Beckman tabletop centrifuge at 1000 rpm for 15 minutes. Pour off supernatant, and resuspend pellet in 10 ml of phosphate buffered saline (PBS). Centrifuge at 1000 rpm for 15 minutes again, then resuspend in PBS to achieve a cell concentration of 1-2 X 107 cells/ml (refer to methods in tissue culture regarding hemocytometer for cell counting procedure).
2. Mix an equal volume of cells (1-2 ml) with premelted 1% low gelling agarose which has been cooled to 45-50°C, using a pasteur pipette; immediately distribute the mixture to plug molds for the CHEF apparatus. Place the mold on ice for 15 minutes.

3. Remove plugs, then incubate in 6 well tissue culture plates (5 ml/well, several plugs/well) in ESP solution at 50°C with gentle shaking for two days; be sure to seal the 6 well plate securely with parafilm to prevent evaporation of the samples. Plugs can now be stored in ESP at 4°C, or one can proceed as follows.

Day 3

1. For each digest condition, use approximately 1/4 of a plug, roughly the size that would fill a well on the CHEF DR apparatus. Place each in 1 ml of 1 mM PMSF (caution: extremely toxic to mucous membranes, do not inhale or allow contact with skin) diluted from 0.1 M stock in TE (pH 8), and slowly rotate plates at room temperature for 2 hours. Replace with new PMSF aliquot and repeat. Remove PMSF, and wash in 1 ml of TE (pH 8) three times for 1-2 hours each by slow rotation at room temperature.
2. Place each plug in 250 µl of appropriate restriction buffer (1X) in eppendorf tubes at room temperature for 30 minutes. Aspirate buffer carefully, then replace with fresh buffer and hold at room temperature for 30 minutes. Replace with 250 µl of fresh buffer, then add 25 µg of BSA to each tube. Add 20 units of enzyme per µg of DNA to tubes containing the appropriate buffer (1/4 of one plug should be approximately equal to 5-10 µg of DNA if the cell counts are correct). Incubate overnight at the appropriate temperature; additional enzyme can be added the next morning and incubation carried out for 4 more hours if so desired.

Day 4

1. Replace buffer with 1 ml of ES solution, and incubate at 50°C for 2 hours with gentle shaking.
2. Replace ES with 250 µl ESP solution and incubate at 50°C for an additional 2 hours.
3. Load the plugs into a gel poured for the CHEF DR apparatus. Run under conditions appropriate for the fragment sizes expected. Fragments in the 200-2200 kb size range can be effectively separated on a 1.0% agarose gel run at 200V with a switch time of 60 seconds for 15 hours, and 90 seconds for 9 hours.

Solutions

- *ESP solution:*
- 0.5 M EDTA, pH 9.0-9.5
- 1% sodium laurylsarcosine

  1 mg/ml Proteinase K
- *PMSF stock:*

  0.1 M PMSF in 2-Propanol
- *ES solution:*
- 0.5 M EDTA, pH 9.0-9.5

  1% sodium laurylsarcosine

M.3: Restriction Enzyme Digestion of DNA

Materials

10X restriction enzyme buffer (see manufacturer's recommendation)

DNA

sterile water

restriction enzyme

phenol : chloroform (1 : 1)

Procedure

1. Add the following to a microfuge tube:

   2µl of appropriate 10X restriction enzyme buffer. 0.1 to 5 µg DNA sterile water to a final volume of 19 µl (Note: These volumes are for analytical digests only. Larger volumes may be necessary for preparative digests or for chromosomal DNA digests.
2. Add 1 to 2 µl (3 to 20 units) enzyme and mix gently. Spin for a few seconds in microfuge.
3. Incubate at the appropriate temperature (usually 37EC) for 1 to 2 hours.
4. Run a small aliquot on a gel to check for digestion.
5. If the DNA is to be used for another manipulation, heat inactivate the enzyme (if it is heat labile) at 70EC for 15 min, phenol/chloroform extract and ethanol precipitate, or purify on Qiagen DNA purification column.

EXERCISE 5.4. ELECTROELUTION OF DNA FRAGMENTS FROM AGAROSE INTO DIALYSIS TUBING

Purpose

To retrieve and purify any specific DNA fragment from an agarose gel slice; the expected yield is from 50-75% of the amount in the gel slice.

Time Required

1. Restriction digest-minimum of 2 hours
2. Electrophoresis-approx. 4 hours
3. Elution-2-5 hours depending on fragment size
4. DNA Purification-2 hours.

Special Equipment

- SPECTRA/POR dialysis tubing (VWR, B, F 10 mm × 50 ft., cat.#132645) and clips.

Procedure

The restriction digest. Use 5-10 µg DNA with the appropriate enzyme digest (see Restriction Digest Methods) for a minimum of 2 hours, maximum of overnight. If available, use a restriction enzyme map of the DNA to be digested to determine what enzyme gives the best fragment separation and what the size of the fragments of interest should be. For example, cutting a plasmid into two similar size fragments makes separating the fragments difficult. The restriction maps can identify another enzyme which only cuts one of these fragments (usually the vector). By cutting with both enzymes, the unwanted fragment can be cut into smaller pieces giving better isolation of the wanted fragment. Note: It is always helpful to run a small aliquot of the digest (250-500 ng) on a mini-gel to visualize the fragment pattern and determine how long to run the gel to get the desired separation, as well as to test for complete digestion.

Electrophoresis

1. Generally, a 0.8% TA gel will give sufficient separation for most digested fragments. Increase agarose to 1.2% to isolate small fragments of 100 to 500 base pairs, decrease agarose to 0.6% for fragments larger than 8 kb. 10 µg of digested DNA will need to be spread out over approximately 6 cm of wells. Some combs have 6 cm "slots" or wells or this can be achieved by taping up 8-10 individual wells. Individual wells can be used but some DNA will be lost due to the trailing up effect from the well edges.
2. Run the gel at 60-80 volts until desired separation has occurred (depending on fragment sizes). Stain the gel for 30 minutes with ethidium bromide, but do not photograph gel yet because short wave UV can damage the DNA.
3. Place Saran Wrap on the FOTODYNE Model 3-3500 UV light box. Transfer the gel to the Saran Wrap and turn on "PREPARATIVE" UV source (long wave). The fragments should become visible.
4. Use a clean razor blade to cut above and below each fragment of interest minimizing the amount of agarose in the slice. Then free the slice by cutting the ends. Leave a small amount of the fragment ends in the gel to verify fragment sizes in the photograph. Minimize the amount of time the DNA is exposed to UV by having sterile labelled tubes ready for the slices.
5. Photograph the gel after the fragment slices have been removed. Sometimes the fragment bands are too faint to be seen with long-wave and you must use short-wave. Again, try to minimize any UV exposure.

Elution

1. Cut a piece of dialysis tubing approximately 3 cm longer than the gel slice and clip one end. Gently push gel slice into open end and down to the clip. Add 300-500 ml of buffer (same as the gel, *e.g.*, if the gel is 1 X TA, then use 1 X TA Buffer) so that the gel slice is completely immersed and there are no bubbles. Clip the open end.

   Note: Gel slices will sit better in the electrophoresis box if clips are positioned identically.
2. Place gel slices parallel to the electrodes and fill the electrophoresis box with buffer (again, same as original gel) until all tubing is submerged. Then remove some buffer until clip edges stabilize and rest on bottom. Electroelute at 80-100 volts for 2-5 hours, longer for large fragments. Monitor the movement of the DNA with a hand held UV source (long-wavelength).

3. DNA may stick to dialysis tubing after current flow. Either reverse the electrodes and run for 15 seconds to dislodge the DNA or be careful to resuspend the DNA before removing the buffer. To resuspend DNA inside the tubing, first open one clip and carefully remove the agarose slice (which can be restained to verify elution) without losing any buffer. Then replace the clip and with your fingers, press along the length of tubing to mix the DNA with the buffer. Open the clip and remove the buffer with a pipetman. If low yields are suspected, add another 100 µl 1X buffer to the empty tubing and repeat these steps to rinse out the tubing. Final volume should be less than 600 µl. If the agarose slice is much longer than 4 cm, the amount of buffer retrieved may be much more than 600 µl. In this case split the DNA buffer into two or three tubes so that no tube has more than 600 µl.

DNA Cleaning and Precipitation

1. Add 1 volume (500 µl) of phenol and mix (Note: hard vortexing will shear large fragments). Spin in microcentrifuge at 14,000 rpm (at 4°C or at room temperature) for five minutes. Phenol sinks and the DNA (in the aqueous phase) will be on top. Remove the top aqueous phase without pulling any debris at the interphase (don't try to get it all - the DNA quality will be better). Place the aqueous phase in clean eppendorf tubes.
2. Add 1 volume (500 µl) of chloroform, mix, and spin in microcentrifuge and 14,000 rpm for two minutes. Chloroform will sink and the DNA will be in the aqueous phase. Remove the top aqueous phase leaving behind any debris at the interphase and transfer to clean eppendorf tubes.
3. Add 1/10 volume (50 µl) 3M Na Acetate and 800 µl 95% EtOH (EtOH must be at –20°C, or place tubes in the –20°C freezer for 15 minutes after adding EtOH). Spin in microcentrifuge at 14,000 rpm, 4°C for 30 minutes. Decant supernatant and add 500 µl 70% EtOH (must be cold!) to the pellet. Spin again for five minutes, decant the ethanol wash and invert the tubes to dry. For low yields or barely visible pellets, speed vacuum drying is recommended.
4. When the DNA pellet is dry, add 20-50 µl 1X TE ; the pellets should resuspend fairly easily. Quantitate on a mini-gel with several lambda DNA standards and 1kb ladder for sizing.

Safety Precautions

Ethidium bromide is a powerful mutagen and moderately toxic. Gloves should be worn when handling gel and gel slices. UV can cause severe burns. Always wear eye protection. Phenol can cause severe burns and gloves should be worn. Phenol waste must be contained and disposed of through WUMS Hazardous Waste Department. Chloroform is a carcinogen.

EXERCISE 5.5. ISOLATION OF RESTRICTION FRAGMENTS FROM AGAROSE GELS BY COLLECTION ONTO DEAE CELLULOSE

Principle

A DNA restriction fragment is isolated by collection on DEAE cellulose paper during electrophoresis, then washed from the DEAE with a high salt buffer, cleaned, precipitated, and resuspended in a small volume. Recovery of 50-90% of the bound DNA can be expected, however fragments larger than 7 kb have lower yields. DNA prepared this way is suitable for subcloning.

A DNA fragment of a given size migrates at different rates through gels containing different concentrations of agarose. By using a gel at the appropriate gel concentration, it is possible to resolve well the DNA of interest. Use the following table as a guide for determining the agarose concentration to use.

% agarose in the gel	*Efficient range of separation of linear DNA molecules (kb)*
0.3	60-5
0.6	20-1
0.7	10-0.8
0.9	7-0.5
1.2	6-0.4
1.5	4-0.2
2.0	3-0.1

Time Required

1. 3-4 hours on Day 1
2. 2-4 hours on Day 2

Special Materials

Schleicher & Schuell NA-45 DEAE membrane.

Procedure

Day 1. Isolating the fragment:

1. Run the restriction digest and the appropriate size markers on a 1X Tris-Borate agarose gel with ethidium bromide. Be sure to leave at least 1-2 wells between samples. Run the gel until the DNA bands are well separated (visualize on the long-wavelength UV lightbox.
2. Cut a slit just ahead of the band of interest using a sharp sterile razor blade or scalpel. Using blunt-edged forceps (such as Millipore forceps) carefully insert an NA-45 paper into the slit (prewet and cut NA-45 to the width of the band, see preparation of thc NA-45 bclow).
3. Place the gel in fresh 1X Tris-Borate buffer, and run the gel until the fragment has moved out of the gel and stopped by the NA-45. Monitor the progress of the band with the hand-held long wavelength UV light. Do not allow other bands of higher molecular weight to run onto the NA-45.
4. Remove the NA-45 paper, rinse in NET buffer and place in a labelled eppendorf tube. Add sufficient high-salt NET buffer to cover most of the membrane (typically 150-300 μl). Spin 5 seconds in a microcentrifuge to submerge the entire strip. Place at 65 degrees C for 1 hour, mixing frequently, and respinning if the membrane rides up the side of the tube.
5. Transfer the buffer (+DNA fragment) to a clean labelled tube. Wash the membrane (in the original tube) with 50 μl high salt NET buffer and add the wash to the DNA fragment tube.

Cleaning the DNA

1. To remove ethidium bromide, extract twice with 3 volumes water-saturated n-butanol.
2. Precipitate the DNA with 2.5 volumes of ethanol at –20°C for at least 1 hour (can sit overnight in the freezer).

Day 2

1. Pellet the DNA (for 20 minutes at high speed in a microcentrifuge) and resuspend in 50 µl TE. Reprecipitate with sodium acetate to remove any residual NaCl: Add 5 µl 3M Na-acetate, and 120 µl ethanol, hold at –20°C for 2 hours or more, pellet as before and resuspend in an appropriate amount of TE.

Solutions

- *Preparation of the DEAE cellulose membrane.* Schleicher and Schuell NA-45 can be used as supplied by the manufacturer, with prewetting in sterile dH_2O. However, the binding capacity of the membrane is increased with the following: 10 minute soak in 10 mM EDTA pH 7.6, then 5 minutes in 0.5 N NaOH, followed by several rapid washes in sterile dH_2O. Membranes can be stored for several weeks in sterile dH_2O at 4°C.
- *NET buffer (500 ml).* 25 ml 3 M NaCl, 150 mM NaCl, 100 ml 0.5 M EDTA, 100 mM EDTA,10 ml 1 M Tris pH 7.5, 20 mM Tris pH 7.5, 365 ml dH_2O. Autoclave to sterilize.
- *High salt NET buffer (500 ml).* 166.7 ml 3M NaCl, 1 M NaCl, 100 ml 0.5 M EDTA, 100 mM EDTA,10 ml 1M Tris pH 7.5, 20 mM Tris pH7.5, 223.3 ml dH_2O. Autoclave to sterilize.

EXERCISE 5.6. LIGATION OF INSERT DNA TO VECTOR DNA

Introduction

DNA cloning requires the DNA sequence of interest to be inserted in a vector DNA molecule. For this, both the vector as well as insert DNA is prepared by digestion with compatible restriction enzymes, so that the ends produced during digestion is complimentary in both. When setting up ligation it is important to consider the permutations that can occur and bias the relative concentration of DNA accordingly.

Usually a 5 to 10 fold excess of insert over the vector DNA is the norm. This ensures that enough ligated product will be produced in the right orientation.

Requirements

(*a*) Vector digest

(*b*) Insert DNA

(*c*) T4 DNA ligase

(*d*) Ligation buffer: 50 mM Tris-HCI, 10 mM $MgCl_2$, 20 mM DTT, 1 mM ATP, 50/µg/ml.

(*e*) Nuclease free BSA, pH 7.4 (generally supplied with the enzyme).

Procedure

1. Set up ligation in 0.5/ll microfuge tubes as follows:
2. Voetoriligest (DHa.5)
3. Insert DNA
4. Ligation buffer
5. Ligase
6. Mix well by flicking the tube.
7. Spin briefly in a microfuge to put all the liquid to the bottom of the tube. Incubate at 16 degrees overnight.
8. Stop the reaction by placing at 70 degrees for 10 minutes.
9. Run an aliquot on a minigel and verify.

Notes

The ligation buffer supplied by the vector along with the enzyme usually contains ATP. However, one should verify this and in case, their is not ATP or if the buffer is little old it should be supplemented with ATP.

The incubation temperature in step 4 depends on the ends. In case the ends are sticky and there is a four base overhang than 16 degrees is optimum. In case of 2 base overhang and blunt ends the optimum temperature is 12 and 4° respectively. This is because to anneal shorter fragment lower temperature is required. However, lower temperature reduces enzyme activity and enzyme concentration should be adjusted as such.

T4 DNA ligase is much more efficient than *E.Coli* DNA ligase, specially in blunt end ligation and preferred accordingly.

Most preparations of T4 DNA ligase are calibrated in Weiss units. 0.015 Weiss unit of T4 DNA ligase will ligate 50% of Hind III fragments of bacteriophase lambda (5 /-tg) in 30 minutes at 16 degrees. Joining of blunt ended molecule to one another is improved greatly by addition of noncovalent cation (150 to 200 mM NaCl) and low concentration of PEG.

Observations

Run aliquots of ligated DNA side by side with unligated DNA on a minimize and observe the difference. Being circular ligated plasmid DNA _will tend to run faster than unligated plasmid DNA.

6. PCR METHODS (POLYMERASE CHAIN REACTION)

Who would have thought a bacterium hanging out in a hot spring in Yellowstone National Park would spark a revolutionary new laboratory technique? The polymerase chain reaction, now widely used in research laboratories and doctor's offices, relies on the ability of DNA-copying enzymes to remain stable at high temperatures. No problem for Thermus aquaticus, the sultry bacterium from Yellowstone that now helps scientists produce millions of copies of a single DNA segment in a matter of hours.

In nature, most organisms copy their DNA in the same way. The PCR mimics this process, only it does it in a test tube. When any cell divides, enzymes called polymerases make a copy of all the DNA in each chromosome. The first step in this process is to "unzip" the two DNA chains of the double helix. As the two strands separate, DNA polymerase makes a copy using each strand as a template.

The four nucleotide bases, the building blocks of every piece of DNA, are represented by the letters A, C, G, and T, which stand for their chemical names: adenine, cytosine, guanine, and thymine. The A on one strand always pairs with the T on the other, whereas C always pairs with G. The two strands are said to be complementary to each other.

To copy DNA, polymerase requires two other components: a supply of the four nucleotide bases and something called a primer. DNA polymerases, whether from humans, bacteria, or viruses, cannot copy a chain of DNA without a short sequence of nucleotides to "prime" the process, or get it started. So the cell has another enzyme called a primase that actually makes the first few nucleotides of the copy. This stretch of DNA is called a primer. Once the primer is made, the polymerase can take over making the rest of the new chain.

A PCR vial contains all the necessary components for DNA duplication: a piece of DNA, large quantities of the four nucleotides, large quantities of the primer sequence, and DNA polymerase. The polymerase is the Taq polymerase, named for Thermus aquaticus, from which it was isolated.

The three parts of the polymerase chain reaction are carried out in the same vial, but at different temperatures. The first part of the process separates the two DNA chains in the double helix. This is done simply by heating the vial to 90-95 degrees centigrade (about 165 degrees Fahrenheit) for 30 seconds.

But the primers cannot bind to the DNA strands at such a high temperature, so the vial is cooled to 55 degrees C (about 100 degrees F). At this temperature, the primers bind or "anneal" to the ends of the DNA strands. This takes about 20 seconds.

The final step of the reaction is to make a complete copy of the templates. Since the Taq polymerase works best at around 75 degrees C (the temperature of the hot springs where the bacterium was discovered), the temperature of the vial is raised.

The Taq polymerase begins adding nucleotides to the primer and eventually makes a complementary copy of the template. If the template contains an A nucleotide, the enzyme adds on a T nucleotide to the primer. If the template contains a G, it adds a C to the new chain, and so on to the end of the DNA strand. This completes one PCR cycle.

The three steps in the polymerase chain reaction - the separation of the strands, annealing the primer to the template, and the synthesis of new strands - take less than two minutes. Each is carried out in the same vial. At the end of a cycle, each piece of DNA in the vial has been duplicated.

But the cycle can be repeated 30 or more times. Each newly synthesized DNA piece can act as a new template, so after 30 cycles, 1 million copies of a single piece of DNA can be produced! Taking into account the time it takes to change the temperature of the reaction vial, 1 million copies can be ready in about three hours.

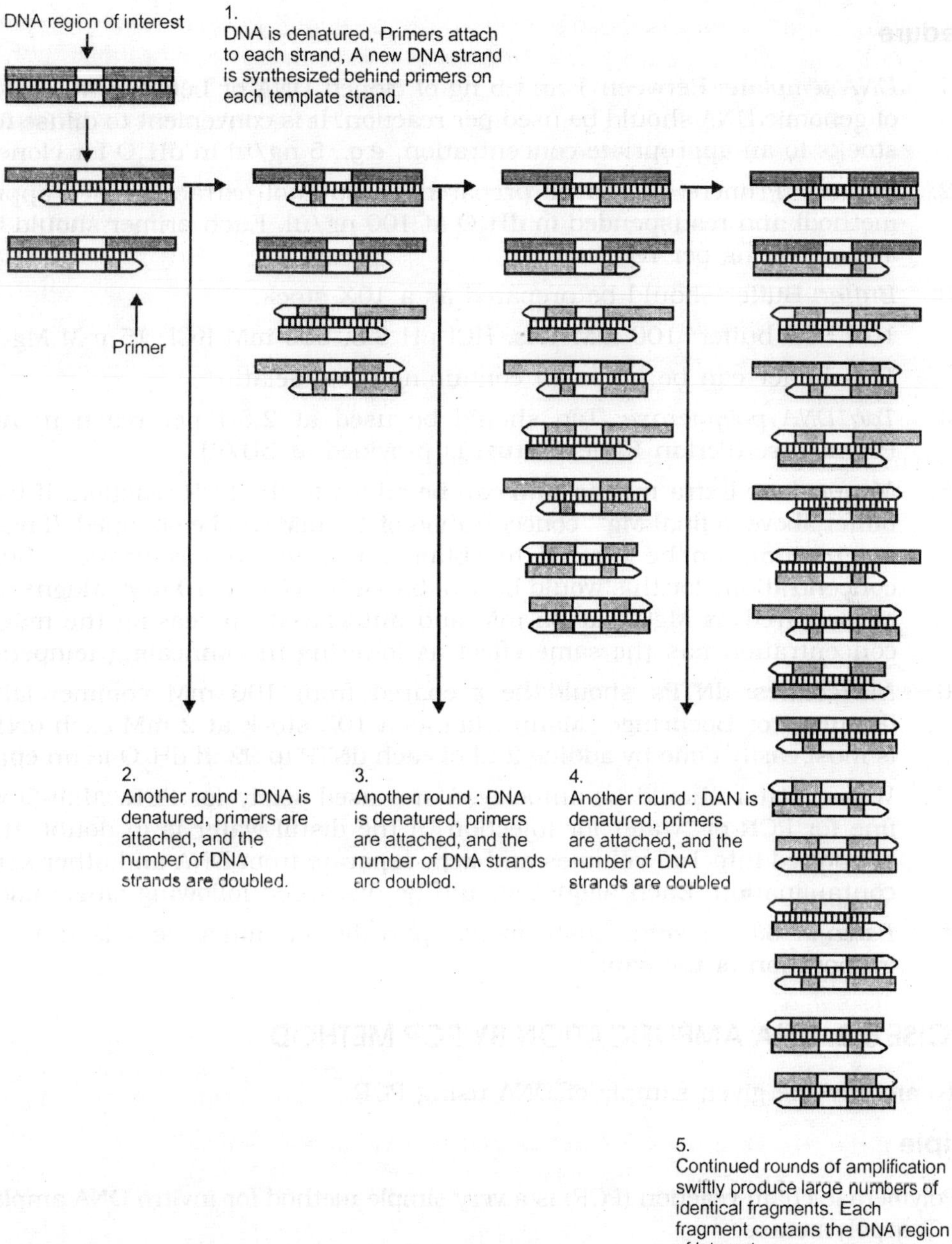

In one application of the technology, small samples of DNA, such as those found in a strand of hair at a crime scene, can produce sufficient copies to carry out forensic tests.

EXERCISE 6.1. POLYMERASE CHAIN REACTION

Contamination of reactions is an often encounted problem with PCR. It is best avoided by using dedicated reagents, Gilsons, and plugged tips. Setting up a PCR reaction in a different location to where template or products are worked with is also a good precaution.

Procedure

1. *DNA template:* Between 1 and 5 ng of cloned DNA or between 40 and 100 ng of genomic DNA should be used per reaction. It is convenient to diluse template stocks to an appropriate concentration, *e.g.*, 5 ng/ul in dH_2O for cloned DNA.
2. *Primers:* Primers should be prepared by butanol extraction (see appropriate method) and resuspended in dH_2O at 100 ng/ul. Each primer should be used at ca. 100 ng per reaction.
3. *Buffer:* Buffer should be prepared as a 10X stock.
   10X PCR buffer: 100 mM Tris. HCl pH 8.3, 500 mM KCl, 15 mM $MgCl_2$.
   This buffer can be prepared containing 0.1% gelatin.
4. *Taq DNA polymerase:* Taq should be used at 2.5U per reaction. Amplitaq Polymerase (Perkin Elmer Cetus) is provided at 5U/ul.
5. *Magnesium:* Extra magnesium can be added to the PCR reaction. If using the buffer above, a final Mg^{2+} concentration of 1.5 mM will be obtained. If necessary, magnesium can be titrated to obtain an optimal concentration. Suggested concentrations for this would be 1.5, 3.0, 4.5, 6.0 and 10 mM. Magnesium can be prepared as $MgCl_2$ at 25 mM and autoclaved. Increasing the magnesium concentration has the same effect as lowering the annealing temperature.
6. *Nucleotides:* dNTPs should be prepared from 100 mM commercial stocks (Promega or Boehringer Mannheim) as a 10X stock at 2 mM each dNTP. This is most easily done by adding 2 ul of each dNTP to 92 ul dH_2O in an eppendrof.
7. *Water:* Water should be autoclaved and used solely for PCR. Milli-Q water is fine for PCR or "water for injection" if the distill water is in doubt. It can be aliquotted into 1ml volumes and kept separate from DNA and other sources of contamination. Each aliquot should be discarded following single use.
8. *Parrafin oil:* In some instruments parrafin oil must be added to prevent evaporation of the sample.

EXERCISE 6.2. DNA AMPLIFICATION BY PCR METHOD

To amplify the given sample of DNA using PCR.

Principle

Polymerase chain reaction (PCR) is a very simple method for invitro DNA amplification using Taq polymerase.

This technique innolves DNA synthesis in 3 simple steps.

Step 1. Denaturation of the template into single strands.

Step 2. Annealing of primers to the template.

Step 3. Extension of new DNA.

Materials Required

Millique water or autoclaned double distilled water, sterile microfuge (0.2 ml), 10 x Taq polymerase assay buffer with MgClz, or NTP mix solution, template DNA, forward and serverse primers, Taq DNA polymerase enzyme, sterile mineral oil and PCR machine.

Procedure

1. 38 ml of sterile millique water (or autoclaved double distilled water) was added to a sterile.
2. To the microfuge 5 of 10 X Taq poly merase assay buffer with MgCl2 was added.
3. 3 ml of 2.5 mm dNTP mixed solution was also added to microfuge.
4. 1 ml of control template DNA was added
5. 1 mol of each forward and senerse primers were also added.
6. Then 1-2 units (0.5-0.7 ml) of Taq DNA polymerase was added.
7. The solution was gently mixed
8. The reaction mixture was layered with 50 ml of mineral oild to avoid enapouration.
9. The amplification was then carried out using the following reaction conditions:
10. Initial denaturation at 940C for 1 min.
11. Denaturation at 940C for 30 sec.
12. Annealing at 480C for 30 sec.
13. Extension at 720C for 1 min
14. Final extension at 720C for 2 min.

Result

The DNA was amplified.

9 TISSUE CULTURE TECHNIQUES

1. TISSUE CULTURE METHODS

I. TYPES OF CELLS GROWN IN CULTURE

Tissue culture is term that refers to both organ culture and cell culture. Cell cultures are derived from either primary tissue explants or cell suspensions. Primary cell cultures typically will have a finite life span in culture whereas continuous cell lines are, by definition, abnormal and are often transformed cell lines.

II. WORK AREA AND EQUIPMENT

(a) *Laminar Flow Hoods.* There are two types of laminar flow hoods, vertical and horizontal, and both types of hoods are available in the Applied Molecular Biology and microbiology laboratory. The vertical hood is best for working with hazardous organisms since the aerosols that are generated in the hood are filtered out before they are released into the surrounding environment. Horizontal hoods are designed such that the air flows directly at the operator hence they are not useful for working with hazardous organisms but are the best protection for your cultures.

Both types of hoods have continuous displacement of air that passes through a HEPA (high efficiency particle) filter that removes particulates from the air. The hoods are equipped with a short-wave UV light that can be turned on for a few minutes to sterilize the surfaces of the hood, but be aware that only exposed surfaces will be accessible to the UV light. Do not put your hands or face near the hood when the UV light is on as the short wave light can cause skin and eye damage. The hoods should be turned on about 10-20 minutes before being used. Wipe down all surfaces with ethanol before and after each use.

(b) *Microscopes.* Inverted phase contrast microscopes are used for visualizing the cells. Microscopes should be kept covered and the lights turned down when not in use. Before using the microscope or whenever an objective is changed, check that the phase rings are aligned.

(c) *CO_2 Incubators.* The cells are grown in an atmosphere of 5-10% CO_2 because the medium used is buffered with sodium bicarbonate/carbonic acid and the pH must be strictly maintained. Culture flasks should have loosened caps to allow for sufficient gas exchange. The humidity must also be maintained for those cells growing in tissue culture dishes so a pan of water is kept filled at all times.

(*d*) *Preservation.* Cells are stored in liquid nitrogen.

(*e*) *Vessels.* Anchorage dependent cells require a nontoxic, biologically inert, and optically transparent surface that will allow cells to attach and allow movement for growth. The most convenient vessels are specially-treated polystyrene plastic that are supplied sterile and are disposable. These include petri dishes, multi-well plates, microtiter plates, roller bottles, and screwcap flasks.

III. PRESERVATION AND STORAGE

Liquid N_2 is used to preserve tissue culture cells, either in the liquid phase or in the vapor phase. Freezing can be lethal to cells due to the effects of damage by ice crystals, alterations in the concentration of electrolytes, dehydration, and changes in pH. To minimize the effects of freezing, several precautions are taken. First, a cryoprotective agent which lowers the freezing point, such as glycerol or DMSO, is added.

The freezing medium we typically use is 90% serum, 10% DMSO. In addition, it is best to use healthy cells that are growing in log phase and to replace the medium 24 hours before freezing. Also, the cells are slowly cooled from room temperature to –80EC to allow the water to move out of the cells before it freezes. The optimal rate of cooling is 1E-3EC per minute. Some labs have fancy freezing chambers to regulate the freezing at the optimal rate by periodically pulsing in liquid nitrogen.

The filled with 200 ml of isopropanol at room temperature and the freezing vials containing the cells are placed in the container and the container is placed in the –80EC freezer. The effect of the isopropanol is to allow the tubes to come to the temperature of the freezer slowly, at about 1EC per minute.

Once the container has reached –80EC the vials are removed and immediately placed in the liquid nitrogen storage tank. Cells are stored at liquid nitrogen temperatures because the growth of ice crystals is retarded below –130EC.

IV. MAINTENANCE

Cultures should be examined daily, observing the morphology, the colour of the medium and the density of the cells. A tissue culture log should be maintained. The log should contain, the name of the cell line, the medium components and any alterations to the standard medium, the dates on which the cells were split and/or fed, a calculation of the doubling time of the culture (this should be done at least once during the semester), and any observations relative to the morphology, etc.

(*a*) *Growth Pattern.* Cells will initially go through a quiescent or lag phase that depends on the cell type, the seeding density, the media components, and previous handling. The cells will then go into exponential growth where they have the highest metabolic activity. The cells will then enter into stationary phase where the number of cells is constant, this is characteristic of a confluent population (where all growth surfaces are covered).

(*b*) *Harvesting.* Cells are harvested when the cells have reached a population density which suppresses growth. Ideally, cells are harvested when they are in a semi-confluent state and are still in log phase. Cells that are not passaged and are allowed to grow to a confluent state can sometime lag for a long period of time and some may never recover. It is also essential to keep your cells as happy as possible to maximize the efficiency of transformation. Most cells are passaged (or at least fed) three times a week.

1. *Suspension cultures.* Suspension cultures are fed by dilution into fresh medium.
2. *Adherent cultures.* Adherent cultures that do not need to be divided can simply be fed by removing the old medium and replacing it with fresh medium. When the cells become semi-confluent, several methods are used to remove the cells from the growing surface so that they can be diluted:

Mechanical. A rubber spatula can be used to physically remove the cells from the growth surface. This method is quick and easy but is also disruptive to the cells and may result in significant cell death. This method is best when harvesting many different samples of cells for preparing extracts, *i.e.*, when viability is not important.

Proteolytic enzymes. Trypsin, collagenase, or pronase, usually in combination with EDTA, causes cells to detach from the growth surface. This method is fast and reliable but can damage the cell surface by digesting exposed cell surface proteins. The proteolysis reaction can be quickly terminated by the addition of complete medium containing serum.

EDTA. EDTA alone can also be used to detach cells and seems to be gentler on the cells than trypsin.

The standard procedure for detaching adherent cells is as follows:

1. Visually inspect daily
2. Release cells from monolayer surface
   (*a*) Wash once with a buffer solution
   (*b*) Treat with dissociating agent
   (*c*) Observe cells under the microscope. Incubate until cells become rounded and loosen when flask is gently tapped with the side of the hand.
   (*d*) Transfer cells to a culture tube and dilute with medium containing serum.
   (*e*) Spin down cells, remove supernatant and replace with fresh medium. Count the cells in a hemacytometer, and dilute as appropriate into fresh medium.

(c) *Media and Growth Requirements*

1. *Physiological parameters*
   A. Temperature –37°C for cells from homeotherms
   B. pH–7.2–7.5 and osmolality of medium must be maintained
   C. Humidity is required
   D. Gas phase-bicarbonate conc. and CO_2 tension in equilibrium
   E. Visible light can have an adverse effect on cells; light induced production of toxic compounds can occur in some media; cells should be cultured in the dark and exposed to room light as little as possible.
2. *Medium requirements: (often empirical)*
   A. Bulk ions—Na, K, Ca, Mg, Cl, P, Bicarb or CO_2
   B. Trace elements iron, zinc, selenium
   C. Sugars glucose is the most common
   D. Amino acids-13 essential

E. Vitamins-B, etc.

F. Choline—inositol

G. Serum contains a large number of growth promoting activities such as buffering toxic nutrients by binding them, neutralizes trypsin and other proteases, has undefined effects on the interaction between cells and substrate, and contains peptide hormones or hormone-like growth factors that promote healthy growth.

H. Antibiotics although not required for cell growth, antibiotics are often used to control the growth of bacterial and fungal contaminants.

For our purposes, we will use the following media components:

Basal medium—IMDM

Serum-10% fetal calf

1. Glutamine-1% - an essential amino acid that tends to be unstable - it is typically stored frozen and added separately; its half-life in medium at 4EC is 3 weeks, at 37EC-1 week.
2. Antibiotic/antimycotic-1% (streptomycin, amphotericin B, penicillin; spectrum: bacteria, fungi and yeast).
3. Feeding—2-3 times/week.
4. Measurement of growth and viability. The viability of cells can be observed visually using an inverted phase contrast microscope. Live cells are phase bright; suspension cells are typically rounded and somewhat symmetrical; adherent cells will form projections when they attach to the growth surface. Viability can also be assessed using the vital dye, trypan blue, which is excluded by live cells but accumulates in dead cells. Cell numbers are determined using a hemocytometer.

V. SAFETY CONSIDERATIONS

Assume all cultures are hazardous since they may harbor latent viruses or other organisms that are uncharacterized. The following safety precautions should also be observed:

1. Pipetting: use pipette aids to prevent ingestion and keep aerosols down to a minimum
2. No eating, drinking, or smoking
3. Wash hands after handling cultures and before leaving thc lab
4. Decontaminate work surfaces with disinfectant (before and after)
5. Autoclave all waste
6. Use biological safety cabinet (laminar flow hood) when working with hazardous organisms. The cabinet protects worker by preventing airborne cells and viruses released during experimental activity from escaping the cabinet; there is an air barrier at the front opening and exhaust air is filtered with a HEPA filter
7. Make sure cabinet is not overloaded and leave exhaust grills in the front and the back clear (helps to maintain a uniform airflow)
8. Use aseptic technique
9. Dispose of all liquid waste after each experiment and treat with bleach.

VI. TISSUE CULTURE METHODS

Each student should maintain his own cells throughout the course of the experiment. These cells should be monitored daily for morphology and growth characteristics, fed every 2 to 3 days, and subcultured when necessary. A minimum of two 25 cm^2 flasks should be carried for each cell line; these cells should be expanded as necessary for the transfection experiments. Each time the cells are subcultured, a viable cell count should be done, the subculture dilutions should be noted, and after several passages, a doubling time determined. As soon as you have enough cells, several vials should be frozen away and stored in liquid N_2. One vial from each freeze down should be thawed 1-2 weeks after freezing to check for viability. These frozen stocks will prove to be vital if any of your cultures become contaminated.

Procedures

1. *Media preparation.* Each student will be responsible for maintaining his own stock of cell culture media; the particular type of media, the sera type and concentration, and other supplements will depend on the cell line. Do not share media with you partner (or anyone else) because if a culture or a bottle of media gets contaminated, you have no back-up. Most of the media components will be purchased prepared and sterile. In general, all you need to do is sterily combine several sterile solutions. To test for sterility after adding all components, pipette several mls from each media bottle into a small sterile petri dish or culture tube and incubate at 37EC for several days. Use only media that has been sterility tested. For this reason, you must anticipate your culture needs in advance so you can prepare the reagents necessary. But, please try not to waste media. Anticipate your needs but don't make more than you need. Tissue culture reagents are very expensive; for example, bovine fetal calf serum cost ~ $200/500 ml. Some cell culture additives will be provided in a powdered form. These should be reconstituted to the appropriate concentration with double-distilled water (or medium, as appropriate) and filtered (in a sterile hood) through a 0-22 μm filter.

All media preparation and other cell culture work must be performed in a laminar flow hood. Before beginning your work, turn on blower for several minutes, wipe down all surfaces with 70% ethanol, and ethanol wash your clean hands. Use only sterile pipettes, disposable test tubes and autoclaved pipette tips for cell culture. All culture vessels, test tubes, pipette tip boxes, stocks of sterile eppendorfs, etc. should be opened only in the laminar flow hood. If something is opened elsewhere in the lab by accident, you can probably assume its contaminated. If something does become contaminated, immediately discard the contaminated materials into the biohazard container and notify the instructor.

2. *Growth and morphology.* Visually inspect cells frequently. Cell culture is sometimes more an art than a science. Get to know what makes your cells happy. Frequent feeding is important for maintaining the pH balance of the medium and for eliminating waste products. Cells do not typically like to be too confluent so they should be subcultured when they are in a semi-confluent state. In general, mammalian cells should be handled gently. They should not be vortexed, vigorously pipetted or centrifuged at greater than 1500 g.

3. *Cell feeding.* Use prewarmed media and have cells out of the incubator for as little time as possible. Use 10-15 ml for T-25's, 25-35 ml for T-75's and 50-60 ml for T-150's.

(*a*) *Suspension cultures.* Feeding and subculturing suspension cultures are done simultaneously. About every 2-3 days, dilute the cells into fresh media. The dilution you use will depend on the density of the cells and how quickly they divide, which only you can determine. Typically 1:4 to 1:20 dilutions are appropriate for most cell lines.

(*b*) *Adherent cells.* About every 2-3 days, pour off old media from culture flasks and replace with fresh media. Subculture cells as described below before confluency is reached.

4. *Subculturing adherent cells.* When adherent cells become semi-confluent, subculture using 2 mM EDTA or trypsin/EDTA.

Trypsin-EDTA:

(*a*) Remove medium from culture dish and wash cells in a balanced salt solution without Ca^{++} or Mg^{++}. Remove the wash solution.

(*b*) Add enough trypsin-EDTA solution to cover the bottom of the culture vessel and then pour off the excess.

(*c*) Place culture in the 37EC incubator for 2 minutes.

(*d*) Monitor cells under microscope. Cells are beginning to detach when they appear rounded.

(*e*) As soon as cells are in suspension, immediately add culture medium containing serum. Wash cells once with serum containing medium and dilute as appropriate (generally 4-20 fold).

EDTA alone:

(*a*) Prepare a 2 mM EDTA solution in a balanced salt solution (i.e., PBS without Ca^{++} or Mg^{++}).

(*b*) Remove medium from culture vessel by aspiration and wash the monolayer to remove all traces of serum. Remove salt solution by aspiration.

(*c*) Dispense enough EDTA solution into culture vessels to completely cover the monolayer of cells.

(*d*) The coated cells are allowed to incubate until cells detach from the surface. Progress can be checked by examination with an inverted microscope. Cells can be gently nudged by banging the side of the flask against the palm of the hand.

(*e*) Dilute cells with fresh medium and transfer to a sterile centrifuge tube.

(*f*) Spin cells down, remove supernatant, and resuspend in culture medium (or freezing medium if cells are to be frozen). Dilute as appropriate into culture flasks.

5. *Thawing frozen cells*

(*a*) Remove cells from frozen storage and quickly thaw in a 37EC waterbath by gently agitating vial.

(*b*) As soon as the ice crystals melt, pipette gently into a culture flask containing prewarmed growth medium.

(*c*) Log out cells in the "Liquid Nitrogen Freezer Log" Book.

6. *Freezing cells*

(*a*) Harvest cells as usual and wash once with complete medium.

(*b*) Resuspend cells in complete medium and determine cell count/viability.

(*c*) Centrifuge and resuspend in ice-cold freezing medium: 90% calf serum/10% DMSO, at 10^6–10^7 cells/ml. Keep cells on ice.

(*d*) Transfer 1 ml aliquots to freezer vials on ice.

(*e*) Place in a Mr. Frosty container that is at room temperature and that has sufficient isopropanol.

(*f*) Place the Mr. Frosty in the –70EC freezer overnight. Note: Cells should be exposed to freezing medium for as little time as possible prior to freezing

(*g*) Next day, transfer to liquid nitrogen (DON'T FORGET) and log in the "Liquid Nitrogen Freezer Log" Book.

7. *Viable cell counts.* Using a Hemocytometer to determine total cell counts and viable cell numbers.

Trypan blue is one of several stains recommended for use in dye exclusion procedures for viable cell counting. This method is based on the principle that live cells do not take up certain dyes, whereas dead cells do.

1. Prepare a cell suspension, either directly from a cell culture or from a concentrated or diluted suspension (depending on the cell density) and combine 20 µl of cells with 20 µl of trypan blue suspension (0.4%). Mix thoroughly and allow to stand for 5-15 minutes.
2. With the cover slip in place, transfer a small amount of trypan blue-cell suspension to both chambers of the hemocytometer by carefully touching the edge of the cover slip with the pipette tip and allowing each chamber to fill by capillary action. Do not overfill or underfill the chambers.
3. Starting with 1 chamber of the hemocytometer, count all the cells in the 1 mm center square and four 1 mm corner square. Keep a separate count of viable and non-viable cells.
4. If there are too many or too few cells to count, repeat the procedure either concentrating or diluting the original suspension as appropriate.
5. The circle indicates the approximate area covered at 100X microscope magnification (10X ocular and 10X objective). Include cells on top and left touching middle line. Do not count cells touching middle line at bottom and right. Count 4 corner squares and middle square in both chambers and calculate the average.
6. Each large square of the hemocytometer, with cover-slip in place, represents a total volume of 0.1 mm^3 or 10^{-4} cm^3. Since 1 cm^3 is equivalent to approximately 1 ml, the total number of cells per ml will be determined using the following calculations:

   Cells/ml = average cell count per square × dilution factor × 10^4;

   Total cells = cells/ml × the original volume of fluid from which the cell sample was removed; % Cell viability = total viable cells (unstained)/total cells × 100.

PLANT GROWTH REGULATORS COMMONLY USED IN PLANT TISSUE CULTURE

Class	*Name*	*Abbreviation*	*Comments*
Auxin	Indole-3-acetic acid	IAA	Use for callus induction at 10-30 μM. Lowering to 1-10 μM can stimulate organogenesis. Is inactivated by light and readily oxidized by plant cells. The synthetic auxins below have largely superceded IAA for tissue culture studies.
	Indole-3-butyric acid	IBA	Use for rooting shoots regenerated via organogenesis. Either maintain at a low concentration (1-50 μM) throughout the rooting process, or expose to a high concentration (100-250 μM) for 2-10 days and then transfer to hormone-free medium. Can also use as a dip for *in vitro* or *ex vitro* rooting of shoots.
	2,4-Dichlorophenoxy aceticacid	2,4-D	Most commonly used synthetic auxin for inducing callus and maintaining callus and suspension cells in dedifferentiated states. Usually used as sole auxin source (1-50 μM), or in combination with NAA.
	p-Chlorophenoxyacetic acid 1-Naphthalene acetic acid	pCPA NAA	Similar to 2,4-D, but less commonly used. Synthetic analogue of IAA. Commonly used either as sole auxin source (2-20 μM for callus induction and growth of callus and suspension cultures; 0.2-2 μM for root induction), or in combination with 2,4-D.
Cytok-inin	6-Furfurylamino-purine (kinetin)	K	Often included in culture media for callus induction, growth of callus and cell suspensions, and induction of morphogenesis (1-20 μM). Higher concentrations (20-50 μM) can be used to induce the rapid multiplication of shoots, axillary/adventitious buds, or meristems.
	6-Benzylaminopurine	BAP, BA	Included in culture media for callus induction, growth of callus and cell suspensions (0.5-5.0 μM), and for induction of morphogenesis (1-10 μM). More commonly used than kinetin for inducing rapid multiplication of shoots, buds, or meristems (5-50 μM).
	N-Isopentenylaminopurine	2iP	Less commonly used than K or BAP for callus induction and growth (2-10 μM), induction of morphogenesis (10-15 μM), or multiplication of shoots, buds, or meristems (30-50 μM).
	Zeatin	Zea	Seldom used in callus or suspension media. Can be used for induction of morphogenesis (0.05-10 μM). Zea is thermolabile and must not be autoclaved.

Gibber ellin	Gibberellin A_3	GA_3	Seldom used in callus or suspension medium (one exception being potato). Can promote shoot growth when added to shoot induction medium at 0.03-14 μM. Also used to enhance development in embryo/ ovule cultures (0.3-48 μM). GA_3 is thermolabile and must not be autoclaved.
Abscisic acid	Abscisic acid	ABA	Used at concentrations of 0.4-10 μM to prevent precocious germination, and promote normal development of somatic embryos.

PLANT TISSUE CULTURE

Introduction

Tissue culture refers to the growth and maintenance of plant in nutrient medium *in-vitro.* The term 'plant tissue culture' is used for culturing of unorganized tissues or callus.

It is now used as a blanket term for protoplast, cell, tissue, organ or whole plant culture under asceptic conditions. The methodology of tissue culture consist of separation of cells tissues organs of plant called 'Explants' and growing them aseptically on a nutrient medium under controlled condition of temperature and light. The explants give rise to an unorganized, proliferative mass of differentiated cells called callus, which later produces shoots.

Plant tissue culture techniques are now being used as powerful tools for the study of various kinds of basic problems not only in plant physiology, cell biology and genetics but also in agriculture, forestry, horticulture and industry. An important contribution made by this technique is the revelation of a unique capacity of plant cells that is the cellular totipotency which means that all leaving cells in a plant body can potentially give rise to whole plant.

Plant tissue culture technology can be divided into five classes based on the type of materials used.

(*i*) *Callus Culture.* Culture of callus on agar medium produced from explants.

(*ii*) *Cell Culture.* Culture of cell in liquid media usually aerated by agitation.

(*iii*) *Organ Culture.* Aseptic culture of embryos, anthers, roots, shoot and ovaries, etc.

(*iv*) *Meristem Culture.* Aseptic culture of shoots meristem or explant tissue in nutrient media.

(*v*) *Protoplast Culture.* The aseptic isolation and culture of plant protoplast from cultured cell or plant tissue.

Materials Required

Balance, Autoclave, Hot air oven, Magnetic stirrer, Refrigerator, Heater, pH meter, Distillation unit, microscope, spirit lamp, fore shapes with blunt end [to inoculate and subculture] forceps with fine tips [to dissect leaves], needles, culture bottles, racks, conical flasks, beakers, measuring cylinders, scalpels, screw cap bottles.

Preparation of the Media

The best-suited media is Mushige and Skoogs Media or MS Media. It has following ingredients:

(*i*) Stock A-Having major elements
(*ii*) Stock B-Having minor elements
(*iii*) Stock C-Having iron elements
(*iv*) Stock D-Having vitamin stores
(*v*) Sucrose [as carbon source] and Agar [as solidifying agent].

Preparation of Stock Solution

The stock A, B, C and D are prepared by weighing the respective chemicals written in the following chart and dividing them in minimum amount of distilled water. The final volume in each stock is made up to 1000 ml by adding distilled water. While preparing Stock - 'C' slight heating is done to dissolve sodium ethylene diamino tetra acetate and ferrous sulphate in water. These stock solutions are stored in refrigerator.

Composition of Stock Solutions

- **Stock A**

	Composition	*mg/litre*
1.	Ammonium nitrate (NH_4NO_3)	1650.00
2.	Potassium nitrate (KNO_3)	1900.00
3.	Magnesium sulphate ($MgSO_4$, $7H_2O$)	370
4.	Potassium dihydrogen phosphate (KH_2PO_4)	170
5.	Calcium chloride ($CaCl_2$, $7H_2O$)	440

- **Stock B**

	Composition	*mg/litre*
1.	Boric acid (H_3BO_3)	6.200
2.	Manganese sulphate ($MnSO_4$, $4H_2O$)	2.300
3.	Zinc sulphate ($ZnSO_4$, $7H_2O$)	8.600
4.	Potassium iodide (KI)	.830
5.	Sodium molybdate (Na_2MoO_4, $5H_2O$)	0.250
6.	Cobalt chloride ($CoCl_2$, $6H_2O$)	0.250
7.	Cupric sulphate ($CuSO_4$, $5H_2O$)	0.025

- **Stock C**

	Composition	*mg/litre*
1.	Disodium ethylene diamino tetra acetate (Na_2EDTA)	37.3
2.	Ferric sulphate ($FeSO_4$, $2H_2O$)	27.8

- **Stock D**

	Composition	*mg/litre*
1.	Glycine	2.00
2.	Nicotinic acid	0.50

3.	Thiamine hydrochloride	0.50
4.	Pyridoxine hydrochloride	0.50
5.	Sucrose	20 gm/litre
6.	Dyco-bacto agar	8 gm/litre

Preparation of Growth Regulators

- *Auxins.* The different types of Auxins used are as follows:
  (*i*) Indole acetic acid-IAA
  (*ii*) Indole-3-Butric acid-IBA
  (*iii*) Naphthalene acetic acid-NAA
  (*iv*) 2,4-Dichloro phenoxy acetic acid-2,4-D.

The known quantities of these hormones are first dissolved in 5 ml of NaOH or KOH and the final volume is made up by adding distilled water.

- *Cytokinins:*
  (*i*) 6-Benzyl amino purine-BAP and
  (*ii*) 6-perfuryl amino purine (Kinetin).

The required quantities of hormones are well dissolved in 5 ml of 0.1 N HCl and the final volume is made by water.

Coconut Milk

The liquid endosperm from the coconut is collected. It is filtered through clean cloth and known quantity of coconut milk is added to the medium before autoclaving.

The medium is prepared depending upon the quantity required to prepare 1 litre of the medium. The quantity of the stock required is 100 ml for Stock A, 1 ml for Stock B, 10 ml for Stock C and 1 ml for Stock D to the boiling water. Known amount of sucrose (2%) and agar (0.8%) are added. Then the stock solution that is the boiling solution is made up to 1000 ml by adding distilled water. PH is adjusted to 5.6-5.8.

(*i*) 25 ml of the media is dispensed into culture bottles and then the mouths of the bottles are closed with lid.

(*ii*) The culture bottles are then transferred to autoclave.

(*iii*) The instrument necessary for the inoculation and distilled water is also kept inside the autoclave.

(*iv*) The media is autoclaved at 15 inch/inch^2 pressure and temperature of 121°C for 15 minutes.

(*v*) It is allowed to cool and the autoclaved media is removed. The culture vessels with media are allowed to solidify before they are transferred to the inoculation chamber for inoculation.

Procedure

Collection of explant materials, pieces of seedlings, swelling of buds, stem, leaves, immature embryos, anthers are collected from a suitable plant material.

Sterilization of Plant Material

Surface of plant parts gets a wide range of microbial contaminants like fungi, bacteria. To avoid the microbial growth in a culture, the explants must be surface sterilized in disinfectant solutions before planting to the medium. The procedure is as follows:

(*i*) The explants are washed with tap water.

(*ii*) They are washed thoroughly by keeping them under running tap water for an hour. Through washing of the material is necessary because it removes superficial contaminants.

(*iii*) After washing, the material is treated with freshly prepared saturated chlorine water for 2 minutes.

(*iv*) It is then rinsed with sterile distilled water.

(*v*) The material is surface sterilized with freshly prepared 0.1% mercuric chloride solution for one minute followed by 4-5 times with sterilized distilled water to remove all traces of contamination.

Inoculation

The surface sterilized explants are transferred to inoculation chamber where asceptic conditions are maintained. The chamber is cleaned with ethanol and subjected to UV radiation for 1 hour. The outer surfaces of the culture bottles are also cleaned with ethanol before placing in the chamber. The materials placed on the sterilized petriplates.

Flame sterilization of the forceps is done by dipping them in ethanol and holding into the spirit lamp. With the help of the flamed forceps, the explants are inoculated on to the media. Flaming of outlets of the bottles is done and the caps are replaced tightly.

Incubation of Cultures

The culture bottles are transferred to the incubator where controlled condition of temperature and light are maintained. The temperature inside the culture room is maintained at 25°C + –2°C. The culture is exposed to 15 hours of light and 5 hours of darkness. Fluorescent tubes are used as light source. The cultures are allowed to grow and periodical observations are made. Subculturing is done once in 6 weeks.

2. PLANT TISSUE CULTURE

INTRODUCTION

Plant Tissue Culturing Techniques have become especially important in the agricultural community over the past ten years. During this time period, plant tissue culture has effectively moved from the confines of small laboratories and has taken its place among some of the more mainstream, broad-scale techniques employed by the agriculture industry.

Plant Tissue Culture more technically known as micropropagation, can be broadly defined as a collection of methods used to grow large numbers of plant cells, *in vitro*, in an aseptic and closely controlled environment. This technique is effective because almost all plant cells are totipotent - each cell possesses the genetic information and cellular machinery necessary to generate an entire organism. Micropropagation, therefore, can be used to produce a large number of plants that are genetically identical to a parent plant, as well as to one another.

The standard protocol for performing plant tissue culture experiments is fairly basic. First, it is essential that a sterile environment be created. The medium used to grow the plant tissue, the plant tissues themselves, and the environment surrounding the tissue culture, must be free of all possible contaminants. The presence of any bacterial, fungal, algal, or viral contaminants could potentially rob the desired plants of the nutrients provided by the culture medium and have devastating effects upon their growth (Kyle, 23002). Once a sterile environment has been established, tissue can be collected from the plant's leaf, shoot, bud, stem, or root (see figure 1). Because each of these cells is totipotent, each has the potential to express an entire organism. The tissue sample can then be placed on an aseptic (free of microorganisms), nutrient-rich medium where its cells will begin to grow and develop into the desired plant product (Thorpe, 1981). The nature of the medium and the nutrients that it contains is dependent upon the type of plant being grown and the properties that the grower wishes to express. Finally, the developing tissue should be maintained in a closely controlled chemical and physical environment, such as a greenhouse, to achieve the best results.

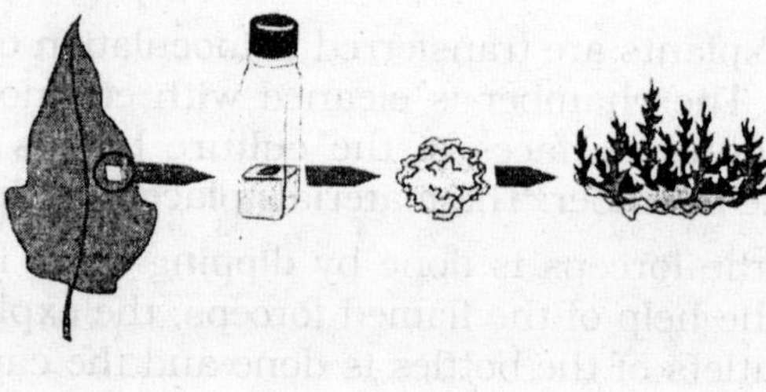

Fig. 1. The basic steps of micropropagation.

The benefits of plant tissue culture are extensive in the agricultural world. Micropropagation is favorable to traditional crop breeding methods in many respects, the first being that it allows for the production of huge numbers of plants in a very short period of time. In India over 100,000,000 plants are produced using micropropagation each year. Plant tissue culture is also advantageous to growers because the overwhelming number of plants can be produced using the tissue collected from a single parent plant - a plant which itself remains unharmed in the tissue harvesting process. Crop production through micropropagation also eliminates the possibility of any interruption in the growing season because it can be carried out inside the carefully regulated environment of a greenhouse. Because the chemical and physical environment inside a greenhouse can be closely monitored, any lull in production that might typically occur as a result of seasonal change can be avoided.

Micropropagation will be crucial to the agriculture industry in the future because it is used to produce plants which have been genetically modified and selected for their ability to resist certain indigenous environmental stresses. Currently, scientists and members of the agricultural community have joined forces to investigate the possibility of creating lines of tomatoes that possess increased salt tolerance (to be grown in areas in which the soil is high in salinity), plants that are completely resistant to various viral, bacterial, algal, and fungal infections, tobacco plants whose leaves can withstand freezing temperatures, and crops that are entirely resistant to harmful and destructive insects.

MANY DIMENSIONS OF PLANT TISSUE CULTURE RESEARCH

Introduction

The practice of plant tissue culture has changed the way some nurserymen approach plant propagation. In the recent past, the applicability of this technology to the propagation of trees and shrubs has been documented. Some firms have established tissue culture facilities and commercial scale operations are presently in operation for the mass propagation of apples, crabapples, rhododendrons, and a few other selected woody species. The intent of this research update is to briefly examine "what is being done" and to explore "what can be done" with regard to the tissue culture of ornamental plants. Such a consideration necessarily includes an overview of tissue culture as a propagation tool. The major impact of plant tissue culture will not be felt in the area of micropropagation, however, but in the area of controlled manipulations of plants at the cellular level in ways which have not been possible prior to the introduction of tissue culture techniques.

The Art and Science of Micropropagation

"Micropropagation" is the term which best conveys the message of the tissue culture technique most widely in use today. The prefix "micro" generally refers to the small size of the tissue taken for propagation, but could equally refer to the size of the plants which are produced as a result.

Micropropagation allows the production of large numbers of plants from small pieces of the stock plant in relatively short periods of time. Depending on the species in question, the original tissue piece may be taken from shoot tip, leaf, lateral bud, stem or root tissue. In most cases, the original plant is not destroyed in the process — a factor of considerable importance to the owner of a rare or unusual plant. Once the plant is placed in tissue culture, proliferation of lateral buds and adventitious shoots or the differentiation of shoots directly from callus, results in tremendous increases in the number of shoots available for rooting. Rooted "microcuttings" or "plantlets" of many species have been established in production situations and have been successfully grown on either in containers or in field plantings. The two most important lessons learned from these trials are that this methodology is a means of accelerated asexual propagation and that plants produced by these techniques respond similarly to any own-rooted vegetatively propagated plant. Micropropagation offers several distinct advantages not possible with conventional propagation techniques. A single explant can be multiplied into several thousand plants in less than one year. With most species, the taking of the original tissue explant does not destroy the parent plant. Once established, actively dividing cultures are a continuous source of microcuttings which can result in plant production under greenhouse conditions without seasonal interruption. Using methods of micropropagation, the nurseryman can rapidly introduce selected superior clones of ornamental plants in sufficient quantities to have an impact on the landscape plant market.

Plant Improvement Through Tissue Culture

In introducing this research update, it was mentioned that the major impact of tissue culture technology would not be in the area of micropropagation, but rather in the area of controlled manipulations of plant germplasm at the cellular level. The ability to unorganize, rearrange, and reorganize the constituents of higher plants has been demonstrated with a few model systems to date, but such basic research is already being conducted on ornamental trees and shrubs with the intent of obtaining new and better landscape plants.

Selection of Plants with Enhanced Stress or Pest Resistance

Perhaps the most heavily researched area of tissue culture today is the concept of selecting disease, insect, or stress resistant plants through tissue culture. Just as significant gains in the adaptability of many species have been obtained by selecting and propagating superior individuals, so the search for these superior individuals can be tremendously accelerated using in vitro systems. Such systems can attempt to exploit the natural variability known to occur in plants or variability can be induced by chemical or physical agents known to cause mutations.

All who are familiar with bud sports, variegated foliage and other types of chimeras have an appreciation for the natural variability in the genetic makeup or expression in plants. Chimeras are the altered cellular expressions which are visible, but for each of these which are observed many more differences probably exist but are masked by the overall organization of the plant as a whole. For example, even in frost-tender species, certain cells or groups of cells may be frost hardy. However, because most of the organism is killed by frost, the tolerant cells eventually die because they are unable to support themselves without the remainder of the organized plant. Plant tissues grown in vitro can be released from the organization of the whole plant through callus formation. If these groups of cells are then subjected to a selection agent such as freezing, then those tolerant ones can survive while all those which are susceptible will be killed. This concept can be applied to many types of stress as well as resistance to fungal and bacterial pathogens and various types of phytotoxic chemical agents. Current research in this area extends across many interests including attempts to select salt tolerant lines of tomato, freezing resistant tobacco plants, herbicide resistant agronomic crops, and various species of plants with enhanced pathogen resistance. Imagine, if you will, the impact of a fireblight-resistant Bartlett pear, a clone of pin oak for alkaline soils, or a selection of southern magnolia hardy to zone 4.

Tissue Culture and Pathogen Free Plants

Another purpose for which plant tissue culture is uniquely suited is in the obtaining, maintaining, and mass propagating of specific pathogen-free plants. The concept behind indexing plants free of pests is closely allied to the concept of using tissue culture as a selection system. Plant tissues known to be free of the pathogen under consideration (viral, bacterial, or fungal) are physically selected as the explant for tissue culture. In most cases, the apical domes of rapidly elongating shoot tips are chosen. These are allowed to enlarge and proliferate under the sterile conditions of the *in vitro* culture with the resulting plantlets tested for presence of the pathogen (a procedure called indexing). Cultures which reveal the presence of the pathogen are destroyed, while those which are indexed free of pathogen are maintained as a stock of pathogen-free material. Procedures similar to these have been used successfully to obtain virus-free plants of a number of species and bacteria-free plants of species known to have certain leaf spot diseases. The impact of obtaining pathogen-free nursery stock can only be speculative, since little research documenting viral, bacterial, or fungal diseases transmitted through propagation of woody ornamentals is available.

Somatic Hybridization

The ability to fuse plant cells from species which may be incompatible as sexual crosses and the ability of plant cells to take up and incorporate foreign-genetic codes extend the realm of plant modifications through tissue culture to the limits of the imagination. Most such manipulations are carried out using plant "protoplasts". Protoplasts are single cells which have been stripped of their cell walls by enzymatic treatment. A single leaf treated under these conditions may yield tens of millions of single cells, each theoretically capable of eventually producing a whole plant. This concept has fueled speculation as diverse as the possibilities of obtaining nitrogen-fixing corn plants on the one extreme to discovering a yellow-flowered African violet on the other extreme.

The observation that has provided the impetus for most of this research is that when cells are stripped of their cell walls and brought into close contact, they tend to fuse with each other. This "somatic hybridization" is not subject to the same incompatibility problems that limit traditional plant breeding strategies. It is conceivable then that one could hybridize a Juneberry with a crabapple or a plum, but the fundamental research required to demonstrate such an event has yet to be conducted.

Summary

Plant tissue culture research is multi-dimensional. While most nurserymen have been introduced to the techniques and advantages of micropropagation, few have ventured to use it as a propagation tool. The applicability of micropropagation for woody trees has been demonstrated as feasible since all aspects of the technology have confirmed the fact that trees produced by this method look like and grow like their counterparts produced by traditional methods of cloning.

Other dimensions of tissue culture research have been less well publicized. The potential for selecting pathogen free plants, for selecting stress-tolerant and pathogen-resistant clones of plants, and the novel genetic combinations to be achieved through somatic hybridization are all lines of research which can have a profound impact on the nursery industry.

WHAT IS PLANT TISSUE CULTURE?

Plant tissue culture, the growth of plant cells outside an intact plant, is a technique essential in many areas of the plant sciences. Cultures of individual or groups of plant cells, and whole organs, contribute to understanding both fundamental and applied science.

It relies on maintaining plant cells in aseptic conditions on a suitable nutrient medium. The culture can be sustained as a mass of undifferentiated cells for an extended period of time, or regenerated into whole plants.

Designing a strategy to culture cells from a plant for the first time can still seem like a matter of trial and error, and luck. However, the commercial production of valuable horticulture crops by micropropagation, which relies on tissue culture, shows that it exists in the routine, as well as experimental, world.

Plant cells can be grown in isolation from intact plants in tissue culture systems. The cells have the characteristics of callus cells, rather than other plant cell types. These are the cells that appear on cut surfaces when a plant is wounded and which gradually cover and seal the damaged area.

Pieces of plant tissue will slowly divide and grow into a colourless mass of cells if they are kept in special conditions. These are:

- initiated from the most appropriate plant tissue for the particular plant variety
- presence of a high concentration of auxin and cytokinin growth regulators in the growth media
- a growth medium containing organic and inorganic compounds to sustain the cells
- aseptic conditions during culture to exclude competition from microorganisms.

The plant cells can grow on a solid surface as friable, pale-brown lumps (called callus), or as individual or small clusters of cells in a liquid medium called a suspension culture. These cells can be maintained indefinitely provided they are sub-cultured regularly into fresh growth medium.

Tissue culture cells generally lack the distinctive features of most plant cells. They have a small vacuole, lack chloroplasts and photosynthetic pathways and the structural or chemical features that distinguish so many cell types within the intact plant are absent. They are most similar to the undifferentiated cells found in meristematic regions which become fated to develop into each cell type as the plant grows. Tissue cultured cells can also be induced to re-differentiate into whole plants by alterations to the growth media.

Plant tissue cultures can be initiated from almost any part of a plant. The physiological state of the plant does have an influence on its response to attempts to initiate tissue culture. The parent plant must be healthy and free from obvious signs of disease or decay. The source, termed explant, may be dictated by the reason for carrying out the tissue culture. Younger tissue contains a higher proportion of actively dividing cells and is more responsive to a callus initiation programme. The plants themselves must be actively growing, and not about to enter a period of dormancy.

The exact conditions required to initiate and sustain plant cells in culture, or to regenerate intact plants from cultured cells, are different for each plant species. Each variety of a species will often have a particular set of cultural requirements. Despite all the knowledge that has been obtained about plant tissue culture during the twentieth century, these conditions have to be identified for each variety through experimentation.

USES OF PLANT TISSUE CULTURE

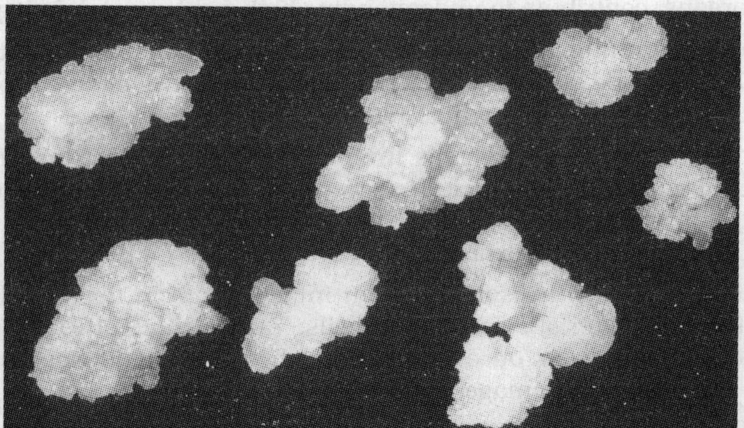

Fig. 2. Callus cultures derived from onion

Plant tissue culture now has direct commercial applications as well as value in basic research into cell biology, genetics and biochemistry. The techniques include culture of cells, anthers, ovules and embryos on experimental to industrial scales, protoplast isolation and fusion, cell selection and meristem and bud culture. Applications include:

- micropropagation using meristem and shoot culture to produce large numbers of identical individuals
- screening programmes of cells, rather than plants for advantageous characters
- large-scale growth of plant cells in liquid culture as a source of secondary products
- crossing distantly related species by protoplast fusion and regeneration of the novel hybrid
- production of dihaploid plants from haploid cultures to achieve homozygous lines more rapidly in breeding programmes
- as a tissue for transformation, followed by either short-term testing of genetic constructs or regeneration of transgenic plants removal of viruses by propagation from meristematic tissues.

1. PLANT TISSUE CULTURE DEMONSTRATION BY USING SOMACLONAL VARIATION TO SELECT FOR DISEASE RESISTANCE

Plant tissue cultures isolated from even a single cell can show variation after repeated subculture. Distinct lines can be selected with their own particular morphology and physiology. It suggests that the tissue culture contains a population of genotypes whose proportion can be altered by imposing an appropriate selection pressure. This variation can be transmitted to plants regenerated from the tissue cultures, and is called somaclonal variation. It provides an additional source of novel variation for exploitation by plant breeders.

The carrot cultivar Fancy was used in our laboratory to generate a series of 197 regenerant progeny lines. These plants showed considerable morphological variation. They were tested for resistance to the leaf spot pathogen *Alternaria dauci*, which can cause total necrosis of mature leaves. They had a greater degree of variation in response than the parental cultivar, including some more resistant lines.

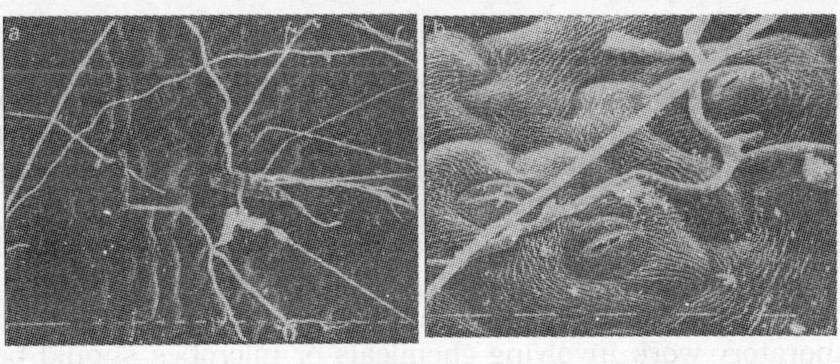

Fig. 3. Scanning electron micrographs of surface of carrot leaf 3 days after inoculation with A. dauci. (LHS) germination from multiseptate conidiospore (RHS) penetration of hyphae through epidermal surface rather than through stomata. Scale bar = 10 micrometres.

One symptom of the disease is loss of chlorophyll and total soluble polyphenol compounds. These reduce to a low level 6 days after inoculation of excised leaves with *A. dauci* spores when compared with uninoculated leaves. Regenerant progeny with high chlorophyll levels maintained higher chlorophyll levels after challenge with *A. dauci.* After self-pollinating selected high- and low-chlorophyll regenerant plants, this characteristic was inherited by their progeny, suggesting that the capacity to resist this infection is inherited.

2. DEMONSTRATION OF TISSUE CULTURE FOR TEACHING

Introduction

The starting point for all tissue cultures is plant tissue, called an explant. It can be initiated from any part of a plant - root, stem, petiole, leaf or flower - although the success of any one of these varies between species. It is essential that the surface of the explant is sterilised to remove all microbial contamination. Plant cell division is slow compared to the growth of bacteria and fungi, and even minor contaminants will easily over-grow the plant tissue culture. The explant is then incubated on a sterile nutrient medium to initiate the tissue culture. The composition of the growth medium is designed to both sustain the plant cells, encourage cell division, and control development of either an undifferentiated cell mass, or particular plant organs.

The concentration of the growth regulators in the medium, namely auxin and cytokinin, seems to be the critical factor for determining whether a tissue culture is initiated, and how it subsequently develops. The explant should initially form a callus, from which it is possible to generate multiple embryos and then shoots, forming the basis for plant regeneration and thus the technology of micropropagation. The first stage of tissue culture initiation is vital for information on what combination of media components will give a friable, fast-growing callus, or a green chlorophyllous callus, or embryo, root or shoot formation.

There is at present no way to predict the exact growth medium, and growth protocol, to generate a particular type of callus. These characteristics have to be determined through a carefully designed and observed experiment for each new plant species, and frequently also for each new variety of the species which is taken into tissue culture. The basis of the experiment will be media and protocols that give the desired effect in other plant species, and experience.

The Demonstration

The strategy for designing a medium to initiate tissue culture, showing how growth regulators and other factors modulate development, can be demonstrated using the African Violet, a popular house plant. Leaf sections are the source of explants. This demonstration is regularly carried out by a student class, and gives reliable results. Sterile supplies are provided from central facilities, and provision of sterile working areas (for example, in laminar flow hoods) is an advantage, although cultures can be initiated in an open laboratory with careful aseptic technique. The standard precautions used during any laboratory work involving chemicals or microbes should be adopted. If you are in any doubt about safety hazards associated with this demonstration, you should consult your local safety adviser.

Step 1. Selection of the Leaves

Leaves are cut from healthy plants, leaving a short length of petiole attached. They should be selected to each yield several explants of leaf squares with approximately 1 cm sides. The youngest and oldest leaves should be avoided.

Wash the dust off the leaves in a beaker of distilled water, holding the leaf stalk with forceps.

Step 2. Surface Sterilisation and Preparation of the Explants

This part of the procedure should be carried out in a sterile working area, or with meticulous aseptic technique.

The leaf, with the petiole still attached, should be immersed in 70% ethanol for 30 seconds, then transferred to a sterile petri dish. Sterile scissors and forceps are then used to cut squares from the leaf as explants, each with approximately 1 cm sides.

The explants are transferred into a 10% hypochlorite bleach solution for 5 minutes, gently agitating once or twice during this time. They are then washed free of bleach by immersing in four successive beakers of sterile distilled water, leaving them for 2-3 minutes in each.

Three explants are placed on each petri dish of growth medium, with the upper epidermis pressed gently against the surface of the agar to make good contact.

The petri dishes are sealed with plastic film to prevent moisture loss, and incubated at 25°C in 16h light/8h dark.

Step 3. Assessment of Tissue Culture Development

The explants are incubated for 4-6 weeks, and inspected at weekly or fortnightly intervals. The growth of obvious bacterial or fungal colonies indicates contamination, and data from such cultures is obviously suspect. The development of dark brown tissue cultures can also be a consequence of contamination.

The media used in the demonstration are designed to show the effects of auxin, cytokinin, sucrose and mineral salts on development. The media were based on the well-known Murashige and Skoog inorganic medium.

Typical Results

These pictures show typical results, after about 8 weeks on each medium. To summarise, multiple adventitious buds form on the *control* medium, leading to many small shoots on the upper surface where the leaf is not in contact with the medium.

Absence of *sucrose* inhibits this production. Shoot production is also limited on the low sucrose concentration, but comparable with the control at high sucrose.

At zero and low levels of *cytokinin*, callus forms where the leaf surface is in contact with the medium, while at high levels, shoot formation is stimulated.

At zero and low levels of *auxins* there is a stimulus to shoot formation, but at high concentrations, large numbers of roots are formed.

At low and zero levels of *MS* salts, there is no growth at all.

These very obvious variations demonstrate the importance of a carbon and inorganic salt source for plant growth, as well as the effect of the auxin:cytokinin ratio on the control of plant development.

3. PREPARATION OF PLANT TISSUE CULTURE MEDIA

Since the pioneer plant tissue culture studies, one of the main objectives was design a proper medium which supports sufficient growth of the explants in a totally artificial environment. The first medium formulations used for plant culture work were based on experience of microorganisms. These media contained several distinct classes of compounds:

- Carbon source
- Organic supplements
- Inorganic salts
- Trace elements

As works on plant cell cultures progressed, attention was paid in order to define specifically media for an optimal growth of the variable plant cells, tissue and organs. Today the most commonly used culture media are based on following components:

- Macroelements-N, P, K, Ca, Mg and S
- Microelements-Fe, Cu, Mn, Co, Mo, B, I, Zn, Cl and some times Al, Ni and Si
- Carbon source-Sucrose, glucose, fructose or sorbitol
- Organic compounds (Vitamins)-thiamine, niacin, pyridoxine, biotin, folic acid, ascorbic acid, tocopherol
- Myo-inositol or casein hydrolysate
- Complex organics-Coconut milk and water, yeast extract, fruit juices or pulps
- Plant growth regulators-Auxins, cytokinins, Gibberellins, Abscisic acid
- Gelling agents-Agar, agarose, gellan gum
- Other components-MES, Activated charcoal, antibiotics, fungicides.

The number of plant species that have been cultured and the different media used are extensive. Once the plant species to be cultured is decided upon, a suitable medium formulation should be selected. One of the most successful media, is the MS medium, was devised by Murashige and Skoog, (1962), based on the constitution of tobacco plants. The choice or formulation of the media depends basically on the requirements of the cultivated explant and species.

Media Formulation

As mentioned above, MS medium was developed for culture of tobacco explants and the formulation was based on the analysis of the mineral compounds present in the tobacco tissue itself (Table 1). It is noted for its high salt levels, particularly K and N salts. LM medium (Linsmaier and Skoog, 1965) is a revision of the MS medium, in which the organic constituents have been modified and only thiamine and myo-inositol retained. White's medium (White, 1963) was devised for the culture of tomato roots and has lower concentrations of salts than MS medium. Gamborg's B5 medium (Gamborg et al. 1968) was devised for soybean callus culture and contains a much greater proportion of nitrate compared to the ammonium. SH medium (Schenk and Hildebrandt, 1972) was developed for callus culture of both monocotyledons and dicotyledons. Nitsch's medium (Nitsch & Nitsch; 1969) was developed for another culture, it contains lower salt concentrations than that of MS, but not as low as that of White's medium. KM medium (Kao & Michayluck, 1975) was designed to grow cells and protoplasts of Vicia at a very low cell density in liquid media. The medium of Chu (N6) is defined to improve the formation, growth and differentiation of pollen in rice. The concentration of ammonium proved to be crucial for the development of callus. Recently other media have been developed for specific explants and species.

Table 1. Composition of commonly used tissue culture media in molar concentration

Compounds	*MS*	*LM*	*Gamborg's*
KNO_3	18.79 mM	18.79 mM	24.73 mM
NH_4NO_3	20.61 mM	20.61 mM	—
$(NH_4)_2SO_4$	—	—	1.01 mM
KH_2PO_4	1.26 mM	1.26 mM	—
NaH_2PO_4	—	—	1.09 mM
$CaCl_2$	2.99 mM	2.99 mM	1.02 mM
$MgSO_4 7H_2O$	1.50 mM	1.50 mM	1.01 mM
$CoCl_2$ $6H_2O$	0.11 µM	0.11 µM	0.11 µM
$CuSO_4$ $5H_2O$	0.10 µM	0.10 µM	0.10 µM
FeNa EDTA	0.10 mM	0.10 mM	0.10 mM
H_3BO_3	0.10 mM	0.10 mM	38.52 mM
KI	5.00 µM	5.00 µM	4.52 µM
$MnCl_2$ H_2O	0.10 mM	0.10 mM	59.16 mM
Na_2MoO_4 $2H_2O$	1.03 µM	1.03 µM	1.03 µM
$ZnSO_4 7H_2O$	29.91 µM	29.91 µM	6.96 µM
Glycine	26.64 µM	—	—
Nicotinic acid	4.06 µM	—	8.12 µM
Pyridoxin-HCl	2.43 µM	—	4.86 µM
Thiamine-HCl	0.30 µM	1.19 µM	29.65 µM
Myo-inositol	0.56 mM	0.56 mM	0.56 mM

Table 2. Composition of commonly used for tissue culture media in molar concentration

Compounds	*Nitsch's*	*KM*	*CHU (N6)*	*SH*
KNO_3	9.40 mM	18.79 mM	27.99 mM	24.73 mM
NH_4NO_3	9.00 mM	7.70 mM	—	—
$(NH_4)_2SO_4$	—	—	3.50 mM	—
NH_4H_2 PO_4	—	—	—	2.61 mM
KH_2PO_4	0.50 mM	1.25 mM	2.94 mM	—
KCl	—	4.02 mM	—	—
$CaCl_2$	1.50 mM	4.08 mM	1.13 mM	1.36 mM
$MgSO_4 7H_2O$	0.75 mM	1.22 mM	0.75 mM	1.62 mM
$CoCl_2$ $6H_2O$	—	0.11 µM	—	0.42 µM
$CuSO_4$ $5H_2O$	0.10 µM	0.10 µM	—	0.80 µM
FeNa EDTA	0.10 mM	0.10 mM	0.10 mM	53.94 µM
H_3BO_3	0.16 mM	48.52 µM	25.88 µM	80.86 µM

$MnSO_4H_2O$	0.11 mM	59.17 μM	19.70 μM	59.17 μM
KI	—	4.52 μM	4.81 μM	6.02 μM
Na_2MoO_4 $2H_2O$	1.03 μM	1.03 μM	—	0.41 μM
$ZnSO_47H_2O$	34.78 μM	6.96 μM	5.22 μM	3.48 μM
Glycine	26.64 μM	—	26.64 μM	—
Nicotinic acid	40.62 μM	—	4.06 μM	40.61 μM
Pyridoxin-HCl	2.43 μM	—	2.43 μM	2.43 μM
Thiamine-HCl	1.48 μM	—	2.96 μM	14.82 μM
Folic acid	1.13 μM	—	—	—
Biotin	0.21 μM	—	—	—
Myo-inositol	0.56 mM	—	—	5.55 mM

4. PLANT TISSUE CULTURE MEDIA

Major Constituents

- Salt Mixtures
- Organic Substances
- Natural Complexes
- Inert Supportive Materials
- Growth Regulators

1. Salt Mixtures

- M.S. (Murashigi and Skoog)
- Gamborg
- Nitsch and Nitsch (Similar to M.S.)
- White
- Knudson

Macro-nutrient salts

- NH_4NO_3 Ammonium nitrate
- KNO_3 Potassium nitrate
- $CaCl_2$-$2H_2O$ Calcium chloride (Anhydrous)
- $MgSO_4$-$7H_2O$ Magnesium sulfide (Epsom Salts)
- KH_2PO_4
- Too much NH_4 + may cause vitrification, but is needed for embryogenesis and stimulates adventitious shoot formation.

Micro-nutrient salts

- FeNaEDTA or (Na_2EDTA and $FeSO_4$)
- H_3BO_3 Boric Acid
- $MnSO_4$-$4H_2O$ Manganese sulfate

- $ZnSO_4$-$7H_2O$ Zinc sulfate
- KI Potassium iodide
- Na_2MoO_4-$2H_2O$ Sodium molybdate
- $CuSO_4$-$5H_2O$ Cupric sulfate
- $CoCl_2$-H_2O Cobaltous sulfide

2. Organic Compounds

Carbon Sources

1. Sucrose (1.5 to 12%, 3% I reference point)
2. Glucose (Sometimes used with monocots)
3. Fructose

White Vitamins

- Thiamine 1.0 mg/l
- Nicotinic Acid and Pyroxidine 0.5 mg/l
- Glycine 2.0 mg/l
- Vitamin C (antioxidant) 100.0 mg/l

Organic Compounds (cont.)

- Amino Acids and Amides
- Amino acids can be used as the sole source of Nitrogen but normally too expensive

  2 amino acids most commonly used
- L-tyrosine, enhances adventitious shoot form.
- L. Glutamine, may enhance adventitious embryogenesis.

Other Organic Compounds

Hexitol-we use I-inositol

- Stimulates growth but we don't know why
- use at the rate of 100 mg/l

Purine/pyrimidine

- adenine stimulates shoot formation
- can use adenine sulfate at 100 mg/l
- still other organics

Organic Acids

- Citric acid (150 mg/l) typically used with ascorbic acid (100 mg/l) as an antioxidant.
- Can also use some of Kreb Cycle acids

Phenolic compounds

- Phloroglucinol-Stimulates rooting of shoot sections
- L-tyrosine-stimulates shoot formation

3. Natural Complexes

- Coconut endosperm
- Fish emulsion
- Protein hydrolysates
- Tomato juice
- Yeast extracts
- Potato agar

4. Nutritionally inert complexes

- Gelling agents
- Charcoal
- Filter Paper Supports
- Other materials

Gelling Agents

Agar-extract from Marine red agar

- Phytagar
- Taiyo
- Difco-Bacto
- TC Agar
- Agarose
- Hydrogels
- Gelatin

Gelrite

- Bacterial polysaccharide
- Same gelling and liquifying
- Better quality control and cleaner than agar
- Gel firmness related to osmolarity starting point about 2g/l
- Sugar content higher the osmotic concentration, the firmer used in low concentration very low concentration of gelrite enhances vitrification Charcoal.

Activated charcoal is used as a detoxifying agent. Detoxifies wastes from plant tissues, impurities. Impurities and absorption quality vary concentration normally used is 0.3 % or lower.

Charcoal for tissue culture

Acid washed and neutralized never reuse

Filter paper supports

Heller platforms

- filter paper should be free of impurities
- filter paper should not dissolve in water
- Whatman # NO 50 or 42

Other inert materials

- polyurethane sponge
- vermiculite
- glass wool
- techiculture plugs
- peat/polyurethane plugs to root cuttings

5. Growth regulators

- Auxin
- Cytokinin
- Gibberellin
- Abscisic acid
- Ethylene

1. Auxins

Order of effectiveness in callus formation, rooting of cuttings, and the induction of adventive embryogenesis

- IAA
- IBA
- NAA
- 2,4-D
- 2,4,5-T
- Pictoram

2. Cytokinins

Enhances adventitious shoot formation

- BA
- 2iP
- Kinetin
- Zeatin
- PBA

3. Gibberellin

- Not generally used in tissue culture
- Tends to suppress root fromation and adventitious embryo formation

4. Abscisic Acid

- Dormin - U.S.
- Abscisin - England
- Primarily a growth inhibitor but enables more normal development of embryos, both zygotic and adventitious

5. Ethylene

Question is not how much to add but how to get rid of it in-vitro

- Natural substance produced by tissue cultures at fairly high levels especially when cells are under stress
- Enhances senescense
- Supresses embryogenesis and development in general

6. Hormone Combinations

- Callus development
- Adventitious embryogenesis
- Rooting of Shoot cuttings
- Adventitious shoot and root formation
- Callus development

Auxin alone

- Pictoram 0.3 to 1.9 mg/l
- 2,4-D 1.0 to 3.0 mg/l

Auxin and cytokinen

- IAA 2.0 to 3.0 mg/ l
- 2iP 0.1 mg/l
- NAA 0.1 mg/l
- 2iP 0.1 mg/l

Adventitious embryogenesis

- Induction is the first step (biochemical differentiation) High auxin in media.
- Development is the second step which includes cell and tissue organization, growth and emergence of organ or embryo
- no or very low auxin. Can also add ABA 10 mg/l, NH_4 and K

Rooting of shoot cutttings

- Induction: need high auxin, up to 100 mg/l for 3-14 days .
- Development: no auxin, in fact auxin may inhibit growth
- Can also add phloroglucinol and other phenolics but we don't know for sure how they fit in

Adventitious shoot and root development

Skoog and Miller's conclusions

- formation of shoots and roots controlled by a balance between auxin and cytokinin
- high auxin/low cytokinin = root development

- low auxin/high cytokinen = shoot development
- concept applies mainly to herbaceous genera and easy to propagate plants
- we lump together induction and development requirements

Enhancing adventitious shoot formation

• adenine	40 to 160 mg/l
• L-tyrosine	100 mg/l
• NaH_2PO_4-H_2O	170 mg/l
• Casein hydrolysate	1-3 g/l
• phynylpyruvate	25-50 mg/l
• NH_4	some

Not sure if these additives belong in induction or development.

Summary

	auxin	cytokinen
• callus	high (2-3 mg/l)	low .1 mg/l (2iP)
• axillary shoots	low to none	very high 10-100 2iP or BA
• adventi.	shoots	equal (2-3 mg/l) equal (2-3 mg/l kinetin)
• rooting	high (10 mg/l IAA)	low .1 mg/l 2iP or none
• embryogenesis	high	low

Physical Quality of Media

1. pH

- Usual range of 4.5 to 6.0
- Liquid 5.0
- Solid 5.6 to 5.8
- Above 6.0 many of the salts precipitate out
- All pH adjustment are made prior to adding gelling agents
- PH adjusted by adding KOH or HCL
- M.S media has a high buffering capacity
- pH changes as soon as you put explant in physical quality of media.

2. Volume/quantity of media

- related to kind of vessel
- growth of tissue depends on medium volume
- related to shape and volume

3. Liquid or gel

- gels use slants to get more growing area, better light
- liquid

- stationary with or without supports
- agitated: rotation less than 10 ppm

Preparation of Media

Pre-Mixes, Individual Stock solutions, Concentrated individual

Prepare 750 ml of media containing 1 mg/l Kinetin

1. VS × CS = VM × CF
2. VS × 100 mg/l = 750 ml × 1 mg/l
3. VS = 750 ml × 1 mg/l/100 mg/l
4. VS = 7.5 ml

Prepare 500 ml of media that contains .4 mg/l of Thiamin HCl

1. VS × CS = VM × CF
2. VS × 100 mg/l = 500 ml × .4 mg/l
3. VS = 500 ml × .4 mg/l/100 mg/l
4. VS = 2 ml
5. Pipette 2 ml of stock solution of Thiamine HCl and add to 498 ml of media.

5. PREPARATION OF PROTOPLASTS

Method for isolating large numbers of metabolically competent protoplasts from leaves of monocotyledons (grasses), dicotyledons (such as spinach and sunflower) or from hypocotyl tissue (e.g. Brassica napus).

1. Leaf slices of monocots and dicots are prepared by cutting the leaves with a sharp razor blade into segments 0.5-1 mm in size. In the case of dicots the epidermis can be scraped off before cutting by rubbing with fine carborundum powder or with a fine nylon brush.
2. Set up 50 ml of digestion medium (for 10-15 g of plant tissue) according to the recipe listed below (Solution A).
3. Incubate the leaf slices or pieces in a 19 cm-diameter dish containing the digestion medium for 3 hours at 25°C, covered with a plastic film. It may be advantageous to replace the digestion medium at intervals of 1 hour, as the enzymes might become inactivated by substances released from broken cells.
4. After completion of the incubation the digestion medium is carefully removed and discarded. It normally contains very few protoplasts. The plant tissue is then washed 3 times by shaking gently with 20 ml wash medium (Solution B).
5. After each wash the tissue is collected by pouring through a tea strainer (0.5 to 1 mm pore size) and the combined washes are then filtered through nylon mesh (100-200 μm pore size) to remove vascular tissue and undigested material.
6. The protoplasts are collected by centrifugation of the combined filtered washes for 3 minutes at 50-100 x g and the supernatant is aspirated and discarded.
7. This crude protoplast preparation also contains some cells and chloroplasts and it is important to purify the protoplasts to remove these contaminants. This can be done with solutions of sucrose and sorbitol of different densities.

8. The protoplast pellet is gently resuspended in 40 ml of Solution C and this suspension is divided among two 100 ml centrifuge tubes.
9. To each tube add slowly 5 ml of Solution D and then overlay this with 5 ml of wash medium (Solution B) to make a three step gradient.
10. Centrifuge at 300 g for 5 minutes.
11. The protoplasts now collect as a band at the interface between the 2 top layers. Carefully remove them with a pasteur pipette.
12. The protoplasts should be examined with a light microscope to ensure that the preparation is free of cells and chloroplasts.
13. When a large portion of the protoplasts is pelleted in this sucrose/sorbitol gradient the density of the 2 layers can be increased by adding 5-10 % Dextran (15,000-20,000 Mr) or 10-20% Ficoll to increase the percentage of the floating protoplasts.
14. The purified protoplasts can be concentrated by diluting with 10 ml of Solution B, cenrifuging at 100 g for 3 minutes and then resuspending the pellet in a small amount of medium by gently shaking the tubes.
15. The protoplasts are stable for up to 24 h when stored on ice. Photosynthetic activity of the protoplasts can be determined by measuring co-dependent 20 evolution with an oxygen electrode, provided rapid stirring is avoided as this will break some of the protoplasts. A suitable medium is listed as Solution E. Protoplasts exhibit a relatively broad pH optimum butat more acidic pH values the bicarbonate concentration should be lowered.

Solutions

Solution A (Digestion medium)

Composition:

For 50 ml use:

500 mM

D-Sorbitol

4.56 g

1 mM

$CaCl_2$

7.35 mg

5 mM

MES-KOH, pH 5.5

49.00 mg

2%

Cellulase Onozuka R10

1.00 g

0.3%

Macerozyme R10

0.15 g

Solution B (Wash medium)

Composition

For 100 ml use:

500 mM

D-Sorbitol

9.11 g

1 mM

$CaCl_2$

14.70 mg

5 mM

MES-KOH, pH 6.0

98.00 mg

The pH of the solution must be adjusted to 6.0 with KOH

The pH must be adjusted to 5.5 with KOH before adding the enzymes

Solution C	Solution D
Composition	Composition
For 100 ml use:	For 100 ml use:
500 mM	400 mM
Sucrose	Sucrose
8.56 g	6.80 g
1 mM	100 mM
$CaCl_2$	D-Sorbitol
7.40 mg	0.90 g
5 mM	1 mM
MES-KOH, pH 6.0	CaCl2
49.00 mg	7.40 mg
The pH of the solution must be adjusted to 6.0 by adding KOH	5 mM
	MES-KOH, pH 6.0
	49.00 mg
	The pH of the solution must be adjusted to 6.0 by adding KOH.

Solution E

Composition

For 100 ml use:

500 mM

D-Sorbitol

9.10 g

1 MM

CaCl2

15.00 mg

30 mM

Tricine-KOH

538.00 mg

5 mM

$NaHCO_3$

42.0 mg

6. PROTOPLAST ISOLATION, CULTURE AND FUSION

Background information

Protoplasts are cells which have had their cell wall removed, usually by digestion with enzymes. Cellulase enzymes digest the cellulose in plant cell walls while pectinase enzymes break down the pectin holding cells together. Once the cell wall has been removed the resulting protoplast is spherical in shape.

Digestion is usually carried out after incubation in an osmoticum (a solution of higher concentration than the cell contents which causes the cells to plasmolyse). This makes the cell walls easier to digest. Debris is filtered and/or centrifuged out of the suspension and the protoplasts are then centrifuged to form a pellet. On resuspension the protoplasts can be cultured on media which induce cell division and differentiation. A large number of plants can be regenerated from a single experiment - a gram of potato leaf tissue can produce more than a million protoplasts, for example.

Protoplasts can be isolated from a range of plant tissues: leaves, stems, roots, flowers, anthers and even pollen. The isolation and culture media used vary with the species and with the tissue from which the protoplasts were isolated.

Protoplasts are used in a number of ways for research and for plant improvement. They can be treated in a variety of ways (electroporation, incubation with bacteria, heat shock, high pH treatment) to induce them to take up DNA. The protoplasts can then be cultured and plants regenerated. In this way genetically engineered plants can be produced more easily than is possible using intact cells/plants.

Plants from distantly related or unrelated species are unable to reproduce sexually as their genomes/modes of reproduction etc. are incompatible. Protoplasts from unrelated species can be fused to produce plants combining desirable characteristics such as disease resistance, good flavour and cold tolerance. Fusion is carried out by application of an electric current or by treatment with chemicals such as Polyethylene Glycol (PEG). Fusion products can be selected for on media containing antibiotics or herbicides. These can then be induced to form shoots and roots and hybrid plants can be tested for desirable characteristics.

Materials Required

- Lettuce (round, green lettuce not iceberg)
- scalpel or sharp knife
- forceps
- tile
- glass Petri dish
- 10 cm^3 syringe
- 1 cm^3 syringe
- 13% sorbitol solution
- Viscozyme enzyme
- Cellulase enzyme
- Dropper
- small filter funnel
- 60 μm gauze square (approximately 12 cm × 12 cm)
- tape
- centrifuge tube
- slide and coverslip.

A centrifuge, high power microscope and incubator set at approximately 35°C must be available in the lab.

Preparation of Materials

It is best to use limp lettuce which has been left in an incubator at about 35°C for an hour. This causes the cells to plasmolyse slightly.

Viscozyme, pectinase and cellulase enzymes can be bought from NCBE. Do not dilute the enzymes. The enzyme solution should be stored in a fridge until use.

Sorbitol solution (70%) can be bought from Boots the Chemist. To prepare a 13% solution add 18.5 cm^3 stock to 100 cm^3 distilled water.

The 60 mm gauze can be ordered from Lockertex. Folly Lane, Warrington, WA5 0NP. Telephone: 01925 644371. £18 per square metre.

Water proof sticky tape can be used to secure the gauze mesh in a small filter funnel. This can be washed and reused.

For Protoplast Fusion

A microcentrifuge can be used to spin the protoplasts. A spin of 3 minutes at the lowest voltage is sufficient. Care should be taken when resuspending the pellet as the protoplasts are very fragile.

Making the Fusion Mix

PEG (12 ml. of 50% solution)

HEPES buffer 0.9 g

10 ml. water

pH 8.0

Starter Experiment

- Lettuce leaf (2 squares, each approximately 3 cm × 3 cm)
- scalpel or sharp knife
- forceps
- tile
- glass Petri dish
- 10 cm^3 syringe
- 1 cm^3 syringe
- 13% sorbitol solution
- 1.2 cm^3 Viscozyme enzyme
- 0.6 cm^3 Cellulase enzyme
- Dropper
- small filter funnel
- 60 μm gauze
- tape
- centrifuge tube
- slide and coverslip
- centrifuge

- incubator
- microscope

Procedure

1. Place 12.5 cm^3 of 13% sorbitol solution into the Petri dish.
2. Place the squares of lettuce leaf bottom side in the Petri dish.
3. Cut up the lettuce leaf into pieces approximately 5 mm × 5 mm.
4. Place the lid on the dish and incubate in the oven at 35°C for 5 minutes.
5. Add the Viscozyme and Cellulase enzymes to the dish. Swirl gently.
6. Incubate for a further 20 minutes; swirl gently at regular intervals.
7. Place the 60 μm mesh into the funnel and secure with tape.
8. Hold the funnel over the lid of the Petri dish.
9. Pour the digested lettuce into the filter funnel. Try to minimise the distance the protoplasts have to fall!
10. Place the funnel over a centrifuge tube and gently pour the contents of the Petri dish lid through the funnel again.
11. Balance your tube with another and centrifuge for 5-10 minutes at 2,000 rpm.
12. The protoplasts will have formed a pellet at the bottom of the tube. Use the dropper to remove the liquid supernatant from the tube, without disturbing the pellet. Remove all but 0.5 cm^3 of the liquid and then resuspend the pellet by gently tapping the tube.
13. Place a drop of the suspension on a slide, cover with a coverslip and examine under high power.
14. If your suspension is too concentrated to see the protoplasts clearly you can add several drops of the sucrose solution to the suspension in the centrifuge tube.
15. You should be able to try to fuse your protoplasts together if you have time.
16. Protoplast Fusion (Somatic Hybridisation).

    Glass slide

    Coverslip

    Glass dropper

    Fusion mix

    Protoplast suspension

    Microscope

Instructions

1. Place a drop of the protoplast suspension on the glass slide.
2. Observe under medium power of microscope.
3. Add one drop of the fusion mix.
4. Observe over a period of ~3 minutes.

7. AGROBACTERIUM CULTURE AND AGROBACTERIUM MEDIATED TRANSFORMATION

Material

- Infiltration Media Selection Plates
- 5X MS salts 1X MS salts
- 1X B5 vitamens .3% sucrose
- 5% sucrose 8g TC agar
- 044 uM benzylamino purine
- 03% Silwet L-77
- Autoclave and add appropriate antibiotics (for pCGN add Kan 50 ug/ml ; I also add timentin 200 ug/ml)
- Silwet L-77 is sold by OSI Specialty 800-523-5862. Speak with Linda Ayers at ext 3. 1 gallon costs approximately $240 and it is not sold in smaller amounts. Lehle seeds has repackaged Silwet L-77 in smaller amounts.

Procedure

Plant Growth

1. Plant seeds of appropriate genotype (I have only used Ler) on top of cheesecloth covered soil, making sure soil fills pot. I use a rubber band to secure the cheesecloth.
2. After plants have bolted, clip off the primary bolt to encourage the growth of secondary bolts. Perform infiltration 4-8 days after clipping.

   Vacuum Infiltration 1. Grow a large liquid culture of Agrobacterium (I have used ASE strain) carrying the appropriate construct. Start a 25 ml overnight (LB+ antibiotics) two to three days ahead of infiltration. One day prior to infiltration use this culture to inoculate a 400 ml culture.
3. After 24 hours of growth, cells are usually at a density of at least 1 OD. Harvest cells by centrifugation and resuspend to an OD of .8 in infiltration media.
4. I use a crystallization dish for the actual infiltration but the pots do not fit exactly, so any ideas/improvements in this area are encouraged. Invert the pot and stick into the infiltration solution and put into vacuum oven at RT. Infiltrate for 10-15 mins at 10-15 in^3 Hg.
5. Release vacuum and remove pot, lay it on its side in tray and cover with Saran wrap. Remove cover the next day and stand upright.
6. The Agro infiltration solution can be reused for multiple pots.

Selection of Transformants

7. Pour selection plates.
8. Sterilize seeds in 50% water/50% bleach/drop of Tween. Rinse 3-4 times with sterile water.
9. Plate seeds by resuspending in room temp .1% agarose and pipetting onto selection plates. Dry plates in laminar flow hood until set. Use 1ml agarose for

every 500-1000 seed. Plate 2000-4000 seed per 150 × 15 mm plate. Put a small number of control seeds from a known transformed plant on one of the plates for comparison.

10. Vernalize plants for two nights in cold room. Move plates to growth chamber.
11. After about 7 days, transformants should be clearly visible at dark green plants with roots which extend over and into the selection media.
12. Transfer seedlings to soil.

8. ISOLATION OF CHLOROPLASTS FROM SPINACH LEAVES

Procedure

1. Prepare an ice bath and pre-cool all glassware to be used (including a mortar and pestle).
2. Select several fresh spinach leaves. Remove the large veins by tearing them loose from the leaves. Weigh out 4 g of de-veined leaf tissue.
3. Chop the tissue as fine as possible with a knife and chopping board.
4. Add the tissue to an ice-cold mortar containing 15 ml of Grinding Solution and grind the tissue to a paste.
5. Filter the ground up tissue solution through double layered cheesecloth into a beaker and squeeze the tissue pulp to recover all of the liquid suspension.
6. Transfer the green suspension to a cold 50 ml centrifuge tube and centrifuge for 1 minute at 200 X g at 4°C to pellet the unbroken cells and cell fragments.
7. Decant the supernatant into a clean centrifuge tube and centrifuge again for 7 minutes at 1,000 X g to pellet the chloroplasts. Decant and discard the supernatant.
8. Resuspend the chloroplasts in 5 ml of cold Suspension Solution (or 0.35 M NaCl). Use a cold glass stirring rod to gently disrupt the packed pellet.
9. Enclose the tube in aluminum foil and place at 0 to 4°C (see Hint #1).
10. Using a hemacytometer, determine the number of chloroplasts per ml of suspension media.
11. To determine the chlorophyll content see Protocol on Determination of Chlorophyll Content in Chloroplasts.

Solutions

Spinach Leaves (Fresh) Suspension Solution

0.33 M Sorbitol

pH 7.6 Adjust with NaOH

1 mM $MgCl_2$

50 mM HEPES

2 mM EDTA

Grinding Solution

0.33 M Sorbitol

pH 6.5 Adjust with HCl

4 mM $MgCl_2$

2 mM Ascorbic Acid (Vitamin C)

10 mM Sodium Pyrophosphate

0.35 M NaCl

Bio Reagents and Chemicals

Sodium Hydroxide

Sodium Pyrophosphate

Sodium Chloride

Spinach Leaves (Fresh)

Magnesium Chloride

Ascorbic Acid

Sorbitol

HEPES

Cheesecloth

Hydrochloric Acid

EDTA

Protocol

The isolation procedure used in this protocol leaves the chloroplast outer membrane intact. If you wish to study the enzymes for photophosphorylation: wash the chloroplasts and rupture the outer membranes. To rupture the outer membranes, resuspend the chloroplasts in diluted suspension solution (1:25). Immediately centrifuge the chloroplast suspension for 5 minutes at 8,000 X g to collect the chloroplasts. Remove the diluted suspension media and resuspend the chloroplasts in isotonic media (0.35 M NaCl or undiluted Suspension Buffer).

9. PREPARATION OF PLANT DNA USING CTAB

Alternatively, the nonionic detergent cetyltrimethylammonium bromide (CTAB) is used to liberate and complex with total cellular nucleic acids. This general procedure has been used on a wide array of plant genera and tissue types. Many modifications have been published to optimize yields from particular species. The protocol is relatively simple, fast, and easily scaled from milligram to grams of tissue; it requires no cesium chloride density gradient centrifugation.

Additional Materials

CTAB extraction solution

2% (w/v) CTAB

100 mM Tris·Cl, pH 8.0

20 mM EDTA, pH 8.0

1.4 M NaCl

Store at room temperature (stable several years)

CTAB Precipitation Solution

1% (w/v) CTAB

50 mM Tris·Cl, pH 8.0

10 mM EDTA, pH 8.0

Store at room temperature (stable several years)

Extraction Buffer

100 mM Tris·Cl, pH 8.0

100 mM EDTA, pH 8.0

250 mM NaCl

100 ug/ml proteinase K (add fresh before use)

Store indefinitely at room temperature without proteinase K

High-salt TE Buffer

10 mM Tris·Cl, pH 8.0

0.1 mM EDTA, pH 8.0

1 M NaCl

Store at room temperature (stable for several years).

CTAB/NaCl Solution (10% CTAB in 0.7 M NaCl)

Dissolve 4.1 g NaCl in 80 ml water and slowly add 10 g CTAB (hexadecyltrimethyl ammonium bromide) while heating and stirring. If necessary, heat to 65°C to dissolve. Adjust final volume to 100 ml.

2% (v/v) 2-mercaptoethanol (2-ME)

High-salt TE buffer

24:1 (v/v) chloroform/isoamyl alcohol

80% ethanol

Pulverizer/homogenizer: mortar and pestle, blender, Polytron (Brinkmann), or coffee grinder

Organic solvent-resistant test tube or beaker

65°C water bath

Beckman JA-20 rotor or equivalent or microcentrifuge.

Procedure

1. Add 2-ME to the required amount of CTAB extraction solution to give a final concentration of 2% (v/v). Heat this solution and CTAB/NaCl solution to 65°C. Approximately 4 ml of 2-ME/CTAB extraction solution and 0.4 to 0.5 ml CTAB/NaCl solution are required for each gram of fresh leaf tissue. With lyophilized, dehydrated, or dry tissues such as seeds, 2-ME/CTAB extraction solution should be diluted 1:1 with sterile water. 2-ME should be used in a fume hood.

2. Chill a pulverizer/homogenizer with liquid nitrogen (–196°C) or dry ice (–78°C). Pulverize plant tissue to a fine powder and transfer the frozen tissue to an organic solvent-resistant test tube or beaker.

   Use young tissue and avoid larger stems and veins to achieve the highest DNA yield with minimal polysaccharide contamination.

3. Add warm 2-ME/CTAB extraction solution to the pulverized tissue and mix to wet thoroughly. Incubate 10 to 60 min at 65°C with occasional mixing.

   A 60-min incubation results in larger DNA yields. If maximum yield is not important, 10 min should be adequate. If the tissue contains large amounts of phenolic compounds, 1% (v/v) polyvinylpyrrolidone (mol. wt. = 40,000) may be added to absorb them.

4. Extract the homogenate with an equal volume of 24:1 chloroform/octanol or chloroform/isoamyl alcohol. Mix well by inversion. Centrifuge 5 min at 7500 µg (8000 rpm in JA-20; ~10,000 rpm in a microcentrifuge, for smaller samples), 4°C. Recover the top (aqueous) phase.

   Octanol, rather than isoamyl alcohol, is used because it may enhance isolation of nuclei. Slower centrifugation speeds are possible if centrifugation time is increased accordingly; a microcentrifuge may be used for small-scale preparations (150 mg starting tissue). After centrifugation, two phases should be evident with tissue debris at the interface.

5. Add 1/10 vol 65°C CTAB/NaCl solution to the recovered aqueous phase and mix well by inversion.

6. Extract with an equal volume of chloroform/octanol. Mix, centrifuge, and recover as in step 4 above.

   The aqueous phase may still be light yellow-brown in colour.

Precipitate Nucleic Acids

7. Add exactly 1 vol CTAB precipitation solution. Mix well by inversion. If precipitate is visible, proceed to step 8. If not, incubate mixture 30 min at 65°C.

8. Centrifuge 5 min at 500 µg (2000 rpm in JA-20; ~2700 rpm in microcentrifuge), 4°C.

   Do not increase the speed or time of centrifugation as the pellet may become very difficult to resuspend. If there is no pellet, add more CTAB precipitation solution (up to 1/10 the total volume). Incubate 1 hr to overnight at 37°C. Centrifuge 5 min at 500 µg, 4°C.

9. Remove but do not discard the supernatant and resuspend pellet in high-salt TE buffer (0.5 to 1 ml per gram of starting material). If the pellet is difficult to resuspend, incubate 30 min at 65°C. Repeat until all or most of pellet is dissolved.

   Polysaccharide contamination may make it excessively difficult to resuspend the pellet. Read the A260 of the supernatant and discard pellet if nucleic acids are present in the supernatant.

10. Precipitate the nucleic acids by adding 0.6 vol isopropanol. Mix well and centrifuge 15 min at 7500 µg, 4°C.

    Ethanol can be used for the precipitation, but isopropanol may yield cleaner pellets.

11. Wash the pellet with 80% ethanol, dry, and resuspend in a minimal volume of TE (0.1 to 0.5 ml per gram of starting material).

    Residual CTAB is soluble and is removed by the 80% ethanol wash. Further purification of the DNA with RNase A and proteinase K may be done using standard methods

10. SUSPENSION CULTURE AND PRODUCTION OF SECONDARY METABOLITES

Introduction

Plant cell suspension cultures are mostly used for the biochemical investigation of cell physiology, growth, metabolism and for large or medium scale production of secondary metabolites. For such purposes normally suspension cultures are used which are propagated in Erlenmeyer flasks on a gyratory shaker and are maintained by regular subculturing after short intervals (usually 1 or 2 weeks). For such use callus cultures are also maintained under a mineral oil layer which reduces the subculture frequency. However, from the callus phase it takes additional time to establish a new suspension culture again and this new suspension may exhibit changed biochemical traits.

For storage purposes, several strategies have been developed, all aiming at reduction of labour and costs of maintenance, while preserving all properties of the cells. Long-term conservation of suspension cultures is usually successful by cryopreservation [1,2]. However, it requires special and expensive equipment and is not suitable for routine work as reactivation of cells requires prolonged incubation and has to pass a callus phase. Therefore, for practical work there is a need for an easy and a time-saving method that could allow the storage of actually investigated suspension cultures for relatively longer terms.

A simple procedure is described here to maintain suspended plant cell cultures for medium-terms. With this method suspension cultures of Agrostis tenuis, Nicotiana tabacum, Nicotiana chinensis, Oryza sativa and Solanum marginatum, could be maintained viable under reduced temperatures for more than 4 months without transfer to fresh medium [3,4]. The suspension cultures were kept without shaking at 10C (in dark or in dim light at about 50 lux) in screw-cap plastic bottles (tissue culture flasks with membranes, Fig. 1) that permitted sterile air to pass through easily. Some effective adsorption materials or stabilizers such as activated charcoal, gelatine, glutamic acid, starch, sugar etc. were added to the cell suspensions. In the case of sensitive microorganisms and unicellular green algae such substances have already shown protection during maintenance and against known freezing and drying injuries [5,6,7]. In the case of plant cell suspensions only with Nicotiana tabacum the addition of 0.01% charcoal +1.0% gelatine resulted in slightly improved cell survival as compared to the cells grown without additives. However, the simple maintenance conditions used proved effective to prolong the storage life of suspension cultures as compared to control samples grown under normal conditions. The stability of the stored cells usually appeared unchanged, as evident from HPLC-fingerprints) and from growth characteristics after re-establishment of the cultures.

The method is convenient and the equipment described here is easily available. It is useful for routine work as during such long-term maintenance a ready inoculum can continuously be obtained (from the same batch of cell suspension) for immediate use.

Material and Methods

Preparation of cultures

Plant cell cultures are grown under normal conditions using routine methods. The suspension cultures should be propagated in Erlenmeyer flasks on a gyratory shaker (100 rpm) at 24C under continuous light at 600 lux. These should be transferred to fresh medium after every 7-10 days.

Storage of cell suspensions

For storage 100 ml of the cell suspension from the early exponential growth phase is transferred to a 650 ml screw caps plastic bottle with membranes which permits sterile air to pass through. Depending on the volume of cell suspension to be maintained, smaller 250 ml bottles with 43 ml cell suspension can also be used. However, the surface to volume ratio should remain the same as with big bottles. These bottles are supplied as tissue culture flasks with closure secure against contamination. Alternatively wide mouth Erlenmeyer flasks with new cotton plugs can also be used. The volume of the added cell suspension in the vessels should lead to a high surface to volume ratio in order to assure sufficient oxygen supply for the cells. The suspension cultures are stored without shaking in the dark and at a reduced temperature of 10C. To prolong cell survival it is recommended to add some stabilizers such as 0.01% activated charcoal + 1.0% gelatine, or doubled to tripled concentration of sugar.

Recultivation and Viability Assay

For recultivation an aliquot of 5 ml of the stored cell suspension is transferred with a sterile pipette from the plastic bottles to 100 ml Erlenmeyer-flasks containing 25 ml of medium. The cells are cultured in these flasks for 1-2 weeks at 24C on a gyratory shaker (100 rpm) under continuous light (600 lux). To determine regrowth on solid media 1ml of cell suspension is placed on a Petridish (60 mm dia.).

For a periodic viability assay of the preserved cultures, the ability of regrowth is determined. After appropriate intervals of storage (every 2-3 weeks or longer depending on the sensitivity of the cell suspension) aliquots of the stored cells and the control samples (grown under normal conditions) are recultivated in equal amount of liquid media. After a definite period of growth (about 1 week when the cells are in exponential growth phase), the cells are harvested by suction filtration. The dry weight of the re-grown cells is determined which is taken as a measure of the inoculum quantity (the cell density developed from the surviving cells during storage). The resulted cell mass from the stored cells is compared after different storage intervals and with the control samples.

Staining of a very small aliquot with fluorescence in diacetate (FDA) on a microscope slide and counting of a fraction of positive cells under the fluorescence microscope is also an easy tool to obtain a quick impression of the viability of the cells [8].

Stability Checking

For stability checking by HPLC, 2g (fresh weight) of the harvested cells are extracted in boiling methanol for one hour. 20 µl of the resulting extracts are analyzed by reverse phase high pressure liquid chromatography (HPLC). The result of the chromatographic procedure is a characteristic "fingerprint" of cell metabolites. Methanol is used as an organic solvent. Substances are eluted by a standard gradient system which runs from

0% methanol to 100% methanol within 30 minutes. The majority of cell cultures show most characteristic "fingerprints" when a slightly acidic solvent system is used. Comparability of such "fingerprints" has to be assured by extracting cells from the same growth stage of the cell culture.

High Pressure Liquid Chromatography (HPLC) Protocol

Column :	Nuleosil 100-7 C18, 7 μm, 4 × 100 mm
Flow rate :	1 ml/min
Solvent A :	Water + 0.1 ml H_3PO_4
Solvent B :	Methanol + 0.1 ml H_3PO_4
Gradient :	0 min : 100% Sol. A / 0% Sol B
	5 min : 100% Sol. A / 0% Sol B
	35 min : 0% Sol. A / 100% Sol B
	40 min : 0% Sol. A / 100% Sol B
	45 min : 100% Sol. A / 0% Sol B
	Detector wavelength : 280 nm

Notes

Plant cell suspension cultures are normally cultivated in Erlenmeyer-flasks on gyratory shakers. At normal growth temperature when gyration of these flasks is not maintained, plant cells normally settle down very fast. Under such conditions a lack of oxygen supply leads to cell death or at least a cell damage occurs within few hours for most species. However, during a systematic study we observed that such plant suspension cultures are able to survive at reduced temperatures and without gyration much longer (more than 16 weeks). The influence of doubling or tripling of the sugar content of the media (up to 6% sucrose or glucose) tested showed that one of the Nicotiana tabacum lines resulted in a re-established culture even after 26 maintenance free weeks at 10C (with 6% sucrose instead of normal 3% concentration). For Cinchona robusta cell suspensions (recalcitrant to any tested cryopreservation protocol) after 8 weeks of standing culture at 10C survival was still high, but no regrowth was obtained after 16 weeks. Nevertheless after 16 weeks with FDA staining 60% "viability" was detected. This indicates that optimized conditions may improve the regrowth.

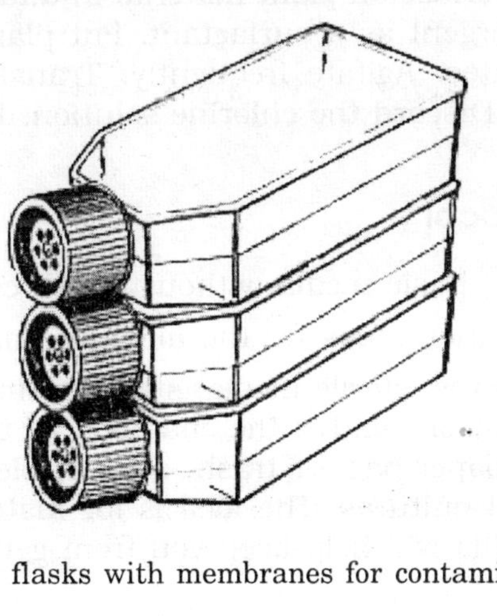

Fig. 4. Tissue culture flasks with membranes for contamination free aeration.

The onset of regrowth normally occured within a few days and after one week normal growth parameters were restored. Nevertheless with Oryza sativa it took about 2 weeks before rapid cell growth resumed and the re-establishment of the normal growth pattern took one or two subculture stages.

One should be prepared that during maintenance of the standing culture much of the medium water may evaporate. Depending on the dryness of the air in the incubator or room climate, up to 1/3 of the volume may disappear within 16 weeks. The bottles or flasks should however, not be sealed or closed air tight to minimize evaporation because of disturbances in the aeration. To avoid this one may start with more volume if frequent samples of inoculum are required.

To optimize the maintenance conditions in one's own laboratory and to establish a maximum storage period for the cell suspension under study it is recommended to conduct perodic viability assay and the stability check of the chemical traits of the preserved cultures. To check whether cell metabolism is affected by these storage conditions or the added substances, methanolic extracts should be analyzed by reversed phase HPLC. The analysis will show characteristic peak patterns (which need not be identified chemically). The storage conditions and reduced incubation temperature should not cause any changes in the characteristic of such peak patterns.

For rice cultures regeneration experiments were carried out. Dependent on the amount of extra sugar during standing culture, regeneration was higher or lower than in the controls, but high regeneration capacity appeared to be correlated with a high incidence of aberrant plants : 20% was albino. Therefore it is recommended as with other preservation methods to check for the occurrence of somaclonal variations.

Although the stabilizers showed no remarkable increase in survival of few of the tested cultures, they might cause some kind of protection in other cultures and result in better stability and viability of such cell suspensions at least over few passages.

PROTOCOLS FOR PLANT TISSUE CULTURE

Tissue culture involves several stages of plant growth, callus formation, shooting, roots and flowering. It starts with sterilizing a piece of plant tissue:

Sterilizing Plant Materials

Inside the glove box, sterilize all plant material in diluted domestic bleach solution containing one drop of detergent as a surfactant. Put plant pieces in a jar containing the bleach for 10-20 minutes. Agitate frequently. Transfer the tissue in the bleach solution to the sterile area. Discard the chlorine solution. Rinse plant pieces twice with sterile water.

1. Callus Formation Protocol

1. Prepare a batch of basic media without added coconut milk or malt extract.
2. Follow the sterilization steps for the medium, instruments and chamber.
3. Organize the cultures, sterile media, sterilized tools and sterile paper towels, and sterile water at one end of the chamber. At the other end of the chamber organize a small paper bag for trash, a jar of bleach solution, and a spot for freshily inoculated cultures. The idea is for materials to always move in the same direction. This will help keep you from getting confused.

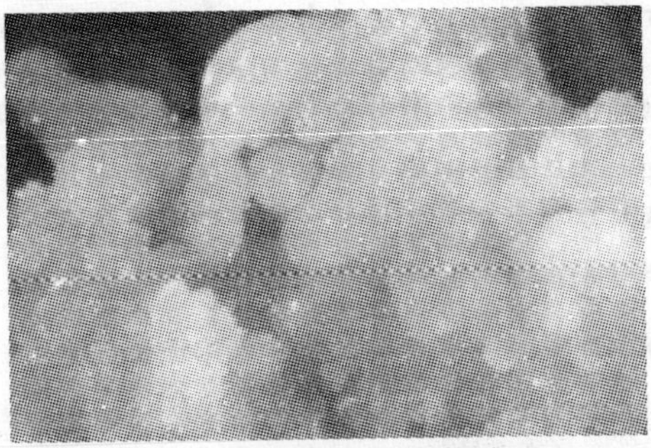

Fig. 5. Callus development.

4. Sterilize your plant tissue, following the above protocol.
5. Take a sterilized piece of tissue from the jar with a pair of forceps (do not touch the plant material with your hands). Working on a dampened sterile paper towel, use the scalpal or a razor blade to cut the tissue sample into 2-3 cm long pieces.
6. Put one piece of the tissue into each container (it is important to have only one shoot per container at this stage so that if the shoot is contaminated it cannot spread to the others). Shut the lids of the containers.
7. Store jars at room temperature away from direct sunlight. Callus will start to become visible after about 7 days post initiation.
8. Friable type II callus should be separated from more organized callus and/or watery unorganized callus 2-3 weeks after initiation.
9. Friable type II callus should be visually selected at each subsequent transfer to maintain an optimal phenotype.
10. Callus should be transferred to fresh medium at 2-4 week intervals depending on growth rate. Switching to different types of media also helps maintain vigor.
11. Callus can be maintained at room temperature in the dark.

2. Shoot Multiplication Protocol

Fig. 6. Source.

1. Prepare a batch of media containing 5% to 10% coconut milk.
2. Follow the sterilization steps for the medium, instruments and chamber.
3. Organize the cultures, sterile media, sterilized tools and sterile paper towels, and sterile water at one end of the chamber. At the other end of the chamber organize a small paper bag for trash, a jar of bleach solution, and a spot for freshly inoculated cultures. The idea is for materials to always move in the same direction. This will help keep you from getting confused.

4. With a pair of forceps, remove a culture from its container. Moisten the sterile paper with some sterile water. It is important to do all the manipulation on a damp paper towel as these plants are very soft and can desiccate readily.
5. Cut stems off the culture and transfer to new jars. At this stage, up to five stems may be put inside each jar.
6. Store cultures as explained in the previous stage.

3. Root Formation Protocol

Once you have established enough shoots, let them grow to at least 2 cm before beginning the rooting process

1. Prepare a batch of media containing malt extract.
2. Follow the sterilization steps for the medium, instruments and chamber.
3. Organize the cultures, sterile media, sterilized tools and sterile paper towels, and sterile water at one end of the chamber. At the other end of the chamber organize a small paper bag for trash, a jar of bleach solution, and a spot for freshly inoculated cultures. The idea is for materials to always move in the same direction. This will help keep you from getting confused.
4. With a pair of forceps, remove a stem from its container. Wash all of the media off the culture and transfer to malt media. (Malt contains auxins, which promote root formation.)
5. Up to five shoots may be put in each culture vessel. Store containers in their usual place as before. Roots should form within two to four weeks.

4. Acclimatization Protocol

1. Fill plastic "vegetable" bags with a potting mix that contains no fertilizer.
2. Autoclave for 15 minutes in the pressure cooker.
3. When cool, inoculate with mycorrhizae.
4. Remove the rooted plants from agar medium using a pair of forceps.
5. Wash off the agar thoroughly from the roots using lukewarm water.
6. Poke a hole in the middle of the potting mix, using a sterile instrument, gently insert the roots in that hole.
7. Dampen the potting mix with Basic Nutrient Mix.
8. Spray the foliage with a hand spray containing sterile water.
9. Keep these bags inside larger plastic containers with a glass cover, out of dirцct sunlight. Gradually remove the glass cover but watch for signs of

desiccation and if needed use the hand spray to spray water on the foliage. Gradually increase the light intensity for the plants also.

10. When the roots are well established and the plants are acclimatized (this should take about 4-6 weeks), they can be given fertilizer and be treated like any other plant.

Flowering Protocol

1. Prepare a batch of jars with peroxidated media.
2. Follow the sterilization steps for the instruments and chamber.
3. Organize the cultures, sterile media, sterilized tools and sterile paper towels, and sterile water at one end of the chamber. At the other end of the chamber organize a small paper bag for trash, a jar of bleach solution, and a spot for freshly inoculated cultures. The idea is for materials to always move in the same direction. This will help keep you from getting confused.
4. With a pair of forceps, remove a plantlet from its container. Wash all of the media off the plantlet and transfer it to peroxidated media.

Pseudospikelets growing in vitro

Bamboo inflorescence

5. Watch the cultures carefully for growth of flower buds.

5. Anther/Ovule Culture Protocol (For Producing Haploid Lines)

1. Collect immature flower buds from plants grown either in the field, the greenhouse or in vitro.
2. Preculture the flower buds in dry test tubes in the refrigerator for 2 days.
3. Dip the flower buds in 70% ethanol for 30 seconds, and then surface sterilize with 0.5-1.0 percent sodium hypochlorite for 20 min in your glove box.
4. Rinse the flower buds with sterile distilled water three times, and aseptically excise the anthers and/or ovules using scalpel and needle.
5. Culture the anthers/ovules on semisolid nutrient medium.
6. Incubate the cultures under complete darkness at room temperature.
7. Once calli or embryos are initiated, transfer the cultures to daylight.
8. Culture the developing embryoids on standard propagation medium.

6. Protoplast Preparation Protocol

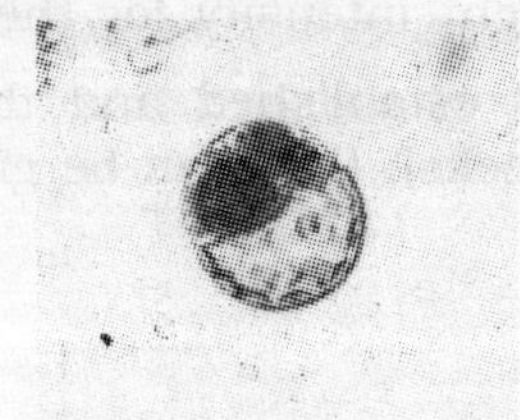

Plant Protoplast

1. Perform all of the following in the glovebox.
2. Buffer Solution. Dissolve 56.94 g of mannitol in 200 ml of distilled water. Add distilled water to bring the final volume to 500 ml.
3. Digestion Solution. Measure 10 ml of buffer solution into a 15 ml test tube or small beaker. Measure 0.1 g pectinase and 0.2 g cellulace onto weighing paper. Drop the pre-measured, powdered enzymes into this solution. Swirl the beaker or cap the test tube and shake it back and forth until the enzymes are completely dissolved.
4. Carefully pour all 10 ml of enzyme solution into the bottom of a sterile jar.
5. Use forceps to float each tissue sample on the surface of the enzyme solution.
6. Seal the jar with Saran Wrap or tape and leave it at room temperature (approximately 25°C) overnight. If proper equipment is available, gentle agitation of the dishes will be helpful.
7. The next day, gently swirl and shake the solution in the Petri dish. (If no protoplasts are observed with the microscope, let the solution stand for another 15-30 minutes then look again for protoplasts.)
8. At the end of the digestion period, gently shake the petri dish to release the protoplasts.
9. Filter the enzyme-protoplast suspension through successively smaller filters, starting with the 100 μm sieve. One sieve will trap mostly protoplasts.

7. Chemical Protoplast Fusion Protocol

1. Mix two droplets of protoplasts from two genetically different strains along with a droplet of PEG (polyethylene glycol) in a test tube.
2. Flood the protoplasts with baking soda dissolved in basic growing mix.
3. After 20 minutes, wash with basic growing mix and centrifuge in the salad spinner centrifuge.
4. Transfer the fused protoplasts onto callus culture medium.

Electrical Protoplast Fusion Protocol

1. Place two droplets of protoplasts from two genetically different strains between two electrodes.
2. Apply 10 volts per millimeter of electrode separation for one minute.
3. Apply a 10 to 20 μsec pulse at 100 volts per millimeter. This will require building an electrical pulse apparatus. A 555 timer discharging through an automotive coil might do the trick.

4. Wash with basic growing mix and centrifuge in the salad spinner centrifuge.
5. Transfer the fused protoplasts onto callus culture medium.

8. Polyploidization Protocol

1. Prepare a batch of media containing 0.001 to 0.01% trifluralin (Preen or Treflan brand name).
2. Follow the sterilization steps for the medium, instruments and chamber.
3. Organize the cultures, sterile media, sterilized tools and sterile paper towels, sterile water and bleach solution in the glove box.
4. Place a clean culture or fused protoplast on trifluralin media.
5. Store containers in a cool, dark location. Within two to four weeks polyploid shoots should begin developing.

Sand Supported Culture Protocol

The purpose of agar in growth media is solely to support the tissue. Instead of agar, use sand to support the culture.

1. Boil sand in water on your stovetop.
2. Store the clean sand in a plastic bag.
3. Put ½ inch of sand in a babyfood jar.
4. Sterilize jars in your pressure cooker.
5. Place sterilized jars, prepared caps and sterile media solution in your glovebox.
6. In the glovebox, add enough sterile media solution to just saturate but not cover the sand.
7. Close the jar with a prepared cap.

Note: Do not microwave the sand, it will splatter all over. When transfering cultures, make sure they are in good contact with the sand but not "swimming" in the media.

STERILE METHODS IN PLANT TISSUE CULTURE

Plants in tissue culture are very vulnerable to disease. Everything has to be sterile, the tissue, the containers, the tools, the medium. Here's how:

Using a Pressure Cooker as an Autoclave

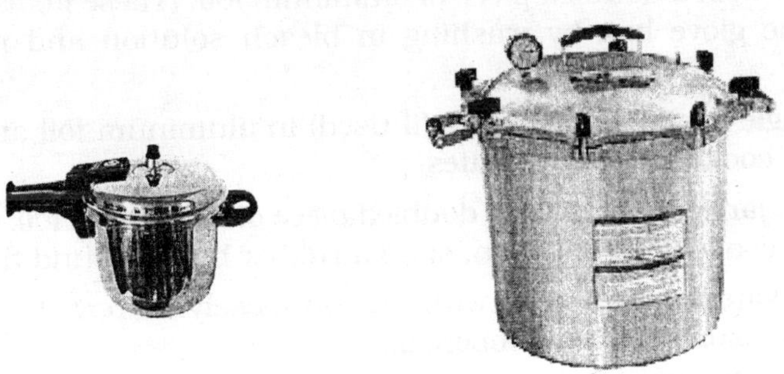

Scientists and housewives use similar instruments.

1. Check the pressure cooker seal. If stiff or damaged, do not use the pressure cooker!
2. Place items on a rack inside the pressure cooker.
3. Add water to a level below the rack.
4. Greese the lugs with petroleum jelly.
5. Close the pressure cooker, making sure that the seal is well seated.
6. Place the pressure cooker on high heat.
7. Begin timing when a steady stream of steam emerges from the vent.
8. After 15 minutes, turn the heat off and let cool before opening the pressure cooker.
   Or, Follow the manufacturer's directions for "canning" fruit and other low-acid food.

Warning

- Opening a pressure cooker while still warm is very dangerous.
- High pressure and hot steam can kill or maim

Sterilizing Instruments and Other Items

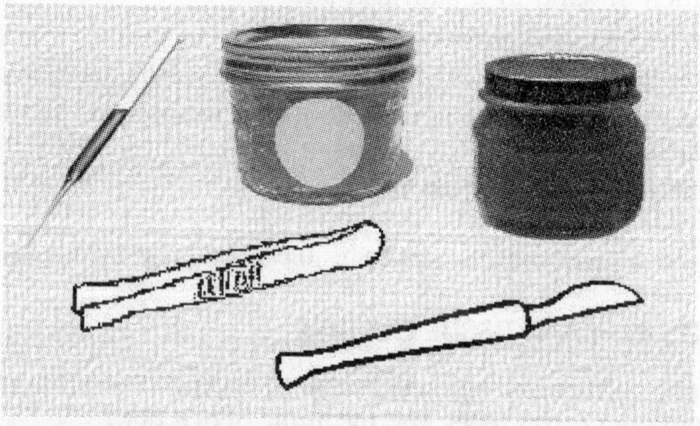

- Place tweezers, forceps and scalpels in a wide mouthed jar, business end down and cover with a doubled piece of aluminum foil. (These items can be resterilized inside the glove box by washing in bleach solution and rinsing with sterile water.)
- Wrap single-edged razor blades (if used) in aluminum foil and autoclave in the pressure cooker for 15 minutes.
- Autoclave jars covered with a doubled piece of aluminum foil. When the pressure cooker is cool enough to open, snap a rubber band around the foil immediately.
- Sterilize water in glass jars with lids on loosely. Screw the lids down as soon as the pressure cooker is opened.
- Place paper inside a paper bag and autoclave in the pressure cooker above the water level. After sterilization dry the bag in an oven set at 80°C. Do not unwrap the papers until needed.

Sterilizing Your Homemade Glove Box

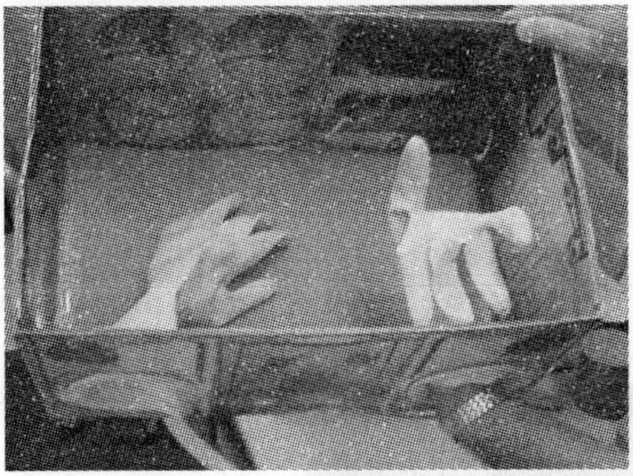

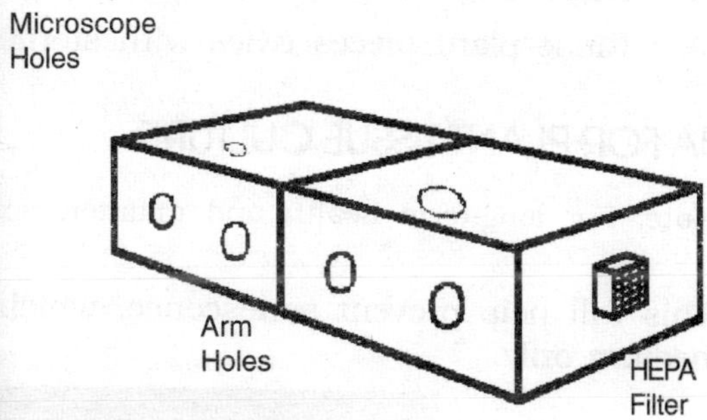

These directions will work for both the inexpensive cardboard-plastic-film glove box and the more expensive plastic glove box.

1. Tie back your hair, roll up your sleeves and remove your watch and other jewelry.
2. Collect and organize all the items you will need inside the cabinet.
3. Close and seal the homemade glovebox. (Plug in the HEPA filter fan, if used.)

Note: The active cabinet will not be tightly sealed like the passive one. (Air needs to leak out.)

4. Wash your hands thoroughly with the Pert (or other disinfectant).
5. Put on a pair of surgical gloves.
6. Insert your gloved hands through the sleeves.
7. Sterilize the inside of the cabinet, including your gloved hands, by spraying with 70% alcohol and wiping dry with sterile tissue.
8. Sterilize all jars and implements by spraying with 70% alcohol and wiping dry with sterile tissue.
9. Discard the used tissues in a small bag or box inside the glove box.
10. If using the active glove box, do the following:
    - o Sterilize the inside of the inner cabinet, including your gloved hands, by spraying with 70% alcohol and wiping dry with sterile tissue.
    - o Turn on the HEPA filter fan.
    - o Pass the items through to the inner chamber.

Note:

- From this point on, do not remove your gloved hands from the glove box.
- If you must, close all jars first and resterilize everything before starting to work on cultures again.

Sterilizing Plant Materials

1. Inside the homemade glove box, sterilize all plant material in diluted domestic bleach solution containing one drop of detergent as a surfactant.
2. Put plant pieces in a jar containing the bleach for 10-20 minutes. Agitate frequently.

3. Transfer the tissue in the bleach solution to the sterile area.
4. Discard the chlorine solution.
5. Rinse plant pieces twice with sterile water.

MEDIA FOR PLANT TISSUE CULTURE

Note: For long-term health and maintenance of strains, alternate two or more types of media.

This will help prevent senescence, which can occur when cultures are grown on one medium only.

Fertilizer Stock Mix

- ½-tablespoon all-purpose 18:18:18 (N.P.K.) water soluble fertilizer.
- 2 Cups of water.
- Gently boil to dissolve all ingredients and store in the refrigerator.

Basic Nutrient Mix

- Two cups of distilled water.
- Between two tablespoonfuls and ¼ cup of sugar.
- ½ Tbs of fertilizer stock.
- One-half Inositol tablet (500 mg).
- ½ vitamin tablet with thiamine.
- Gently boil to dissolve all ingredients and store in the refrigerator.

Rooting Mix

- ½ teaspoon of malt.
- Add basic nutrient mix to make two cups.
- Check the pH using narrow range pH indicator tape.
- Adjust pH to between 5 and 6 with vinegar or sodium bicarbonate.
- Add several drops of yellow food colouring to identify the formula.
- Gently boil to dissolve all ingredients and cool.

- Cover and put in the glove box.
- Add 100 mg, silver nitrate.
- Pour into sterile jars and cap.
- Store in the refrigerator until use.

Multiplication Mix

- ¼ to ½ cup of coconut milk.
- Add basic nutrient mix to make two cups.
- Check the pH using narrow range pH indicator tape.
- Adjust pH to between 5 and 6 with vinegar or sodium bicarbonate.
- Add several drops of red food colouring to identify the formula.
- Gently boil to dissolve all ingredients and cool.
- Cover and put in the glove box.
- Add 100 mg, silver nitrate.
- Pour into sterile jars and cap.
- Store in the refrigerator until use.

Peroxidated Flowering Mix

- (Optional) ½ teaspoon of malt.
- Add basic nutrient mix to make two cups.
- Add several drops of blue food colouring to identify the formula.
- Sterilize and cool to around 130°F.
- Cover and put in the glove box.
- Stir in ½ teaspoon peroxide.
- Add 100 mg, silver nitrate.
- Pour into sterile jars and cap.
- Store in the refrigerator until use.

Sand Supported Culture Protocol

The purpose of agar in growth media is solely to support the tissue. Instead of Agar, use sand to support the culture.

1. Boil sand in water on your stovetop.
2. Store the clean sand in a plastic bag.
3. Put one ounce of sand (volume) in a babyfood jar.
4. Sterilize jars in your pressure cooker.
5. Place sterilized jars, prepared caps and sterile media solution in your glovebox.
6. In the glovebox, add enough sterile media solution to just saturate but not cover the sand.
7. Close the jar with a prepared cap.
8. Store in the refrigerator until use.

Note. Do not microwave the sand, it will splatter all over.

The following are bacterial or fungal agars, use them for ideas.

Amaranth Soy Agar

- 40 grams amaranth flour
- 40 grams soy flour
- 19 grams agar
- 1 litre distilled water

Entheo Genesis No. 442

- 20 grams amaranth flour
- 20 gm brown rice flour
- 20 gm potato flour
- 20 gm soy flour
- 4 gm malted barley
- 19 gm agar
- 1 litre distilled water

Cornmeal Dextrose Agar

- 50 gm yellow cornmeal
- 5 gm dextrose
- 19 gm agar
- 1 litre distilled water

Barley Malt Extract Agar

- 40 gm barley flour
- 2 gm malt extract
- 1-2 gm yeast extract (optional)
- 9.5 gm agar
- 500 litre distilled water

Dr. Pollock's Modified Agar

- 20 gm dried dog food
- 20 gm amaranth flour
- 4 gm dextrose or malt extract
- 1 litre distilled water

Malt Extract Peptone Agar

- 30 gm Malt extract
- 2 gm Soya peptone
- 15 gm Agar

- 1 litre distilled water
- Adjust pH to 5.6. Sterilize for 16 min.

Malt Agar

- 20 gm of agar
- 20 gm of malt extract
- 2 gm of yeast
- water to one litre

Potato Dextrose Agar

- 20 gm of agar
- the strained broth produced by boiling 300 grams of diced potatoes in one litre of water for one hour
- 10 grams of dextrose
- clean water back up to one litre volume

In one experiment, tapioca starch produced the greatest number of shoots and nodes.

Chemical		*Abbreviations*	*Molecular Weight*
Cytokinins	Benzyladenine	BA	225.3
	Isopentil Adenine	2-iP	203.2
	Kinetin	KIN	215.2
	Zeatin	ZEA	219.2
Auxins	Indole-3-acetic acid	IAA	175.2
	Indole-3-butyric acid	IBA	203.2
	1-Naphtaleneacetic acid	NAA	186.2
	2,4-Dichlorophenoxyacetic acid	2,4-D	221.04
	2,4,5-Trichlorophenoxyacetic acid	2,4,5-T	255.5
	Picloram	PIC	241.5
Other	Thidiazuron	TDZ	220.2
	Silver Nitrate	$AgNO_3$	169.9

SAFETY IN PLANT TISSUE CULTURE

- Sterile tissue culture technique eliminates or almost eliminates the danger of loose seeds, pollen broadcast, vegetative escape, etc.
- Bamboo is unlikely to cause gene flow due to long intervals between flowering.
- Handling and growing cultures inside a locked and filtered plastic box isolates the organisms from infection and prevents escape.
- Choosing clumping bamboos limits the possibilities of vegetative escape, even if planted in the ground.

Therefore, looking at standard practices for biosafety level 1-P, I conclude that:

1. A locked apartment door and secure windows are more than enough access restriction, in this case.
2. I can inform myself of hazards and safeguards.

3. Given an organism with low potential for gene flow, my procedures are more than adequate.
4. No special containment is required other than keeping cultures inside the plastic glove box/growing chamber. If removed they can be sealed inside a small plastic carrying box that is passed into the chamber using aseptic procedures.
5. Records can be kept in a lab notebook.
6. Supplies can be sterilized as they are passed into the glovebox/growing chamber and sterilized again as they are passed out.
7. All organisms can be inactivated by placing inside a pressure cooker which is sealed before removing from the chamber. The pressure cooker is not unsealed until after an adequate processing time.
8. Maintenance in tissue culture and inactivating through pressure cooking provide adequate pest control.
9. Since there will be no motile organisms, sterile tissue culture is adequate containment.
10. Given the built-in safety of sterile tissue culture and the organisms selected, no signage is needed.
11. Working inside a air-tight glovebox minimizes the creation of aerosols.
12. Use of gloves eliminates the need for special clothing. Hand washing before and after procedures is adequate.
13. Any accidents and cleanup procedures can be recorded in a lab notebook.

3. PREPARATION OF MEDIA FOR ANIMAL CELL CULTURE

Introduction

The growing interest in products from animal cells has caused an extensive research effort forward development of media for cell cultivation. The basic components in the media used for cultivation of animal cells vary depending upon the characters of the cells, and the cultivation method. Basic components consist of an energy source, nitrogen source, vitamins, fats, fatty acids and fat soluable components, inorganic salts, nucleic acid, antibiotics, oxygen, pH buffering systems, hormones, growth factors serum and extensive efforts are directed toward developing serum-free media, protein media or (synthetic media) chemically defined media (eg: MEM- minimum essential media).

Almost 50% of the biological produced today or planned to be produced in the future are from animal cells origin. Therefore there is an increasing interest in developing technologies for cultivation of animal cells and production of a wide spectrum of biologicals. The world wide activities and market of biologicals from mammalian cells were recently received.

Although the major achievements in the field of animal cells cultivation have been accomplished in the last three decades, it has along history of about one hundred years. Apart from the development of various types and size of culturing vessels. Research and development of optimal media for cell division is also carried out among most groups involved in the field of animal cell cultivation and production of biologicals. Media used for cell cultivation is considered to include two major parts.

1. Essential basal ingredients that full-fill all cellular requirements, for nutrients and known as the basal growth medium.
2. A set of supplements that satisfy other types of cellular growths requirements and make it possible to growth medium.

Water for Animal Cell Media

Out of the most important points for consideration when preparing media is the required high quality of the water. Water used for culture media should be pyrogen free (especially if the product is for human or animal use it should have resistant of 1.5-2.0 m. Ohms indicating a low salt content.)

It is highly recommended to use fresh ultra-pure water and not store water since in some storage tanks, organic materials or irons from plastic or glass may dissolve in the water.

Purity of Chemicals, Stability and Self Life

Chemicals of the highest purity are required for preparation of media. Commercial chemicals although being pure inevitably contain traces of contaminants. Some of the traces may be toxic (like Hg). With regards to stability of media ingredients, inorganic chemicals are indefinitely stable. Vitamins are the least Hormones, several antibiotics and growth factors are recommended to be stored frozen (–20°C) or refrigerated (0-4°C)

Several ingredients used in animal cell culture media are known for their instability. eg. Ascorbic acid and Glutamine. Most factors affect the self life if media among them are the following Natural decay rates of unstable compounds, pH, moisture, storage temperature, access of oxygen and an exposure to near ultra-violet, day light or inflorescence light. Most media should be stored at 41°C and in a dark place. Storage of media by freezing may cause loss of some purely soluble ingredients. Powdered media may be stored for several years.

Basic Components in Media

1. Energy sources - Glucose, Fructose, Amino acids.
2. Nitrogen sources - Amino acids.
3. Vitamins - Mainly water soluble vitamins - B and C.
4. Fat and Fat soluble components: Fatty acids, cholesterols.
5. Inorganic salts: Na^{+}, K^{+}, Ca^{2+}, Mg^{2+}
6. Nucleic acid precausers.
7. Antibiotics.
8. pH and buffering systems.
9. Oxygen.
10. Hormones and growth factors.

Sera in Animal Cell Media

Sera is the most important and most problematic component in animal cell media. During more than three decays sera has been an essential medium component with the following functions:

1. Provides nutrients.
2. Provides proteins that solubilize essential nutrients that do not dissolve readily.
3. Binds essential nutrients that are toxic when present in excessive amounts and releasing then slowly in a controlled manner.
4. Provides hormones and growth factors.
5. Modulates the physical and chemical properties of the medium (Viscosity, rate of diffusion) - protect cells in agitated culture.
6. Has a pH buffering function.

Despite these advantages, there are several problems associated with the use of serum for cell cultures.

1. Serum is the most expensive component.
2. Being highly viscosis, sera sloes down the sterilization by filtration of the media.
3. From time to time there is a storage in world supply of sera.
4. Possible availability of contaminants.
   Eg: Mycoplamas, Viruses.
5. Availability of serum in media, increases the complicity of the downstream processing of the desired biological media.

1. ASEPTIC TECHNIQUE

This protocol describes basic procedures for aseptic technique for the novice in cell culture technology. One basic concern for successful aseptic technique is personal hygiene. The human skin harbors a naturally occurring and vigorous population of bacterial and fungal inhabitants that shed microscopically and ubiquitously. Most unfortunately for cell culture work, cell culture media and incubation conditions provide ideal growth environments for these potential microbial contaminants. This procedure outlines steps to prevent introduction of human skin flora during aseptic culture manipulations.

Every item that comes into contact with a culture must be sterile. This includes direct contact (e.g., a pipette used to transfer cells) as well as indirect contact (e.g., flasks or containers used to temporarily hold a sterile reagent prior to aliquoting the solution into sterile media). Single-use, sterile disposable plastic items such as test tubes, culture flasks, filters, and pipettes are widely available and reliable alternatives to the laborious cleaning and sterilization methods needed for recycling equivalent glass items. However, make certain that sterility of plastic items distributed in multiunit packages is not compromised by inadequate storage conditions once the package has been opened.

Ideally, all aseptic work should be conducted in a laminar however, work space preparation is essentially the same for working at the bench. Flame sterilization is used as a direct, localized means of decontamination in aseptic work at the open bench. It is most often used (1) to eliminate potential contaminants from the exposed openings of media bottles, culture flasks, or test tubes during transfers, (2) to sterilize small instruments such as forceps, or (3) to sterilize wire inoculating loops and needles before and after transfers. Where possible, flame sterilization should be minimized in laminar-flow environments as the turbulence generated by the flame can significantly disturb the sterile air stream.

Materials

Antibacterial soap
70% ethanol or other appropriate disinfectant
95% ethanol
Clean, cuffed laboratory coats or gowns
Latex surgical gloves
Clean, quiet work area
Shallow discard pans containing disinfectant
Bunsen burner or pilot-activated burner (*e.g.*, Touch-o-Matic, VWR)

Take Personal Precautions

1. Just prior to aseptic manipulations, tie long hair back behind head. Vigorously scrub hands and arms at least 2 min with an antibacterial soap.

   Superficial lathering is more prone to loosening than removing flaking skin and microbial contaminants. Loosely adhering skin flora easily dislodge and can potentially fall into sterile containers.

2. Gown appropriately. For nonhazardous sterile-fill applications, wear clean, cuffed laboratory coats and latex gloves.

   Greater stringencies may be necessary depending upon laboratory regulatory requirements. Work with potentially hazardous agents certainly mandates additional considerations for safety. Front-closing laboratory coats are not recommended for work with hazardous biological agents. Safety glasses should be worn by laboratory personnel when manipulating biological agents outside the confines of a biosafety cabinet.

3. Frequently disinfect gloved hands with 70% ethanol while doing aseptic work.

   Although the gloves may initially have been sterile when first worn, they will no doubt have contacted many nonsterile items while in use.

   Note that 70% ethanol may not be an appropriate agent for latex glove disinfection when working with cultures containing animal viruses, as studies have shown that ethanol increases latex permeability, reducing protection for the wearer in the event of exposure. In this case, quarternary ammonium compounds are more appropriate.

4. Dispose of gloves by autoclaving after use. Do not reuse. Bag and autoclave single-use laboratory coats after use. Bag, autoclave (if necessary), and wash other laboratory coats within the laboratory facility or send out for cleaning at a laundry certified for handling biologically contaminated linens.

   Never take laboratory clothing home for washing.

5. Thoroughly wash hands after removing protective gloves.

   Prepare and maintain the work area.

6. Perform all aseptic work in a clean work space, free from contaminating air currents and drafts. For optimal environmental control, work in a laminar-flow cabinet.

7. Clear the work space of all items extraneous to the aseptic operation being performed.

8. Wipe down the work surface before and after use with 70% ethanol or other appropriate disinfectant.
9. Wherever feasible, wipe down items with disinfectant as they are introduced into the clean work space. Arrange necessary items in the work space in a logical pattern from clean to dirty to avoid passing contaminated material (*e.g.*, a pipette used to transfer cultures) over clean items (*e.g.*, flasks of sterile media).
10. Immediately dispose of any small contaminated items into a discard pan.
11. When the aseptic task has been completed, promptly remove any larger contaminated items or other material meant for disposal (*e.g.*, old culture material, spent media, waste containers) from the work space and place in designated bags or pans for autoclaving. Disinfect the work space as in step 8.

Flame Sterilize the Opening of a Vessel

12. For a right-handed person, hold the vessel in the left hand at ~45° angle (or as much as possible without spilling contents) and gently remove its closure. Do not permit any part of the closure that directly comes in contact with the contents of the vessel to touch any contaminating object (*e.g.*, hands or work bench).

    Ideally, and with practice, one should be able to hold the closure in the crook of the little finger of the right hand while still being able to manipulate an inoculating loop or pipettor with the other fingers of the hand.

    Holding the vessel off the vertical while opening will prevent any airborne particulates from entering the container.
13. Slowly pass the opening of the vessel over the top of (rather than through) a Bunsen burner flame to burn off any contaminating matter.

    Be careful when flaming containers of infectious material. Any liquid lodged in the threads of a screw cap container will spatter as it is heated. Aerosols thus formed may actually disseminate entrapped biological agents before the heat of the flame is hot enough to inactivate them.
14. While still holding the vessel at a slant, use a sterile pipette and pipettor to slowly add or remove aliquots to avoid aerosol formation.
15. Flame-sterilize again as in step 13, allow the container to cool slightly, and carefully recap the vessel.

    Flame sterilize small hand instruments.
16. Dip critical areas of the instrument (*i.e.*, those that come into contact with the material of concern) in 95% ethanol.

    Make certain that the alcohol is in a container heavy enough to support the instrument without tipping over.

    Caution: 95% ethanol is flammable; keep the container at a safe distance from any open flame.
17. Remove the instrument from the alcohol, being careful not to touch the disinfected parts of the instrument. Allow excess ethanol to drain off into the container.

18. Pass the alcohol-treated part of the instrument through the flame of a Bunsen burner and allow residual alcohol to burn off.
19. Do not let the sterilized portion of the instrument contact any nonsterile material before use. Let the heated part of the instrument cool for ~10 sec before use.
20. After use, return the instrument to the alcohol disinfectant until needed again. Flame sterilize inoculating loops and needles.
21. Hold the inoculating wire by its handle and begin in the centre of the wire to slowly heat the wire with the flame of a Bunsen burner. Proceed back and forth across the wire's full length until it glows orange.
22. While still holding the handle, allow the inoculating wire to cool back to room temperature (~10 sec) before attempting any transfer of material.

    If transfers are made while the inoculating wire is hot, cells will be killed by the hot wire, and aerosols created from spattering material can disperse biological material throughout the work space.
23. After the transfer is made, reheat the inoculating wire as in step 21 to destroy any remaining biological material. Let cool to room temperature before putting aside for next use.

2. CULTURE AND MAINTENANCE OF CELL LINES

Purpose

Cell and organ cultures are used to maintain living animal cells and groups of cells outside the body (in vitro). With separate, living cell cultures, it is possible to see and study the behavior of animal cells in greater detail than when they are in the animal (in vivo). Cell culture also frees the cells from some of the controls that normally regulate their activities. Cells, or tissues, are kept alive for varying periods of time, at times undergoing repeated divisions over many generations.

Cells grown and cultured for study have been taken from a wide variety of species, such as humans, monkeys, mice, dogs, cats, frogs, insects, fish and many others. The cultures have come from a number of organs - heart, lungs, liver, kidney, blood, skin, etc. In cancer research it is common practice to grow cells from normal and cancerous tissues to compare their properties. In fact, cell cultures have become one of the best means of testing potential anti-cancer drugs; utilizing cell cultures is more cost-effective and faster than experiments using animals and, with this method, isolated human cancerous tissue can be ested.

General Principles and Techniques

Tissue fragments used in the preparation of cell cultures must be handled with care in the laboratory to avoid microbial contamination; sterile, or aseptic, technique must be employed at all times. All instruments, culture vessels, etc. that come in contact with the cells or medium must be sterile. The tissues are kept at 37°C and suspended in a physiologically balanced salt solution. Addition of a small amount (1-15%) of blood serum helps protect the cells during these preliminary manipulations. Antibiotics may also be added as well as pH indicators such as phenol red. Tissues are cut into very small fragments, called explants, which are put into vessels and bathed with nutrient medium. Sometimes the fragments are attached to the surface of the vessels by opposite charge attraction or with blood plasma, which is then allowed to clot. In the case of the cultures of this type, cells migrate from the explanted tissue into the medium and undergo division to produce a "halo" of outgrowth around the original tissue.

Cell cultures are also started by treating tissue fragments with chemicals (enzymes such as trypsin, or chelating agents such as versene) to dissociate them into a suspension of single cells. These cells are then placed in a nutrient medium and a portion of the suspension is put into a suitable culture vessel. The culture vessel is incubated without being disturbed, which allows the cells to settle out from the suspension, attach to the wall of the vessel, and grow. The complete sheet of cells covering the vessel wall is called a monolayer.

"Cell lines", which are capable of continuous growth, are usually grown as monolayers. To transplant such cultures, the cells are either scraped carefully from the glass with a rubber spatula or the medium is taken off and the cells are removed with a chemical (enzyme or chelating agent). The cells are then suspended and counted, and the suspension is diluted with fresh medium to obtain the number of cells desired.

Objectives

Students will demonstrate one of the methods used to initiate primary cell cultures from fresh embryonic tissue and observe patterns and rate of cell growth in a mixed culture (the culture prepared representing a mixture or many cell types). Students will also become acquainted with the initiation and subculture of tissue cells in "in vitro" culture and with fixation and staining techniques used in the field of cytology, and will clearly differentiate between nuclear material and cytoplasmic material in animal tissue cells.

Required Materials

2 Curved Dissecting Forceps

1 Sterile Culture Tube

1 Sterile Petri Dish per Group

25 Sterile Pipettes

10 Sterile Culture Flasks

1 Sterile Glass Rod

2 Sterile Versene Tubes

1 Sterile Medium Tubes 199

5 Sterile Alcohol Pads

1 Staining Jar

1 Bottle of Hanks Balanced Salt Solution, 100 ml

Methanol 30 ml

Hematoxylin Stain 30 ml

Eosin Stain 30 ml

Isopropyl Alcohol 99% 100 ml

Histoclear

Piccolyte II Mounting Medium 15 ml

Fertile Hen's Egg, Incubated for Seven Days

Isopropyl Alcohol, 70%

Incubator

Compound Microscope

Safety Goggles

Lab Aprons

Pre-Lab Preparation

Seven days before the experiment obtain a fertile hen's egg and incubate according to instructions accompanying your incubator. See Appendix I for additional information on egg incubation.

A. Establishment of the Primary Cell Culture

Procedure Notes

Though the embryo is relatively underdeveloped, working with a chick embryo may be a sensitive issue for some students. If this is the case, you may perform steps 1 through 12, involving maceration of the embryonic tissue, in advance. One fertile egg, incubated for seven days, provides sufficient material for five lab groups. Do not remove sterile materials from their protective packages until you are ready to use them.

Procedure

Note: Sterile technique is essential when obtaining primary cell cultures. Do not touch any instruments or culture vessel openings, or any other contaminated surface. Consider all surfaces not specifically sterilized to be contaminated. When removing a cap from a vessel, hold the cap with your little finger. Your hand will still be free, yet the cap is protected from contamination. For more information on sterile technique.

1. Obtain a fertile hen's egg which has been incubated for seven days. Candle the egg to verify the presence of a developing embryo and to locate its position. To candle the egg, use a box containing a 150-watt bulb, with a 2"-diameter hole cut in the box. The box should be in a dark room. Place the egg, large end up, over the hole. If the egg is fertile, the developing blood vessels will be visible.
2. Place the egg, large end up, in an egg carton. The large end contains the air sac. Wipe the egg thoroughly with a sterile alcohol pad to sterilize the surface.
3. Sterilize a pair of forceps by passing them slowly through an open flamc; allow to cool before using.

   Note. Once used, items are no longer considered sterile.
4. Gently crack the shell over the air sac. Remove the shell surrounding the air sac with the sterile forceps, taking care not to rupture the shell membrane.
5. Tear off the shell membrane.

   Note. Do not let bits of shell fall into the egg.
6. Carefully pour the contents of the egg into a sterile Petri dish. With two pairs of sterile forceps, remove the membrane from around the embryo.

   Note. Be very careful when removing the membrane; the embryo explant.

   Materials can be damaged easily if pressed too tightly or pulled too hard.

7. Place the embryo in the sterile culture tube. Cap the tube.
8. With a sterile pipette, add 1 ml versene solution to the explant material. Discard the pipette.

   Note. Set aside the remainder of the versene for use in part B of the investigation.
9. Gently grind the explant material and versene with a sterile glass-grinding rod. Replace the cap on the culture tube and set aside for approximately 20 to 30 minutes.

   Note. You may also macerate embryonic tissue with a sterile, large-capacity. *Syringe.* The process is similar to grinding with less chance of contamination. Add 1 ml versene solution to the tissue in the culture tube. Place the embryo and versene in a syringe with at least a 50 ml capacity. Reset the plunger, position the syringe over the culture tube, and press, using one fluid motion. Pushing the tissue through the syringe will facilitate the action of the versene.
10. With a sterile pipette, draw the tissue up and expel it from the pipette, several times to homogenize it further. Discard the pipette.
11. With a sterile pipette, add one tube of Medium 199 to the culture tube. Discard the pipette.
12. Shake to mix the suspension well. Allow to stand for five minutes.
13. With a sterile pipette, carefully transfer 3 ml of the cell suspension to a sterile culture flask. Do not draw up large particulate matter. Cap the flask.
14. Label the flask with the date of the culture and the tissue source. Incubate the flask with its largest, flat surface facing downward at 37°C (99°F).
15. Once the tissue has been incubated, invert the flask and place it on the stage of a compound microscope. The cells will be growing on the surface that was facing down during incubation. Examine the cells at 100X. At this magnification, an individual cell will appear to be about the size of a grain of rice. Focus carefully to view the nucleus within each of the cells.

    Note. You should be able to observe a network of cells. Since there will be a mixture of cell types, some cells will be elongated and others compact. On occasion, some cells may regenerate into a large organized mass.
16. If your cultures turn yellow (acidic) or the fluid becomes cloudy, it is an indication that they have become contaminated; they should be discarded.

    Note. Disposed of biologic materials should be autoclaved. If an autoclave is not available, place the unopened cultures in a pan of water, bring to a vigorous boil for 20 minutes, then discard. Instruments used in the investigation should likewise be sterilized by either autoclaving or boiling.

B. Propagation of Chick Cells onto Coverslips

Once tissue cells have been established in primary culture (first isolate), they can usually be maintained for some time by serial subculture. However, primary cultures, as developed in part A, can usually be transplanted, or subcultured, only two or three times. When a coverslip is added to the culture medium in which the cells are suspended, some of the cells will settle out and grow on its surface, creating a monolayer that can be viewed under a microscope.

Procedure

1. Obtain a culture flask that has been incubated for six to seven days.
2. Pour off the medium, taking care not to contaminate the flask opening or its contents. Replace the cap. Discard the medium.
3. With a sterile pipette, add 1 ml versene solution to the flask. Swirl the flask and pour off the versene. Replace the cap. Discard the pipette.
4. With a sterile pipette, add 1 ml versene solution to the flask. Replace the cap.
5. Place the flask on a flat surface so the solution completely covers the layer of cells. Leave at room temperature for 15 minutes.

   Note. Mark the side of the culture flask on which the cells are growing.
6. The cells will be loosened by the versene solution and will float off the surface. If necessary, you may extend the incubation period, or you may agitate the culture flasks vigorously to help dislodge the cells.
7. With a sterile pipette, draw up and eject the suspension several times to break up any larger cell clumps that may be present.
8. Draw up the full amount of cell suspension in the pipette and transfer it to the unused vial of Medium 199. Pool the cells from five flasks into one fresh medium vial. Discard the pipette.
9. Swirl the vial gently to mix the transferred cells with the fresh medium.
10. Obtain a sterile vial containing a coverslip.
11. With a sterile pipette, add 2 ml of the cell suspension to the vial. Cap the culture vial tightly. Discard the pipette.
12. Label the vial with the name of the culture, its original initiation date, and the subsequent transfer.
13. Place the vial in the incubator; stand it upright to ensure the coverslip remains perfectly flat on the bottom of the vial. Incubate the culture vial at 37°C (99°F) for four to five days. The cells should settle and attach to the coverslip within a few hours.

C. Fixation and Staining of Chick Cell Monolayer

Cells, which are normally colourless, are fixed and stained for viewing under a microscope. Fixing is a process that stabilizes the chemical and structural characteristics of the cells. Cells are then stained with a dye or combination of dyes to highlight structural details. Different fixation and staining procedures highlight different structural features. The fixative methanol is good for monolayers. Hematoxylin, a powerful stain commonly used by cytologists, stains nuclear material bright blue to dark blue. Hematoxylin does not stain cytoplasmic material, however, so another stain must be employed, such as eosin, which stains cytoplasmic material light pink to red.

Procedure

Safety. Wear safety goggles at all times. Use care in handling stains

Note. This procedure should be performed three to four days after preparing the monolayer tissue culture.

1. Heat Hanks Balanced Salt Solution to 37°C (99°F.) Place in a staining jar.

2. Carefully remove the coverslip from the culture vessel: tip the coverslip away from the bottom of the vessel and gently remove it with forceps.

   Note. The coverslip is extremely thin; be very careful when handling it with forceps.

3. Immediately place the coverslip in warm Hanks Balanced Salt Solution in the staining jar. Orient the coverslip so the side with cells attached is easily identified.
4. Taking care to keep the coverslip in the staining jar, pour off and discard the used balanced salt solution. Immediately add fresh, warm balanced salt solution for a second washing. Repeat, for a total of three rinses in the salt solution.
5. Pour enough methanol into the staining jar to just cover the coverslip. Allow cells to fix for seven to 10 minutes. Agitate the staining jar occasionally to ensure thorough fixation.

   Note. Be careful when using methanol; it is poisonous if taken internally.

6. Pour off and discard the methanol. Rinse the coverslip with water to remove excess methanol.
7. Add enough Hematoxylin stain to completely submerge the coverslip. Allow the cells to stain in this solution for 10 minutes.
8. Pour off and discard the stain. Rinse the coverslip with water.
9. Add balanced salt solution and set aside for two minutes.
10. Pour off and discard salt solution. Rinse the coverslip with tap water.
11. Add enough eosin stain to completely submerge the coverslip. Allow to the cells to stain for five minutes.
12. Pour off and discard the stain. Rinse the coverslip in 99% isopropyl alcohol three times. Allow one minute for each rinse, agitating the staining jar occasionally.
13. Add Histoclear to the staining jar and allow to set for two minutes, agitating occasionally.
14. Pour off and discard the Histoclear. Add fresh Histoclear and set aside for two minutes.
15. During these two minutes, clean a glass microscope slide with a lint-free cloth.
16. Just before the end of the two-minute period, add one drop of Piccolyte II mounting medium to the center of the clean microscope slide.
17. Remove the coverslip from the Histoclear and place it cell-side down onto the centre of the microscope slide.

    Note. Be sure not to trap air bubbles beneath the coverslip. This can be avoided by placing one side of the coverslip against the microscope slide and gently lowering the coverslip onto the slide. If an air bubble is tapped, it can be forced out by gently pressing on the coverslip with a sharp instrument.

18. Allow the slide to dry for at least 24 hours before handling it.
19. Examine the slide under 100X magnification. Note the pattern and rate of cell growth. The increase in cell rate can be roughly calculated by selecting a representative microscope field and counting the number of cells in that field. Mark the field by circling it with a glass marking pencil.

TRYPSINIZING AND SUBCULTURING CELLS FROM A MONOLAYER

A primary culture is grown to confluency in a 60 mm petri plate or 25 cm^2 tissue culture flask containing 5 ml tissue culture medium. Cells are dispersed by trypsin treatment and then reseeded into secondary cultures. The process of removing cells from the primary culture and transferring them to secondary cultures constitutes a passage, or subculture.

Materials

Primary cultures of cells

HBSS without Ca^{2+} and Mg^{2+}, 37°C

Trypsin/EDTA solution

Complete medium with serum with 10% to 15% (v/v) FBS

Sterile Pasteur pipettes

37°C warming tray or incubator

Tissue culture plasticware or glassware including pipettes and 25 cm^2 flasks or 60 mm petri plates, sterile

Note. All culture incubations should be performed in a humidified 37°C, 5% CO_2 incubator unless otherwise specified. Some media (*e.g.*, DMEM) may require altered levels of CO_2 to maintain pH 7.4.

1. Remove all medium from primary culture with a sterile Pasteur pipet. Wash adhering cell monolayer once or twice with a small volume of 37°C HBSS without Ca^{2+} and Mg^{2+} to remove any residual FBS that may inhibit the action of trypsin.

   Use a buffered salt solution that is Ca^{2+} and Mg^{2+} free to wash cells. Ca^{2+} and Mg^{2+} in the salt solution can cause cells to stick together.

   If this is the first medium change, rather than discarding medium that is removed from primary culture, put it into a fresh dish or flask. The medium contains unattached cells that may attach and grow, thereby providing a backup culture.

2. Add enough 37°C trypsin/EDTA solution to culture to cover adhering cell layer.

3. Place plate on a 37°C warming tray 1 to 2 min. Tap bottom of plate on the countertop to dislodge cells. Check culture with an inverted microscope to be sure that cells are rounded up and detached from the surface.

   If cells are not sufficiently detached, return plate to warming tray for an additional minute or two.

4. Add 2 ml 37°C complete medium. Draw cell suspension into a Pasteur pipette and rinse cell layer two or three times to dissociate cells and to dislodge any remaining adherent cells. As soon as cells are detached, add serum or medium containing serum to inhibit further trypsin activity that might damage cells.

   If cultures are to be split 1/3 or 1/4 rather than 1/2, add sufficient medium such that 1 ml of cell suspension can be transferred into each fresh culture vessel.

5. Add an equal volume of cell suspension to fresh plates or flasks that have been appropriately labelled.

Alternatively, cells can be counted using a hemocytometer or Coulter counter and diluted to the desired density so a specific number of cells can be added to each culture vessel. A final concentration of ~5 × 104 cells/ml is appropriate for most subcultures.

For primary cultures and early subcultures, 60 mm petri plates or 25 cm^2 flasks are generally used; larger vessels (*e.g.*, 150 mm plates or 75 cm^2 flasks) may be used for later subcultures.

Cultures should be labelled with date of subculture and passage number.

6. Add 4 ml fresh medium to each new culture. Incubate in a humidified 37°C, 5% CO_2 incubator.

   If using 75 cm^2 culture flasks, add 9 ml medium per flask.

   Some labs now use incubators with 5% CO_2 and 4% O_2. The low oxygen concentration is thought to stimulate the in vivo environment of cells and to enhance cell growth.

   For some media it is necessary to adjust the CO_2 to a higher or lower level to maintain the pH at 7.4.

7. If necessary, feed subconfluent cultures after 3 or 4 days by removing old medium and adding fresh 37°C medium.

8. Passage secondary culture when it becomes confluent by repeating steps 1 to 7, and continue to passage as necessary.

4. CELLULAR BIOLOGY TECHNIQUES

1. MEDIA AND SOLUTIONS REQUIRED FOR ROUTINE ES CELL CULTURE

Media used to prepare 100 ml medium

DMEM	80 ml
FCS	15 ml
Non-essential amino acids (100X)	1 ml
Pen/strep (5,000 1U/ml, 5000 ug/ml)	1 ml
L-Glutamine 200 mM	1 ml
Nucleosides stock (100X)	1 ml
BME 0.1M	0.2 ml

- Dulbecco's Modification of Eagles Medium (DMEM)
- 1X without L-glutamine with 4.5 g/L Glucose. Store at 4°C
- Cytosystems 500 ml Cat. No. 11.016.0500V
- Non essential amino acids 100X
- Cytosystems 100 ml Cat No. 21-145-0100V. Store at 4°C
- Penicillin - streptomycin (5,000 1U/ml, 5,000 ug/ml)
- Cytosystems 100 ml Cat No. 21-140-0100V. Store at –20°C
- Foetal Calf Serum (FCS)

- Needs to be tested for ability to support growth of ES Cells. Serum should be stored frozen and heat inactivated before use by heating at 56°C for 30 mins. It may be stored at 4°C following inactivation. The heat inactivation removes complement activity which along with natural heterophile antibodies, may be toxic for ES cells.
- 100 × Nucleoside Stock
- To prepare 100 ml (100X)

Adenosine	80 mg	3 mM	Sigma A 4036
Guanosine	85 mg	3 mM	Sigma G 6264
Cytidine	73 mg	3 mM	Sigma C 4654
Uridine	73 mg	3 mM	Sigma U 3003
Thymidine	24 mg	1 mM	Sigma T 1895

- Add to 100 ml Travenol water and dissolve by warming to 37°C. Filter sterilize and aliquot while warm. Store at 4°C for months. The nucleosides will come out of solution. Warm to 37°C before use to re-solubilize. Solution seems to be stable at 4°C.
- Other solutions required.
  - 1X Dulbecco's Phosphate Buffered Salt Solution (PBS) without calcium and magnesium.

Cytosystems	100 ml	11.075.0100V
	500 ml	11.075.0500V

- Trypsin 2.5% Cytosystems 100 ml 21-159-0100V
- Trypsin - EDTA (1:250) Cytosystems (0.05% Trypsin) 100 ml 21-160-0100V
- Solutions to be made up
- PBS/EGTA (0.5 mM)

To 100 mls of PBS add 1 ml of 0.05M stock EGTA (Sigma E4378) to give a final concentration of 0.5 mM.

- Stock EGTA in H_2O 0.05M add conc. NaOH to dissolve
- Trypsin 0.25%.
- To 20 mls Cytosystems Trypsin/EDTA 0.05%, add 1.6 mls 2.5% Trypsin
- 0.1% Gelatin in PBS
- To 100 mls PBS add 0.1 gm Gelatin Sigma G-1890 and autoclave
- 0.1M 2-mercaptoethanol (§ME) sigma M-6350.
- Add 0.1 ml §ME (14.4M) to 14.3 mls PBS and filter through a 0.2 uM ACRODISC. Store at –20°C for up to one month.
- LIF (if no feeders are used)
- LIF ESGRO AMRAD (murine Lif in PBS/BSA solution) 1 ml ampoule
- 107 U/ml. Dilute 1/100 in DMEM 10% FCS, aliquot into 10 × 10 ml tubes,
- store at –20°C until use.
- 107 U/ml dilute 1/100 = 105 U/ml. (100X conc)
- Dilute 1/100 for use = 1000 U/ml

2. ROUTINE CULTURING OF ES CELLS

Cell are normally passaged every 2-3 days, this is important to avoid differentiation. *Signs of differentiation are:*

- colonies are surrounded by flattened differentiated cells.
- large colonies with necrotic centres, these appear as cells with defined boundaries.
- colonies appear as individual cells rather than as a syncial mass.
- colonies are more "rounded" than "flat", they also have a clearly defined boundary, worse than this, they have formed free floating embryoid bodies.

Cells are passaged as followed:

1. All reagents are warmed to 37°C

   Medium: PBS, PBS/EGTA, TRYPSIN/EDTA
2. Remove medium
3. Wash with PBS (5 ml/25 cm^2 flask, 10 ml/80 cm^2 flask), aspirate
4. Wash with PBS/EGTA, aspirate
5. Place flask on 37°C warming tray for approx. 1 min or until individual cells can be seen in colonies.
6. Add 0.5 ml (25 cm^2), 1 ml (80 cm^2) TRYPSIN/EDTA and rock flask backwards and forwards until colonies float off, this should take ~1 min.
7. Using a 1 ml pipette, pipette up and down, (avoid making bubbles as these kill the cells), for approx. 1 min. Check that all colonies have been dispersed and that a single cell suspension has been achieved. Don't leave cells in TYRPSIN/EDTA for longer than 3 mins as it is quite toxic.
8. Neutralise trypsin by adding an equal volume of medium, mix by gentle pipetting.
9. Aspirate media from feeder flask, as this is different media to ES cell media. Seed feeder* flasks with an aliquot of cell suspension, a 1:10 and 1:20 split is appropriate for a well growing culture with medium - large colonies that are not touching each other but are reasonably close together.

   *Feeder flasks contain Mitomycin C treated (see protocol) Primary Mouse Embryo Fibroblasts (PMEFs) at a concentration of 0.3 × 106/25 cm^2 flask, 1 × 106/80 cm^2 flask, or Mitomycin C treated STO cells at a concentration of 1.25 × 106/25 cm^2 flask, 4 × 106/80 cm^2 flask. STO cells are smaller than PMEF's. (see charts).

   If cells are to be grown in the presence of LIF only *i.e.*, no feeder layer, flasks or plates must be treated with 0.1% gelatin in PBS at 37°C for at least 1 hr–2 hrs. This is removed before the medium is added.

3. ISOLATION OF PRIMARY MOUSE EMBRYO FIBROBLASTS

- You will need 13.5 day pregnant mouse (we use MTK NEO inbred white mice)
- 2 sets sterile instruments
- One containing a pair of curved forceps and a pair of iris scissors

- One containing two pairs curved forceps, one pair iris scissors and a #3 size scalpel handle
- Phosphate buffer saline (PBS)
- Sterile medium size petri dishes (tissue culture standard)
- 18 gauge needle
- Luer lock syringe (about 6cc should suffice)
- #11 size flat-edged scaple blade
- Trypsin/EDTA
- Dulbecco's Modification of Eagles Medium (with 10% fetal calf serum, 1% penicillin/streptomycin, 1% L-glutamine 0.2% 0.1 m BME)
- Large flasks (tissue cultured standard about 154 cm^2 area)
- Class II Laminar Flow hood.

Before starting, pour out 2X petri dishes of PBS in the hood. Pregnant mouse is killed by cervical dislocation. (This is not done in the hood but on clean benchcote). Lay mouse out on its back and swab belly with 70% ethanol. With a pair of scissors (not sterile) nip a small cut across the belly. Grasping the skin above and below the nip with your finger, tear the skin apart and draw back over the head and hind legs to expose the viscera of the gut. This method is cleaner than cutting through the fur and enable you to reach the uterus with no risk of touching the fur (cutting through dry fur creates a bacterial aerosol).

Using sterile forceps and iris scissors dissect out the uterus, taking care not to touch the fur or the benchcote with the uterus or instruments. Place the uterus into a petri dish of sterile PBS and swirl around to remove blood. Transfer uterus to second petri dish of sterile PBS and move dish to hood.

Using the second set of sterile instruments and a fresh sterile petri dish, isolate the embryos. Be sure to remove the placement and embryonic sacs. Using the scalpel handle with the #11 blade on it, cut off embryo heads and scoop out the liver with a pair of forceps. The head and forelimb should be cut off as shown below.

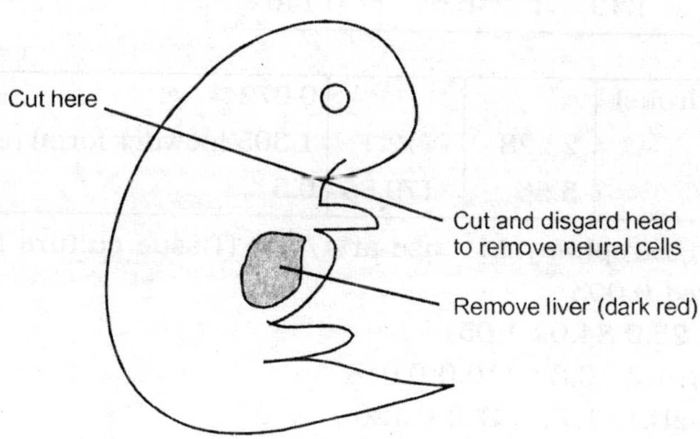

Discard the head and liver and leave the bodies in fresh PBS in a fresh petri dish. Take 6cc luer lock syringe with 18 gauge needle attached and remove plunger. KEEP PLUNGER STERILE. Drop the embryo bodies inside the syringe and add 3 ml trypsin/EDTA. Put plunger back in syringe and squirt contents of syringe into a large tissue culture flask. Place the flask onto a warming tray (37°C) for 2-3 minutes. Back to the hood and add 20 ml DMEM. When adding the DMEM, try to wash any tissue off the walls of the flask. Pipette the tissue/medium up and down a few times to help break up the tissue. Transfer flask to an incubator at 37°C with 5% CO_2. Do not put less than 7 embryo in one large flask.

Important. Loosen lid of flask in incubator to allow gas exchange in medium. This is the primary isolation or passage one. PMEF's should attach and begin to divide in 1-3 days. During this time do not disturb, so as to allow PMEF's to settle and attach. After 2 days change the medium. It will be very acidic. After 3-4 days the culture will need splitting. Remove media and gently wash the monolayer with 2 X 10 ml PBS. Add 2 ml trypsin EDTA and split 1:4. After a further 2-4 days the culture will be ready for freezing. The number obtained from each flask will be between 5-10 × 106 cells. Freeze cells in 10% DMSO at 3 X 106/ampule.

When recovering the cells from LN2 put all the cells into a medium flask. When confluent these are split into 1 medium and one large flask (1:3). The large flask can be treated with mitomycin C and the medium flask split again. Do not passage beyond P6.

4. MEDIA FOR EMBRYO CULTURE AND MANIPULATION

M16 Medium (Protocol obtained from Karen Austen-Reed from SS Tan Laboratory, Anatomy Department)

For oocyte maturation and routine culture of mouse embryos, M16 culture medium is used. This medium is unable to maintain its own pH and must therefore by used in conjuction with an incubator buffered with 5% CO_2. The CO_2 maintains the required pH level of the medium.

Compound	mM	Mol. Wt.	g/500 ml
NaCl	94.66	58.45	2.766
KCl	4.78	74.557	0.178
KH_2PO_4	1.19	136.091	0.081
$MgSO_4$-$7H_2O$	1.19	246.5	0.146

A OR

$MgSO_4$ (anhydrous)			0.072
Na Lactate	23.28	112.1	1.305 (powder form) or 1.69 ml (60% syrup
Glucose	5.56	179.86	0.5

- Penicillin/Streptomycin - use at 1/100 (Tissue culture Pen/Strep)
- Phenol red 0.005
- $NaHCO_3$ 25.0 84.02 1.051
- C Na Pyruvate 0.33 110.0 0.018
- D $CaCl_2$-$2H_2O$ 1.71 147.2 0.126
- BSA (Bovine Serum Albumin) 4 mg/ml

Method

1. Weigh out all of stock "A" (except Pen/Strep and Na lactate) into a measuring cylinder and make up to 90 ml with "Travenol" water. Add Pen/Strep to cylinder

and then Na lactate (Note: the Na lactate is quite viscous - by heating it up to ~37°C prior to use it can be more easily and accurately pipetted). Make up to 100 ml with "Travenol" water and pour into 500 ml bottle.

2. Weigh out stock "B" (Phenol red, $NaHCO_3$) components into a measuring cylinder. Make up to 100 ml with "Travenol" and pour into 500 ml bottle. Weigh stock "C" (Na Pyruvate) into measuring cylinders, make up to 100 ml with "Travenol" water and pour into 500 ml bottle. Weigh stock "D" ($CaCl_2$-$2H_2O$) into measuring cylinder, make up to 100 ml with "Travenol" water and pour into 500 ml bottle. Add about 100 ml of "Travenol" water to bring the total volume to 500 ml. Be sure that the components of each stock have dissolved properly before adding to the 500 ml bottle. Also be sure to add stocks in their alphabetical order to avoid precipitation of some ingredients.
3. Make 50 ml aliquots of this solution and freeze them. Use one 50 ml aliquot at a time by aliquoting it into tubes of 9 ml each, then refrigerate until use. The osmolarity should be 288-292 m osmol.

   Before use, lightly gas with CO_2, then add 1 ml of FCS (foetal calf serum) to a 9 ml aliquot of M16. Then sprinkle 36 mg of BSA (4 mg/ml M16) on top and allow to dissolve - do not shake up or stir. Then Filter-sterilize this mix using a 12 ml syringe with an acrodisc.

DMEM with HEPES

- This medium is used for manipulations that are performed on the mouse embryos whilst out of the incubator. Since there is not 5% CO_2 present to help maintain the pH, this medium contains HEPES to keep the pH constant.
- Ingredient % of Final Volume
- 1 X DMEM with 20 mM HEPES buffer 80.8%
- Penicillin/Streptomycin (5,000 i.u/ml/5,000 ug/ml) 1%
- L-glutamine (200 mM, 100X) 1%
- non-essential amino acid (100X) 1%
- nucleosides (100X) 1%
- b-mercaptoethanol (0.1M) 0.2%
- Foetal calf serum 15%
- This recipe is made up fresh each week (the same stocks of ingredients may be used). An alternative to DMEM with HEPES is M2, the recipe for which can be found in Bridgette Hogan's Book, "Manipulating the Mouse Embryo".

5. COLLECTION OF MORULAE AND EARLIER

1. Before you start, have ready the following:
   - Sterilize 2 pairs of curved forceps, 1 pair of iris scissors and 2 pairs of fine watchmaker forceps.
   - 6 ml syringe filled with DMEM with HEPES, with an 18 gauge needle attached. To the needle attach a 20 cm length of clear vinyl tube and into the end of this tubing insert a sharp flusher.
2. At 2.5 days p.c. (post coitus) the morulae are present in the oviducts. For this reason, it is only necessary to remove the oviducts from the mouse.

3. Kill the 2.5 day pregnant mouse and lay on its back on benchcloth or absorbent paper. Swab belly with ethanol (70%) and nick the skin with a pair of scissors. Pull the skin back over the head and towards with tail. Cut and pull back the body wall to expose the contents of the abdomen. Push aside the guts to expose the reproductive tract. Using a pair of curved forceps, grasp the uterus just below the oviduct and carefully separate the oviduct from the ovary using the iris scissors. Cut through the uterus just below the uterotubal junction and place the oviduct in a petri dish.

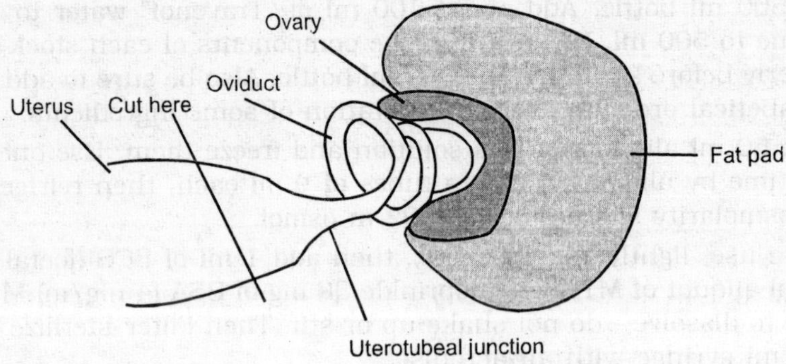

4. Transfer the dish to the stage of a stereo dissecting microscope. Using the watchmaker forceps, manipulate the oviduct so that you have the end of the oviduct (close as possible to the uterotubal junction) between the forceps and insert the flusher so that it points away from the uterotubal junction. Be careful not to pierce through both sides of the oviduct tube. Keeping the flusher inserted into the oviduct, pick up the syringe in the other hand and give a couple of short, sharp squeezes on the plunger. This should flush all morulae from the oviduct. Sometimes, if no morulae are flushed through, it is hard to tell whether the oviduct has been correctly flushed. If there is much debris floating around, then the oviduct has been properly flushed. The morulae are then collected using a mouth pipette and deposited into a petri dish in microdrops of M16 medium. These drops are covered with fluid 200 (Dow Corning, viscosity 50CS) and placed in a 37°C incubator buffered with 5% CO_2.

6. BLASTOCYST TRANSFER

1. Blastocyst transfer is usually performed 24 hours after aggregation when the morulae have become expanded blastocysts and on the same day as injection. A little time is given between injection and transfer to allow blastocysts to re-expand.
2. The Recipient
3. Careful selection of the recipient is most important as the pups are the end result of a lot of hard work. Two strains of mice are used. RB Swiss and (CBA*C57BL6/J)f1. RB Swiss are quiet and make excellent mothers but they become overweight quickly and do exhibit bad planes of anaesthesia. CBA*C57 mice are hardy and display hybrid vigour. They do not carry excess weight and go under anaesthetic well. This strain can be very nervous when housed separately which could be dangerous to their young. They are most suitable

if a young RB Swiss is placed into the cage as a companion which can be removed as soon as the pups are seven days old. By this age destruction of the litre is very unlikely. If the CBA mother is to be housed alone she must not be disturbed for ten days.

Prior to surgery, sterilise the following:

- 3 pairs of curved forceps
- 2 pairs of iris scissors
- 1 pair suture clamps
- 1 serafin clip
- sterile suture with needle attached (small - for mouse surgery)
- Michelle clips (small size)
- Michelle clip applicators
- 1 mouth pipette and flame-polished transfer pipette
- You will also need an anaesthetic. Rompun / Ketavet is found to be quite effective. To make up 10 ml
- 0.5 ml 2% Rompun (20 mg/ml Xylazine)
- 0.5 ml 100 mg/ml Ketavet 100 - Delta Veterinary Lab, 8 Rosemead Rd Hornsby NSW 2077
- Make up to 10 ml with PBS
- The dosage is 0.02 ml/g body weight
- Store wrapped in tin foil at 4°C.
- Shake well before use as it tends to separate in the fridge
- The Transfer

1. Select a mouse that is 2.5 days pseudo pregnant and weigh. Do not use anything lighter and 25 g or anything heavier than 30 g. Underweight mice tend to re-absorb the embryos as they are not physically ready to support a pregnancy. Overweight mice make surgery difficult in the absorption of anaesthetic into the fat reduces the potency of the anaesthetic. Also, the presence of fat means the presence of blood vessels, and cutting through all the extra fat causes a lot of unnecessary bleeding. This makes it difficult to see what you are doing and may also clog up the tip of your transfer pipette.
2. Anaesthetise mouse with Rompun / Ketavet, administered intraperitoneally. After administering the anaesthetic, put the mouse back into the box from which it came. The mouse will be more relaxed when placed in a familiar environment and the anaesthetic will act more quickly than it would on a distressed mouse.
3. To check that the mouse is fully anaesthetised, press or squeeze the pads of the feet. If the mouse can feel this it will try to withdraw its leg from your grasp. Do not commence surgery until there is no reflex reaction to this test.
4. Take the anaesthetised mouse and shave its lower back. Lay mouse on belly on a petri dish lid taking care to keep the airway clear by resting the teeth on the edge of the petri dish. This makes it easier to move the mouse around

without having to actually touch it. Swab the shaven area with hibitane or 70% ethanol.

5. Instruments should have been laid out. Use one pair of iris scissors and one pair of forceps for cutting the skin - call these "outside" instruments. Use one pair of iris scissors and two pair of forceps for working inside the mouse - called the "inside" instruments.

6. Using the outside forceps and scissors, make a small cut (about 1cm long) along the dorsal midline of the lower back. Through the shaven, moistened skin it is fairly easy to see blood vessels. Try to avoid these vessels when making the incision (see below).

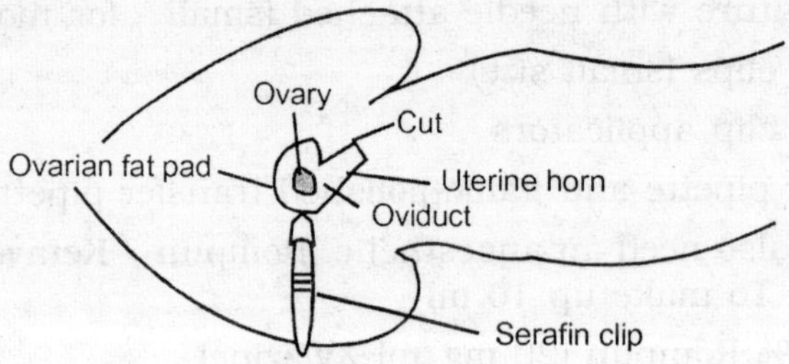

7. Using a pair of outside forceps and inside forceps, pick up the skin and separate skin from the body wall and cut the body wall as indicated about 5 mm long, avoiding blood vessels.

8. If the cut has been made in the right place, the ovarian fat pad is easily visible. If not, the fat pad can be located by lifting the edge of the body wall and scouting around with the other pair of inside forceps. Once you have located the fat pad attach serafin clip to it, taking care not to clip ovary. It is important not to damage the ovary as it is responsible for hormone production throughout pregnancy. Gently ease the ovary, oviduct and part of the uterus out through the incision in the body wall. DO NOT PULL. A traumatised uterus will contract and move quite violently, making surgery difficult and may cause expulsion of the transferred blastocysts. When the tip of the uterus is visible, rest the serafin clip across the mouse's back to hold the uterus in place. If the uterus or uterine horn continually slip back into the cavity it may be necessary to gently lie the mouse on the side being careful not to block the airway.

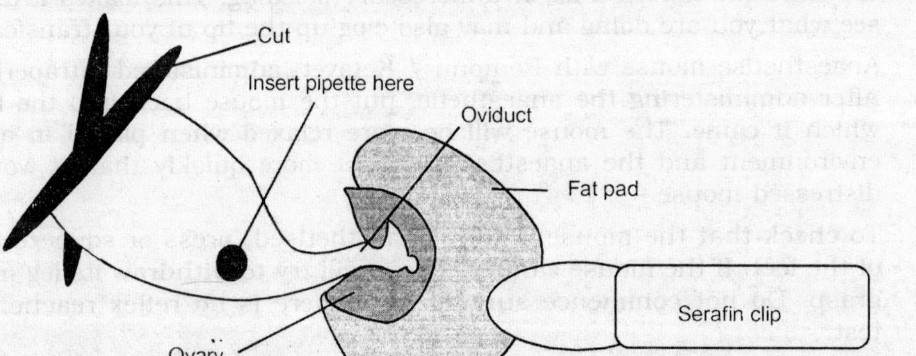

9. The transfer pipette should now be loaded. Five or six embryos must be transferred to each horn, any less than this and the chances of a pregnancy resulting are serverely reduced. It may be loaded to suit yourself but this is a method that is popular. Take up a minute amount of DMEM with Hepes in the very tip of the transfer pipette, then make a small bubble by taking up a little air. Then take up some more medium - roughly the same volume as you hope to transfer the blastocysts in. Take up another bubble, same size as before. Then take up your blastocysts in the smallest possible volume of medium, lining them up side by side in the transfer pipette. This is how your transfer pipette should look when loaded.

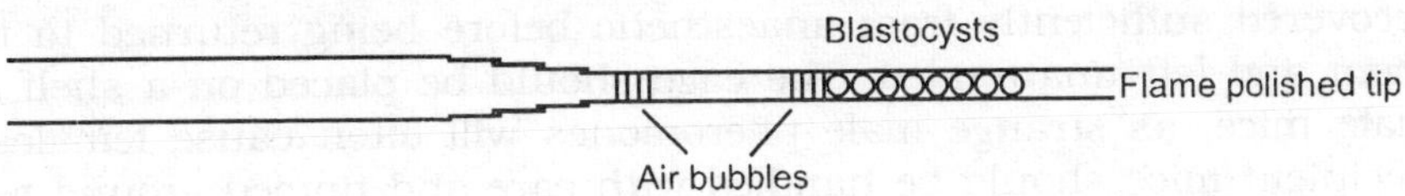

10. This will take some practise. Make sure that you are competent at loading a transfer pipette before any attempt at a blastocysts transfer. During surgery is not the time to learn how to load a pipette. If it is likely to take you more than a few minutes to load the transfer pipette, then do not expose the uterus until the pipette has been loaded. This prevents drying out and further trauma to the uterus. Alternatively, the uterus, ovary, etc. may be moistened repeatedly with a sterile cotton bud and saline.

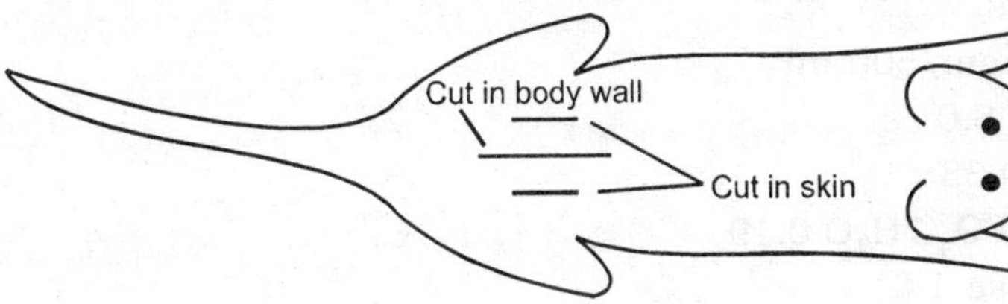

11. Once the pipette is loaded and the uterus positioned, move the petri dish lid supporting the mouse to the microscope and turn on the overhead light source. Once the lights and focus have been adjusted and the mouse positioned to suit yourself, gently grasp the top of the uterine horn with a pair of inside forceps. Whilst still holding the horn with one hand, use the other hand to gently insert a 26 gauge hypodermic needle through the uterine wall (close to the oviduct) and into the lumen. Choose an area of the uterus that is relatively devoid of blood vessels as blood will clot in the tip of your pipette and block it. Remove the needle and carefully (so as not to lose the site of the hole), without averting your eyes, pick up the loaded transfer pipette. Gently insert the transfer pipette about 3 mm into the uterine lumen. Gently blow the blastocysts into the uterus, using the air bubbles in your transfer pipette to monitor the transfer. Be careful not to blow any air into the uterus. Once transfer is complete, quickly rinse out transfer pipette in some HEPES buffer medium (M2) and check to see if there were any blastocysts stuck in the transfer pipette. If there were, transfer these blastocysts again.
12. With the transfer complete, the serafin clip can now be removed and the uterus gently eased back into the body. Do not touch the uterus, but ease it back by the edges of the incision in the body wall and allowing the uterus to fall back in, without actually handling it. This procedure is then repeated on the other uterine horn. The incision in the body wall is not sutured. The skin

is closed with Michelle clips - two per incision is usually sufficient. Michelle clips are used on the skin instead of suture as the mice will chew at the suture thread and effectively open their wounds up.

13. Once surgery is complete, the mouse is placed in a box of clean autoclaved sawdust. Under anaesthetic, mammals are unable to retain heat as effectively as when conscious. For this reason, and also because the animal has been shaven, the mouse should be wrapped in a tissue to help keep it warm. This will also be used as bedding until the fur has grown back. It is also very important that the mouse by house alone as anaesthetised animals are often buried by cage mates who think they are dead. All animals should have recovered sufficiently from anaesthetic before being returned to the animal room and left unattended. The cage should be placed on a shelf away from male mice, as strange male pheromones will often cause females to abort. Recipient mice should be handled with care and tiptoed around as pregnant mice are easily upset, sometimes leading to abortion, or even cannibalism of pups.

7. DNA TRANSFECTION OF EUKARYOTIC CELLS USING CALCIUM PHOSPHATE

Stock Solutions

1. 2M $CaCl_2$ Merck. Filter (to remove particulates), autoclave and store at –20° in aliquots.
2. 2X HEBS gm/500 ml
   - NaCl 8.0
   - KCl 0.38
   - $Na_2HPO_4.7H_2O$ 0.19
   - glucose 1.0
   - Hepes 5.0
   - Adjust pH to 7.05 ± .05 with NaOH, Autoclave, Store frozen in aliquots
3. 1 X HEBS/15% glycerol. 50 ml 2X HEBS)
   - 15 ml glycerol
   - 35 ml DDW, Autoclave and store at –20° in aliquots
4. Carrier DNA. Mouse liver DNA, sheared at 90 ug/ml in 0.2X SSC by passing 2X through a 20 gauge needle; filtered through a 4.45 um filter.

Transfection Protocol

1. Plate cells at 1X 106 in 10 cm dishes
2. Next day perform transfection.

Preparation of $CaPO_4$ Precipitates

1. Set up two rows of tubes - A & B
   - In tube A place : 15 ug of plasmid DNA (linearized, phenol extracted, precipitated and resuspended at 1 ug/ul in sterile, 0.2X SSC, 69 ul 2M $CaCl_2$, 460 0.2X SSC
   - In tube B place 550 ul 2X HEBS

2. Using two autopipettes and 1 ml pipettes have tube B bubbling whilst slowly adding contents of tube A.
3. Stand 15-20 mins, while precipitate forms, giving a milky fine pipette.
4. Carefully add precipitate dropwwise to 10 cm dish of cells whilst maintaining the pH of cultures.
5. Leave 3-4 hrs in CO_2 incubator (Can be left o/n).
6. Glycerol Shock:

   Aspirate medium. Wash by adding ~3 ml medium then aspirate, Add 2.5 ml 15% glycerol in HEBS.

   Sit 4 mins at room temp. then aspirate, Wash with 5 ml medium per plate. Add fresh medium.

   Note. Cell are fragile and only loosely attached at this stage. Hand very gently. If pipette was left on cells o/n do the splits next day.

Next day

7. Trypsinise to harvest cells. Recover cells in medium containing antibiotic and divide directly in the ratio 3/5th's and 1/20th into two 10 cm plates, respectively.
8. Feed every 2-3 days with medium containing antibiotic (once a week when most cells have been killed).

8. CELL CULTURE FROM WHOLE MICE EMBRYO (DAY 10-11 PC)

Materials and Equipment

- Sterile technique
- Autoclave 1 X PBS
- Dissection tools (watchmaker forceps X 2-3, Surgical scissors)
- Sterile dishes (10 cm)
- Prepare 6 well dishes to culture the embryo cell culture and mark the well.
- Prepare syringe contain required amount (300 ul) of trypsin and keep in hood.
- Stopcock 3-way transfusion

Procedure

1. Kill mice
2. Open abdomen and remove the both side of the uterus — sterile from now!!
3. Open uterus and separate embryo (include decidua)
4. Take single embryo into fresh dish, open decidua, open and remove york sack for PCR genotype (soak in PBS of the 12 well dish, wash twice and dip into DNA buffer).
5. Rinse the forceps, take the embryo into a fresh syringe, carry the syringe contain the embryo into tissue culture hood. Place on a 3-way stopcock and add on a syringe containing 300 ul trypsin on the other one side. Start timing before push the syringe, push the syringe from one side at a time, carefully

close the stopcock bit by bit (not too much as the cells will be lysed), this requires about 20 seconds (shorter is better), and place syringe in one well then place dish on warm plate up to 3 minutes in total.

6. Rinse the syringe with 1 ml media and add total media of 3 - 3.5 ml. Do not use pipette take up and push down the culture since the embryo cells tend to be happier when the cells are aggregated.
7. Incubate the culture an 37°C, 5% CO_2 for 3-5 days till the cell reach to confluence, transfer the cell into 10 cm dish by 1 ml trypsin and 8-10 ml media.

9. PCR GENOTYPING FROM THE EMBRYO YOKE SACK

Material

- DNA buffer : 1 X PCR buffer
- 0.45% NP - 40
- 0.45% Tween - 20
- Proteinase K 100 mg/ml

Prepare the DNA for PCR

1. In PCR tube contain 50 ul DNA buffer,
2. Dig washed york sack into tube, spin down,
3. Incubate at 50°C for 30 minutes,
4. Boil the lysate for 5 minutes, spin down,
5. Take 5ul for PCR.

PCR Reaction (50 ul) Contains

1 X PCR buffer

2 mM $MgCl_2$

100 nM dNTPs

Primers

(*a*) 0.25 uM

(*b*) 0.05 uM

(*c*) 0.2 uM

a c 400bp (targeting region)

a b 200bp (wild type)

2.5 unite/100 ul Taq DNA polymerase

PCR Cycle

Initial denaturing by 94°C for 4 minutes

28 cycles of 94°C for 1 minute

60°C for 1 minute

72°C for 1 minute

Final extension by 72°C for 10 minutes.

INVITRO METHODS

1. CELL CULTURE ASSAYS

Introduction

Cell cultures of various sorts are under wide consideration and development in toxicology to replace the use of animals, supplement the use of animals, and perhaps provide information that can be obtained by no other convenient means. An Institute of Alternatives to Animal Testing operates at Johns Hopkins University on funding from a variety of private and public sources to further the application of in vitro methods to toxicology testing and research.

A variety of in vitro systems based on cell cultures or bacterial cultures are available. Perhaps the best known, and the oldest, is the Ames Assay, developed by Bruce Ames to screen chemicals for mutagenic potential. This bacterial system, based on specially designed strains of *S. typhimurium*, is supplemented with rat microsomes and the back mutation rate scored as a function of dose of chemical. Because chemicals are identified by mutagenic action, and the system uses mammalian microsomes, the screen is also popularly considered to screen for carcinogenic potential.

Simple cell cytotoxicity assays are entering the market. One system is based on the release of a cytoplasmic enzyme into the culture medium. The enzyme, lactate dehydrogenase, is measured by a simple calorimetric assay. Another test monitors dead cells by a failure to exclude Trypan blue. These tests are scored manually by counting cells on a microscope slide and are therefore less popular.

In this experiment, we will examine the effect of heat shock on the cell's ability to resist exposure to cadmium, a known lung toxin. We will use a simple test for cell activity (not cell viability) to determine a dose response curve for cadmium, in the presense and absence of heat shock. The assay is the uptake of a vital stain, Neutral red. Most cells will accumulate NR in lysosomes. The process requires intact membranes and active metabolism in the cell. Failure to take up NR therefore indicates that the cells have suffered damage. The dye taken up by the cells is subsequently extracted and measured. Since the amount of dye taken up by the cells is a function of cell number, some indication of cell mass is also necessary to interpret the results. In this experiment, cell mass will be determined by a sensitive assay for protein.

Procedure

The response of the cultures to cadmium chloride solutions will be examined. Two plates are set up, so divide yourselves into teams, one team per plate. The cultures were exposed prior to the laboratory (about 9 am today) to a range of $CdCl_2$ concentrations: 0.0, 5, 10, 20 and 40 uM. At the beginning of the class period, the medium will be replaced with medium containing Neutral red, to a final concentration of 50 ug/ml. After an additional 1 hour exposure, you will remove the medium, rinse the cells and determine the Neutral red uptake. Follow the protocol below. The amount of NR will be determined spectrophotometrically using a microplate reader (Directions attached).

Cell Culture

A culture of rat lung epithelial cells are grown on solid substrate in sterile flasks. The cells are released from the substrate and suspended in medium. About 50,000 cells are plated per well in 24 well plates. The cells are plated and grown at 37° in a medium of 5% newborn calf serum/defined medium (F12/DMEM) under 5% CO_2. Typically, the cultures reach confluence after 24 hrs. Only cultures at, or near, confluence are used in this experiment.

Heat Shock

Twelve hours before the experiment, 1 plate of cells will be transferred to a 44° incubator for 1 hr. The plate is then returned to the regular 37° incubator.

Exposure to Cadmuim Chloride

Replicates of four wells were exposed to the following concentrations of $CdCl_2$ solution by the addition of appropriate volumes of mM $CdCl_2$.

Column 1: Blank, no cells

Column 2: 5 uM Cd

Column 3: 10 uM Cd

Column 4: 20 uM Cd

Column 5: 40 uM Cd

Column 6: Control, no exposure.

2. NEUTRAL RED ASSAY

Stock solutions

- Neutral Red (0.5% solution in F12/DMEM medium, prepared for student).
- Destaining solution (1% glacial acetic acid, 49% dH_2O and 50% ethanol, V/V).
- Dulbecco's phosphate buffered saline (PBS)

Preparation of the NR medium

1. Dissolve NR stock solution in ethanol to final concentration of 5 mg/ml. Incubate at 37° for 24 hrs.
2. Dilute NR stock solution 1 to 100 in cell medium (F12/DMEM) 2 hr. prior to application to the medium, and hold at 37°.
3. Filter the NR/medium through a 0.45 u filter to remove filtrate and insure sterility.

Assay

1. Remove the plates from the incubator (normally, this step would be done in a laminar flow safety hood; however, the short period of incubation required for this step allows us to do this on the bench top). Remove the existing medium by aspiration. Replace with 1 ml of the NR medium. Use a sterile pipette or sterile pipette tips.

2. Return the plates to the incubator for 1 hrs at 37°.
3. Observe the plates in the inverted microscope. Describe the cultures by reference to the control plate.
4. Remove medium from cells by gentle aspiration. Add 1.0 ml of PBS to each well *gently*. Avoid blasting cells free from bottom of plate. Rotate the plates several times over a 5 min period.
5. Aspirate the PBS. Repeat with another identical volume of PBS.
6. Carefully add 1.0 ml of destaining solution. Gently rotate the plates several times every 5 minutes for 15 min. A plate shaker is available.
7. Read the plates on the Bio-Tech microplate reader (Instructions attached). The microplate reader is in Rm 432.

Protein Assay

Carefully aspirate the extraction medium. Add 0.1 ml of 1 N NaOH to each well. Swirl the plates to dissolve the protein. Add 1.5 mL Biuret reagent. Read in the plate reader after 15 min. (same settings as the NR assay).

Interpretation of Results

The absorbance value at 540 nm is the amount of NR taken up by the cells. As cells lose viability, they lose the ability to take up NR.

Obtain and attach copies of both 24 well plate reports as tables in your report.

Plot your values of NR uptake, average of 4 wells, *vs.* concentration and construct a dose response curve. You may wish to convert the NR uptake values to a percent response for purposes of plotting. Plot results of both plates on the figure. As a separate figure, plot the Biuret absorbance *vs.* concentration to determine if the toxin caused a lose of cell protein from the well. Comment specifically on whether heat shock altered the dose response of the cells to cadmium exposure.

HUMAN CELL CULTURE METHODS

EXERICSE 1. LOGGING IN SPECIMENS AND RECORD KEEPING

Purpose

To keep a written and computerized record of all cell lines, the dates when cell lines were received and frozen, freezer locations, and any other important information such as dates of birth, sex, etc.

Procedure

1. Refer to cell line growth record sheet. When a cell line arrives or is established from whole blood, information such as the cell line number, family position in the pedigree, sex, date of birth, date arrived, etc., is recorded on a cell line growth record sheet in a binder which corresponds with the study group to which the cell line belongs. Other information is recorded, such as the dates the cell pellet is frozen for DNA extraction and for permanent storage. The freezer locations (in stainless steel racks) of the cell line aliquots are recorded to facilitate locating the cell line at a later date.

2. To locate a cell line frozen for DNA extraction, first look for the location of the cell line on the master list of all the cell lines in the study group. This master list is located in the front of the study group binder to which the cell line belongs. After the rack location is identified, locate the position of the cell line in the rack. A separate binder labelled '–80 Revco' has forms representing all the racks with in the freezer; refer to –80 Revco sheet. When a cell line is removed it is erased from the –80 rack sheet and crossed off the master list in the study group binder.
3. A binder to locate frozen cell lines for permanent storage is labelled '–135 Cryostar'. The –135 freezer has the capacity to hold 20 racks with 10 boxes in each rack. Each box has the capacity for 81 cryotubes with one empty space used for rack orientation. In the –135 binder, 20 dividers separate the 20 racks, and between each divider are 10 sheets (each labelled –135 Cryostar sheet), corresponding to the 10 boxes in that rack. The –135 Cryostar sheet is used to record the information of what cell lines are in ach box. On these sheets data such as kindred number, cell line number, and date the cell line is frozen is recorded. When a vial is removed, it is erased off the sheet and off the master list in the front of the study group binder.

EXERCISE 2. LYMPHOCYTE TRANSFORMATION

Principle

Lymphocytes are transformed to establish cell lines. Mononuclear cells (lymphocytes) from anticoagulated venous blood are isolated by layering onto histopaque. During centrifugation, erythrocytes and granulocytes are aggregated by ficoll and rapidly settle to the bottom of the tube; lymphocytes and other mononuclear cells remain at the plasma-histopaque interface. Erythrocyte contamination is neglible. Most extraneous platelets are removed by low speed centrifugation during the washing steps.

Special Reagents

- Cyclosporin A (CSA, from the Sandoz Research Institute; East Hanover, New Jersey 07936)
- Request CSA several months in advance in order to receive it when needed.
- Send a statement of investigator form to cover the release. It is an experimental drug and is used only in research work and not intended for human use.

Time Required

2-2.5 hours to prepare two transformations. Cell lines will require 3-4 weeks in a T-25 cm^2 flask before passaging to a T-75 cm^2. After passaging to the larger flask, each cell line requires several more weeks to reach a cell density of 1 X 108 cells/ 100 ml.

Procedure

1. Collect 27 ml of anticoagulated blood in 3 yellow top tubes (citrate), 9 ml each. The blood should be set up in culture as soon as possible for best results. Blood should be kept at room temperature prior to use in this procedure.

2. Wipe the exterior of the tubes of blood with EtOH, divide evenly and transfer the blood into 2-50 ml tubes. Bring the volume of each tube up to 40 ml with wash media.
3. Place 10 ml of histopaque –1077 into two other 50 ml tubes. Overlay the blood and wash media mixture onto the 10 ml of histopaque. Do this very slowly making sure not to mix the two layers.
4. Centrifuge tubes for 30 minutes at 1500 rpm at room temperature (no brake), in the TJ-6 centrifuge. Aspirate the top layer down to within 1/4" of the white blood cell layer.
5. Collect the WBC layer using a 10 ml pipette, moving the pipette in a circular motion around the inside of the tube just below the surface of the WBC layer. Transfer the WBC layer to another 50 ml tube.
6. Bring the volume of each tube up to 50 ml with wash media, gently invert tubes to mix.
7. Centrifuge the tubes for 20 minutes at 1200 rpm at room temperature (no brake) in the TJ-6 centrifuge. Aspirate supernatant.
8. Add 12 ml wash media, resuspend the cell pellette and transfer to 15 ml centrifuge tube.
9. Centrifuge 8 minutes at 1000 rpm at room temperature (no brake) in the TJ-6 centrifuge. Aspirate supernatant.
10. Cell counts can be done to determine the appropriate volume of media to be added to the cells. Cells should be set up in culture using a minimum of 2.6×10^6 cells/ml and not more than 7 ml per 25 cm^2 flask. The average WBC count of whole blood ranges from 1×10^6 cells/ml to 3×10^6 cells/ml. An ideal primary culture should contain between 5 and 7 ml of cells/25 cm^2 flask. Cells are resuspended in 10 ml of RPMI (without serum) for counting. If cell counts cannot be done, set up cultures in 5-7 ml of fresh RPMI, 20% FBS, and 2 µg/ml cyclosporin A.
11. Inoculate cells with an equal volume of virus (refer to procedure describing the maintainence of the B95-8 cell line). Incubate cells at 37 degrees C with 5% CO_2 and loose caps on flasks.
12. Feed the culture after 24-48 hours if the media has turned yellow. This is done by removing 1/2 of the media and replacing it with 1 µg/ml cyclosporin A media. Discontinue cyclosporin A after 3-4 weeks. Do not overfeed cells. Do not increase the volume of media for at least 2 weeks. If cells do not seem to be growing, reduce the volume of media and feed only once a week. After 2 weeks, when cells are very clumpy and the media is changing colour from orange to yellow within 3 days incubation, increase the volume of media by 10-20 ml. Cultures are always fed by removing 1/2 of the old media and replacing it with a slightly increased volume of fresh media. Cultures can be split when they reach 25-30 ml of media in a 25 cm^2 flask.

Solutions

- *Growth media:* (600 ml)

  To 500 ml of sterile RPMI 1640 with 2 mM L-glutamine, add: 90.0 ml FBS, 6.0 ml 200 mM (100X) L-glutamine, 0.6 ml 50 mg/ml gentamicin reagent.

  Note. Filter sterilize through a 0.22 µm filter and store up to 2 weeks at 4°C.

- *Wash media:* (1 litre)

  1 litre of sterile RPMI-1640 with 2 mM L-glutamine, add: 10.0 ml 2.5M (100X) Hepes buffer 1.2 ml 50 mg/ml gentamicin reagent.

  Note. Filter sterilize through a 0.22 μm filter and store up to 2 weeks at 4°C.
- *2X Cyclosporin A media:* (1 μg/ml)

  To 100 ml of growth media add 2 ml of 100X Cyclosporin A.
- *100X Cyclosporin A:* (100 ml)

  Dissolve 1 mg CSA in 0.1 ml ethanol, add 0.02 ml Tween 80 and mix well. While continually stirring, add 1 ml RPMI, drop by drop. Quantitate to a final volume of 100 ml with RPMI. Filter sterilize and store at 4°C for up to 4 months.

EXERCISE 3. PREPARATION OF LYMPHOBLASTOID CELL LINES FOR LONG TERM STORAGE

Purpose

To store cell lines in a form that will insure recovery with high viability. A culture in logarithmic phase of growth with a total volume of 80-100 ml/T-75 flask should yield enough cells to freeze 10 ampules (1.0 ml/ampule). Cells should have a count of 4 X 106 cells/ampule to 9 X 106 cells/ampule. Too high or too low a cell count lowers recovery viability. Cell are frozen in RPMI-1640 with 15% Fetal Bovine Serum +10% DMSO. Cultures are frozen slowly using a Model 700 Controller freezing chamber. This precision electronic device automatically controls the injection of liquid nitrogen into the freezing chamber to provide a 1°C/minute freezing rate from +4°C to –45°C (with automatic heat of fusion compensation), then a 10°C per minute freezing rate to –90°C. Frozen ampules should be stored in liquid nitrogen for long term storage or in a –135°C Cryopreservation System.

Note: Cryotubes should be labelled with cell line number and date prior to beginning this procedure.

Time Required

2.5-3.0 hours to freeze 10 aliquots from each of 6 cell lines. Only 6 cell lines or 60 cryotubes should be frozen at one time. It is essential to keep the time the cells are exposed to the DMSO at a minimum. The freezing chamber can hold up to 120 tubes so two people can freeze samples at the same time to save liquid nitrogen.

Procedure

1. Aspirate media from the T-75 flask down to the 50 ml mark.
2. Resuspend cells by shaking gently and transfer 40 ml of the cell suspension to a 50 ml centrifuge tube.
3. Add 10 ml of fresh media to the culture flask and reincubate at 37°C. Keep the culture flask growing until a test thaw is done on one cryotube (done to determine if the cells were successfully frozen. Refer to reactivating cell line for DNA growth and extraction procedure. The cell line will begin growing within days if the freezing conditions were correct). Greater than 99% of the cell lines are successfully frozen using this procedure.

4. Remove 200 µl of the cell suspension from centrifuge tube for a cell count. (Refer to cell counting procedure.) Use the cell count to adjust the cell concentration to between 4×10^6 and 9×10^6 cells/ampule. Too high or too low a cell concentration decreases the viability of the cell line when the cryotube is thawed for growth.
5. Centrifuge the 50 ml tube for 10 minutes at 1200 rpm, no brake, room temperature, in the TJ-6 centrifuge.
6. Aspirate supernatant down to 1/4 inch above the cell pellet.
7. Place a control sample (freezing media in 1.0 ml cryotube) into the freezing chamber in a central location with the thermocouple probe placed equidistant from side to bottom. It will take approximately 6 minutes for the sample temperature to reach start temperature of 4°C on the chart drive.
8. Resuspend cell pellet with 10 ml of freezing media. Pipette 1.0 ml into each of 10 cryotubes on ice. DMSO is toxic to cells, therefore begin freezing immediately after transferring the cells to cryotubes.
9. Load the cryotubes into the chamber when the sample temperature is +4°C on the chart drive paper.
10. Again allow the chamber and cells to cool to the start temperature of +4°C.
11. Place the selector switch to the freeze ampule position. The controller will automatically cycle through the freezing program until the end temperature is reached. This takes approximately 55 minutes.
12. Remove samples after the recorder has reached –90°C and transfer to a permanent storage container. Samples should be moved quickly to prevent thawing or warming and sample deterioration.

    Warning. Wear cryoprotective gloves when working with the freezing chamber and other permanent storage containers. Also, protective eyeglasses are necessary in case of explosion of a cryotube.

Solutions

- *Freezing media:* (1 litre)
  Prepare a 1 litre volume and divide into 25-50 ml, centrifuge tubes containing 40 ml each. Store the tubes at –80°C for up to one year. 700 ml RPMI-1640 with 2 mM L-Glutamine
- 200 ml fetal bovine serum (FBS)
- 100 ml dimethyl sulfoxide (DMSO, sigma)
- 1000 ml total volume

  Note. Filter sterilize media and FBS with a 0.22 µm cellulose acetate filter. Do Not filter DMSO, it will dissolve the cellulose acetate membrane.

EXERCISE 4. REACTIVATING CELL LINES AND CELL GROWTH FOR DNA PREPARATION

Purpose

Cell lines are reactivated and grown to a count of 1×10^8 cells. The cells are pelleted and stored frozen at –80°C prior to DNA extraction.

Time required

15-20 minutes to begin growing 2-4 cryovials.

Procedure

1. Frozen cells should be thawed quickly. Remove the cryovial from its long term storage container in the –135°C Cryostar, and place immediately in a 37°C water bath for 2 minutes.
2. Remove the cells from the vial and place in 10 ml wash media. This is necessary to remove traces of dimethyl sulfoxide from the cells.
3. Centrifuge cells for 10 minutes at 1200 rpm (no brake) at room temperature using the TJ-6 centrifuge.
4. Remove the supernatant above the cell pellet.
5. Resuspend the cell pellet in 7-10 ml of 1X Cyclosporin A media.
6. Aspirate half of culture media within 3-4 days. Add growth media and slightly increase volume by 5 ml. Increase the volume of media by 5-10 ml two times a week by aspirating off half of media from culture flask (do not to suction off cells from bottom of flask) and replacing it with fresh growth media. Cells can be harvested for extraction when a T-75 cm^2 flask reaches a volume of 100 ml of media and there is a monolayer of cells on bottom of flask.

Solutions

- *Wash media:* (1 litre)
- To 1 litre of sterile RPMI 1640 with 2 mM L-glutamine, add: 10.0 ml 2.5 M (100X) Hepes buffer and 1.2 ml 50 mg/ml gentamicin reagent

  Filter sterilize through a 0.22 μm cellulose acetate filter and store up to 2 weeks at 4°C.

- *Growth media:* (1 litre)

  To 1 litre of sterile RPMI 1640 with 2 mM L-glutamine, add:

- 165.0 ml fetal bovine serum, heat inactivated at 50-60°C for one half hour 12.0 ml 200 mM (100 X) L-glutamine
- 1.2 ml 50 mg/ml gentamicin reagent

  Filter sterilize through a 0.22 μm cellulose acetate filter and store up to 2 weeks at 4°C.

- *1X Cyclosporin media:* (100 ml)
- To 100 ml of growth media add: 1.0 ml 100X cyclosporin A
- *100X Cyclosporin A:* (100 ml)
- Dissolve 1 mg CSA in 0.1 ml ethanol in a sterile 15 ml centrifuge tube with a small magnetic stirrer. Add 0.02 ml (= 20 μl) of Tween 80 and mix well. While continually stirring, add 1 ml RPMI drop by drop. Bring to a final volume of 100 ml with RPMI.
- Filter sterilize with a 0.22 μm filter. Store at 4°C for up to 4 months.

EXERCISE 5. PREPARATION OF LYMPHOCYTE CELL PELLET FOR STORAGE

Purpose

Following propagation to 1 X 108 cells, lymphoblastoid cells are conveniently stored at –80°C to preserve the high molecular weight DNA in the cells until the DNA is purified. This procedure describes the steps required to harvest and freeze the cells for long term storage.

Time Required

2-3 hours to prepare 12-15 cultures for storage.

Procedure

1. Aspirate the growth media from the lymphoblastoid cell culture to the 40 ml mark on the T-75 cm^2 flask.
2. Resuspend the cells in the flask by shaking gently. Remove 200 µl of the cell suspension and determine the cell count (refer to cell counting procedure). Transfer the cell suspension either by decanting or pipetting to a 50 ml conical centrifuge tube.
3. Centrifuge the tubes containing cells for 10 minutes, 1200 rpm, at room temperature using the T-J6 centrifuge. Do not apply the brake at the end of the centrifuge run.
4. Aspirate the supernatant above the cell pellet. Resuspend the cells with 10 ml of PBS.
5. Label a 15 ml tube with the date, kindred#, cell line#, and cell count. Transfer the cell suspension to the labelled 15 ml centrifuge tube, centrifuge again for 10 minutes and aspirate the supernatant.
6. Transfer the tube to a –80°C Revco freezer and record the rack location on the cell line growth record sheet.

EXERCISE 6. MAINTENANCE OF B95-8 CELL LINE AND OBTAINING VIRUS FOR LYMPHOCYTE TRANSFORMATION

Principle

The B95-8 cell line was initiated by exposing marmoset blood leukocytes to Epstein-Barr virus (EBV) extracted from a human leukocyte line. B95-8 is a continuous line and releases high titres of transforming EBV. Thus it provides a source of EBV to establish continuous lymphocytic cell lines from human donors.

Safety Considerations

B95-8 must be handled with precautions since EBV can infect primates. A Biological Safety cabinet must be used when passaging the culture and harvesting the virus. Use bleach to kill unused virus. All material that comes in contact with virus must be autoclaved. In addition, the door of the room should remain closed to prevent outside contaminants from entering the room and to prevent any harmful viruses from leaving the area. Gloves should always be worn in dealing with any human or hybrid cell line because latent virus genomes can be present.

Special Reagents

The B95-8 cell line. [available from American type culture collections CAT NO. ATCC CRL 1612].

Time Required

5 minutes twice a week to feed and split the culture to maintain the correct cell density of 1.0-2.0 X 106 cells/ml.

Procedure

1. B95-8 should be grown in growth media (RPMI-1640 + 16% fetal bovine serum). The culture should be passaged twice a week: on Mondays and Thursdays, or on Tuesdays and Fridays. Passaging (subculturing) cells denotes the transplantation of cells from one culture vessel to another. Aspirate half of old media and replace it with an equal volume of new media.
2. To maintain a culture at a density of around 1 X 106 cells/ml it is necessary to split it 1:4 once a week. For example, to a culture with a cell density greater than 1.5 X 106 cells/ml, one fourth is diluted with 3 parts growth media (10 ml cells +30 ml media). Save the old flask as a backup in case the new culture becomes contaminated. When the subculture is passaged the next time, dispose of the old flask.
3. Media containing fresh virus can be prepared at the same time the culture is passaged: Using a 10 ml or 25 ml disposable pipette, remove and transfer the media (above the cells) to a 50 ml centrifuge tube. Always be careful not to pull up any cells at bottom of the culture flask. Reserve 25 ml of media in the flask and add a equal amount of new growth media to maintain the culture.
4. Centrifuge the tube with the media-containing-virus for 10 minutes at 1200 rpm (no brake) at room temperature, using the TJ-6 centrifuge. Centrifuging the media will pellet any marmoset cells to the bottom of the centrifuge tube.
5. With a 10 ml pipette, transfer all but the bottom 10 ml of virus in the centrifuge tube to a 150 ml 0.22 µm cellulose acetate filter. Filter and store the virus at 4°C for up to 7 days.

Solutions

- *Growth media:*

  To 1 litre of sterile RPMI-1640, add: 165.0 ml fetal bovine serum, 1.2 ml gentamicin reagent, 12.0 ml L-glutamine.

  Filter sterilize and store at 4°C, for up to 2 weeks.

EXERCISE 7. CELL COUNTS USING A HEMOCYTOMETER

Purpose

The purpose of this procedure is to determine the cell density of the culture. Cell cultures always have some dead cells; the viable and non-viable cells can be distinguished with the use of trypan blue dye and a hemocytometer. Living cells will not take up the dye while dead cells do.

Time Required

5 minutes for 2 two cell counts

Procedure

1. Transfer 200 µl of the cell suspension into a 15 ml centrifuge tube.
2. Add 300 µl of PBS and 500 µl of trypan blue solution to the cell suspension (creating a dilution factor of 5) in the centrifuge tube. Mix thoroughly and allow to stand 5 to 15 minutes.

   Note: If cells are exposed to trypan blue for extended periods of time, viable cells may begin to take up dye as well as non-viable cells, thus, try to do cell counts within one hour after dye solution is added.
3. With a cover-slip in place, use a pasteur pipette and transfer a small amount of the trypan blue-cell suspension to a chamber on the hemocytometer. This is done by carefully touching the edge of the cover-slip with the pipette tip and allowing the chamber to fill by capillary action. Do not overfill or underfill the chambers.
4. Count all the cells (non-viable cells stain blue, viable cells will remain opaque) in the 1 mm centre square and the four corner squares. Refer to diagram below. Keep a separate count of viable and non-viable cells. If greater than 25% of cells are non-viable, the culture is not being maintained on the appropriate amount of media; reincubate culture and adjust the volume of media according to the confluency of the cells and the appearance of the media. A culture growing well will have many clumps of cells and will turn the media from an orange to a yellow colour within several days (increase the amount of media). A culture not growing well will have few clumps of cells and the media will not change to the yellow colour (it may even turn pink, if so, decrease the volume of media). Cells may be frozen if greater than 75% of the cells are viable. Note: If greater than 10% of the cells appear clustered, repeat entire procedure making sure the cells are dispersed by vigorous pipetting in the original cell suspension as well as the trypan blue suspension. If less than 20 or greater than 100 cells are observed in the 25 squares, repeat the procedure adjusting to an appropriate dilution factor. Repeat the count using the other chamber of the hemocytometer.
5. Each square of the hemocytometer (with cover slip in place) represents a total volume of 0.1 mm^3 or 10^{-4} cm^3. Since 1 cm^3 is equivalent to 1 ml, the subsequent cell concentration per ml (and the total number of cells) will be determined using the following calculations.

   Cells per ml = the average count per square × the dilution factor × 104 (count 10 squares)

   Example: If the average count per square is 45 cells × 5 × 104 = 2,250,000 or 2.25 × 10^6 cells/ml.

   Total cell number = cells per ml × the original volume of fluid from which cell sample was removed.

   Example: 2.25 × 10^6 (cell per ml) × 10 ml (original volume) = 2.25 × 10^7 total cells.

EXERCISE 8. REMOVAL OF YEAST CONTAMINATION FROM LYMPHOBLAST CULTURES

Purpose

This method is advantageous for saving the occasional cultures that become contaminated. Yeast contaminated cultures will appear cloudy when slightly shaken and lymphocytes will not cluster together as much as normal. If cultures are suspect, a drop of culture can be streaked on a YPD media plate to check for growth of yeast colonies, or a 5 ml sample can be taken to Barnes Diagnostic Centre for identification of yeast strain.

Procedure

1. Pipette 5 ml histopaque into a 15 ml centrifuge tube.
2. Carefully layer on top of histopaque 10 ml of contaminated culture (or concentrated cells/yeast resuspended in growth media).
3. Centrifuge tube for 25 minutes at 2500 rpm (no brake) at room temperature, using the TJ-6 centrifuge.
4. The yeast cells will pellet to the bottom of the histopaque gradient and the lymphoblast cells will be located on top of histopaque gradient. Remove the lymphoblast cells with a 10 ml disposable pipette, and transfer to a 15 ml centrifuge tube.
5. Wash cells by adding 10 ml of wash media to cells. Centrifuge 10 minutes at 1200 RPM, no brake, at room temperature. Aspirate off the wash media and resuspend in RPMI-growth media containing 1X antimycotic/antibiotic. This will remove the rest of the yeast cells.
6. Transfer the cells to a 25 cm^2 Tissue Culture Flask and feed the culture twice a week with 1X antimycotic/antibiotic media until all traces of contamination are gone. This will depend on the severity of the contamination (usually for cultures moderately contaminated, 2 weeks or 4 feedings will suffice). After contamination is no longer visible, feed the cultures with growth media containing only antibiotic and not the antimycotic.

Solutions

- *Wash media:* (1 litre)

  To 1 litre of sterile RPMI 1640 with 2 mM L-glutamine, add:

  10.0 ml 2.5 M (100X) Hepes buffer, 1.2 ml 50 mg/ml gentamicin reagent.

  Filter sterilize through a 0.22 μm cellulose acetate filter and store up to 2 weeks at 4°C.

- *Growth media:* (1 litre)

  To 1 litre of strerile RPMI 1640 with 2 mM L-glutamine, add:

  165.0 ml fetal bovine serum, heat inactivated at 50-60°C for one half hour. 12.0 ml 200 mM (100 X) L-glutamine, 1.2 ml 50 mg/ml gentamicin reagent as added. Filter sterilize through a 0.22 μm cellulose acetate filter and store up to 2 weeks at 4°C.

- *1X Cyclosporin media:* (100 ml)

  To 100 ml of growth media add: 1.0 ml 100X cyclosporin A

- *100X Cyclosporin A:* (100 ml)

  Dissolve 1 mg CSA in 0.1 ml ethanol in a sterile 15 ml centrifuge tube with a small magnetic stirrer. Add 0.02 ml (or 20 µl) of Tween 80 and mix well. While continually stirring, add 1 ml RPMI drop by drop. Quantitate to a final volume of 100 ml with RPMI. Filter sterilize with a 0.22 µm filter. Store at 4°C for up to 4 months.
- *Antimycotic/antibiotic media:*

  To 1 litre of sterile RPMI 1640 with 2 mM L-glutamine, add:

  165.0 ml fetal bovine serum, heat inactivated

  12.0 ml 200 mM (100X) L-glutamine

  12.0 ml antimycotic/antibiotic (100X), liquid,

  Filter sterilize through a 0.22 µm cellulose acetate filter and store up to 2 weeks at 4°C.

EXERCISE 9. MAINTAINING LYMPHOBLASTOID CELL LINES

Purpose

To grow lymphoblastoid cells for permanent storage and for DNA extraction.

Safety Considerations

All cultured animal and human cells have the potential for carrying viruses, latent viral genomes, and other infectious agents. Cell cultures should be handled very carefully by trained persons under laboratory conditions which afford adequate biohazard containment. A Biological Safety Cabinet must be used when passaging cell lines. Use bleach in a suctioning apparatus to kill unused virus. All material used in passaging the cell lines must be autoclaved. Gloves are always worn to protect hands from contamination. A laboratory coat should be worn to protect clothes from contamination. Doors of the tissue culture room should remain closed to decrease the amount of airborn contaminants entering the incubators and the room. Equipment (incubators, centrifuges, microscopes, tabletops, etc.) should be cleaned routinely to help maintain a sterile work environment.

Time Required

3-4 weeks to grow a cell line from a frozen stock, or to grow an established cell line arriving in a T-25 cm^2 flask to 1 × 100 million cells.

Allow 6-8 weeks establish and grow a lymphoblastoid cell line from whole blood to 1×10^8 cells.

Procedure

Maintaining lymphoblastoid cultures is fairly simple if two important characteristics are taken into consideration: 1) the cell cycle (primary culture and established cell line) and 2) the cell concentration.

Cell Cycle

Every lymphoblastoid culture is unique and should be treated accordingly. For example, some cultures will grow very rapidly, while others may require twice the amount of time. Cultures which require the media to be changed every other day are rapidly dividing and will form many clumps. Cell lines which grow slowly will change the colour of the media every 3-4 days, and may require the use of cyclosporin A and less media with each feeding.

Lymphoblast cultures grow in clumps and do best if periodically shaken up to break up the clumps. The cells will usually settle to the bottom of the flask, but do not attach unless the culture is in the primary stage of transformation or is lacking in nutrients. Cultures growing well will turn the media acidic within 12-24 hours after being fed. The colour is a good indication of cell growth and concentration (yellow: growing well; orange or pink: not growing well).

Lymphoblasts can be grown in T-25 cm^2 flasks or T-75 cm^2 flasks. Occasionally it is necessary to use a 24 or 96 well plate if a culture is not growing well. Primary cultures are set up in a 25 cm^2 flask and maintained until a volume of 15-20 ml is reached. The culture is then transferred to a larger flask to continue growth.

Cell Concentration

The cell concentration of the suspension is important. A cell count above a certain number means decreased viability (the dead cells stain blue with Trypan blue, see cell counting procedure). When the cell count is too low, cultures will show little growth. The absolute lo west cell concentration for any cell line should be 1.5-2.0 X 100 thousand viable cells/ml. Cultures can be split when the cell count is 2.0 X million viable cells/ml.

Cultures are grown upright in T-flasks. They are maintained with RPMI-1640 (supplemented with 1% of a 200 mM L-glutamine solution) plus 15% fetal bovine serum (heat inactivated) plus an antibiotic, such as gentamicin reagent or penicillin/streptomicin. Incubation conditions are 37°C and 5% CO_2. Cultures are fed every 3 to 4 days. If a cell line is not fed frequently enough, the majority of the cells will not be in the logarithmic phase of growth; therefore the optimum growth of the cell line is never reached. Cultures are fed by removing half of the media from the flask and replacing it with a slightly increased volume of new media. If a culture is not growing well, half of the media is removed, and the volume of added media is decreased slightly.

EXERCISE 10. LYMPHOBLASTOID CELL LINES FROM FROZEN WHOLE BLOOD

Purpose

Blood Samples can be stored frozen as a backup in case an LCL is needed at a later date.

Time Required

15 minutes to freeze 1-4 cryotubes placing them directly into the −135°C freezer; or one hour to freeze the tubes using the Cryomed freezing chamber. Cells have been shown to be more viable if temperature is lowered gradually with the freezing chamber.

Procedure

Freezing cells:

1. Pipette 1.0 ml of whole blood into 2 cryotubes (1.25 ml).
2. Add to each cryotube 100 µl dimethyl sulfoxide (DMSO), or 10% of the volume of blood.

3. Immediately begin freezing the whole blood in the cryomed freezing chamber until the chart drive printer reads –90°C.
4. Quickly transfer the frozen sample to long term storage in a –135°C freezer or a liquid nitrogen storage container. These whole blood samples have been shown to be viable for as long as 5 months by G. Chenevix-Trench, et al.

Thawing cells for transformation:

5. When an LCL is needed, the cells are thawed rapidly in a 37°C water bath: Place the cryotubes in a bubble rack. Shake the rack to help thaw the cells, usually for 1-2 minutes.
6. With a 1 ml disposable pipette, transfer the sample to a 15 ml conical centrifuge tube filled with 10 ml of wash media.
7. Centrifuge the cells for 10 minutes at 1200 rpm, no brake, at room temperature.
8. Aspirate the wash media to just above the cell pellet. Wash the pellet again with 10 ml wash media and centrifuge as in step 7. Repeat the wash a total of 4 times or until the red cell contamination is minimal. If the red cell contamination is not eliminated, several days in culture will decrease the amount of red cells substantially.
9. Aspirate the wash media and resuspend the cell pellet in 300 µl filtered supernatant from a B95-8 marmoset culture containing Epstein Barr virus. Transfer the cell suspension to a T-25 cm^2 flask and incubate for 2 hours at 37°C with 5% CO_2. If there is a very small volume of cells, leave the cells in the 15 ml centrifuge tube for the incubation.
10. After incubation, add 800 µl RPMI-1640 containing 20% fetal bovine serum and 2X Cyclosporin A.
11. Using a 5 ml disposable pipette, plate out cells in serial dilution in a 96-well microtiter plate: Transfer half of the cells in the first well into a second well. Add enough media to fill the second well. Then take half of this cell/media mixture and transfer to a third well. Fill up the third well with media. Incubate at 37°C with 5% CO_2.
12. Feed the cells twice-weekly by removing half of the old media and replacing with fresh media until transformed colonies are apparent (usually 2-3 weeks). The new media should contain 1X CSA.
13. Subculture cells to a 24-well plate before transferring to culture flasks. Maintain the subcultures on growth media (no CSA).

Solutions

- *Growth media:* (600 ml)
- To 500 ml of sterile RPMI 1640 with 2 mM L-glutamine, add: 90.0 ml FBS, 6.0 ml 200 mM (100X) L-glutamine, 0.6 ml 50 mg/ml gentamicin reagent.

  Note: Filter sterilize through a 0.22 µm filter and store up to 2 weeks at 4°C.
- *Wash media:* (1 litre)
- To 1 litre of sterile RPMI 1640 with 2 mM L-glutamine, add: 10.0 ml 2.5 M (100X) hepes buffer, 1.2 ml 50 mg/ml gentamicin reagent.

  Note. Filter sterilize through a 0.22 µm filter and store up to 2 weeks at 4°C.
- *2X Cyclosporin A media:* (1 µg/ml)

  To 100 ml of growth media add 2 ml of 100X Cyclosporin A.

- *100X Cyclosporin A:* (100 ml)

  Dissolve 1 mg CSA in 0.1 ml ethanol, add 0.02 ml Tween 80 and mix well. While continually stirring, add 1 ml RPMI, drop by drop. Quantitate to a final volume of 100 ml with RPMI. Filter sterilize and store at 4°C for up to 4 months.

EXERCISE 11. CELL CULTURE MEDIA AND SOLUTIONS

Antimycotic/Antibiotic Media

To 1 litre of sterile RPMI 1640 with 2 mM L-glutamine, add:

165.0 ml fetal bovine serum, heat inactivated

12.0 ml 200 mM (100X) L-glutamine

12.0 ml antimycotic/antibiotic (100X), liquid, Gibco, Cat. No. 600-5240AG

Filter sterilize through a 0.22 µm cellulose acetate filter and store up to 2 weeks at 4°C.

1X Cyclosporin Media: (100 ml)

To 100 ml of growth media add:

1.0 ml 100X Cyclosporin A

2X Cyclosporin A Media: (1 µg/ml)

To 100 ml of growth media add 2 ml of 100X Cyclosporin A.

100X Cyclosporin A: (100 ml)

Dissolve 1 mg CSA in 0.1 ml ethanol in a sterile 15 ml centrifuge tube with a small magnetic stirrer. Add 0.02 ml (= 20 µl) of Tween 80 and mix well. While continually stirring, add 1 ml RPMI drop by drop. Bring to a final volume of 100 ml with RPMI. Filter sterilize with a 0.22 µm filter. Store at 4°C for up to 4 months.

Freezing Media: (1 litre)

Prepare a 1 litre volume and divide into 25-50 ml centrifuge tubes containing 40 ml each.

Store the tubes at –80°C for up to one year.

700 ml RPMI-1640 with 2 mM L-Glutamine

200 ml fetal bovine serum (FBS)

100 ml dimethyl sulfoxide (DMSO, Sigma)

1000 ml total volume

Note. Filter sterilize media and FBS with a 0.22 µm cellulose acetate filter. Do NOT filter DMSO, it will dissolve the cellulose acetate membrane.

Growth Media: (600 ml)

To 500 ml of sterile RPMI 1640 with 2 mM L-glutamine, add:

90.0 ml FBS

6.0 ml 200 mM (100X) L-glutamine

0.6 ml 50 mg/ml gentamicin reagent

Note. Filter sterilize through a 0.22 µm filter and store up to 2 weeks at 4°C.

Growth Media: (1 litre)

To 1 litre of sterile RPMI 1640 with 2 mM L-glutamine, add:

165.0 ml fetal bovine serum, heat inactivated at 50-60°C for one half hour

12.0 ml 200 mM (100 X) L-glutamine

1.2 ml 50 mg/ml gentamicin reagent

Filter sterilize through a 0.22 µm cellulose acetate filter and store up to 2 weeks at 4°C.

Wash Media: (1 litre)

To 1 litre of sterile RPMI 1640 with 2 mM L-glutamine, add:

10.0 ml 2.5 M (100X) Hepes buffer

1.2 ml 50 mg/ml gentamicin reagent

Filter sterilize through a 0.22 µm cellulose acetate filter and store up to 2 weeks at 4°C.

APPENDIX

A. UNITS AND MEASURES

Size

Cell biology deals with things which are relatively small. The units of measurement typically used are the micron at the light microscope level, and the nanometer at the electron microscope level. For molecular measurements, the norm is the Angstrom. These units are defined within the following table:

Measure	*Symbol*	*Relative Length*	*Exponential Notation*
Meter	M	1	10^{0}
Decimeter	dm	.1	10^{-1}
Centimeter	cm	.01	10^{-2}
Millimeter	mm	.001	10^{-3}
Micrometer or micron	µ	.000001	10^{-6}
Nanometer	nm	.000000001	10^{-9}
Angstrom	Å	.0000000001	10^{-10}

From this table it is apparent that:

10 Å = 1 nm

1000 nm = 1 mm

10 mm = 1 cm

Not apparent are that:

1 inch = 2.54 cm = 25.4 mm = 25,400 µ = 25,400,000 nm

1 inch = 2.54 cm = 2.54×10^{1} mm = 2.54×10^{4} µ = 2.54×10^{7} nm

1 mm = 0.04 inches.

B. BASIC UNITS USED IN BIOCHEMISTRY

Mole Units

1 mole

1 millimole (mmole) = 10^{-3} moles

1 micromole (umole) = 10^{-6} moles

1 nanomole (nmole) = 10^{-9} moles

1 picomole (pmole) = 10^{-12} moles

Gram Units

1 gram (g)

1 milligram (mg) = 10^{-3} g

1 microgram (μg) = 10^{-6} g

Concentration = weight/volume

1 M (molar) = 1 mole/litre or 1 mmole/ml

1 mM = 1 mmole/litre etc.

Volume

Volumes are measured relative to a litre, with the most commonly used measurements, the millilitre and the microlitre. The following table gives the relative volumes:

Measure	*Symbol*	*Relative Volume*	*Exponential Notation*
Litre	L	1	10^{7}
Decilitre	Dl	.1	10^{-1}
Millimeter	Ml	.001	10^{-3}
Microlitre	μl	.000001	10^{-6}

There are 1,000 μl in 1 ml.

1 ml = 1 cm^3

1 gallon = 3.8 litres

1 quart = 0.95 litres

1 liquid ounce = 29.6 ml

Weight

The most common measurements of weight at the gram, milligram and microgram.

Measure	*Symbol*	*Relative Weight*	*Exponential Notation*
Kilogram	Kg	1000	10^{3}
Gram	G	1	10^{0}
Milligram	mg	.0001	10^{-3}
Microgram	μg	.0000001	10^{-6}

C. CONCENTRATION

Most concentrations used throughout cell biology are those of a solute disolved or suspended within a solvent, and in most cases the solvent is water. There are two general methods of identifying the concentration of a solution; as molarity or as a percent. Molarity is based on the number of moles of solute in the solvent, while percent is based on the number of parts, either grams (for a solid solute) or millilitres (for a liquid solute).

Molarity equals the number of moles of solute in 1 litre of solution. A mole is equal to the gram molecular weight (or formula weight) of the solute. Sodium Chloride (NaCl), for example has a formula weight of 58.43 (22.98 for Na and 35.43 for Cl). Thus, if 58.43 grams of NaCl are dissolved in 1 litre of water, the result would equal a 1 molar solution of NaCl. This is designated as 1 M NaCl, or as simple M NaCl.

We often deal with solutions of less volume than 1 litre, and the following should be noted:

1 M NaCl = 58.43 grams/litre = 58.43 mg/ml = 58.43 μUg/μUl.

A 0.002 M NaCl solution contains 0.002 moles of NaCl or 0.1168 grams (0.002 × 58.43) in one litre of solvent. Note that molar is abbreviated as M, but that there is no abbreviation for moles. The 0.002 M solution contains 0.002 moles (or 2 millimoles) or solute in one litre. A 0.002 M solution would contain 0.001 moles of solute in a half litre.

The number of moles = Volume (in litres) × Molar Concentration

The number of millimoles = Volume (in ml) × Molar Concentration

Note that chemical equations are always balanced via moles. Moreover, note that for dilutions of known concentrations, one can use the simple formula:

Molarity × Volume = Molarity × Volume

If you have a 0.002 M solution of NaCl and you wish to obtain 100 ml. of a 0.001 M solution,

0.002 M × Needed Volume = 0.001 M × 100 ml

Needed Volume = 0.001 M × .1 litres/0.002 M = .050 litres = 50 ml.

Measure 50 ml of the 0.002 M solution, and dilute it to 100 ml with the solvent (usually water, or an appropriate buffer).

Molarity is appropriate for use when chemical equations are to be balanced. When we deal with physical properties of solutions, molarity is not as valuable as a similar measurement of concentration, molality. For colligative properties of solutions (freezing point depression, boiling point elevation, osmotic pressure, density, viscosity), there is a better correlation between the property and molality.

Molality (designated with a lower case m) is equal to the number of moles of solute in 1000 gm of solvent. At first this may not appear any different from molarity, since a ml of water equals 1.0 gm. Indeed, for dilute solutions in water, there is little or no practical difference between a molar solution and a molal solution. In concentrated solutions, with temperature fluctuations and with changes in solvent, there is appreciable difference.

For example: A 2 m (2 molal) solution of sucrose contains 684.4 gm of sucrose (twice the molecular weight or two moles of sucrose) dissolved in 1000 gm (approximately 1 litre) of water. The weight of this solution is 684.4 gm +1000 gm or 1684.4 gm.

This solution (2 m sucrose) has a density of 1.18 gm/ml or 1180 gm/litre.

Since there are 1684.4 gms, division by the density (1180 gm/litre) would indicate that there are 1.43 litres of solution. That is, 684.4 gm of sucrose dissolved in 1000 gm of water would yield 1.43 litres of solution. This solution would contain 2 moles of sucrose, however and would have a molarity = 2 moles/1.43 litres or 1.40 M. So, a 2 m sucrose solution equals a 1.4 M sucrose solution.

D. PERCENT SOLUTIONS

In the example above of 2 m sucrose, there were 684.4 gm of sucrose in the final solution which weighed 1684.4 gm (684.4 gm sucrose + 1000 gm water). The percent of sucrose on the basis of weight is therefore 684.4/1684.4 × 100, or 40.6%.

There are three means of expressing concentration in the form of a percent figure:

1. Percent by weight (w/w); gm solute/100 gm solvent
2. Percent weight by volume (w/v); gm solute/100 ml solvent
3. Percent by volume (v/v); ml solute/100 ml solution.

For dilute solutions, these differences are not significant, but at higher concentrations, they are. Chemists (when they use Percent designations) usually use w/w. Biochemists and physiologists more often use w/v. Both use v/v if the solute is a liquid. It is important to distinguish among these alternatives.

Using ethanol as an example, consider a 20% solution of ethanol in water, mixed according to the three designations of w/w, w/v and v/v.

1. w/w would contain 20 g of absolute ethanol mixed with 80 gm of water to yield a 20% (w/w) solution.
2. w/v would contain 20 g of absolute ethanol mixed with water to form a final volume of 100 ml.
3. v/v would contain 20 ml of absolute ethanol diluted to 100 ml with water.

The three solutions are not the same. First, the density of alcohol is not equal to that of water, and thus conversion of g to ml is not equivalent. A 20% (w/w) solution of ethanol, for example, has a density of 0.97 g/ml and 20 gm of ethanol plus 80 gm of water would have a volume of 103 ml. The % (w/v) for this solution would be 20 gm ethanol / 103 ml, or 19.4% (w/v). Similarly, absolute ethanol has a density of 0.79 gm/ml and thus 20 ml of ethanol would weigh 15.8 gm. A 20% (v/v) solution would contain 15.8 gm of ethanol in 100 ml and be a 15.8% (w/v) solution.

So, for ethanol:

20% (w/w) = 19.4 % (w/v)
20% (w/v) = 20.0 % (w/v)
20% (v/v) = 15.8 % (w/v)

In cell biology, the most common use of Percent solution is as (w/v). In practice, these are simple solutions to mix. For a 20% (w/v) sucrose solution, for example, simply weigh 20 gm of sucrose and dissolve to 100 ml with water.

Unless specifically stated otherwise, solutions lacking the appropriate designation should be assumed to be (w/v).

E. SCIENTIFIC NOTATION AND DILUTIONS

I. Scientific Notation

When doing scientific calculations or writing, scientific notation is commonly used. In scientific notation, one digit (a number between 1 and 9) only is found to the left of the decimal point. The following examples are written in scientific notation:

- 3.17×10^3
- 5.2×10^{-2}

Note that exponents (the powers of 10) are used in these conversions.

Multiples of 10 are expressed in positive exponents :

- $10^0 = 1$
- $10^1 = 10$
- $10^2 = 100 = 10 \times 10$
- $10^3 = 1000 = 10 \times 10 \times 10$
- $10^4 = 10{,}000 = 10 \times 10 \times 10 \times 10$
- $10^5 = 100{,}000 = 10 \times 10 \times 10 \times 10 \times 10$
- $10^6 = 1{,}000{,}000 = 10 \times 10 \times 10 \times 10 \times 10 \times 10$

Fractions of 10 are expressed as negative exponents :

- $10^{-1} = 0.1$
- $10^{-2} = 0.01 = 01 \times 0.1$
- $10^{-3} = 0.001 = 0.1 \times 0.1 \times 0.1$
- $10^{-4} = 0.0001 = 0.1 \times 0.1 \times 0.1 \times 0.1$
- $10^{-5} = 0.00001 = 0.1 \times 0.1 \times 0.1 \times 0.1 \times 0.1$
- $10^{-6} = 0.000001 = 0.1 \times 0.1 \times 0.1 \times 0.1 \times 0.1 \times 0.1$

A. Procedure for converting numbers that are multiples of 10 to scientific notation

1. Convert 365 to scientific notation.

(*a*) Move the decimal point so that there is only one digit between 1 and 9 to the left of the point (from 365.0 to 3.65).

(*b*) 3.65 is a smaller number than the original. To equal the original you would have to multiply 3.65 by 100. As shown above, 100 is represented by 10^2. Therefore, the proper scientific notation of 365 would be 3.65×10^2.

(*c*) A simple way to look at these conversions is that you add a positive power of 10 for each place the original decimal is moved to the *left.* Since the decimal was moved two places to the left to get 3.65, the exponent would be 10^2, thus 3.65×10^2.

2. Convert 6,500,000 to scientific notation.

(*a*) Move the decimal point so there is only one digit to the left of the point (6,500,000 becomes 6.5).

(*b*) To equal the original number, you would have to multiply 6.5 by 1,000,000 or 10^6. (Since you moved the decimal point 6 places to the *left,* the exponent would be 10^6).

(*c*) Therefore, the proper scientific notation of 6,500,000 would be 6.5×10^6.

B. Procedure for converting numbers that are fractions of 10 to scientific notation

1. Convert 0.0175 to scientific notation.

(*a*) Move the decimal so there is one digit between 1 and 9 to the left of the decimal point (0.0175 becomes 1.75).

(*b*) To equal the original number, you would have to multiply 1.75 by 0.01 or 10^{-2}. Therefore, the proper scientific notation for 0.0175 would be 1.75×10^{-2}.

(c) A simple way to look at these conversions is that you add a negative power of 10 for each place you move the decimal to the right. Since the decimal point was moved 2 places to the right, the exponent becomes 10^{-2}, thus 1.75×10^{-2}.

3. Convert 0.000345 to scientific notation.

(a) Move the decimal point so only one digit (between 1 and 9) appears to the left of the decimal (0.000345 becomes 3.45).

(b) To equal the original number, you would have to multiply 3.45 by 0.0001 or 10^{-4}. (Since you moved the decimal point 4 places to the *right*, the exponent becomes 10^{-4}).

(c) Therefore, the proper scientific notation of 0.000345 is 3.45×10^{-4}.

C. Other examples

- $12{,}420{,}000 = 1.242 \times 10^{7}$
- $21{,}300 = 2.13 \times 10^{4}$
- $0.0047 = 4.7 \times 10^{-3}$
- $0.000006 = 6.0 \times 10^{-6}$

II. Dilutions: Examples

A. 1 ml of bacteria is mixed with 1 ml of sterile saline. The total ml in the tube would be 2 ml, of which 1 ml is bacteria. This is a 1:2 dilution (also written 1/2, meaning 1/2 as many bacteria per ml as the original ml).

B. 1 ml of bacteria is mixed with 3 ml of sterile saline. The total ml in the tube would be 4 ml, of which 1 ml is bacteria. This is then a 1:4 dilution (also written 1/4, meaning 1/4 as many bacteria per ml as the original ml).

C. 1 ml of bacteria is mixed with 9 ml of sterile saline. The total ml in the tube would be 10 ml, of which 1 ml is bacteria. This is then a 1:10 dilution (also written 1/10 or 10^{-1}, meaning 1/10 or 10^{-1} as many bacteria per ml as the original ml).

D. For dilutions greater than 1:10, usually serial dilutions (dilutions of dilutions) are made. The following represents a serial ten-fold dilution (a series of 1:10 dilutions):

- The dilution in tube #1 would be 1/10 or 10^{-1}.
- The dilution in tube #2 would be 1/100 or 10^{-2} (1/10 of 1/10).
- The dilution in tube #3 would be 1/1000 or 10^{-3} (1/10 of 1/100).
- The dilution in tube #4 would be 1/10,000 or 10^{-4} (1/10 of 1/1000).

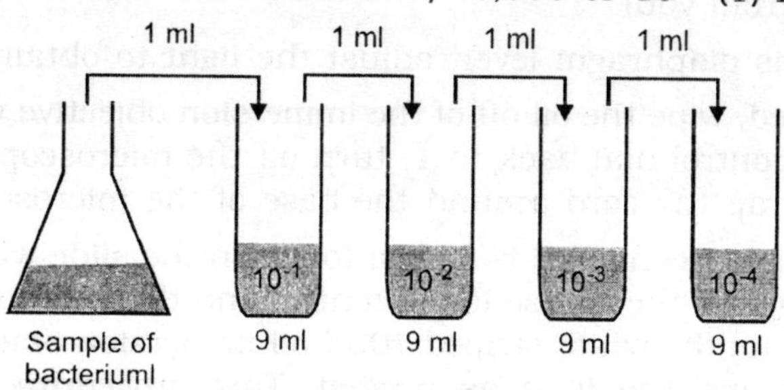

The *dilution factor* is the *inverse* of the dilution. (Inverse means you flip the two numbers of the fraction; with scientific notation you use the *positive* exponent).

- For a dilution of 1/2, the dilution factor would be 2/1 or 2.
- For a dilution of 1/4, the dilution factor would be 4/1 or 4.
- For a dilution of 1/10 or 10^{-1}, the dilution factor would be 10/1 or 10 or 10^1.
- For a dilution of 1/1,000,000 or 10^{-6}, the dilution factor would be 1,000,000/1 or 1,000,000 or 10^6.

On other words, the dilution factor tells you what whole number you have to multiply the dilution by to get back to the original 1 ml.

FOCUSING USING OIL IMMERSION MICROSCOPY

Microscope

1. Before you plug in the microscope, turn the voltage control dial on the right-hand side of the base of the microscope to 1. Now plug in the microscope and turn it on.
2. Place the slide in the slide holder, centre the slide using the two mechanical stage control knobs under the stag on the right-hand side of the microscope, and place a rounded drop of immersion oil on the area to be observed.
3. Rotate the white-striped 100X oil immersion objective until it is locked into place. This will give a total magnification of 1000X.
4. Turn the voltage control dial on the right-hand side of the base of the microscope to 9 to 10. Make sure the iris diaphragm lever in front under the stage is almost wide open, (toward the left side of the stage), and the knob under the stage on the left-hand side of the stage controlling the height of the condenser is turned so the condenser is all the way up.
5. Watching the slide and objective lens carefully from the front of the microscope, lower the oil immersion objective into the oil by raising the stage until the lens just touches the slide. Do this by turning the coarse focus (larger knob) away from you until the spring-loaded objective lens just begins to spring upward.
6. While looking through the eyepieces, turn the fine focus (smaller knob) towards you at a slow steady speed until the specimen comes into focus. (If the specimen does not come into focus within a few complete turns of the fine focus control and the lens is starting to come out of the oil, you missed the specimen when it went through focus. Simply reverse direction and start turning the fine focus away from you).
7. Using the iris diaphragm lever, adjust the light to obtain optimum contrast.
8. When finished, wipe the oil off of the immersion objective with lens paper, turn the voltage control dial back to 1, turn off the microscope, unplug the power cord, and wrap the cord around the base of the microscope.

An alternate focusing technique is to first focus on the slide with the yellow-striped 10X objective by using only the coarse focus control and then without moving the stage, add immersion oil, rotate the white-striped 100X oil immersion objective into place, and adjust the fine focus and the light as needed. This procedure is discussed in the introduction to the lab manual.

ABBREVIATIONS AND ACRONYMS

- A260: Absorbance at 260 nm wavelength (UV light)
- ABI: Applied Biosystems, Inc.
- ALARA: as low as reasonable achievable
- amp or AMP: ampicillin
- APS: ammonium persulfate
- ATCC: American Type Culture Collection
- ATP: adenosine triphosphate
- BCS: bovine calf serum
- BIOS: The BIOS Corporation
- βMe, 2-ME: β-mercapto ethanol
- bp: base pairs
- BRL: Bethesda Research Laboratories
- BSA: bovine serum albumin
- (CA)n: dC*dA dinucleotide repeat
- CGM: Centre for Genetics in Medicine (WUMS)
- CEPH: Centre d'Etude Polymorphisme Humain
- CHEF: Contour-clamped homogeneous electric fields
- CIP: calf intestinal alkaline phosphatase
- cm: centimetres
- cM: centimorgans
- cntl, CTRL: control key
- CO_2: carbon dioxide
- cpm: counts per minute
- CRI: Collaborative Research, Inc.
- CSA: cyclosporin A
- DAPI: 4,6-diamidino-2-phenylindole-dihydrochloride
- ddH_2O: distilled and deionized water
- ddNTP: dideoxyribonucleoside triphosphatc
- DGGE: denaturing gradient gel electrophoresis
- dH_2O: deionized or distilled water
- DMEM: Dulbecco's modified eagle medium (tissue culture)
- DMSO: dimethyl sulfoxide
- dNTP: deoxyribonucleoside triphosphate (usually one of dATP, dTTP, dCTP, dGTP)
- dpm: disintegrations per minute
- DTT: dithiothreitol
- EBV: Epstein Barr virus

- EDTA: ethylenediaminetetraacetic acid (powder is a disodium salt)
- EIU: electronic interface unit (Biomek)
- EMBL: European molecular biology laboratory
- EtBr: ethidium bromide
- EtOH: ethanol
- FBS: fetal bovine serum
- FF: father's father
- FITC: fluoresce in isothiocyanate
- FM: father's mother
- FW: formula weight
- GTL: G-banded chromosomes using trypsin and Leishman's stain
- HEPES: N-2-hydroxyethylpiperazine-N'-2-ethanesulfonic acid
- HET: heterozygosity
- HGM: human gene mapping
- IM: informative meioses
- IPTG: isoprcpylthio-β-D-galactoside
- ISCN: International System for Human Cytogenetic Nomenclature
- kan: kanamycin
- Kb, kb: kilobases
- Kg, kg: kilograms
- LCP: Linkage Control Package
- mA: milliamperes
- uCi: microCuries
- MEM: minimal essential media (tissue culture)
- MeOH: methanol
- MF: mother's father
- MIM#: Mendelian Inheritance in Man Number
- ug: micrograms
- ul: microlitres
- mg: milligrams
- ml: millilitres
- MM: mother's mother
- MW: molecular weight
- ng: nanograms
- nm: nanometres
- OD260: Optical density at 260 nm wavelength (UV light); = A260
- oligos: Oligonucleotides

- OS: operating system
- e32P: radioactive phosphorus isotope
- PAGE: polyacrylamide gel electrophoresis
- PBS: phosphate buffered saline
- PCR: polymerase chain reaction
- PEG: polyethylene glycol
- PFGE: pulsed field gel electrophoresis
- pfu: plaque forming units
- PHA: phytohemagglutinin
- PI: propidium iodide
- PIC: polymorphism information content index
- PMSF: phenylmethylsulfonyl fluoride
- RAM: radioactive materials
- RFLP: restriction fragment length polymorphism
- rpm: revolutions per minute
- RPMI-1640: Roswell Park Memorial Institute medium (tissue culture)
- RSO: radiation safety officer
- R.T.: room temperature
- e35S: radioactive sulfur isotope
- SDS: sodium dodecyl sulfate
- Solution A (tissue culture support centre): Dulbecco's Ca^{++}/Mg^{++} free medium
- STS: sequence tagged site
- TA, TAE: Tris-acetate buffer
- TB, TBE: Tris-borate buffer
- TC: tissue-culture
- TE: Tris-EDTA buffer
- TEMED: N,N,N',N'-tetramethylethelenediamine
- Tet: tetracycline
- TSC: Tissue Support Center (WUMS)
- U: units
- UV: ultra-violet light
- V: volts
- VNTR: variable number of tandem repeats
- v/v: volume per volume
- W: watts
- WBC: white blood cells
- w/v: weight per volume
- X-gal: 5-bromo-4-chloro-3-indolyl-beta-D-galactoside
- YAC: yeast artificial chromosome

CHEMICAL PREPARATIONS

Appendix A: Chemicals or Reagents Required for Molecular Biology Experiments

The following is a list of the solutions and chemicals required throughout the laboratory manual. It is organized alphabetically, and individual exercises list the materials needed for that exercise. For many solutions, directions are given for a molar solution and the user is left to dilute to the appropriate concentration for their needs. There are many vendors of the chemicals listed, and many of the solutions can also be purchased pre-mixed.

Acetic acid (MW 60.05)

Glacial acetic acid is 99.6% (w/v) acetic acid, and is 17.4 M.

1 M	Add 57.5 ml of glacial acetic acid to 800 ml of water and then make to 1 litre with water.
0.05 N	Add 2.87 ml of glacial acetic acid to 800 ml of water and then make to 1 litre with water.
0.9 M	Add 54 ml of glacial acetic acid to 800 ml of water and then make to 1 litre with water.
7% (w/v)	Add 70 ml of glacial acetic acid to 800 ml of water and then make to 1 litre with water.
45% (w/v)	Add 450 ml of glacial acetic acid to 500 ml of water and then make to 1 litre with water.

Acetic Acid/Butanol/Water (15:60:25)

Combine 150 ml of glacial acetic acid, 600 ml of butanol and 250 ml of water.

Aceto-orcein

Add 2.0 grams of orcein to 45 ml of glacial acetic acid. Bring to a boil and continue to heat until completely dissolved. Cool and add 55 ml of distilled water. Filter prior to use. Note: Some early investigators added of an iron salt (such as ferric citrate) as a mordant. It tends to increase the intensity of the aceto-orcein stain. The same reaction can be had by chopping plant material with an older steel (not stainless) razor blade.

Acid alcohol

Add 1.0 ml of concentrated HCl to 100 ml of 70% (v/v) ethyl alcohol.

Acid Orcinol Reagent (0.1% $FeCl_3$ in 10% HCl)

Add 0.1 grams of $FeCl_3$ to 50 ml of 10% (v/v) HCl and make to 100 ml with 10% HCl.

Acrylamide Solutions

Acrylamide solutions for PAGE are given as total concentration of acrylamide (acrylamide + bisacrylamide) and the amount of cross linker (bisacrylamide). This is listed as the T:C ratio. For example, a 10% gel (10%T:5%C) would contain a total of 10 grams of acrylamide per 100 ml, and would be composed of 5 grams of acrylamide and

5 grams of bisacrylamide. Usually, a stock solution of 30% acrylamide is produced containing 0.8% bis-acrylamide. Many investigators use 30 grams of acrylamide plus 0.8 grams of bis-acrylamide per 100 ml of water, but 29.2 grams of acrylamide plus 0.8 grams of bis would be technically correct. In practice, it makes little difference since the gels are diluted to 10% or less. The 30% stock solution is filtered through a 0.45 µ filter and stored at 4°C in the dark. For use, the stock solution is diluted with an appropriate buffer (usually a 2X Tris-HCl). The stock solution is stable for about one month. Discard after this period.

Acrylamides in their monomeric form are neurotoxic. Polymerize all acrylamide solutions prior to disposal.

SDS, β-mercaptoethanol and a tracker dye (bromophenol blue) are added at various points. Refer to Chapter 4 for more complete details.

Alcian blue (MW ≃ 1300)

0.001 M: Dissolve 0.13 grams of Alcian Blue 8GX (Sigma # A-2899) in 100 ml of water.

Alcohol orcinol reagent (10% orcinol in 95% ethanol)

Dissolve 1.0 gram of orcinol in 95% ethanol to a final volume of 10 ml.

Alkaline Distilled Water

Add one pellet of NaOH to 1 litre of distilled water.

Alkaline Solution for G-banding

Dissolve 2.8 grams of NaOH and 6.2 grams of NaCl to a final volume of 1 litre with water.

ρ-Aminosalicic acid (PAS MW 175.1)

6% (w/v): Dissolve 6.0 grams of PAS to a final volume of 100 ml with water or buffer.

Ammonium acetate (MW 77.08)

0.1 M: Add 7.708 grams of Ammonium acetate to a final volume of 1 litre of water.

Ammonium persulfate (MW 228.2)

10% (w/v): Dissolve 1.0 grams of ammonium persulfate to a final volume of 10 ml with water. Mix fresh, prior to use as a catalyst for PAGE. Normally, about 50 µl of ammonium persulfate is added to each 15 of gel solution for polymerization. Dissolve 0.13 grams of Alcian Blue 8GX (Sigma # A-2899) in 100 ml of water.

Ammonium sulfate (MW 132.14)

2% (w/v): Add 2 grams of ammonium sulfate to a final volume of 100 ml water.

0.001 M: Dissolve 542 grams of ammonium sulfate to a final volume of 1 litre.

n-Amyl alcohol (Pentanol $C_5H_{11}OH$ MW 88.15)

Density = 0.8144 grams/ml

0.38 M: The amyl alcohol can be weighed (33.5 grams) or measured volumetrically by using the density. That is, 33.5 grams ÷ 0.8144 grams/ml or 41.1 ml of n-amyl alcohol. Weigh or measure the appropriate amount and dilute to a final volume of 1 litre with water.

Amylase, buffered pH 7.0

1% (w/v): Dissolve 0.5 grams of amylase to a final volume of 50 ml with 0.01 M sodium phosphate buffer, pH 7.0.

Ascorbic acid (MW 176.12)

2 mM: Dissolve 35.2 mg of ascorbic acid to a final volume of 100 ml with water.

ATP (Adenosine triphosphate, MW 507.21)

5 mM: Dissolve 254 mg of ATP to a final volume of 100 ml with water or buffer. Dissolve 35.2 mg of ascorbic acid to a final volume of 100 ml with water.

Baker's Formalin

Add 1.0 gram of calcium chloride, 1.0 gram of cadmium chloride and 10 ml of concentrated formalin to 75 ml of water. Make to a final volume of 100 ml with water.

Benzoic acid (MW 122.12)

8 mM: Dissolve 98 mg of benzoic acid to a final volume of 100 ml with water or buffer.

Bis-acrylamide (N,N'-Methylene-bis-acrylamide)

Cross linker for acrylamide gels. Refer to Acrylamide solutions or Chapter Four for more details.

Biuret Reagent

Add 1.50 grams of $CuSO_4 \cdot H_2O$ and 6.0 grams of sodium potassium tartrate to 500 ml of water. Separately make 300 ml of 10% (w/v) NaOH by dissolving 300 grams of NaOH to a final volume of 300 ml with water. Combine the two solutions in a 1 litre volumetric, swirl to mix and make up to 1 litre with water. Store the final solution in a dark, plastic bottle. Discard if black or red precipitate forms.

Bovine Serum Albumin (BSA)

There are many grades of BSA available and care should be taken when using this protein. For routine protein concentration standards, a 96-99% pure fraction (Sigma A 2153) may be used. For tissue culture, RIA, or molecular weight standardization, BSA should be obtained which is extracted and purified specifically for that purpose.

1% (w/v): Dissolve 0.5 grams of BSA to a final volume of 50 ml in water or buffer.

Bradford Protein Assay

This procedure uses an absorbance shift in an acidic Coomasie Blue solution. It is commercially available from Pierce Chemical Company, Rockford, Illinois as Protein in Assay Reagent, Cat. # 23200. It contains methanol and solubilizing agents and is very reliable.

If you wish to make your own, dissolve 100 mg of Coomasie Brilliant Blue G-250 in 50 ml of 95% ethanol. Add 100 ml of 85% phosphoric acid, and bring to a final volume of 1 litre with distilled water.

Phosphoric acid is extremely corrosive. Handle with care.

Bromophenol blue (Sodium salt, MW 692.0)

0.01 (w/v): Dissolve 1 mg of Bromophenol blue, sodium salt (Sigma # B7021) to a final volume of 100 ml with either water or buffer.

n-Butanol (C_4H_9OH MW 74.12)

Density = 0.8098 grams/ml

1.1 M: The butanol can be weighed (81.5 grams) or measured volumetrically by using the density. That is, 81.5 grams ÷ 0.8098 grams/ml or 100.7 ml of n-butanol. Weigh or measure the appropriate amount and dilute to a final volume of 1 litre with water.

C. Elegans Ringers

This is a basic saline solution for nematodes. Dissolve 11.36 grams of $Na_2HPO_4 \cdot 7H_2O$, 3.0 grams of KH_2PO_4, 0.5 grams of NaCl and 1.0 gram of NH_4Cl to a final volume of 1 litre. Adjust the pH to 7.0. May be autoclaved for sterilization.

Calcium chloride (MW 110.99)

0.0033 M: Dissolve 0.522 grams of calcium acetate to a final volume of 1 litre with water or buffer.

Calcium chloride (MW 110.99)

0.001	Dissolve 0.111 grams of anhydrous calcium chloride to a final volume of 1 litre with water or buffer.
0.08 M	Dissolve 8.879 grams of anhydrous calcium chloride to a final volume of 1 litre with water or buffer.
2% (w/v)	Dissolve 2 grams of anhydrous calcium chloride to a final volume of 100 ml with water or buffer.

cAMP (Adenosine monophosphate, cyclic MW 329.22)

0.001 M (1mM): Dissolve 33 mg of cAMP to a final volume of 100 ml with water, buffer or media.

Carnoy Fixative

Combine 10.0 ml of glacial acetic acid with 60.0 ml of absolute ethyl alcohol and 30.0 ml of chloroform.

Chloroplast Homogenization Buffer

To 400 ml of distilled water, add 30.058 grams of sorbitol, 2.23 grams of sodium pyrophosphate, 0.407 grams of magnesium chloride, and 0.176 grams of ascorbic acid. Adjust the pH to 6.5 with HCl and dilute to a final volume of 500 ml.

Chloroplast Suspension Buffer

To 400 ml of distilled water, add 30.058 grams of sorbitol, 0.372 grams of EDTA, 0.102 grams of magnesium chloride and 5.958 grams of HEPES buffer. Adjust the pH to 7.6 with NaOH and dilute to a final volume of 500 ml.

Chrome Alum Gelatin (Subbing solution)

Dissolve 5.0 gram of gelatin in 1 litre of boiling water. Cool and add 0.5 grams of potassium chrome alum ($CrK(SO_4)_2 \cdot 12H_2O$). Store in refrigerator. To use, dip clean slides into the solution and dry in vertical position in a dust free location.

Citric Acid ($H_3C_6H_5O_7 \cdot H_2O$ MW 210.14)

0.1 M: Dissolve 21.01 grams of citric acid to a final volume of 1 litre.

Citrate Buffer (Sodium phosphate-Citrate buffer) 0.001 M

PH 4.8	Add 493 ml of 0.2 M Na_2HPO_4 to 507 ml of 0.1 M citric acid.
PH 3.6	Add 322 ml of 0.2 M Na_2HPO_4 to 678 ml of 0.1 M citric acid.
PH 4.2	Add 414 ml of 0.2 M Na_2HPO_4 to 586 ml of 0.1 M citric acid.
PH 5.4	Add 557.5 ml of 0.2 M Na_2HPO_4 to 442.6 ml of 0.1 M citric acid.
PH 6.0	Add 631.5 ml of 0.2 M Na_2HPO_4 to 368.5 ml of 0.1 M citric acid.
PH 6.6	Add 727.5 ml of 0.2 M Na_2HPO_4 to 272.5 ml of 0.1 M citric acid.
PH 7.2	Add 869.5 ml of 0.2 M Na_2HPO_4 to 130.5 ml of 0.1 M citric acid.
PH 7.8	Add 957.5 ml of 0.2 M Na_2HPO_4 to 42.5 ml of 0.1 M citric acid.

Cobaltous Nitrate (MW 182.96)

2% (w/v): Dissolve 2.0 grams of cobaltous nitrate (hexahydrate is very soluble) to a final volume of 100 ml with water. Keep well closed in a cool place.

Colcemid

10 µg/ml: Dissolve 10 g of colcemid per ml of saline or culture medium.

Coomasie blue (Coomasie Brilliant Blue R250)

0.25% (w/v) 0.001 M: Dissolve 2.50 grams of Coomasie Brilliant Blue R250 to a final volume of 1 litre with 20% (w/v) trichloroacetic acid (TCA). Some investigators use a 0.25% solution of Coomasie Blue in methanol-water-glacial acetic acid (5-5-1).

Copper sulfate ($CuSO_4 5H_2O$ MW 249.68)

0.5% (w/v): Dissolve 0.5 grams of copper sulfate to a final volume of 100 ml with water.

Copper tartrate/carbonate (CTC)

Dissolve 0.5 grams of copper sulfate and 1.0 gram of potassium sodium tartrate to a final volume of 100 ml with water. Combine 1.0 ml of this solution with 50 ml of 2% Na_2CO_3 in 0.1 N NaOH. Must be made fresh, prior to use. Stock solutions are stable.

Crystal Violet

Dissolve 0.1 grams of crystal violet and 0.25 ml of glacial acetic acid to a final volume of 100 ml with water.

DCMU (3-(3,4-Dichlorophenyl)-1,1-Dimethylurea MW 233.1)

1×10^{-4} M 0.001 M	Dissolve 2.3 mg DCMU to a final volume of 100 ml with water or buffer.
5×10^{-7} M 0.001 M	Dilute the 1×10^{-4} M solution 1/200 prior to use.

Dichlorophenolindophenol (DCPIP MW 290.1)

0.0025 M	Dissolve 73 mg of DCPIP to a final volume of 100 ml with water or buffer.
0.0001 M	Dissolve 2.9 mg of DCPIP to a final volume of 100 ml with water or buffer.

Dinitrophenol (DNP MW 184.11)

18.4 mg%: Dissolvc 18.4 mg of 2,4-dinitrophcnol to a final volumc of 100 ml with water or buffer.

Dische diphenylamine reagent

Dissolve 500 mg of diphenylamine in 49 ml of glacial acetic acid. Add 1.0 ml of concentrated HCl.

Dithiothreitol (Cleland's Reagent MW 154.3)

0.01 M : Dissolve 154 mg of dithiothreitol to a final volume of 100 ml with water or buffer.

DOPA (3-(3,4-Dihydroxyphenyl)-L-alanine MW 197.19)

8 mM: Dissolve 158 mg of L-DOPA to a final volume of 100 ml with water or buffer. Note that the maximum solubility of DOPA in water is 165 mg/100 ml (8.3 mM).

EDTA (Ethylenediaminetetraacetic acid MW 292.24)

1 M: Dissolve 292.24 grams of EDTA, free acid to a final volume of 1 litre. If the more soluble disodium salt of EDTA is used, adjust the weight accordingly. The pH can be adjusted with acetic acid or NaOH. For corresponding concentration dilutions, multiply the weight in grams by the desired molarity. For example, for 10 mM EDTA, multiply 292.24 X 0.010 to obtain 2.92 grams of EDTA per litre.

EGTA (Ethylene Glycol-bis(β-aminoethyl Ether) N,N,N',N'-Tetraacetic Acid MW 380.4)

1 mM: Dissolve 380 mg of EGTA to a final volume of 1 litre with water or buffer.

Eosin

0.5% (w/v): Dissolve 0.5 grams of Eosin Y in 100 ml of water.

Ethanol (C_2H_5OH MW 46.07)

Density = 0.7893 gm/ml

50-95% (v/v): Since 95% ethyl alcohol is less expensive and easier to store than absolute, these dilutions should be made with 95% ethyl alcohol. Unless otherwise stated, denatured alcohol works as well as the more expensive non-denatured. A simple way to make the % solution is to use the appropriate amount of 95% ethanol and dilute to 950 ml instead of 1 litre. For example, to make a 50% (v/v) solution, measure out 500 ml of 95% ethyl alcohol and dilute to a final volume of 950 ml with water. For a 70% solution, measure 700 ml of ethyl alcohol and dilute to 950 ml with water. Absolute ethanol should be used directly as 100% ethanol. It is important for histology that this be truly 100%. Since it is hydroscopic (it absorbs water from the air), do not assume it is absolute unless it is sealed or treated to ensure no water. To test, add a drop to a sample of xylol. If any cloudiness occurs, the alcohol is not absolute.

8.5 M: The ethanol can be weighed (391.6 grams of absolute, 412.2 grams of 95% (v/v) or measured volumetrically by using the density. That is, 391.6 grams ÷ 0.7893 grams/ml or 496.1 ml of absolute ethanol. Using 95%, 412.2 grams ÷ 0.7893 grams / ml or 522.2 ml. Weigh or measure the appropriate amount and dilute to a final volume of 1 litre with water.

Ethanol-acetic acid fixative for histology (3:1)

To 75 ml of absolute alcohol, add 25 ml of glacial acetic acid. Must be made fresh, just prior to use.

Fetal Calf Serum (FCS)

While it is possible to prepare your own serum from whole blood, it is easier to purchase FCS from a reputable supplier. Commercial sources are free of mycoplasma, pre-sterilized and controlled for the presence of antibodies. There are a number of serum substitutes available on the market and these may be less expensive when storage is considered. Suppliers include Gibco, Flow Laboratories, KC Biological and Sigma Chemical Co.

Folin-Ciocalteu Reagent

This is usually purchased premixed, since it is difficult to make. Also known as 2N Folin-phenol reagent.

Giemsa stain

Prepare a stock solution by dissolving 3.8 grams of giemsa powder in 25 ml of glycerin. Heat gently with stirring for about 2 hours at 60°C. Cool and add 75 ml of methanol (neutral, acetone free).

For a working solution, dilute the stock solution 1/10 with water before use. For chromosome banding, combine 5.0 ml of stock Giemsa, 3.0 ml of absolute methanol, 3.0 ml of 0.1 M citric acid and 89 ml of distilled water. Adjust the pH of the solution to 6.6 with Na_2HPO_4.

Glucose (MW 180.16)

10% (v/v) : Dissolve 10 grams of D-glucose (dextrose) in a final volume of 100 ml with water or buffer.

Glutaraldehyde (GTA)

5 %: GTA is usually supplied as a 25% or 50% (w/v) solution. It is used for electron microscope fixation as a 5% solution in a buffer. For routine use, add 20 ml of 25% GTA to 80 ml of 0.2 M sodium cacodylate buffer, pH 7.4.

Glycerol (MW 92.09)

10% (v/v): To 10 ml of glycerol (glycerine) add enough water to make a final volume of 100 ml.

8 M: Weigh 73.67 grams of glycerol and add to a final volume of 100 ml. Alternatively, measure 499.1 ml of glycerol and make to a final volume of 1 litre (the density of glycerol at room temperature is 1.476) with water or buffer.

For 8 M glycerol in MT buffer, make a 2X MT buffer for use as the diluent.

Glycine (MW 75.07)

0.192 M: Dissolve 1.44 grams of Glycine to a final volume of 100 ml with water or buffer.

Gram's iodine

Dissolve 0.33 grams of iodine and 0.67 grams of potassium iodide to a final volume of 100 ml with water.

HEPES (N-[2-Hydroxyethyl]piperazine-N'-[2-ethanesulfonic acid] MW 238.3)

50 mM : Dissolve 11.92 grams of HEPES, free acid to a final volume of 1 litre. If hemisodium salt is used, adjust weight accordingly. Do not use sodium salts unless specified. Hemisodium salt contains 0.5 moles of sodium for each mole of HEPES.

10 mM, pH 7.6: Dissolve 2.38 grams of HEPES, free acid in 900 ml of water. Adjust the pH with NaOH or HCl to 7.6. Adjust the final volume to 1 litre with distilled water.

Hydrochloric Acid (HCl MW 36.46)

Concentrated HCl has a molarity of approximately 11.6.

HCl is a gas, which is soluble in water and which comes in the form of concentrated reagent grade HCl. This solution is approximately 36-38% (w/v) HCl. To make a 1 N solution, add 86 ml of concentrated HCl to 800 ml of water and dilute to a final volume of 1 litre. For 0.1 N, dilute the 1 N by a factor of 10.

For % solutions, note that liquid HCl is only 38% HCl, thus a 1% solution would require 2.6 ml of concentrated HCl (1/.38) per final volume of 100 ml.

Janus Green B

0.01% (w/v): Dissolve 10 mg of Janus Green B in 2-3 ml of absolute ethanol. Dilute to a final volume of 100 ml with water.

Knudson Media

Knudson X4 Stock:

$Ca(NO_3)_2 \cdot 4H_2O$ 4.0 grams

$(NH_2)SO_4$	2.0 grams
$MgSO_4 \cdot 7H_2O$	1.0 gram
Distilled H_2O	1.0 litre
B5 Minor Elements:	
H_2SO_4	0.5 ml
$MnCl_2 \cdot 4H_2O$	2.5 grams
H_3BO_3	2.0 grams
$ZnSO_4 \cdot 7H_2O$	50 mg
$CoCl_2 \cdot 6H_2O$	30 mg
$CuCl_2 \cdot 2H_2O$	15 mg
$Na_2MoO_4 \cdot 2H_2O$	25 mg
Distilled H_2O	1.0 litre
Ferric Citrate:	
$FeC_6H_5O_7 \cdot 5H_2O$	2.5 grams
Distilled H_2O	100 ml
Stock Phosphate	
K_2HPO_4	25 grams
Distilled H_2O	100 ml
1X Media	Add 250 ml of Knudson X4 to 750 ml of distilled water. Add 0.5 ml of B5 Minor Elements, 0.5 ml of Stock Phosphate, and 0.4 ml of Ferric citrate. Adjust the pH to 5.5 with HCl, add 15 grams of agar and heat to dissolve. Autoclave and pour into plates.
Note	2.50 grams of sucrose may be added prior to adjustment of the pH, if desired. It is not necessary for germination of spores, but adds an organic source for mutants and abnormal fern growths. It also increases the need for subsequent aseptic technique.

Krebs Phosphate Ringers (KPR)

Prepare each of the following separately:

0.90% (w/v) NaCl

1.15% (w/v) KCl

1.22% (w/v) $CaCl_2$

3.82% (w/v) $MgSO_4 \cdot H_2O$

0.1 M phosphate buffer, pH 7.4 (17.8 grams of $Na_2HPO_4 \cdot H_2O$ + 20 ml of 1 N HCl, diluted to 1 litre)

To mix, combine 200 ml of NaCl, 8 ml of KCl, 6 ml of $CaCl_2$ and 2 ml of $MgSO_4$. Carefully, and with constant stirring, add 40 ml of phosphate buffer.

LPS buffer (Lower Pad Solution buffer)

Dissolve 1.5 grams of KCl, 0.5 grams of $MgCl_2$, and 0.5 grams of steptomycin sulfate in 500 ml of water. Add 40 ml of 1 M phosphate buffer, pH 6.5 and dilute to 1 litre with water.

Magnesium chloride ($MgCl_2$ MW 95.23)

1. mM: Dissolve 95.2 mg of magnesium chloride per final volume of 1 litre.
2. 4 mM: Dissolve 0.381 grams of magnesium chloride per final volume of 1 litre.
3. 10 mM: Dissolve 0.952 grams of magnesium chloride per final volume of 1 litre.
4. 0.1 M: Dissolve 9.523 grams of magnesium chloride per final volume of 1 litre.

Note. A single stock solution of 1 M $MgCl_2$ can be mixed by dissolving 95.23 grams of magnesium chloride to a final concentration of 1 litre with water, and all dilutions made appropriately from this stock solution.

Magnesium sulfate ($MgSO_4$ MW 120.39)

5% (w/v): Dissolve 5.0 grams of magnesium sulfate to a final volume of 100 ml with water or buffer.

Mayer's hematoxylin

Purchase commercially or mix with either of the following procedures:

A. Dissolve 1.0 gram of hematoxylin in 10 ml of absolute ethanol. Dissolve 20 grams of potassium alum ($KAl(SO_4)_2 \cdot 12H_2O$) in 200 ml of water. In a chemical hood, with protection against explosion, bring the potassium alum solution to a boil and add hematoxylin/ethanol mixture. Continue to boil for approximately 1 minute. Add 0. 5 grams of mercuric oxide and cool rapidly. Add 0.5 ml of glacial acetic acid. Filter before use. This mixture is stable for about two months.

B. Alternatively: Dissolve 5.0 grams of hematoxylin in 50 ml of absolute ethanol and add to 650 ml of warm water. Heat gently until the hematoxylin dissolves and then add 300 ml of glycerin, 0.3 grams of sodium iodate and 20 ml. of glacial acetic acid. Cool and make volume up to 1 litre with distilled water. Filter before use.

β-Mercaptoethanol (MW 78.13)

0.5 M : Density = 1.2 grams/ml. Use either 3.91 grams OR 3.26 ml of mercaptoethanol in a final volume of 100 ml of water or buffer.

5% (w/v): Use 5.0 grams or 4.167 ml in a final volume of 100 ml of water or buffer.

MES (2-(N-Morphilino)ethanesulfonic acid MW 195.2) 0.1 M

Dissolve 1.952 grams of MES to a final volume of 100 ml with water or buffer.

Methanol (CH_3OH MW 32.04)

Density = 0.7914 grams/ml

22 M	The methanol can be weighed (704.9 grams) or measured volumetrically by using the density. That is, 704.9 grams ÷ 0.7914 grams/ml or 890.7 ml of methyl alcohol. Weigh or measure the appropriate amount and dilute to a final volume of 1 litre with water.

Methanol/Acetic Acid (for fixing proteins in acrylamide gels) 45%:12%:

Add 120 ml of glacial acetic acid to 450 ml of methanol and dilute to a final volume of 1 litre with water.

Methanol/Acetic Acid (for destaining or fixing proteins in acrylamide gels) 5%:7%:

Add 70 ml of glacial acetic acid to 50 ml of methanol and dilute to a final volume of 1 litre with water.

Methyl green

0.2% (w/v): Dissolve 0.2 grams of methyl green to a final volume of 100 ml with 0.1 M acetate buffer, pH 4.2.

Acetate buffer (0.1 M pH 4.2) is prepared by dissolving .361 grams of sodium acetate (trihydrate) in approximately 80 ml of water. Add .42 ml of glacial acetic acid and adjust the volume to 100 ml with water.

Microtubule Buffer (MT buffer)

Dissolve 19.52 grams of MES in 800 ml of distilled water. Add 0.380 grams of EGTA and 47.62 grams of $MgCl_2$. Adjust the pH to 6.4 with HCl or NaOH and dilute to a final volume of 1 litre with distilled water.

Minimum Essential Medium (MEM)

For all purposes of this manual, MEM refers to Eagle's MEM. While it is possible to mix this medium, it is infinitely easier (and less expensive) to purchase the media premixed from any number of commercial sources (Gibco, Flow, KC Biological, Sigma Chemical). The ingredients are listed in Table 12.1. It is essential that chemicals of the highest purity are used throughout.

NG agar (Nematode Growth agar)

Dissolve 3.0 grams of NaCl, 2.5 grams of peptone, and 17 grams of agar in a final volume of 1 litre. Boil to dissolve the agar, autoclave to sterilize.

Meanwhile, prepare separate sterile solutions of:

1 M $CaCl_2$

2 mg/ml uracil

10 mg/ml cholesterol in ethanol

1 M Potassium phosphate buffer, pH 6.0

1 M $MgSO_4$

Using proper sterile technique, cool the agar solution slightly and add 1 ml of $CaCl_2$, 1 ml of uracil, 0.5 ml of cholesterol, 25 ml of phosphate buffer, and 1 ml of $MgSO_4$. Swirl to mix all ingredients and pour plates.

ρ-Nitrophenyl phosphate (MW 263.1)

0.05 M	Dissolve 1.32 grams of nitrophenyl phosphate to a final volume of 100 ml with water or buffer.
0.8% (w/v)	Dissolve 0.8 grams of nitrophenyl phosphate to a final volume of 100 ml of water or buffer.

Osmium tetroxide (OsO_4 MW 254.2)

1% : Osmium tetroxide is a gas which is used in solution for EM preservation. It is best purchased in sealed vials of 2 ml of 4% OsO_4. For use, add 6.0 ml of water or buffer to the 2.0 ml of 4% osmium tetroxide. Seal in a tightly sealed container, wrapped with aluminium foil and keep in the refrigerator. Use of a fume hood is mandatory when using OsO_4. Osmium tetroxide will rapidly fix the nasal passages and exposed cornea if not properly vented. It should be handled with extreme care.

Perchloric acid (PCA MW 100.47)

2% (w/v): Dissolve 2.0 grams of PCA to a final volume of 100 ml with water or buffer

Percoll

Colloidal PVP coated silica.

Periodic acid (Periodate Used for PAS reaction)

Dissolve 0.6 grams of periodic acid in 100 ml of water and add 0.3 ml of concentrated nitric acid.

Phenazine methosulfate (PMS MW 306.34)

Mutagen and irritant.

0.033% (v/v): Add 33 μl of phenazine methosulfate to 90 ml of water or buffer and make up to 100 ml final volume. Must be made immediately prior to use.

Phenol mixture

Combine 555 ml of aqueous phenol (or 500 grams of phenol crystals plus 55 ml of water) with 70 ml of mcresol. Add 0.5 grams of 8-hydroxyquinoline.

Phenol will cause severe burns and readily dissolves all plastic and rubber compounds. Use extreme caution when handling this compound.

ρ-Phenylenediamene oxalate (PPDO MW 198.18)

0.02% (w/v): Dissolve 20 mg of PPDO to a final volume of 100 ml with water or buffer.

Phosphate buffered saline (PBS)

Mix 100 ml Ca^{++}/Mg^{++} free 10X PBSA with 800 ml of distilled water. Separately, dissolve 0.1 gram of magnesium chloride and 0.1 gram of anhydrous calcium chloride to a final volume of 100 ml with water. With constant stirring, slowly add the magnesium/calcium chloride solution to the diluted PBSA. If a precipitate forms, start over, and add slower with continuous stirring.

Ca^{++}/Mg^{++} free Phosphate buffered saline-10X (10X PBSA)

Dissolve 80 grams of NaCl, 2.0 grams of KCl, 15.0 grams of Dibasic sodium phosphate and 2.0 grams of Monobasic potassium phosphate in 1 litre of distilled water. This makes a 10X solution of Ca^{++}/Mg^{++} free phosphate buffered saline. Dilute 1:10 prior to use. Store in a refrigerator.

Phosphate buffered saline-Tween 20 (4.8)

Mix PBS and add 0.1% (v/v) Tween 20.

Phytohemaglutinin (PHA)

Available as kidney bean lectin. It is typically used as a stock solution of 10-20 g/ml in balanced salt solution. For tissue culture it must be cold sterilized prior to use.

Potassium chloride (KCl MW 74.55)

1 M: Dissolve 74.55 grams of KCl to a final volume of 1 litre with water or buffer. For other concentrations, multiply the weight by the required molarity. For example, for 0.150 M (150 mM), use 0.150 X 74.55, or 11.183 grams of KCl in 1 litre of water or buffer. Use half as much to obtain 0.075 M for karyotyping.

Potassium cyanide (KCN MW 65.11)

8 mM: Dissolve 52 mg KCN to a final volume of 100 ml with water or buffer.

Potassium phosphate, monobasic (KH_2PO_4 MW 136.09)

0.01M: Dissolve 1.36 grams of monobasic potassium phosphate to a final volume of 1 litre with water.

Potassium phosphate, dibasic (K_2HPO_4 MW 174)

0.01M: Dissolve 1.74 grams of dibasic potassium phosphate to a final volume of 1 litre with water.

Potassium phosphate buffer

0.01M, pH 7.4: Prepare 500 ml of 0.01 M K_2HPO_4 and 500 ml of 0.01 M KH_2PO_4. Place the K_2HPO_4 onto a magnetic stirrer and insert a pH electrode. Add the KH_2PO_4 slowly to adjust the pH to 7.4.

Potassium hydroxide (KOH MW 56.10)

0.5 N: Dissolve 28.05 grams of KOH to a final volume of 1 litre with water.

10% (w/v): Dissolve 10 grams of KOH to a final volume of 100 ml with water. Store in a plastic container.

Potassium sodium tartrate (Rochelle salt $KNaC_4H_4O_6 \cdot 4H_2O$ MW 282.23)

n-Propanol (C_3H_7OH MW 60.11)

Density = 0.8035 grams/ml

3M: The n-propanol can be weighed (180.3 grams) or measured volumetrically by using the density. That is, 180.3 grams ÷ 0.8035 grams/ml or 224.4 ml of n-propanol. Weigh or measure the appropriate amount and dilute to a final volume of 1 litre with water.

Protein buffer

Dissolve 1.46 grams of KH_2PO_4 and 0.92 grams of K_2HPO_4 in 80 ml of distilled water. Add 2.5 grams of crystalline serum albumin and adjust the volume to a final 100 ml with water.

Pyronin Y (acetone)

0.6% (w/v): Dissolve 0.6 grams of pyronin Y in 100 ml of acetone

Ribonuclease

0.1% (w/v): Dissolve 10 mg of pancreatic ribonuclease type A in 10 ml of water or buffer. Use for enzyme treatment of histological sections by floating 0.5-1.0 ml of this solution onto the section, with the slide set into a covered petri plate.

Safranin

Dissolve 2.5 grams of Safranin O in 10 ml of 95% ethanol and dilute to 100 ml with water.

Saline (NaCl)

0.85% (w/v): Saline refers to a solution of NaCl, with the most common usage for that which is isotonic to mammalian blood cells, notable a 0.85% or 0.9% solution. To mix, dissolve 8.5 grams of NaCl to a final volume of 1 litre with water.

Saline citrate (1/10 dilution of SSC)

Dissolve 0.878 grams of NaCl and 0.294 grams of sodium citrate to a final volume of 1 litre with water.

Saline citrate buffer (SSC)

20X: It is common to prepare this buffer as a stock 20X solution, to be diluted to 2X, 1X or 0.1X prior to use. To prepare a 20X stock solution, dissolve 175 grams of NaCl and 88 grams of sodium citrate in 900 ml of water. Adjust the pH to 7.0 with 1 N HCl and bring to a final volume of 1 litre.

For use, as a 1X SSC, dilute 1 part 20X stock with 19 parts distilled water. For a 2X SSC, dilute 1 part 20X stock with 9 parts water.

Schiff's Reagent

Dissolve 0.8 grams of basic fuchsin in 85 ml of distilled water. Add 1.9 grams of sodium metabisulfite and 15.0 ml of 1 N HCl. Place the solution in separatory funnel and shake at 2 hour intervals for a period of approximately 12 hours. Add 200 grams of activated charcoal, shake for 1 minute and filter the clear solution. If the solution is still pink, add another 100 grams of charcoal and shake for an additional minute. Filter and store in a dark bottle. Solution should be clear (no pink coloration) for use.

Scott solution

Dissolve 2.0 grams of sodium bicarbonate and 20.0 grams of magnesium sulfate in water to a final volume of 1 litre. Add a pinch of thymol to retard the growth of molds.

SDS

Refer to Sodium lauryl sulphate.

1X SDS-Electrophoresis running buffer

Dilute 5X Tris-Glycine buffer to 1X and add 1.0 gram of SDS per litre of 1X Tris-Glycine. The pH should be 8.3 after dilution.

2X SDS Sample buffer

Dissolve 1.52 grams of Tris base, 2.0 grams of SDS, 20 ml of glycerin, 2.0 ml of mercaptoethanol and 1 mg of bromophenol blue to a final volume of 100 ml with water.

Siliconized pipettes

Pasteur pipettes can be siliconized by soaking them in a beaker containing 5% (v/v) dichlorodimethylsilane in chloroform for about 1 minute. Remove, drain and rinse several times with distilled water. Bake the pipettes at 180°C for 2 hours and cool before use.

Dichlorodimethylsilane and chloroform are both toxic and volatile. Use only in proper fume hood and keep all flames away from work area. Insure that all silicone and chloroform are removed from glassware before placing in an oven.

Silver nitrate solution (for electrophoresis staining)

Dissolve 0.15 grams of NaOH in 150 ml of water. Add 3.5 ml of concentrated NH_4OH and bring to a volume of 200 ml. Separately, dissolve 2.0 grams of silver nitrate in a final volume of 10 ml. With constant stirring, add 8.0 ml of the silver nitrate to the 200 ml of $NaOH/NH_4OH$.

This solution should be prepared immediately prior to use, and used within 30 minutes.

Dispose of this solution with copious flushing. It becomes explosive upon drying.

SM agar medium (Slime Mold medium of Sussman)

Dissolve 10.0 grams of glucose, 10.0 grams of peptone, 1.0 gram of yeast extract, 1.0 gram of $MgSO_4$, 1.5 grams of KH_2PO_4, 1.0 gram of K_2HPO_4 and 20.0 grams of agar to a final volume of 1 litre. Heat to dissolve the agar, autoclave and dispense to petri plates.

Sodium acetate (MW 82.04)

1 M: Dissolve 82.04 grams of sodium acetate to a final volume of 1 litre with water or buffer.

0.02 M: Dissolve 1.64 grams of sodium acetate to a final volume of 1 litre with water or buffer.

Sodium acetate buffer

1 M pH 5.7: To 925 ml of 1 M sodium acetate, add 75 ml of 1 M acetic acid.

Sodium azide (MW 65.02)

0.01M: Dissolve 0.065 grams of sodium azide to a final volume of 100 ml with water.

0.39% (w/v): Dissolve 0.39 grams of sodium azide to a final volume of 100 ml with water or buffer.

Sodium barbitol

0.2% (w/v): Dissolve 0.2 grams of sodium barbitol to a final volume of 100 ml with water.

Sodium bicarbonate ($NaHCO_3$ MW 84.0)

0.1 M: Dissolve 0.84 grams of $NaHCO_3$ to a final volume of 100 ml with water or buffer.

Sodium cacodylate buffer

0.2 M pH 7.4 : Prepare a 0.2 M solution of cacodylic acid, sodium salt (MW 159.91). Dissolve 3.20 grams of cacodylic acid, sodium salt to a final volume of 100 ml with water. Adjust the pH to 7.4 with HCl.

Cacodylic acid contains arsenic. Handle properly.

Sodium carbonate (MW 106.0)

2% (w/v): Dissolve 2.0 grams of sodium carbonate to a final volume of 100 ml with 0.1 N NaOH.

Sodium chloride (MW 58.44)

M: For a molar solution of sodium chloride, dissolve 58.44 grams of NaCl to a final volume of 1 litre with water or buffer. For corresponding dilutions, multiply the weight by the molarity required. For example, for 0.05 M, multiply 58.44 by 0.05 or 2.92 grams/litre.

% : For % solutions, they are invariably w/v. For a 1 % (w/v) solution, dissolve 1.0 gram of NaCl to a final volume of 100 ml with water or buffer. Multiply the weight by a corresponding change in % for other concentrations.

200, 300, 400 mOsM: Osmoles for NaCl are calculated as twice the molar concentration. Thus, a 200 mOsM solution would be .100 M NaCl. Likewise, 300 mOsM would be .150 M and 400 mOsM would be .200 M NaCl.

Sodium citrate (MW 294.10)

0.09 M: Dissolve 2.65 grams of sodium citrate to a final concentration of 100 ml with water.

Sodium citrate/formaldehyde (for silver stained proteins)

Dissolve 5.0 grams of sodium citrate in 800 ml of water, add 5.0 ml of concentrated formalin (37% formaldehyde solution) and dilute to 1 litre with water.

Sodium deoxycholate (Deoxycholic acid, sodium salt MW 392.58)

0.15% (w/v): Dissolve 150 mg of deoxycholic acid, sodium salt to a final volume of 100 ml with water.

Sodium dithionite ($Na_2S_2O_6 \cdot 2H_2O$ MW 242.16)

.1 mg/ml : Dissolve 10 mg of sodium dithionite in 100 ml of water just prior to use. Alternatively, to reduce a solution, the dry powder can be added as needed. Sodium dithionite should be stored at –20°C.

Sodium fluoride (NaF MW 42.0)

0.1 M: Dissolve 4.2 grams of NaF to a final volume of 1 litre with water.

Sodium lauryl sulfate (SDS or SLS MW 288.38)

0.1% (w/v): Dissolve 0.1 grams of SDS to a final volume of 100 ml with water or buffer. Mix by gentle stirring, do not shake.

10% (w/v): Dissolve 10 grams of SDS to a final volume of 100 ml with water.

SDS should not be inhaled in its powder form. When weighing, use a mask, or better, a hood.

Sodium malonate (MW 104.0)

0.6 M: Dissolve 2.49 grams of malonic acid, sodium salt, to a final volume of 25 ml with water or buffer.

Sodium perchlorate ($NaClO_4 \cdot H_2O$ MW 140.47)

1 M: Dissolve 14.01 grams of sodium perchlorate to a final volume of 100 ml with water or buffer.

Sodium phosphate, monobasic ($NaH_2PO_4.H_2O$ MW 137.99)

1 M: Dissolve 14.01 grams of sodium perchlorate to a final volume of 100 ml with water or buffer.

0.01 M: Dissolve 1.38 grams of monobasic sodium phosphate to a final volume of 1 litre.

Sodium phosphate, dibasic ($Na_2HPO_4 \cdot 7H_2O$ MW 268.07)

1 M : Dissolve 268.07 grams of dibasic sodium phosphate to a final volume of 1 litre.

0.2 M : Dissolve 53.61 grams of dibasic sodium phosphate to a final volume of 1 litre.

0.01 M : Dissolve 2.68 grams of dibasic sodium phosphate to a final volume of 1 litre.

Sodium phosphate buffer

These are the most common buffers used in biology. They are produced by adding equimolar solutions of KH_2PO_4 and Na_2HPO_4. Equal volumes of the two will yield a pH of 7.0, while sodium phosphate will increase the pH. Increased volumes of potassium phosphate will decrease the pH. The pH can be adjusted from 5.4 to 8.2.

If a pH of 7.0-8.2 is desired, start with 500 ml of sodium phosphate and add potassium phosphate while stirring and monitoring the pH with a pH meter until the desired pH is reached.

If a pH of 5.4-7.0 is desired, start with 500 ml of potassium phosphate and add sodium phosphate until the desired pH is reached.

Typically, the molarity of the buffer will range from 0.01 to 0.1 M. Use the appropriate molarity of KH_2PO_4 and Na_2HPO_4. That is, if 0.05 M buffer is desired, use 0.5 M KH_2PO_4 and 0.5 M Na_2HPO_4 as directed above.

Sodium potassium phosphate buffer

Refer to Sodium phosphate buffer.

Sodium pyrophosphate ($Na_4P_2O_7 \cdot 10H_2O$ MW 446.06)

10 mM: Dissolve 0.446 grams of $Na_4P_2O_7 \cdot 10H_2O$ to a final volume of 100 ml with water.

Sodium succinate (MW 270.16)

0.6 M: Dissolve 16.2 grams of succinic acid, sodium salt to a final volume of 100 ml with water or buffer.

Sorbitol (MW 182.17)

0.33 M: Dissolve 60.12 grams of sorbitol to a final volume of 1 litre with water or buffer.

Sorenson phosphate buffer

Refer to sodium phosphate buffer.

0.2 M pH 7.5: Dissolve 24.14 grams of Na_2HPO_4 and 4.08 grams of KH_2PO_4 in 800 ml of water. Dilute to a final volume of 1 litre.

Subbing solution (Slides)

Refer to chrom alum, gelatin.

Sucrose (MW 342.3)

1.0 M: Dissolve 34.2 grams of sucrose to a final volume of 100 ml with water or buffer. For other molarities, multiply the weight by the required molar concentration. For example, for 0.25 M sucrose, weight 34.2 × 0.25 or 8.55 grams to a final volume of 100 ml.

40% (w/v): Dissolve 40 grams of sucrose to a final volume of 100 ml with water or buffer. Dilute this solution for lower percent requirements. If using for sucrose density gradients, the sucrose should have 0.1 ml of diethylpyrocarbonate added, the solution brought to a boil for 3-5 minutes and cooled before use. This will eliminate RNAase, which would otherwise be a contaminant of the solution. Store all sucrose solutions in a refrigerator.

Sulfuric acid (H_2SO_4 MW 98.08)

Caution: Sulfuric acid is extremely caustic and will cause severe burns. It must always be added to the water, when making dilutions. Upon addition to water or alcohol, heat will be generated while the solution will contract in volume. Use extreme care in handling this acid.

Concentrated H_2SO_4 is 17.8 M or 35.6 N. 1.09 N add 30.6 ml of concentrated sulfuric acid slowly, with constant stirring, and with adequate protection from splashes, to approximately 800 ml of water. Cool and make up volume to 1 litre with water.

Sulfurous acid (for Feulgen Reaction)

Add 1.0 ml of concentrated HCl and 0.4 grams of sodium bisulfite to 100 ml of distilled water. This solution should be made fresh prior to use. It does not store well.

Swabbing detergent

For tissue culture purposes, use a non-toxic detergent designed for surgical scrubbing. e.g. Phisohex, Betadine or equivalent. For most routine swabbing, 70% (v/v) ethanol is sufficient and has the advantage that it will leave no residue.

TEMED (N,N,N',N'-tetramethylethylenediamine)

Catalyst for PAGE. Use directly and add 10 l TEMED per 15 ml of gel solution.

Toluidine blue

0.1% (w/v): Dissolve 0.1 grams of toluidine blue in 10 ml of ethanol and add water or citrate buffer (pH 6.8-7.2) to a final volume of 100 ml.

Trichloroacetic acid (TCA CCl_3COOH MW 163.4)

Extremely caustic acid. Handle with care.

72% (w/v): Dissolve 72 grams of TCA to a final volume of 100 ml. TCA is hydroscopic and will readily absorb water. The solid crystals will become liquid if the stock bottle is placed in warm water, with a loose cap (melting point 57-58°C. It is easier to handle as a liquid. Storage of solutions greater than 30% (w/v) are not recommended as decomposition is rapid. Therefore these solutions should be made as needed.

Tris buffer

There are many variations on the basic Tris-HCl buffer combination, most of which are commercially available. Solutions with EDTA are known as TE buffers, while solutions with EDTA and acetic acid are known as TAE buffers. The terminology varies with the author, with Tris buffer being used to mean Tris-HCl solutions. Sigma Chemical Co., St. Louis, carries a full line of the buffers marketed under the tradename of Trizma (base and HCl).

The basic buffer is a combination of Tris (tris(hydroxymethyl)aminomethane) and HCl acid. These are sometimes referred to as Tris-base and Tris-HCl solutions. Tris buffers should not be used below a pH of 7.2 or above a pH of 9.0. Tris buffers are also extremely temperature senstive. Directions are given for room temperature (25°C). The pH will decrease approximately 0.028 units for each degree decrease in temperature.

1M : Dissolve 121 grams of Tris in 800 ml of distilled water. Adjust the pH with concentrated HCl. Dilute to a final volume of 1 litre. Lower required molarities can be diluted from this stock or mixed as combinations of lower molarities of Tris and HCl. It is important to measure the pH at the temperature and molarity that will be used in the final analysis.

Tri-Glycine buffer

5X: Dissolve 15.1 grams of tris base and 72.0 grams of glycine to a final volume of 1 litre. For use, dilute 1 part 5X buffer with 4 parts water.

Trypan blue

0.2% (w/v): Dissolve 0.2 grams of trypan blue to a final volume of 100 ml with water.

Trypsin

0.25%: Dissolve 0.25 grams of crude trypsin in PBSA to a final volume of 100 ml. Cold sterilize by filtration.

Alternatively, purchase pre-diluted crude trypsin, sold as 1:250 which is pre-sterilized as well.

Note. When using trypsin for tissue disaggregation, it must be subsequently inhibited by the use of serum in the culture media, or by the addition of soyabean trypsin inhibitor.

Trypticase soy broth

Add 17.0 grams of trypticase peptone, 3.0 grams of phytone peptone, 5.0 grams of sodium chloride, 2.5 grams of dipotassium phosphate and 2.5 grams of glucose to 1 litre of water. Adjust the pH to 7.3, and autoclave.

Tween 20 or 80 (Polyoxyethylene sorbitan mono-oleate)

1% (v/v): Add 1.0 ml of Tween to 90 ml of water. Mix and dilute to a final volume of 100 ml with water. Note that Tween is extremely viscous and care must be taken to accurately pipette 1.0 ml. Wipe the outside of the pipette before dispensing.

Uranyl acetate (MW 424.19)

5% (w/v): Dissolve 5.0 grams of uranyl acetate to a final volume of 100 ml in 50% (v/v) ethanol. Store in the dark at room temperature. Allow at least 24 hours for the uranyl acetate to completely dissolve. This solution will keep for about 3 months.

Urea (MW 60.06)

2.5 M : Dissolve 15.02 grams of urea to a final volume of 100 ml with water or buffer.

10 M : Dissolve 60.06 grams of urea to a final volume of 100 ml with water or buffer.

14 M : Dissolve 84.08 grams of urea to a final volume of 100 ml with water or buffer.

Appendix B: Reagents Required for Tissue Culture Experiments (Stock Solutions and Media)

Sterile water bottles

200 ml, 500 ml, or 950 ml water.

Autoclave.

0.05 M $CaCl_2$ (per 200 ml)

$CaCl_2{\cdot}2H_2O$ 1.5 g

Add 198 ml water.

Autoclave.

0.5 M $CaCl_2$ (per 200 ml)

$CaCl_2{\cdot}2H_2O$ 15.0 g

Add 188 ml water.

Autoclave.

0.1 M $MgSO_4$ (per 200 ml)

$MgSO_4$ (anhydrous)	2.4 g

Add 200 ml water.

Autoclave.

10% MSG (for 200 ml)

Glutamic acid (Na)	20.0 g

Add 188 ml water.

Autoclave.

40% Glucose (per 200 ml)

Glucose(dextrose)	80.0 g

Add water to 200 ml

(~145-150 ml).

Autoclave.

10% CAA (per 200 ml)

Casein Acid Hydrolysate	20.0 g

Add 186 ml water.

Autoclave.

10% Yeast (per 200 ml)

Yeast extract	20.0 g

Add 150-160 ml water. Autoclave.

Add 150 ul β-mercaptoethanol

after autoclaving.

20X Tryptone/NaCl (per 200 ml)

Tryptone	40.0 g
NaCl	20.0 g

Add 190 ml water. Autoclave.

Add 150 ul β-mercaptoethanol

after autoclaving.

10X M9 stock (for 200 ml)

Na_2HPO_4 (anhydr.)	12.0 g
KH_2PO_4	6.0 g
NaCl	1.0 g
NH_4Cl	2.0 g

Add 196 ml H_2O. Autoclave.

Note. leave out NH_4Cl to prepare 10X M9-N.

10X M9 - N stock (for 200 ml)

Na_2HPO_4 (anhydr.)	12.0 g

KH_2PO_4	6.0 g
NaCl	1.0 g

Add 196 ml H_2O. Autoclave.

Note. leave out NH_4Cl

100X P (for 200 ml)

KH_2PO_4	18.1 g
K_2HPO_4	5.0 g

pH to 6.3 with 10 N KOH!

Add 192 ml H20. Autoclave.

1X SM (for 200 ml)

NaCl	1.16 g
$MgSO_4$	0.20 g
Gelatin (2%)	1.00 ml or 20 mg

Add 190 ml H_2O, then add:

Tris (1 M, pH 7.5)	10.0 ml

Autoclave.

2% Gelatin (per 500 ml)

Gelatin	10 g

Add 500 ml water.

Autoclave.

20% Succinate (for 500 ml)

Succinic acid	100 g
KOH	90 g

Add to 400 ml **ice cold** water.

Measure pH to 6.4, filter sterilize.

10X TBE (electrophoresis buffer)

Tris base	108.0 g
Na_2EDTA	9.3 g
Boric Acid	55.0 g

Bring volume to 1 L with water.

Use double distilled water.

pH should be 8.3; if not adjust dropwise with conc. NaOH.

Water agar stock (for 190 ml)

Agar	3.0 g

Add 190 ml H_2O.

Autoclave, store in 60C water bath.

R agar stock

	200 ml	500 ml
Tryptone........	2.0 g	5.0 g
NaCl...........	1.0 g	2.5 g
Agar...........	3.0 g	7.5 g
Add H_2O.........	194.0 ml	485.0 ml

Autoclave, store in 60C water bath.

LB agar 200 ml 1000 ml

	200 ml	1000 ml
Tryptone	2 g	10 g
NaCl	1 g	5 g
Yeast extract.	1 g	5 g
Agar	3 g	15 g
Add H_2O	193 ml	990 ml

Autoclave, store in 60C water bath. LB agar can also be made by adding 10 ml 10% yeast extract and 10 ml 20X Tryptone/NaCl to 190 ml WA.

NZ 200 ml 500 ml

	200 ml	500 ml
NZ amine	2.0 g	5.0 g
NaCl	1.0 g	2.5 g
$MgSO_4$ (anhydrous).	0.2 g	0.5 g

Add 10 M NaOH (1 drop/100 ml) after adding water.

Autoclave.

For agar media:

Agar	2.4 g	6.0 g

Stir before and after autoclaving.

GYPC to WA stock: 200 ml 1000 ml

	200 ml	1000 ml
Glucose (40%)	2.0 ml	10.0 ml
Yeast extract (10%)	4.0 ml	20.0 ml
100X P	2.0 ml	10.0 ml
CAA (10%)	4.0 ml	20.0 ml
Succinate (20%)....	1.0 ml	5.0 ml

Add appropriate antibiotics.

Pour plates.

ORS minimal 200 ml 1000 ml

	200 ml	1000 ml
10X M9	10 ml	50 ml
Succinate (20%).	2 ml	10 ml
MSG (10%)	2 ml	10 ml
Glucose (40%) ..	1 ml	5 ml
$MgSO_4$ (0.1 M) ..	1 ml	5 ml

$CaCl_2$ (0.5 M) ..	200 ul	1 ml
1000X vitamins ..	200 ul	1 ml
Water	184 ml	920 ml

Omit succinate and add 10 ml glucose for storage medium.

E. coli minimal	**200 ml**	**1000 ml**
10X M9	20 ml	100 ml
$MgSO_4$ (0.1 M) ...	2 ml	10 ml
$CaCl_2$ (0.5 M) ...	0.2 ml	1 ml
1000X vitamins ..	0.2 ml	1 ml
Glucose (40%) ...	2 ml	10 ml

XC minimal	**200 ml**	**1000 ml**
10X M9	5.0 ml	25 ml
$MgSO_4$ (0.1 M) ...	1.0 ml	5 ml
$CaCl_2$ (0.5 M) ...	0.2 ml	1 ml
200X 7 aa stock..	0.2 ml	1 ml
Glucose (40%) ...	0.5 ml	2.5 ml
Glycerol (50%) ..	4.0 ml	20 ml

Nif plates

10X Nif buffer	100 ml
10X trace elements	100 ml
1000X vitamins	1 ml
Sterile dd H_2O	799 ml
Agarose	6 g
Succinate (20%)	20 ml

(Succinate is already 0.4% from that in 10X nif buffer)

10X Nif buffer 1000 ml

K_2HPO_4	34.02 g
KH_2PO_4	57.06 g
Succinate (20%)	200.00 ml

pH should be 6.4 after 10X dilution.

1000X ORS vitamins

Nicotinic acid	2.0 mg/ml
Pantothenate	1.0 mg/ml
Biotin	0.2 mg/ml

Stir, filter sterilize.

Trace elements

$MgSO_4$ (0.1 M)	25 ml

$CaCl_2$ (0.5 M)	5 ml
Na_2MoO_4 (1 mg/ml)	5 ml
$FeCl_3$ (1 mg/ml)	5 ml
Water	460 ml

YEM (per 500 ml)

K_2HPO_4	0.5 g
$MgSO_4 7H_2O$	0.2 g
NaCl	0.1 g
Mannitol	10.0 g
Yeast extract	0.4 g
Agar	15.0 g
pH to 6.9	

YEM substitute (per 200 ml)

To 180 ml WA stock:

10X M9	8.0 ml
Mannitol (20%)	20.0 ml
Yeast extract (10%)	1.6 ml

SM10 selective media (for storage)

10X versatile Mg	100 ml
Thiamine	5 ml
Threonine	5 ml
Leucine	5 ml
Glucose (40%)	10 ml
Water	875 ml

Amino acids

SM10 stocks:	Thiamine (0.337%)
	Threonine (0.71%)
	Leucine (0.79%)
Klebsiella:	Histidine (0.31%)

10X versatile Mg (for 200 ml)

$MgSO_4$	2.0 ml
Thiamine (1 mg/ml)	2.0 ml
$CaCl_2$	0.2 ml

200X PLT (for 200 ml)

Proline	9.20 g
Leucine	1.60 g

Thiamine	0.67 g	
Autoclave.		

Mg PLT

10X versatile Mg	100 ml
200X PLT	5 ml
40% glucose	10 ml
Water	885 ml

Z buffer (lacZ assays)

	200 ml	**1000 ml**
$Na_2HPO_47H_2O$	3.22 g	16.10 g
$NaH_2PO_4H_2O$	1.10 g	5.50 g
KCl	0.15 g	0.75 g
$MgSO_4$ (0.1 M)	1.00 ml	0.20 ml
β-mercaptoethanol.	2.70 ml	0.54 ml

Filter sterilize.

ONPG stock

4 mg/ml in phosphate buffer (pH 7)

Phosphate buffer (pH 7.0) per 100 ml:

$Na_2HPO_47H_2O$	1.64 g
$NaH_2PO_4H_2O$	1.38 g

0.2 M Na_2CO_3 (100 ml)

Na_2CO_3 (anhydr.)	10.6 g

Jensen's Medium (Sesbania)

10X Macro nutrients (1000 ml):

$K_2HPO_43H_2O$	2.6 g
NaCl	2.0 g
$MgSO_4$	0.97 g
$CaCl_2$	10.0 g
$FeCl_36H_2O$	1.4 g
KNO_3	3.0 g

Jensen's Medium (Sesbania)

1X solution (1000 ml):

$K_2HPO_43H_2O$	0.20 g
NaCl	0.20 g
$MgSO_47H_2O$	0.20 g
$CaHPO_4$	1.00 g
$FeCl_36H_2O$	0.10 g

or Fe EDTA	0.26 g
KNO_3 (only for 10X)	3.00 g

Jensen's Medium (1942)

see Vincent 1970 p. 75.

per litre:

$K_2HPO_43H_2O$	0.2 g
NaCl	0.2 g
$MgSO_47H_2O$	0.2 g
$CaHPO_4$	1.0 g
$FeCl_36H_2O$	0.1 g
or Fe-EDTA	0.26 g

Dilute 1:5 when used.

15 ml diluted medium/pouch.

1000X Micronutrients (200 ml)

H_3BO_3	572 mg
$MnCl_2$	406 mg
$ZnSO_4H_2O$	44 mg
$CuSO_45H_2O$	16 mg
$NaMO_4BH_2O$	18 mg
CO^{2+}	20 mg
Ni^{2+}	20 mg

Nodule grinding buffer (500 ml)

50 mM KH_2PO_4 (pH 7.6), 0.15 M NaCl

K_2HPO_4 (1M)	3.25 ml
KH_2PO_4 (1M)	21.75 ml
NaCl	4.38 gm

Add to 450 ml distilled water, pH to 7.6, and adjust volume to 500 ml.

Rapidyne germicide

For working solution:

Dilute concentrated stock at 0.6 ml/L.

1 M Tris-Cl (500 ml)

To 400 ml, add:

Tris base	60.5 g

Add conc. HCl to pH bring to volume.

Tris buffer mixtures for pH

pH	Tris HCl	Tris base
7.0	.923	.077

7.1	.905	.095
7.2	.90	.10
7.3	.865	.135
7.4	.84	.16
7.5	.82	.18
7.6	.78	.22
7.7	.74	.26
7.8	.67	.33
7.9	.63	.37
8.0	.50	.50

Southern Transfer solutions

I. 0.25 N HCl:

21.5 ml conc. HCl (11.6N) to 1 litre volume.

II. 0.5 M NaOH, 1.5 M NaCl:

87.7 g NaCl, 20.0 g NaOH to 1 litre volume.

III. 1.0 M Tris, 1.5 M NaCl:

121.1 g Tris base, 87.7 g NaCl, pH to 8.0 with conc. HCl (~50 ml), adjust to 1 litre volume.

IV. 20X SSC:

175.3 g NaCl, 88.2 g Na_3 citrate, pH to 7.0 with conc. HCl (~1 drop), adjust to 1 litre volume.

Random-primer oligonucleotide labelling

Mix:

9.00 ul DNA (not more than 200 ng)

1.25 ul hexamers

Boil, then place on ice, then add:

5.0 ul HEPES buffer

5.0 ul nucleotides [in 250 mM Tris (pH 8), 25 mM $MgCl_2$, 50 mM β-mercaptoethanol]

1.0 ul BSA (40 mg/ml)

0.5 ul Klenow

5.0 ul-32P CTP

Let reaction go at room temperature for at least 4 hours, or overnight.

MS Medium for Arabidopsis

To 990 ml H_2O add:

Sucrose .... 10.0 g

MOPS .... 0.5 g

Agar .... 8.0 g

Adjust pH to 5.7 with 1 M KOH.

Autoclave.

When cooled add sterile stock solutions of:

KNO_3...............	20.0 ml 940 mM (95.00 g/L)
NH_4NO_3............	20.0 ml 1000 mM (80.04 g/L)
KH_2PO_4............	12.5 ml 100 mM (13.60 g/L)
$MgSO_4 7H_2O$.........	15.0 ml 100 mM (24.65 g/L)
$CaCl_2 7H_2O$.........	10.0 ml 300 mM (44.00 g/L)
Glycine............	2.67 ml 10 mM
Nicotinate.........	0.4 ml 10 mM
Thiamin............	0.02 ml 10 mM
Pyridoxine HCl.....	0.24 ml 10 mM
Myoinositol........	5.5 ml 100 mM
$FeSO_4 7H_2O$	1.0 ml (2.780 g/100 ml)
Na_2EDTA	1.0 ml 100 mM (3.720 g/100 ml)
$MnSO_4 4H_2O$.........	1.0 ml 100 mM (1.690 g/100 ml)
$ZnSO_4 4H_2O$.........	3.0 ml 10 mM (0.288 g/100 ml)
H_3BO_3	1.0 ml 100 mM (0.618 g/100 ml)
KI................	0.5 ml 10 mM (0.166 g/100 ml)
$CuSO_4 5H_2O$	0.01 ml 10 mM (0.170 g/100 ml)
$Na_2MoO_4 2H_2O$	0.1 ml 10 mM (0.242 g/100 ml)
$CoCl_2 6H_2O$	0.01 ml 10 mM (0.238 g/100 ml)

B5 Medium for arabidopsis

To 990 ml H_2O add:

Sucrose	20.0 g
MOPS	0.5 g
Agar	8.0 g

Adjust pH to 5.7 with 1 M KOH.

Autoclave.

When cooled add sterile stock solutions of:

KNO_3...............	26.3 ml 940 mM (95.00 g/L)
$NaH_2PO_4H_2O$........	11.0 ml 100 mM
$MgSO_4 7H_2O$.........	10.0 ml 100 mM (24.65 g/L)
$(NH_4)_2SO_3$..........	10.0 ml 100 mM
$CaCl_2 7H_2O$.........	3.3 ml 300 mM (44.00 g/L)

Myoinositol........	5.5 ml 100 mM
Thiamine...........	3.0 ml 10 mM
Nicotinate.........	0.8 ml 10 mM
Pyridoxine HCl.....	0.48 ml 10 mM
10X Micronutrients.	10.0 ml (see below)
Sequestrene 330 Fe.	10.0 ml (0.28 g/100 ml)

For Callus Inducing Medium (CIM) add:

2,4-D....................	0.50 ml
Kinetin..................	0.05 ml

For Shoot Inducing Medium (SIM) add:

IAA......................	0.15 ml
6-dimethylaminopurine.....	5.00 ml

10X Micronutrients (per 100 ml)

$MnSO_44H_2O$.........	0.6 ml 100 mM (1.690 g/100 ml)
$ZnSO_44H_2O$.........	10.07 ml 10 mM (0.288 g/100 ml)
H_3BO_3	4.9 ml 100 mM (0.618 g/100 ml)
KI................	4.5 ml 10 mM (0.166 g/100 ml)
$CuSO_45H_2O$	2.3 ml 10 mM (0.170 g/100 ml)
$Na_2MoO_42H_2O$	1.0 ml 10 mM (0.242 g/100 ml)
$CoCl_26H_2O$	1.0 ml 10 mM (0.238 g/100 ml)

LB (Luria-Bertani) Broth

Bacto-Tryptone	10 g
Bacto-Yeast Extract	5 g
NaCl	5 g

H_2O up to 1L

If desired, adjust the pH 7.5 with NaOH.

CA (Complete Agar)

Casein Hydrolysate	10 g
Yeast Extract	5 g
K_2HPO_4	4 g
Agar	15 g

H_2O up to 1 L

Stab Agar

Nutrient Broth	10 g
NaCl	5 g
Agar	6 g

H_2O up to 1 L

Mix well and autoclave for 15 min at 121 C.

S.O.C. (from BRL)

	1X 100 ml
2% Bacto-Tryptone	2.0 g
0.5% Yeast Extract	0.5 g
10 mM NaCl	2.92 g
2.5 mM KCl	3.73 g
20 mM $MgCl_2$	20.2 g
20 mM $MgSO_4$	24.6 g
20 mM Glucose	9.0 g

Mix above, bring to 100 ml, and filter through a sterile 0.45 μm filter. Store the stock solution at –20°C or –70°C.

Bacillus Thuringiensis Medium

Component	G/L
Glucose	3.0
$(NH_4)_2SO_4$	2.0
Yeast Extract	2.0
K2HPO$_4$3H$_2$O	0.5
$MgSO_4$*7H_2O	0.2
$CaCl_2$*2H_2O	0.08
$MnSO_4$*4H_2O	0.05

Add componenets to distilled water. Mix throughly and bring pH to 7.3.

Autoclave for 15 min at 15 psi – 121°C.

Buffers

Buffer Formula (required precision; 2 %)

Always to try to prepare buffer solution at the temperature and concentration before planing to use during the experiment. Stock solutions are acceptable as long as pH adjustment made after temperature and concentration adjustment. Also good buffers, e.g., Tris and Phosphate, are stable for a long time period.

10X TE (10:2)

1X 1L 10X

10 mM Tris-HCl 8.9 g Tris-HCl

5.3 g Tris-Base

2 mM Na_3EDTA 7.4 g

pH 8 at 25°C. Mix and store at 4 C. Periodically dump the 1X TE and make up the fresh 1X TE from 10X stock.

10X TE (10:1)

1X 1L 1X 1L 10X

10 mM Tris-HCl 1.06 g Tris-HCl 10.6 g

0.39 g Tris-Base 3.94 g

1mM Na_2EDTA 0.475 g 4.75 g

pH 8 at 25°C. Mix and store at 4°C.

10X TE (50:50)

1X 1L 10X

50 mM Tris-HCl 44.4 Tris-HCl

26.5 Tris-Base

50 mM Na_3 EDTA 186 g

Store at 4°C.

TE (25:10) 50 mM Glucose

1X 0.25 L 1X

25 mM Tris-HCl 0.55 g Tris-HCl

0.33 g Tris-Base

10 mM Na_3 EDTA 0.92 g

50 mM Glucose 2.25 g

Store at 4°C.

50 mM Tris pH 8.0 10 mM EDTA

50 mM 1 M Tris pH 8.0 5 ml

1 mM 0.5 M EDTA 2 ml

H_2O up to 100 ml

Store at 4°C.

1 M Tris pH 8.0, 1.5 M NaCl

1X 1L 1X 2L 1X

1M Tris-HCl 88.8 g Tris-HCl 178 g

53.0 g Tris-Base 106 g

1.5 M NaCl 87.7 g 175 g

Store at 25°C (room temperature is OK)

1 M Tris pH 7.4, 1.5 M NaCl

1X 1L 1X 2L 1X

1M Tris-HCl 132.2 g Tris-HCl 264.4 g

19.4 g Tris-Base 38.8 g

1.5 M NaCl 87.7 g 175 g

Store at 25°C (room temperature is OK)

0.5 M Tris pH 7.5, 1.5 M NaCl

Conc. 1L 2L 3L

0.5 M Tris Tris-HCl 63.5 g 127.0 190.0

Tris-Base 11.8 g 23.6 g 35.4 g

1.5 M NaCl 87.6 g 175.0 g 262.8 g

Store at 25°C (room temperature is OK)

10 mM Tris pH 7.5, 15 mM NaCl

Conc. Stock Volume

10 mM 1 M Tris pH 7.5 100 µl

15 mM 5 M NaCl 30 µl

H_2O to 10 ml

Store at 4°C.

10X TBE Buffer (pH 8)

1X 1L 10X 4L 10X

89 mM Tris-Base 108 g 432 g

89 mM Boric acid 55 g 220 g

2 mM Na_2 EDTA 9.3 g 37.2 g

EtBr (10 mg/ml) 0.5 ml 2 ml

Mix and store at room temperature without EtBr, 4°C with EtBr in a brown bottle. Use EtBr stock solution (10 mg EtBr/ml) when TBE is made.

50X TAE Buffer (pH 8)

1X 1L 50X 1L 10X

40 mM Tris-acetate 242 g Tris Base 48.4 g

57.1 ml acetic acid 11.4 ml

2 mM Na_2 EDTA 37 g 7.4 g

EtBr (10 mg/ml) 2.5 ml 0.5 ml

Mix and store at room temperature without EtBr, 4°C with EtBr in a brown bottle. Use EtBr stock solution (10 mg EtBr/ml) when TAE is made.

Cotton DNA Extraction buffer (pH 6)

Conc. Stock 1L 1X

100 mM Na_2 citrate.$2H_2O$ 29.4 g

Glucose 63 g

5 mM 0.5 M Na_2 EDTA 10 ml

Na_2 Diethyldithiocarbamic acid 10 g

PVP-40,000 MW 20 g

BSA 10 g

Adjust to pH 6.0 with HCl. Store at 4°C.

RSB (Reaction Stop Buffer)

Conc. Stock Volume (ml)

10 mM 1 M Tris pH 8.0 1.0

2 mM 0.5 M EDTA 0.4 0.2 % 20 % SDA 1.0

H_2O 97.6

Store at room temperature.

STE (Sodium chloride and TE) pH 8.0

Conc. Stock 1X 1L 10X 1L

10 mM 1M Tris pH 8.0 10.0 8.88 g Tris-HCl

5.3 g Tris-Base

1 mM 0.5 M EDTA 2.0 4.65 g Na_2 EDTA

100 mM 5 M NaCl 20.0 5.84 g NaCl

H_2O up to 1 L up to 1 L

pH 8.0 at 25 C. Store at 4°C.

1.5 M NaCl 0.5 N NaOH

1X 1L 1X 2L 1X

0.5 N NaOH 20.0 g 40 g

1.5 N NaCl 87.7 g 175 g

Store at room temperature.

0.2 N NaOH 0.6 M NaCl

Conc. 1L 2L 3L 4L

0.2 N NaOH 8.0 g 16.0 g 24.0 g 32.0 g

0.6 M NaCl 35.04 g 70.08 g 105.2 g 140.2 g

Store at room temperature.

20X SSC (Adjust pH 7.0)

1X 1L 20X 2L 4L 6L

150 mM NaCl 175.3 g 350.6 g 701 g 1052 g

15 mM Na_3 Citrate 88.3 g 176.6 g 353 g 530 g

Mix, adjust pH to 7.0 with HCl, and store at 4°C.

25X SSC (Adjust pH 7.4)

25X 1L 4L

3.0 M NaCl 219 g 876 g

0.3 M Na_2Citrate 110 g 440 g

Mix, adjust pH to 7.4 with HCl, and store at 4°C.

3 M K 5 M Ac

1X 200 ml 1 X

3 m KAc 29.4 g (or 20 ml of 5 M KAc)

2 M Acetic acid 23 ml

Dissolve the KAc in 150 ml of H_2O, bring to 177 ml, and then add the glacial acetic acid. Mix and store at 4°C.

Minimal Hybridization Buffer

Conc. Stock 1L 2L

10% 50% PEG 200 400

7% 20% SDS 350 700

0.6 X 25 X SSC 24 48

10 mM 1 M $NaHPO_4$ 10 20

5 mM 0.5 M EDTA 10 20

100 µg/ml Denature Salmon Sperm 10 20

H_2O 396 792

Aliquote 45 ml into 50 ml PPT, store at –20 C.

Oligo Buffer

After preapring the following stock solutions, mix A, B, and C in a ratio of 1:2.5:1.5, respectively.

Solution A

Stock

2-mercaptoethanol (bME) 18 µl

DXTPs (A, T, G) 5 µl each

Solution O 850 µl

Store at –20°C

Solution O

Conc. Stock 250 ml 100 ml

1.47 M Tris-Base 44.52 g 17.81 g

0.147 M $MgCl_2$ 7.47 g 2.99 g

Adjust pH to 8.0 with conc. HCl.

Solution B

: 2M hepes buffer

Conc. Stock 250 ml 100 ml

2M Hepes 119.15 g 47.66 g

Adjust pH to 6.6 with 4 N NaOH. Store at 4°C.

Solution C

Hexamer or oligo nucleotides. Add 55 µl H_2O directly to the Pharmacia bottle which contains 50 units of lyophilized hexamer. Sotre at –20°C.

Klenow

Dilute to 1 unit klenow fragment by adding klenow buffer. Stock klenow from BRL comes as 500 units in 84 µl. Dilute to 1 unit by adding 420 µl of klenow buffer to the 500 unit/84 µl stock. Store –20°C.

Klenow buffer (250 ml)

Conc Stock

7mM Tris-HCl 0.211 g

7 mM $MgCl_2$ 0.355 g

50 mM NaCl 0.730 g

50% Glycerol 125 ml

Add the above to about 80 ml H_2O. Stir to dissolve. Adjust volume to 250 ml with H_2O. Store at –20 C.

React buffer (See Restriction enzyme buffer formula)

10 X React buffer 2 : for, *e.g.* Hind III and Sst 1 (5 ml)

1 X Conc. Stock

500 mM 1 M Tris (pH 8.0) 2.5 ml

100 mM 1 M $MgCl_2$ 0.5 ml

500 mM NaCl 0.5 ml

25 mM Spermidine (3HCl) 0.032 g

H_2O 1.5 ml

Mix, aliquote 2 ml in screw-cap vial, store at –20°C.

10 X React buffer 3 : for *e.g.* Eco RI (5.0 ml)

1 X Conc Stock

500 mM 1M Tris (pH 8.0) 2.5 ml

100 mM 1M $MgCl_2$ 0.5 ml

1 M 5 M NaCl 1.0 ml

25 mM Spermidine (3HCl) 0.032 g

H_2O 1.0 ml

Mix, aliquote 2 ml in screw-cap vial, store at –20°C.

Phosphate (Sodium) buffer Chart

Stock solution A

2 M monobasic sodium phosphate, monohydrate (276 g/L)

Stock solution B

2 M dibasic sodium phosphate (284 g/L).

Mixing an appropriate volume (ml) of A and B as shown in the table below and diluting to a total volume of 200 ml, a 1 M phosphate buffer of the required pH at room temperature.

A	B	pH
92.0	8.0	0.8
90.0	10.0	5.9
87.7	12.3	6.0
85.5	15.0	6.1
81.5	19.5	6.2
77.5	22.5	6.3
73.5	26.5	6.4
68.5	31.5	6.5

62.5	37.5	6.6
56.5	43.5	6.7
51.0	49.0	6.8
45.0	55.0	6.9
39.0	61.0	7.0
33.0	67.0	7.1
28.0	72.0	7.2
23.0	77.0	7.3
19.0	81.0	7.4
16.0	84.0	7.5
13.0	87.0	7.6
10.5	89.5	7.7
8.5	91.5	7.8

Stock Solution

BSA

Stock BSA from BRL comes as 50 mg/1000 µl. Add 4 ml of H_2O for a final concentration of 10 mg/ml. Store at –20°C.

EtBr (10 mg/ml)

Dissolve 100 mg EtBr in 10 ml H_2O. Stir well and store at 4°C in an amber bottle.

0.5 M EDTA (pH 8.0): 500 ml

Dissolve 95.05 g of Na_2 EDTA in appr. 400 ml H_2O. Adjust pH to 8.0 with 6 N NaOH. Then bring volume to 500 ml and autoclave.

0.5 M Na_3 EDTA: 1L

Add 186.2 g (?) to 900 ml H_2O. Adjust pH with conc HCl (or NaOH ?) to 8.0, and bring up to 1 L.

5 M LiCl 50 mM MOPS (pH 8.0)

Conc 5 M LiCl 21.2 g

50 mM MOPS 1.05 g

H_2O up to 100 ml

Adjust pH 8.0 with NaOH. Store at 4°C.

50 % PEG

Add 100 g PEG (Polyethylene Glycol, appr. MW 8.000) to 50 ml H_2O and stir for about one hour. Then bring up to 200 ml with H_2O.

3 N Perchloric acid

Dilute 49.2 ml of perchloric acid up to 200 ml with H_2O in a hood. Stable at room temperature for several month. Treat with care.

1 N NaOH

Dissolve 40 g of NaOH in 1000 ml H_2O.

1M $NaHPO_4$

Stock

$NaH_2PO_4.H_2O$ (monobasic monohydrate) 47.5 g
$Na_2PO_4.7H_2O$ (dibasic 7-hydrate) 42.2 g
H_2O to 500 ml

10 M NH_4Ac

Add 400 ml H_2O to a 1 Kg bottle NH_4Ac, dissolve, and adjust to 1300 ml.

Tris 1 M ph 8.0

Stock
Tris-HCl 88.8 g 177 g 355.2 g
Tris-Base 53.0 g 106.0 g 212.0 g
H_2O 1L 2L 4L

Tris 1 M pH 7.5

Stock
Tris-HCl 127 g 254 g 508 g
Tris-Base 23.6 g 47.2 g 94.4 g
H_2O 1L 2L 4L

76% EtOH 10 mM NH_4Ac

Conc. Stock Volume
75% 95% EtOH 80.0 ml
0.2 M 5M NH_4OHAc 0.2 ml
H_2O 19.8 ml
5 M NH_4Ac stock = 38.55 g/100 ml

76% EtOH 0.2 M NaAc

Conc. Stock Volume
76% 95% EtOH 160 ml
0.2 M3 M NaAc 13.33
H_2O 26.67
3 M NaO Ac stock = 40.83 g/100 ml

0.2 N NaOH 0.6 M NaCl

1L 2L 3L 4L
NaOH 8.0 g 16.0 g 24.0 g 32.0 g
NaCl 35.04 g 70.08 g 105.2 g 140.2 g

0.5 M Tris (pH 7.5) 1.5 M NaCl

1L 2L 3L
Tris-HCl 63.5 g 127.0 g 190.0 g
Tris-Base 11.8 g 23.6 g 35.4 g
NaCl 87.6 g 175.0 g 262.8 g

25 mM $NaHPO_4$

1M $NaHPO_4$ 100 ml
H_2O to 4L

10 X Treacking Dye

Conc.

50 mM Tris pH 8.0

50 mM Na_3 EDTA

25% Glycerol

5% Ficoll

0.1% Xylenol Cyanol

0.1% Bromophenol Blue

1% SDS

Glycerol Solution

Glycerol 650 ml

1 M $MgSO_4$ 100 ml

1 M Tris (pH 8) 25 ml

H_2O up to 1L

Mix well and autoclave for 15 min at 121°C.

0.1 M Spermidine Solution

Dissolve 255 mg spermidine (Sigma #S2501) in H_2O to a final volume of 10 ml. Store at –20°C.

Saline-MOPS

1X 1L 1X

0.85% NaCl 8.5 g

50 mM MOPS 6.3 g MOPS free acid

4.7 g MOPS Na Salt

pH 7 autoclave and store at room temperature.

Proteinase K–10 mg/ml in TE

Weigh out 10 mg proteinase K (Boeringer Manheim #161 497) and dissolve in 1 ml TE. The solution can be used immediately or aliquoted and stored at –20°C.

RNase A and RNase T1

Dissolve 100 mg RNase A (bovine pancrease: Sigma #R4875), if desired, add 5,00 units of RNase T1 together, in 10 ml of 10 mM Tris 15 mM NaCl. Boil for 15 min and allow to cool slowly to room temperature. Distribute 1 ml aliquote into 1.5 ml MFT, and store at –20°C.

Antibiotic Concentration in Media

Antibiotics should be added to lower than 60°C broth, or filter sterilized: See note. "C" refers to chromosomal resistance: "P" plasmid based resistance, "FS" filter sterilization required. Temperature is required for storage.

Concentration (µg/ml)

Antibiotics for *Pseudomonas, E. Coli, Agrobacterium,* etc.

Ampicillin 1000 (C/P) 50 (P) Stock: 25 mg.ml Na salt:FS: Store at –20°C, carbenicillin is more stable than Amp.

Carbenicillin 1000 (C/P) 50 (P) 100 mg/ml: FS: –20°C

Chlroamphenicol 500 (C) 30 (C/P) 170 μg/ml for plasmid amplification 34 mg/ml in EtOH: –20°C.

D-cycloserine 200 (C) Add dry after autoclaving

Erythromycin 200 (C) 25 mg/ml EtOH: –20°C

Gentamycin 10 (C/P) Add dry after autoclaving

Geneticin 200 for fungi same as kanamycin in mode of action and preparation

Hygromycin 250 for fungi 25 mg/ml: FS: –20°C

Kanamycin 200 (P) 50 (P) 25 mg/ml: FS: –20°C

Kasugamycin 200 (C) 20 mg/ml: FS: –20°C

Mercuric chloride 40 (C/P) 10 μg/ml if in minimal media, 40 mg/ml in EtOH: –20°C

Nalidixic acid 1000 (C) 20 (C) 100 mg/ml in H_2O, add 6 N NaOH until dissolved ; 4C

Neomycin 200 (P) 50 (P) 25 mg/ml: FS: –20°C

Rifampicin 200 (C) 200 mg/ml make fresh in Methanol

Spectinomycin 200 (C/P) 25 (P) 20 mg/ml: FS: –20°C

Streptomycin 200 (C/P) 25 (P) 20 mg/ml: FS: –20°C

Sulfamethoxazole 500(C) Add dry after autoclaving

Triomethoprim 500 (C) Add dry after autoclaving

Tetracycline 20 (C) 50 (P) 15 (P) 12.5 mg/ml 50% EtOH: –20°C

Appendix C. Chemicals Required for Microbiology Experiment

1. Staining Reagents

Acid alcohol

Ethyl alcohol (95%)	97.0 ml
HCL	3.0 ml

Methylene Blue

Methylene Blue	.3 gm
Distilled water	100.0 ml

Capsule Stain

Crystal Violet (1%)

Crystal Violet (85% dye content)	1.0 gm
Distilled water	100.0 ml

Copper sulphate solution (20%)

Coppersulfate ($CuSO_4$)	20.0 gm
Distilled water	80.0 ml

Fungal Stains

Lactophenol-cotton blue solution

Lactic acid	20.0 ml
Phenol	20.0 gm
Glycerol	40.0 ml

Distilled water	20.0 ml
Aniline Blue	0.05 gm

Note. Heat gently to dissolve in hot water. Then add aniline blue dye.

Water-iodine solution

Grams iodine	10.0 ml
Distilled water	30.0 ml

Grams Stain

Crystal Violet

Solution A

Crystal Violet (90% dye content)	2.0 gm
Ethyl alcohol (95%)	20.0 ml

Solution B

Ammonium oxalate	0.8 gm
Distilled water	80.0 ml

Mix Solutions A and B

Grams iodine

Iodine	1.0 gm
Potassium iodide	2.0 gm
Distilled water	300 ml

Ethyl alcohol (95%)

Ethyl alcohol (100%)	95.0 ml
Distilled water	5.0 ml

Safranin

Safranin	0.25 ml
Ethyl alcohol (95%)	10.0 ml
Distilled water	100 ml

Negative Stain

Nigrosin

Nigrosin, Water soluble	10.0 gm
Distilled water	100 ml

Immerse in boiling water bath for 30 minutes

Formalin	0.5 ml

Filter twice through double filter paper

Spore Stain

Malachite green

Malachite green	5.0 gm
Distilled water	100 ml

Safranin

Safranin	0.25 ml
Ethyl alcohol (95%)	10.0 ml
Distilled water	100 ml

Acid fast Stain

Carbol fuchsin

Solution A

Basic fuchsin	0.3 gm
Ethyl alcohol (95%)	10.0 ml

Solution B

Phenol	5.0 ml
Distilled water	100 ml

Mix solutions A and B. Add 2 drops of triton X per 100 ml of stain for use in heatless method.

2. Microbiological Media

The formulas of the media used in the exercises in this manual are listed alphabetically in grams per litre of distilled water unless otherwise specified. Sterilization of the media is accomplished by autoclaving at 15 lb pressure for 15 minutes unless otherwise specified. Most of the media are available commercially in powdered form, with specific instructions for their preparation and sterilization.

Ammonium sulfate broth (pH 7.3)

Ammonium sulphate	2.0
Magnesium sulphate .7H_2O	0.5
Ferric sulphate .7H_2O	0.03
Sodium chloride	0.3
Magnesium carbonate	10.0
Dipotassium hydrogen phosphate	1.0

Bacteriophage broth 10 × (pH 7.6)

Peptone	100.0
Beef extract	30.0
Yeast extract	50.0
Sodium chloride	25.0
Potassium dihydrogen phosphate	80.0

Basal salts agar* and broth (pH 7.0)

0.5 M Sodium diphosphate	100.0 ml
1.0 M Potassium dihydrogen phosphate	100.0 ml
Distilled water	800.0 ml

0.1 M Calcium chloride	1.0 ml
1.0 M Magnesium sulphate	1.0 ml

Note: Swril

Ammonium sulphate	2.0
*Agar	15.0

Note: Swirl until completely dissolved, auto-clave, and cool. Aseptically add 10.0 ml of 1% sterile glucose.

Bile esculin (pH 6.6)

Beef extract	3.0
Peptone	5.0
Esculin	1.0
Oxgall	40.0
Ferric citrate	0.5
Agar	15.0

Blood agar (pH 7.3)

Infusion from beef heart	500.0
Tryptose	10.0
Sodium chloride	5.0
Agar	15.0

Note: Dissolve the above ingredients and auto-clave. Cool the sterile blood agar base to 45°C to 50°C. Aseptically add 50 ml of sterile defibrinated blood. Mix thoroughly, avoiding accumulation of air bubbles. Dispense into sterile tubes or plates while liquid.

Brain heart infusion (pH 7.4)

Infusion from calf brain	200.0
Infusion from beef heart	250.0
Peptone	10.0
Dextrose	2.0
Sodium chloride	5.0
Disodium phosphate	2.5
Agar	1.0

Campy BAP agar (pH 7.0)

Trypticase peptone	10.0
Thiotone	10.0
Dextrose	1.0
Yeast extract	2.0
Sodium chloride	5.0
Sodium bisulphide	0.1
Agar	15.0

Vancomycin	10.0 mg
Trimethoprim lactate	5.0 mg
Polymyxin B sulphate	2500.0 IU
Amphotericin B	2.0 mg
Cephalothin	15.0 mg
Defibrinated sheep blood	10.0%

Note: Aseptically add the antibiotics and defibrinated sheep blood to the sterile, cooled, and molten agar.

Chocolate agar (pH 7.0)

Proteose peptone	20.0
Dextrose	0.5
Sodium chloride	5.0
Disodium phosphate	5.0
Agar	5.0

Note: Aseptically add 5.0% difibrinated sheep blood to the sterile and molten agar. Hcat of 80°C until a chocolate colour develops.

Deoxyribonuclease (DNase) agar (ph 7.3)

Deoxyribonucleic acid	2.0
Phytane	5.0
Sodium chloride	5.0
Trypticase	15.0
Agar	15.0

Endo agar (ph 7.5)

Peptone	10.0
Lactose	10.0
Dipotassium phosphate	3.5
Sodium sulphite	2.5
Basic fuchsin	0.4
Agar	15.0

Eosin-methylene blue agar, Levine (ph 7.2)

Peptone	10.0
Lactose	5.0
Dipotassium phosphate	2.0
Agar	13.5
Eosin Y	0.4
Methylene blue	0.065

Gel diffusion agar

Sodium barbital buffer	100.0 ml
Noble agar	0.8

Glycerol yeast extract agar supplemented with aureomycin (ph 7.0)

Glycerol	5.0 ml
Yeast extract	2.0
Dipotassium phosphate	1.0
Agar	15.0

Note: Aseptically add aureomycin, 10 μg per ml, to the sterile, cooled, and molten agar.

Glucose salts broth (pH 7.2)

Dextrose	5.0
Sodium chloride	5.0
Magnesium sulphate	0.2
Ammonium dihydrogen phosphate	1.0
Dipotassium hydrogen phosphate	1.0

Grape juice broth

Commercial grap or apple juice	
Ammonium biphosphate	0.25%

Note: Sterilization not required when using a large yeast inoculum.

Inorganic synthetic broth (pH 7.2)

Sodium chloride	5.0
Magnesium sulphate	0.2
Ammonium dihydrogen phosphate	1.0
Dipotassium hydrogen phosphate	1.0

KF broth (pH 7.2)

Polypeptone	10.0
Yeast extract	10.0
Sodium chloride	5.0
Sodium glycerophosphate	10.0
Sodium carbonate	0.636
Maltose	20.0
Lactose	1.0
Sodium azide	0.4
Phenol red	0.018

Lactose fermentation broth 1X and 2X* (pH 6.9)

Beef extract	3.0
Peptone	5.0
Lactose	5.0

Note: *For 2X broth use twice the concentration of the ingredients.

Litmus milk (pH 6.8)

Skim milk powder	100.0
Litmus	0.075

Note: Autoclave at 12 lb pressure for 15 minutes.

MacConkey's agar (pH 7.1)

Bacto peptone	17.0
Proteose peptone	3.0
Lactose	10.0
Bile salts mixture	1.5
Sodium chloride	5.0
Agar	13.5
Neutral red	0.03
Crystal violet	0.001

Mannitol salt agar (pH 7.4)

Beef extract	1.0
Peptone	10.0
Sodium chloride	75.0
d-Mannitol	10.0
Agar	15.0
Phenol red	0.025

M-endo broth (pH 7.5)

Yeast extract	6.0
Thiotone peptone	20.0
Lactose	25.0
Dipotassium phosphate	7.0
Sodium sulphite	2.5
Basic fuchsin	1.0

Note: Heat until boiling, do not autoclave.

M-FC broth (pH 7.4)

Biosate peptone	10.0
Polypeptone peptone	5.0
Yeast extract	3.0
Sodium chloride	5.0
Lactose	12.5
Bile salts	1.5
Aniline blue	0.1

Note: Add 10 ml of rosolic acid (1% in 0.2N sodium hydroxide). Heat to boiling with agitation, do not autoclave.

Milk agar (pH 7.2)

Skim milk powder	100.0
Peptone	5.0
Agar	15.0

Note: Autoclave at 12 lb pressure for 15 minutes.

Minimal agar (pH 7.0)

Minimal agar, supplemented with streptomycin and thiamine*

Solution A (pH 7.0)

Potassium dihydrogen phosphate	3.0
Disodium hydrogen phosphate	6.0
Ammonium chloride	2.0
Sodium chloride	5.0
Distilled water	800.0 ml

Solution B (pH 7.0)

Glucose	8.0
Magnesium sulfate . $7H_2O$	0.1
Agar	15.0
Distilled water	200.0 ml

Note: Autoclave solutions A and B separately and combine.

*To solution B add 0.001 gm of thiamine prior to autoclaving. To the combined sterile and molten medium, add 50 mg (1 ml of 50 mg per ml) sterile streptomycin solution before pouring agar plates.

MR-VP broth (pH 6.9)

Peptone	7.0
Dextrose	5.0
Potassium phosphate	5.0

Mueller-Hinton agar (pH 7.4)

Beef, infusion	300.0
Casamino acids	17.5
Starch	1.5
Agar	17.0

Mueller-Hinton tellurite agar (pH 7.4)

Casamino acids	20.0
Casein	5.0

1-Tryptophane	0.05
Potassium dihydrogen phosphate	0.3
Magnesium sulphate	0.1
Agar	20.0

Note: Aseptically add 12.5 ml of tellurite serum to the sterile, 50°C molten agar.

Nitrate broth (pH 7.2)

Peptone	5.0
Beef extract	3.0
Potassium nitrate	5.0

Nitrite broth (pH 7.3)

Sodium nitrite	2.0
Magnesium sulphate . $7H_2O$	0.5
Ferric sulphate . $7H_2O$	0.03
Sodium chloride	0.3
Sodium carbonate	1.0
Dipotassium hydrogen sulphate	1.0

Nitrogen-free mannitol agar* and broth (pH 7.3)

Mannitol	15.0
Dipotassium hydrogen phosphate	0.5
Magnesium sulphate	0.2
Calcium sulphate	0.1
Sodium chloride	0.2
Calcium carbonate	5.0
*Agar	15.0

Nutrient agar* and broth (pH 7.0)

Peptone	5.0
Beef extract	3.0
*Agar	15.0

Nutrient gelatin (pH 6.8)

Peptone	5.0
Beef extract	3.0
Gelatin	120.0

Peptone broth (pH 7.2)

Peptone	4.0

Phenol red dextorse broth (pH 7.3)

Trypticase	10.0
Dextrose	5.0

Sodium chloride	5.0
Phenol red	0.018

Note: Autoclave at 12 lb pressure for 15 minutes.

Phenol red inulin broth (ph 7.3)

Trypticase	10.0
Inulin	5.0
Sodium chloride	5.0
Phenol red	0.018

Note: Autoclave at 12 lb pressure for 15 miuntes.

Phenol red lactose broth (ph 7.3)

Trypticase	10.0
Lactose	5.0
Sodium chloride	5.0
Phenol red	0.018

Note: Autoclave at 12 lb pressure for 15 minutes.

Phenol red sucrose broth (ph 7.3)

Trypticase	10.0
Sucrose	5.0
Sodium chloride	5.0
Phenol red	0.018

Note: Autoclave 12 lb pressure for 15 minutes.

Phenylethyl alcohol agar (pH 7.3)

Trypticase	15.0
Phytane	5.0
Sodium chloride	5.0
β-Phenylethyl alcohol	2.0
Agar	15.0

Sabouraud agar (pH 5.6)

Sabouraud agar supplemented with aureomycin*

Peptone	10.0
Dextrose	40.0
Agar	15.0

Note: *Aseptically add aureomycin, 10 μg per ml, to the sterile, cooled, and molten medium.

Salt medium—Halobacterium

Sodium chloride	250.0
Magnesium sulphate . $7H_2O$	10.0

Potassium chloride	5.0
Calcium chloride . $6H_2O$	0.2
Yeast extract	10.0
Tryptone	2.5
Agar	20.0

Note: The quantities given are for preparation of one litre final volume of the medium. In preparation, make up two solutions, one involving the yeast extract and tryptone, and the other salts. Adjust the pH of the nutrient solution to 7. Sterilize separately. Mix and dispense aseptically.

SIM agar (pH 7.3)

Peptone	30.0
Beef extract	3.0
Ferrous ammonium sulphate	0.2
Sodium thiosulphate	0.025
Agar	3.0

Simmons citrate agar (pH 6.9)

Ammonium dihydrogen phosphate	1.0
Dipotassium phosphate	1.0
Sodium chloride	5.0
Sodium citrate	2.0
Magnesium sulphate	0.2
Agar	15.0
Brom thymol blue	0.08

Synder test agar (pH 4.8)

Tryptone	20.0
Dextrose	20.0
Sodium chloride	5.0
Brom cresol green	0.02
Agar	20.0

Sodium chloride borth, 6.5% (pH 7.0)

Brain heart infusion broth	100.0 ml
Sodium chloride	6.5

Starch agar (pH 7.0)

Peptone	5.0
Beef extract	3.0
Starch (soluble)	2.0
Agar	15.0

Thioglycollate, fluid (pH 7.1)

Peptone	15.0
Yeast extract	5.0
Dextrose	5.0
1-Cystine	0.75
Thioglycollic acid	0.3 ml
Agar	0.75
Sodium chloride	2.5
Resazurin	0.001

Tinsdale agar (pH 7.4)

Proteose peptone, No. 3	20.0
Sodium chloride	5.0
Agar	20.0

Note: Following boiling, distribute in 100 ml flasks. Autoclave, cool to 55°C, add 15 ml of rehydrated Tinsdale enrichment to each 100 ml, and mix thoroughly before dispensing.

Top agar (for Ames test)

Sodium chloride	5.0
Agar	6.0

Tributyrin agar (pH 7.2)

Peptone	5.0
Beef extract	3.0
Agar	15.0
Tributyrin	10.0

Note: Dissolve peptone, beef extract, and agar while heating. Cool to 90°C, add the tributyrin, and emulsify in a Waring blender.

Triple sugar-iron agar (pH 7.4)

Beef extract	3.0
Yeast extract	3.0
Peptone	15.0
Proteose peptone	5.0
Lactose	10.0
Saccharose	10.0
Dextose	1.0
Ferrous sulphate	0.2
Sodium chloride	5.0
Sodium thiosulphate	0.3
Phenol red	0.024
Agar	12.0

Trypticase nitrate broth (pH 7.2)

Trypticase	20.0
Disodium phosphate	2.0
Dextrose	1.0
Agar	1.0
Potassium nitrate	1.0

Trypticase soy agar (pH 7.3)

Trypticase	15.0
Phytane	5.0
Sodium chloride	5.0
Agar	15.0

Tryptone agar* and broth

Tryptone	10.0
Calcium chloride (reagent)	0.01–0.03 M
Sodium chloride	5.0
*Agar	11.0

Tryptone soft agar

Tryptone	10.0
Potassium chloride (reagent	5.0 ml
Agar	9.0

Urea broth

Urea broth concentrate	
(filter-sterilized solution)	10.0 ml
Sterile distilled water	90.0 ml

Note: Aseptically add the urea broth concentrate to the sterilized and cooled distilled water. Under aseptic conditions, dispense 3 ml amounts into sterile tubes.

Yeast extract broth (pH 7.0)

Peptone	5.0
Beef extract	3.0
Sodium chloride	5.0
Yeast extract	5.0.